COLLINS

GCSE D&T

Technology

Written in association with
Lincolnshire County Council

Authors: P. Fowler
M. Horsley
Editor: A. Breckon

Collins Educational
An imprint of HarperCollins*Publishers*

ACKNOWLEDGEMENTS

The publishers are grateful to the following individuals and organisations for the photographs and illustrations in this book:

Acorn Computers Ltd, fig 8.52;
J. Allen Cash, figs 3.21, 6.2, 6.7, 6.16. 6.18, 6.19;
Allsport Photo Library, fig 6.1;
Heather Angel, fig 6.6;
Ann Ronan Photo Library, figs 3.6, 3.13, 3.26;
Argos Distributors Ltd, figs 1.9, 1.20, 1.23, 1.25, 1.26, 1.27, 1.28, 1.29;
Art Directors Photo Library, fig 9.60;
Associated Press, fig 3.32;
Australian Information Service, fig 3.25;
J.C. Bamford Ltd, fig 1.6;
Barnabys Picture Library, figs 10.68, 10.69 (inset), 11.27, 11.38;
St Bartholomew's Hospital, figs 3.27, 3.28;
Alastair Black, fig 9.38;
Black and Decker Ltd, page 148;
Boxford Ltd, figs 3.12; 8.70;
British Telecom plc, figs 1.11, 1.12, 1.13, 1.14, 1.15, 1.16, 1.17, 1.18, 1.19, 8.1
   (cordless telephone);
Charles Brereton, figs 2.11, 3.5, 3.14, 5.21, 7.94, 7.121, 7.122, 7.123, 7.130, 7.131,
   8.138, 8.139, 8.140;
BBC Hulton Picture Library, figs 3.9, 3.18, 3.22, 3.30;
Ray Bush, fig 10.62;
Canon UK Ltd, fig 1.1;
Casio Electronics Co Ltd, fig 8.1 (electronic organ);
Castle Associates, fig 8.75;
Economatics Ltd, figs 8.23, 8.84.
Ever Ready Ltd, figs 1.22, 1.24;
Mike Finney, pages 203, 205, 207, 209, 214
Ford Motor Co Ltd, figs 5.6, 11.26;
John Freeman and Co Ltd, fig 3.20;
Sally and Richard Greenhill, figs 1.8, 1.41, 3.29, 3.31, 5.1, 5.5, 6.6, 6.15, 8.117;
Hamleys Ltd, fig 4.1 (jethopper);
Hoover Ltd, fig 3.7;
Howes Electrical and Automation Service, figs 4.21, 4.23, 4.24, 4.25, 4.26;
IBM UK Ltd, Figs 8.59, 8.61, 8.66, 8.84;
Journal of Applied Pneumatics, page 152;
Kenwood Ltd, page 149;
LEGO UK Ltd, figs 5.24, 5.25, 5.26;
Linear Graphics Ltd, figs 8.67, 9.44;
A.M. Lock and Co Ltd, figs 5.28, 5.30;
Marconi Electronic Devices Ltd, fig 8.60;
NCR Ltd, fig 3.15;
Norgren Martonair Ltd, figs 10.37, 10.38, 10.39, 10.58;
Philips Electronics, figs 3.10, 4.1 (video remote controls), 8.1 (video controls);
R.S. Components Ltd, figs 1.21, 5.35, 5.39, 7.4, 7.7, 7.8, 7.9, 7.11, 7.13, 7.43, 7.54,
   7.56, 7.60, 7.63, 7.64, 7.76, 7.108, 8.53, 8.54, 8.80, 8.90, 8.91, 8.92, 8.93, 8.99;
Rapid Electronics Ltd, figs 5.3, 5.34, 7.5, 8.40, 8.89, 8.113;
Renold Gear Ltd, 9.66, 9.68, 9.69;
Renold Conveyors Ltd, figs 9.61, 9.82;
Research Machines Ltd, figs 5.31, 8.68;
Science Photo Library, figs 3.4, 3.8, 4.1 (robot arm), 4.12, 6.3, 6.5, 6.11, 6.13,
   6.17, 7.79, 7.106, 7.109, 8.1 (microcomputer and robot arm), 8.63, 8.68;
Teletext Ltd, figs 3.24;
University of Durham, School of Education, fig 7.15;
Welwyn Strain Measurement (Basingstoke) Ltd, fig 11.72;
Zanussi Ltd, figs 4.12, 4.13, 4.14, 4.15, 4.16, 4.17, 4.18, 4.19, 4.20;

We would also like to thank the Micro Electronic Development Unit, Lincoln, for their general technical assistance.

© Lincolnshire County Council 1988

First published 1988 by Collins Educational
An imprint of Harper Collins *Publishers*
77-85 Fulham Palace Road
Hammersmith, London W6 8JB

Reprinted 1989, 1990, 1991, 1992, 1993, 1998, 1999, 2000.

ISBN 0 00 322036 2

All rights reserved. No part of this publication may be reproduced, stored in a retrieval system, or transmitted in any form or by any means, electronic, mechanical, photocopying, recording or otherwise, without the prior permission of the copyright holder.

Cover design by Sands-Straker Studios Limited, London
Artwork by Tim Cooke, Sam Denley, Peter Harper, Illustrated Arts
Typeset by Burns and Smith, Derby.

Printed and bound in Great Britain by Martins the Printers Ltd, Berwick upon Tweed.

CONTENTS

ABOUT THIS BOOK

This book seeks to cover the appropriate material for GCSE Design and Technology, in particular D&T: Systems and Control Technology. The book is based on the principle that D&T is an activity-focused area of the curriculum concerned with designing and making quality products to meet a specific purpose. The book provides the information necessary to tackle a wide range of design problems. It does not seek to solve them, but offers support and guidance in the process of transforming your ideas into reality. An important feature of the book is to place D&T in the context of the world in which we live. Most sections end with a range of questions based on the chapter, arranged so that they become progressively more difficult. Finally, the book is intended to stimulate those studying the course to extend their interest in designing and making either by appreciating and/or improving the human-made environment, or by studying the subject further and perhaps pursuing an associated career.

TEACHER'S PREFACE

The material in this book was originally developed as part of a comprehensive curriculum programme for CDT in Lincolnshire. Initially the material was written by teachers in the form of pupil guide books, which were evaluated and completely rewritten by the authors. This Technology book was originally published in 1988. However, following minor revision, it still remains relevant for courses in Systems and Control Technology and meets the appropriate Key Stage 4 programmes of study.

Technology is one of a series of GCSE books for D&T. They have been written by a team of six teachers. The principal authors of *Technology* are, Peter Fowler, Head of Design Technology at Lincoln Christ's Hospital School, and Michael Horsley, Technology Consultant. They were ably assisted by Colin Chapman, Kevin Crampton, Mike Finney, and the late Melvyn Peace. I am indebted to them for their commitment and professionalism in working so successfully as a team on this venture. I would also like to express my gratitude to the wives and children of these teachers for their patience and understanding throughout the absorbing but very demanding task of writing this book.

This book was made possible by the encouragement of the former County Education Officer, Mr F. G. Rickard and the continued support of the former Director of Education, Mr D. G. Esp. Their professional suppport has been encouraged by enthusiastic members of Lincolnshire County Council, in particular Councillors W. J. Speechley and P. Newton, whose positive approach to the development of CDT has greatly benefited the education of many pupils in Lincolnshire and hopefully, through this book, many others.

Finally, I would like to thank the many pupils and teachers who, through their interest and enthusiasm, have made such a positive contribution to the creation of this book.

A. M. Breckon, Chief Executive DATA
Formerly Education Inspector, Lincolnshire County Council

DESIGN AND DESIGNING

fig 1.1. 1987 camera

fig 1.3. Common logos

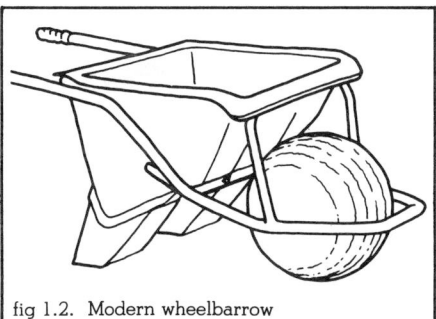

fig 1.2. Modern wheelbarrow

fig 1.5. Food packaging

Design plays a very important part in our lives. It controls and affects much of what we do in our everyday lives, and we are all capable of designing.

You might design the pattern of your day, the layout of your bedroom or the way you travel to see a friend. From this you can see that design is always with you, whether it is the shelter you live in, your means of travelling, or the clothes you wear.

This chapter looks at design and designing which are central to the whole subject of Design and Technology (D&T). It illustrates the different forms of designing related to D&T courses, and offers two frameworks which can help you to solve the design problems central to your course.

The rest of the book provides the necessary information and some helpful ideas for tackling a wide range of design problems.

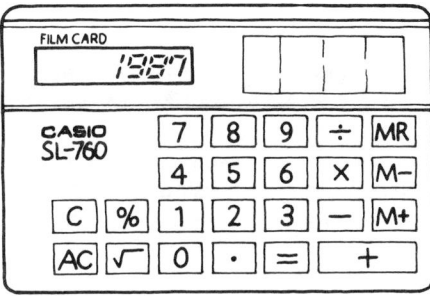

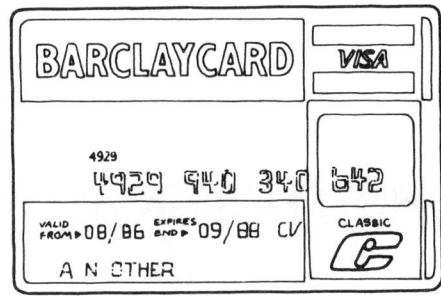

fig 1.4. Calculator and credit card

fig 1.6. JCB excavator

fig 1.7. Lloyd's building

1

DESIGN AND DESIGNING

WHAT IS DESIGNING?

Designing is an activity which uses a wide range of experiences, knowledge, and skills to find the best solution to a problem, within certain constraints.

Designing involves identifying and clarifying a problem, making a thoughtful response, and then creating and testing your solution. You can then usually start to modify your solution, so that the process of designing begins again.

Designing is a creative activity. You may often use known facts or solutions, but the way you combine these to solve your own particular problem requires creative thinking.

Designing is far more than just problem-solving. It involves the whole process of producing a solution from conception to evaluation. This includes elements such as cost, appearance, styling, fashion, and manufacture.

Designers work in almost every area of life — textile design, product design, graphic design, interior design, engineering design and environmental design. Each area requires a different type of knowledge, but they all involve a similar design activity.

fig 1.8. Modern clothing

WHY DESIGNS CHANGE

There are various reasons why designs change. One is the change in the needs of society. An example of this is the fairly recent change to smaller cars. These were designed in response to the demand for economical, easier-to-park vehicles as our cities became more congested. You might argue, however, that the designer creates the change, and society then reacts to this change (as with fashion for example).

A second reason for change is the development of new technology which can be applied to traditional products. The new designs of computer have only been possible because of the development of 'microchip' technology. This was itself stimulated by people's desire for space flight.

HOW DESIGN AFFECTS OUR LIVES

A world without calculators, televisions, trains, aeroplanes, house insulation, advertising, shopping precincts and microwaves is difficult to imagine. These products have by and large benefitted us. Few people would like to do without all the products designed in the last 20 years. However, designers have responsibilities to society because of the way their designs can affect us.

As well as solving the problem, designers may also work under timing, financial, and political pressures. However they must always place people first.

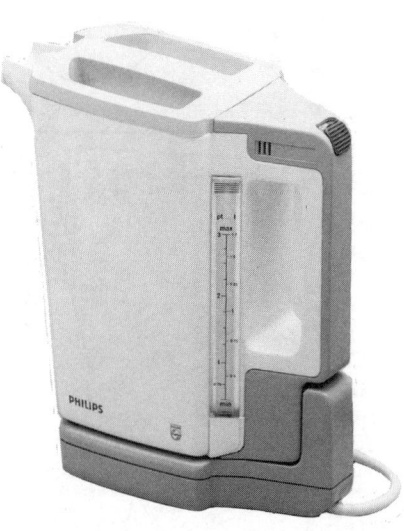

fig 1.9. Cordless kettle

fig 1.10. Interior design of kitchen

LOOKING AT DESIGN

These first two pages show various designs which illustrate the fashion, style and technology of today. Design is all around us and the next three pages illustrate three different aspects. Page 3 looks at the historical development of the telephone, page 4 looks at how designers have developed differing solutions to the same problem, and page 5 looks at how designers communicate ideas, images and information. There are many other examples which you may also wish to investigate.

DESIGN CHANGES THROUGH TIME

The telephone was invented over 100 years ago. It is a vital part of our lives which we use all the time to communicate.

Its function has remained almost identical over the years, but new technology has allowed changes from metal to wood to plastics, and from mechanical switches to electrical switches.

The flexibility of modern electronics allows the designer to create almost any shape of telephone. This page shows some of the external design changes. At the same time there have been many improvements in the telephone's performance, which cannot be shown here.

fig 1.14. Telephone c. 1929

fig 1.18. BT Sceptre — memory and clock

fig 1.11. Ericson Magneta table phone c. 1895

fig 1.15. Telephone 1937

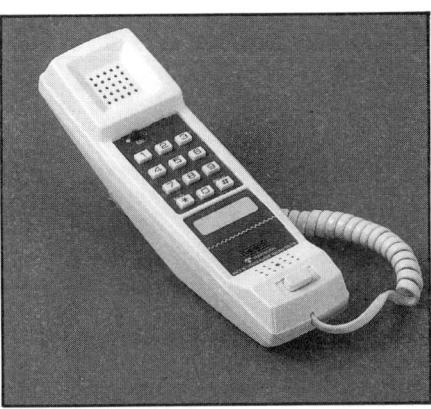

fig 1.19. One piece wall mounted telephone

fig 1.12. Desk telephone c. 1907

fig 1.16. Telephone 1968

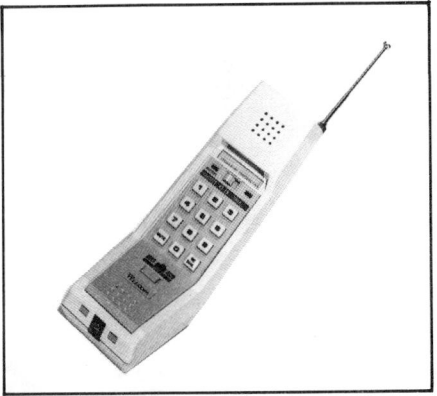

fig 1.20. Remote control telephone

fig 1.13. Strowger calling dial c. 1905

fig 1.17. British telephone 1960s to 1980s

fig 1.21. Mobile phone

DESIGN SOLUTIONS TO SIMILAR PROBLEMS

This page shows some solutions to the problem of providing a portable source of light. Most have been designed to meet other needs as well, but they are all design solutions to the same basic problem.

When looking at a problem, designers think about how their solution will be used. A torch has two principal uses — to provide light to see, and to act as a warning device. Sometimes both these functions are required at once. For example, you cannot cycle safely along country roads at night without a front light which does both.

When developing a new torch, the designer is likely to be constrained by the size of the batteries and bulbs. Another constraint for a cycle lamp is likely to the means of attaching it to the cycle frame. These things are fixed because the designer is unlikely to be able to demand a specially designed battery or bicycle frame. There may be additional constraints, such as making the torch waterproof. Can you think of other possible constraints?

There are many other torch designs besides those shown here. Can you think of some other design solutions to this problem?

fig 1.25. Ever Ready cycle torch

fig 1.26. Motorist's lantern

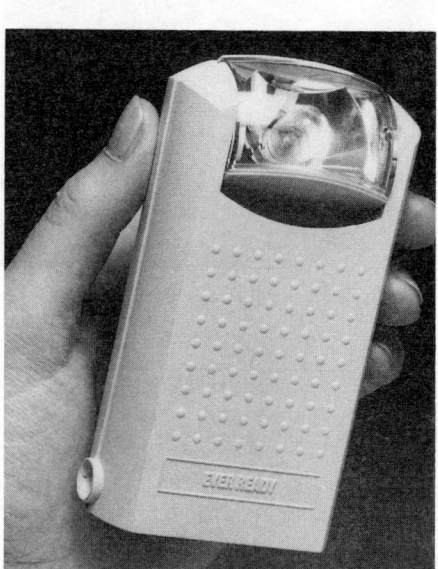

fig 1.22. Ever Ready swivel head torch

fig 1.24. Ever Ready small torch

fig 1.27. Dual purpose torch

fig 1.28. Pifco waterproof torch

fig 1.23. Durabeam torch

fig 1.29. Ever Ready rechargeable torch

4

DESIGN IN COMMUNICATION

You are constantly being bombarded by the communications industry creating images — either television, newspapers, magazines, posters, shop displays or exhibitions.

These images are created to communicate — to get a message across. The message may encourage you to buy a product or service, or it may provide entertainment or information.

The designers who create these images must be very sensitive to people's feelings, because they want to catch the eye of the consumer without offending.

The designer will use colour, texture, cartoons, pictures, and printed letters to get the message across. The development of computer graphics has also made the communication of information more effective.

Look carefully at some of the images on this page to see what messages or information they are getting across.

fig 1.31. Communicating through cartoons

fig 1.30. Company trademarks

fig 1.35. International communication symbols

fig 1.32. Communicating weather forecasts

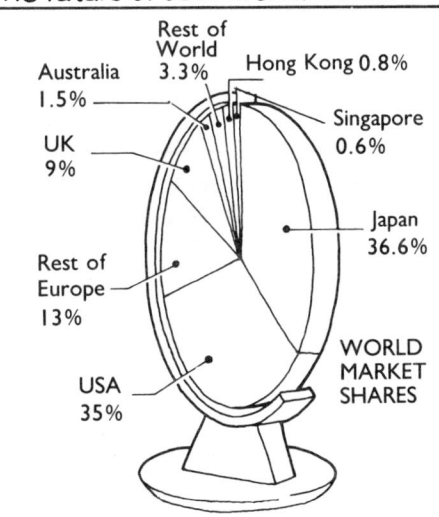
fig 1.34. Communicating data

fig 1.36. Advertising through carrier bags

fig 1.33. Comic strip

FRAMEWORKS FOR DESIGNING

In your D&T course, designing is concerned with the whole process from identifying a problem, through to creating a solution and then testing it. This process requires you to consider many factors, and make a number of decisions.

There are many different methods or routes for doing this, and the one you choose will depend on the nature of the problem. This chapter introduces the idea of a framework which will support you in your design work.

A framework is a series of linked stages which will help you solve your problem. At each stage you will need to refer back to earlier stages. This chapter looks at two frameworks, although it is quite feasible to use others. Equally, these two frameworks can be adapted to suit your own particular problem.

This page looks at the **outline designing framework.** This gives simple guidance on solving straightforward problems, or problems where time is very limited.

The following pages then look at the **integrated designing framework** which gives a lot more detail. This is suitable for more complex problems, and problems for which you have more time.

When designing, points you might consider include:

1. What is the problem I am trying to solve?

2. What is the purpose of my solution?

3. How will my solution be used?

4. How realistic are my suggested solutions?

5. Are the materials available?

6. Is my solution economical?

7. Is there sufficient time to make it?

8. Is this design task a challenge?

NOTE: Specific syllabuses may have a defined design process or framework. You should check this before tackling a design task as the framework may be linked to the assessment scheme.

OUTLINE DESIGNING FRAMEWORK

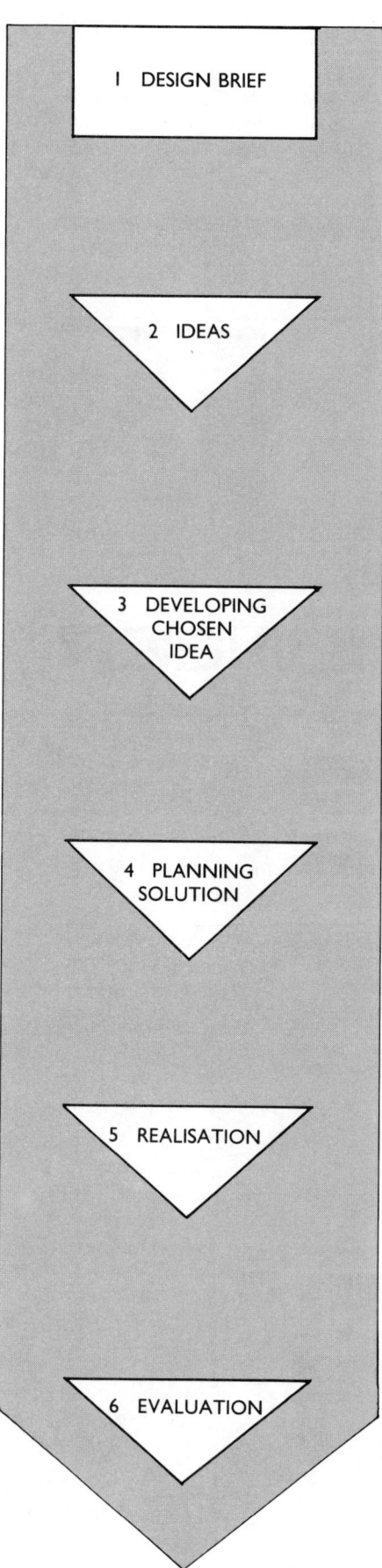

1 DESIGN BRIEF

2 IDEAS

3 DEVELOPING CHOSEN IDEA

4 PLANNING SOLUTION

5 REALISATION

6 EVALUATION

fig 1.37.

1. **The design brief** is a clear description of the problem you are going to solve. You may have to form the brief from either a situation (such as the layout of your kitchen) or a theme (such as storage). Before finalising the brief you may need to do some investigations.

2. From the design brief you should get some initial **ideas** for your design. You may then need to carry out research before beginning to sketch these ideas. It is wise to sketch a number of realistic ideas so you have some choice about the solution. On completion of the ideas you should decide which one to carry forward, giving your reasons.

3. It is important to **develop the chosen idea** into a practical solution. This may involve further drawing of how it fits together, or modelling to see how it will work. At the end of this stage you should have a sound idea of what you are going to make.

4. The **planning of the solution** involves two stages. First, you need to make some drawings from which your solution can be made. These are often called working drawings. Secondly, you should plan how your solution is to be made and which materials and components are required.

5. **Realisation** means making — either a scale model, a prototype or the final product. It is the realised model or product which will be evaluated, so the quality is very important. The realisation is one of the most exciting stages, but if the work in the previous stage has been poor, you may have difficulties.

6. **Evaluation** means finding out how well your solution works, and comparing it with your initial brief. Your evaluation should also suggest possible improvements.

INTEGRATED DESIGNING FRAMEWORK

Stage 1. Brief
Recognition of problem
Identification of needs
Recognition of situation
Formation of design brief

Stage 2. Investigation
Research into topic
Collation of useful information
Analysis of topic
Specification of requirements

Stage 3. Ideas
Generation of realistic ideas to
satisfy design brief

fig 1.39. Thinking and sketching an idea

Stage 4. Evaluating
Evaluation of ideas against the
specification
Identification of a proposed solution

Stage 5. Developing
Modelling, developing and refining
the proposed solution

Stage 6. Planning
Drafting drawings from which it can
be realised
Planning and organising proposed
realisation

Stage 7. Realisation
Realisation of solution in the form of
a model, prototype, artefact or
system

Stage 8. Testing
Testing to see if it works and how
well it works

Stage 9. Evaluation
How does it meet the brief?
How can it be improved?
How did I tackle the problem?

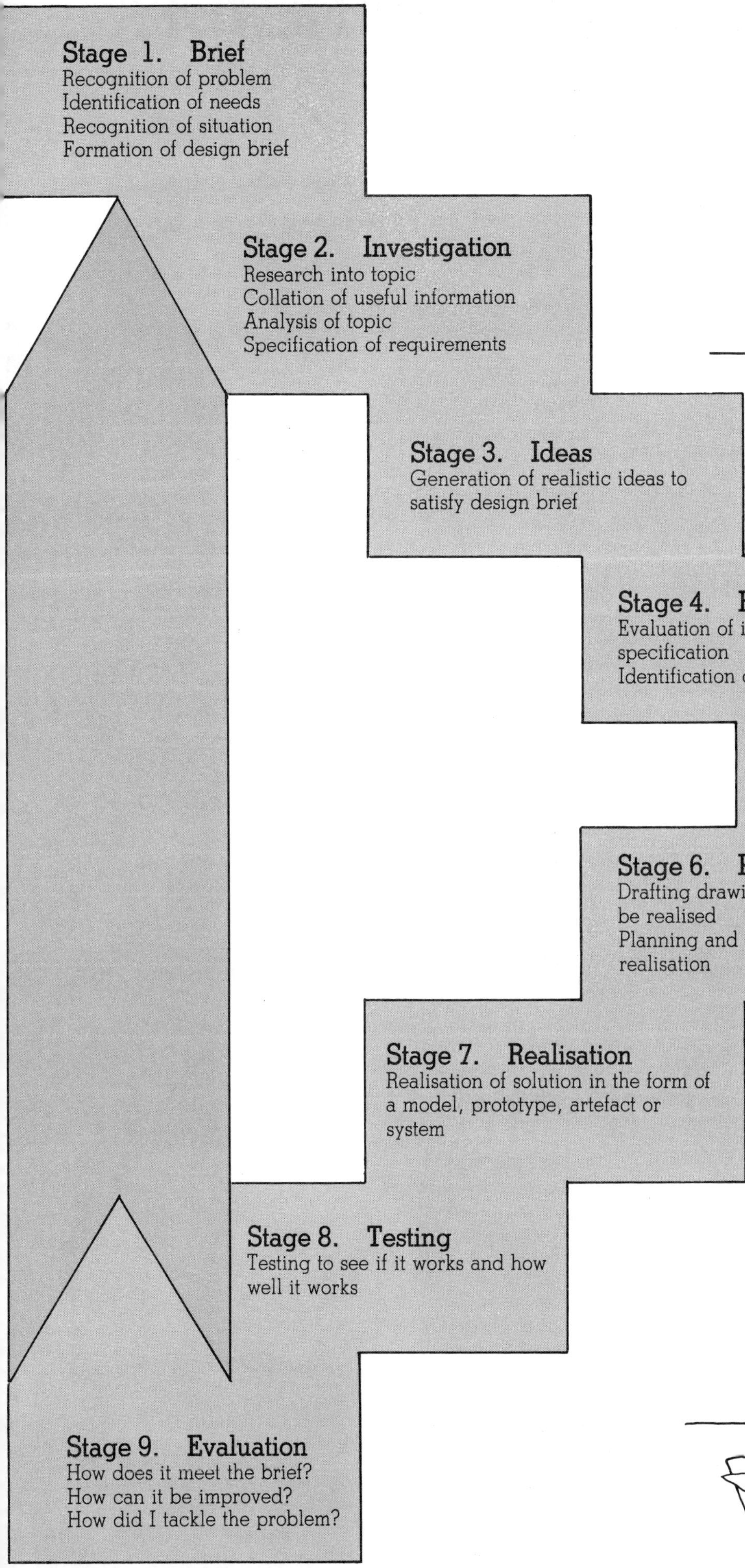

fig 1.38. Integrated designing framework

fig 1.40. Modelling card

INTEGRATED DESIGNING FRAMEWORK

Designing becomes a lot easier if you have a framework to follow. The following pages explain how the integrated designing framework can be used. It is important to recognise that this should not be followed laboriously, but should be modified for specific problems. Various design methods and techniques are covered, but they are only suggestions which might be helpful.

STAGE 1: BRIEF

Where do you begin when you write a design brief? In some cases you might begin by recognising a problem, for example, how to get books to stand upright on shelves.

Another starting point might be seeing the need to improve something, like the instructions for fitting a burglar alarm.

A third starting point might be recognising an area where things are not working well because of layout — for example, in the kitchen. In this case you might have to do some research to find the real need.

From these differing starting points you will need to write a design brief. Design briefs range from the simplistic (*Design a seat*) to the more precise (*Design a garden seat which will be used by elderly people who may wish to sit in pairs*). This second brief is more useful because it sets out clear guidelines. The brief could go on to define more clearly details of colour, or describe the environment in which it would be situated.

The amount of detail given in the brief will decide the amount of freedom the designer has to experiment. If for instance, a design brief stated *'Design a birthday card'*, the designer would have difficulty, because it could be for an 8-year-old or a 90-year-old. However, *'Design a birthday card for a teenager'* states the target without restricting ideas.

It is important not to give too much detail in a design brief, otherwise the designer can do little creative work. There are therefore four key points to remember about your design brief:

1. Identifty a task which you are keen to work on.
2. Make sure the brief has a purpose.
3. Do not begin with such a vague brief that you have no idea where to start.
4. Do not define the brief so precisely that there is no room for innovation.

fig 1.41. Situation with design opportunities

fig 1.42. A need to package these biscuits

STAGE 2: INVESTIGATION

Investigation leads to a clearer understanding of the limits of the design problem. First of all you should read and understand they key words in the brief.

Consider the brief, *'Design a storage unit for kitchen roll, cling film and aluminium foil, which will take up a visible position in the kitchen and will dispense the material easily.* The key points are storage, the three rolls, dispensing ability and appearance. These give a useful starting point for your design. Having analysed the brief you then need to research into the problems. For example:

1. Visit shops or exhibitions to view current products.
2. Draw up a questionnaire to discover further information.
3. Interview people about the problem.
4. Visit libraries and read magazines and books to find further information.
5. Write to and/or visit industry to discover more information.
6. Take a similar existing product or system and analyse it carefully.

Having carefully researched the topic, it is important to sort through the information and decide what is most useful. You should then have a good understanding of the task and can set out the exact limits and constraints for the designer. This is called a **specification,** and helps to focus towards the key aspects of the problem. A simple, systematic way of seeing if you have created a

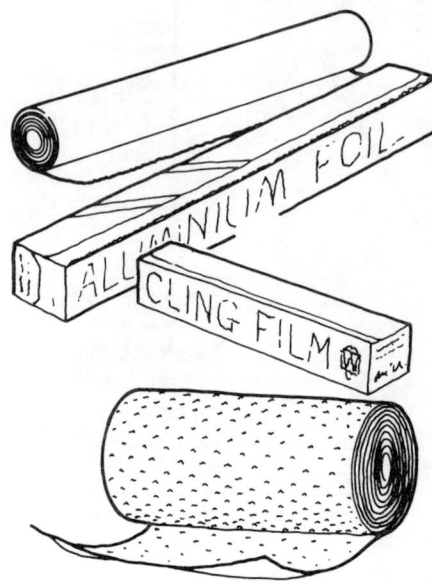
fig 1.43. How can these be stored?

good specification is to check whether key factors are covered. These may not all be appropriate to any one problem, but having checked them, at least you will know the point has been considered. For example:

size	function	appearance	storage
cost	safety	environment	materials
manufacture	ergonomics	shape	reliability
maintenance	finish		

The order and priority will change according to the design problem. For example, the design of a 'pop-up' brochure would place low priority on safety, whereas a child's toy would give high priority to safety.

The design factors will often result in a series of conflicting points being raised. However, balancing these points is the essence of good design and it is now that you move into the third stage of creating ideas.

STAGE 3: IDEAS

Generating ideas which solve the problem is the most creative area of the whole designing activity. The quality of these ideas is one of the key elements in D&T work. Ideas can be generated through thinking and sketching. At this stage you might want to draw complete artefacts very precisely. This is a mistake, as it tends to create rigid, isolated ideas. It is far better to make quick sketches of outlines and rough forms which you can easily modify.

Different problems will lead to differing approaches. For example, if you are designing a car jack, the functional operations will be a key part of the design. However, if you are designing a piece of jewellery, the starting position may be looking at shapes and forms. The sketches should be a means of thinking on paper, using notes where appropriate, and ensuring your rich ideas are recorded. Many syllabuses expect to see between three and five realistic ideas as minimum. With complex problems, ideas may be created for parts of the problem rather than the whole solution.

It should be recognised that ideas do not automatically appear when you wish. Ideas may come at any time, and you must sketch or note them when they occur, because they can easily be mounted in your folder at a later stage. However, solutions to design problems cannot wait for ideas to just arrive, they must be worked at to determine solutions. There are several ways of working at the generation of ideas and different methods can be used, depending on the problem. The following may be helpful.

1. **Observation and adaptation.** Look at existing solutions to similar problems, and then from these you can usually develop ideas. Look at nature to see how it solved the problem and then consider how this can be adapted. Remember that design is about solving problems in the best form, rather than always creating original ideas.

2. **Ideas from drawings.** The creation of ideas for drawing is particularly useful for work in creating the shape and form of a product. This method may begin from shapes such as lines, circles, cubes, prisms or pyramids, which are then cut, rotated or combined to generate a new shape. These visual investigations are a very effective way of creating ideas.

3. **Brainstorming.** This is usually a group activity where everyone thinks of ideas to solve the problem. The ideas are shared, which often stimulates further ideas or adaptations.

4. **Checklists.** The use of a checklist can provide more starting points to stimulate ideas. A word or phrase in a checklist can help you think about the problem from another viewpoint. The checklist may be as at the top of this page or it may be in the form of questions:
What is the purpose of what I am trying to design?
Who may use it?
Where is it going to be used?
Can it be modified?
What is the important part of the design?

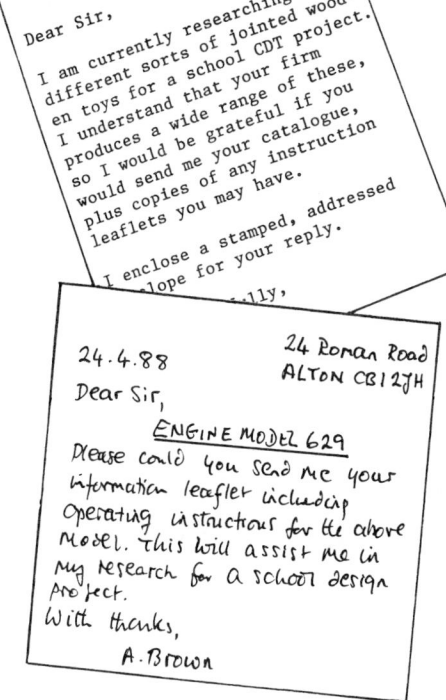

fig 1.44. Letters requesting information

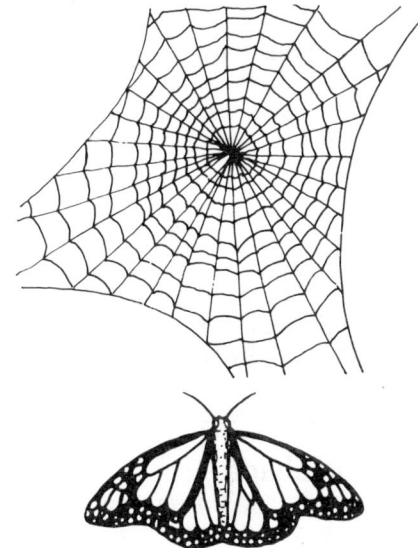

fig 1.45. Use nature as a source for ideas

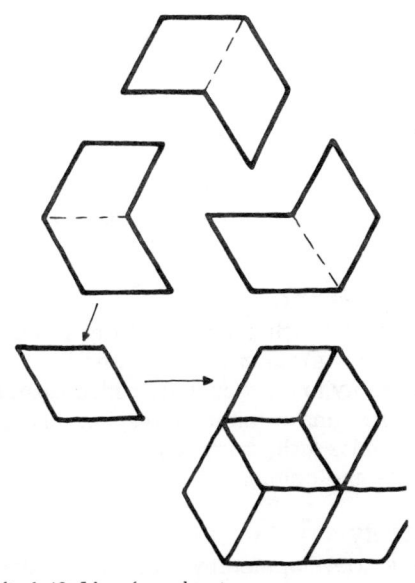

fig 1.46. Ideas from drawing

STAGE 4: EVALUATING

The evaluation of ideas is a critical phase, because it is at this stage that the proposed solution is first identified. It is wise to look carefully at all the ideas, but you need to be clear what you are looking for when choosing an idea to develop. These points may be helpful:

1. Does the idea meet the brief you started with?
2. If not, does it satisfy the need better?
3. Does it meet the specification?
4. Is it possible for the proposed solution to be made with the resources of time, materials and equipment available?
5. Is it financially viable?

In your design work it is wise to write down your reasons for making choices. At the end of the project it may be interesting to know the exact reasons for the decision-making at this critical stage.

STAGE 5: DEVELOPING

Developing and refining the proposed solution is the stage when you convert the idea into reality. A key part of this stage is the modelling of the proposal to see how it works and how it can be improved. The models can use specialist kits or modelling materials. There is no doubt that a good model can be most helpful in developing and refining a proposal. At this stage, a number of factors are likely to arise and require you to make a decision.

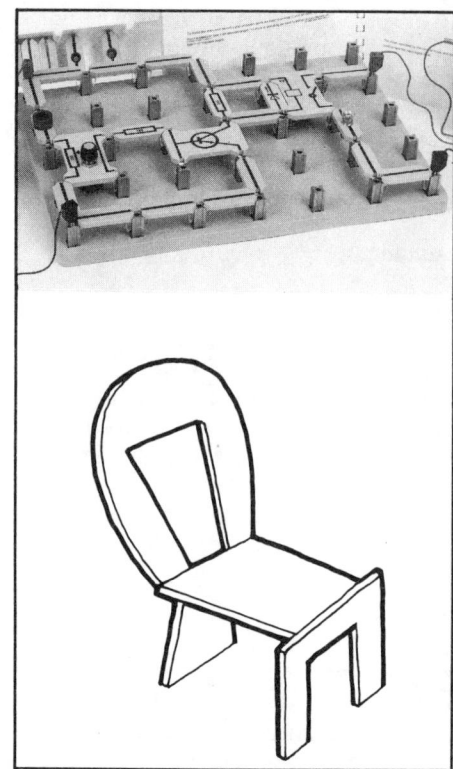

fig 1.47. Modelling solutions

Materials

Materials provide a major constraint. First, you must know how you wish the various parts of your solution to behave. You can then begin to identify an appropriate material for each part. Properties of materials which you might consider are:

weight   strength   toughness   feel   resistance to heat/corrosion

colour   hardness   conductivity   appearance   flexibility

Having identified the properties of the material, you should then consider its availability. This will depend on both the material (e.g. acrylic, mild-steel, beech, card), and also its form (e.g. sheet, tube or block). When you specify the form you should give accurate sizes. In choosing materials you should also consider the cost. For example the cost of precious metals or specialist electronic components might lead you to reconsider.

Sizes and shape

These key points will greatly influence your final design. You will almost certainly need to consider how your proposed solution will come into contact with people. This will affect sizes, ranging from printed letters that can be read easily, to the height of a table.

The study of how objects, systems, and the environment can be designed to fit in with people is called **ergonomics.** It is important for deciding such things as the best height for a computer screen, or the smallest size for calculator buttons which can easily be pressed one at a time.

In order to design solutions which fit in with people, you will need to know human body measurements. These are called anthropometric data.

It is also important to consider how your proposed solution will look in its environment. When designing your solution you will therefore need to consider its overall size in relation to other objects, and to assess this constantly.

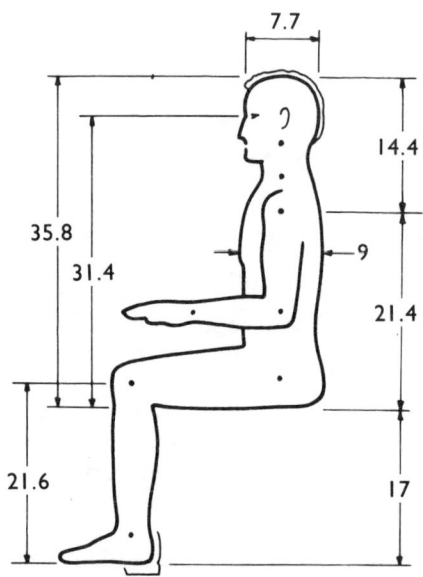

fig 1.48. Anthropometric data

Appearance

A highly functional product which looks awful is unlikely to sell, as is an attractive product which doesn't function. The visual qualities which give a sense of beauty to a product are called **aesthetics.** Aesthetic values vary with different cultures and within cultures, and fashion and styles change. It is wise to recognise fashion and style, but remember that there will always be differing tastes and opinions.

Safety

When developing any product, it is important that safety is considered throughout. Appropriate safety standards must always be applied to the design.

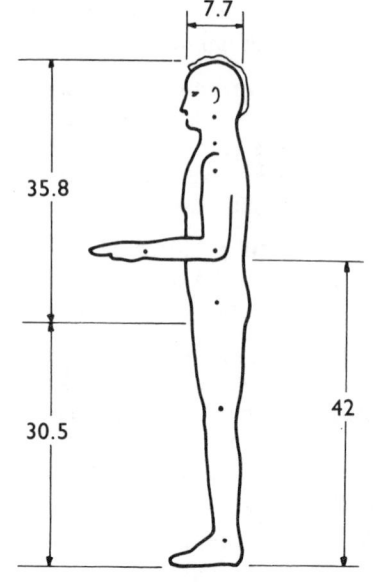

fig 1.49.

STAGE 6: PLANNING

Once the developing and refining of the proposed solution is completed, it is necessary to plan the realisation. Planning is done in two parts.

The first is concerned with the creation of working or production drawings, and the second involves planning to realisation. The working drawings are usually of a formal nature, perhaps in orthographic projection. They should show each part and how it fits together, and give details of all dimensions. This detailed planning through drawing is crucial, and it is essential that such drawings can be understood by others. As well as the working drawings there may sometimes be a presentation drawing which helps to convey the complete idea in its final form. With the final drawings there should be a list of materials to be used, and their sizes. This was traditionally called a **cutting list** and is now called a **parts** or **component list**.

Planning and organising the realisation requires considerable thought. At this stage it is wise to recognise the amount of time available for realisation, because working to a time limit is crucial in stage 7. It is important to write out a procedural flow chart for realisation and a time schedule. These may then need adjusting to fit your time limit. It may help to set a short time limit for each part. Planning can help you identify in advance materials and specialist equipment required.

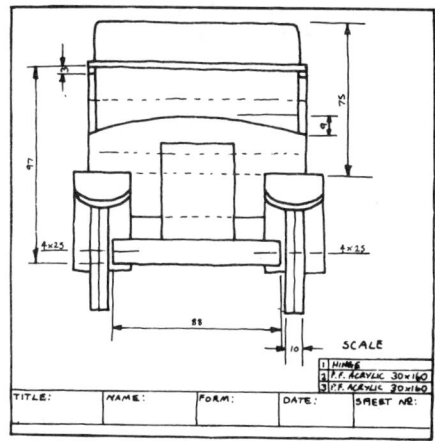

fig 1.50. A working drawing

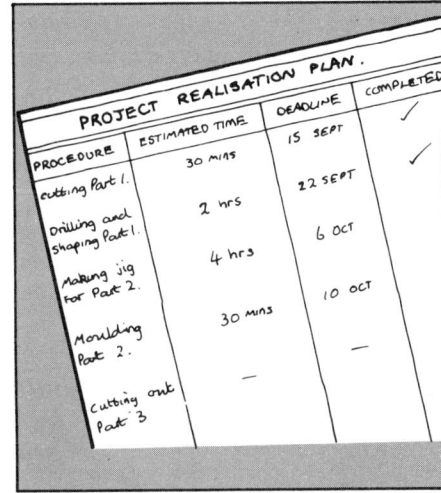

fig 1.51. Project plan

STAGE 7: REALISATION

This is one of the most exciting but time-consuming stages. It can also be one of the most frustrating if the planning stage has not been done thoroughly. The methods of realisation in D&T will vary depending on the particular course or problem. The medium will vary from paint and ink on some dimensional graphics, to heavy construction in perhaps timber, to an electronic system using mass manufactured components. Good design will often involve the use of several different media together to create a sound solution.

The realisation of the solution could include models, prototypes, artefacts or systems. Throughout the realisation there are likely to be problems which were not envisaged, and this can often lead to redesigning parts of the solution.

In the realisation stage it may become necessary to design special tools to help you make the solution. The following are common examples:

Templates are made so that they can be marked around or cut around to repeat a shape several times.

Jigs are tools which are made to allow an operation to be repeated accurately, for example drilling or sawing.

Formers are shapes which are made to allow materials to form around them. This is especially common in plastics.

Moulds are shapes into which materials are usually poured to repeat the shape. It is often necessary to make a pattern first from which the mould can then be made. The pattern is almost identical to the required component.

The making of these specialist tools takes time, but if they are well made it is usually worthwhile. If a project is part of the course work for your examination, you must keep these tools for assessment.

When carrying out the realisation there are some common ground rules:

1. Use tools, equipment and processes in a careful, safe manner.
2. Use materials economically, for example, do not cut material from the centre of a sheet.
3. Always measure and mark out materials accurately.
4. Aim for a high quality finish at all times.
5. On completion, always treat the end product with care.
6. Record briefly the procedures for making in a diary. This will help you learn from your errors and should help you in the future.
7. Try to set yourself small tasks within the project, so that you can assess your progress more easily.
8. Realisation of your solutions should be exciting, so enjoy it.

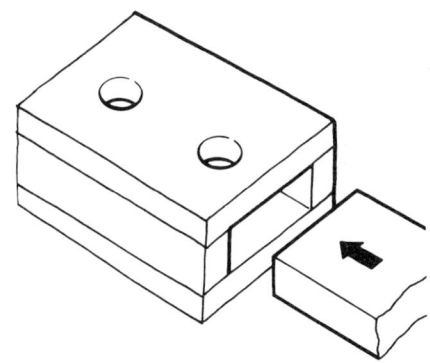

fig 1.52. A jig

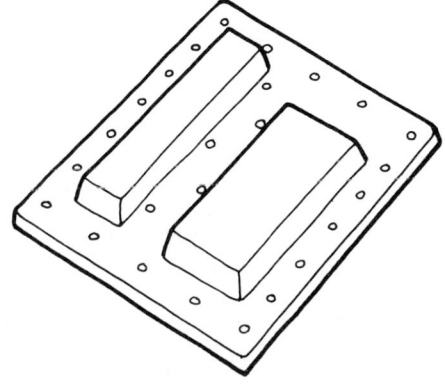

fig 1.53. Former for moulding acrylic

STAGE 8: TESTING

Testing the solution is an important part of designing. It will discover whether your solution works, and if so, how well. Testing may be functional, seeing how well your solution works by trying it out, or it may involve repeated tests, checking your solution's reliability. This can be done by carefully setting up an experiment to discover whether the solution meets the requirements in the brief. Testing should place the solution in its intended environment, and observe how it works.

However, if the test could cause an accident, it may be wise to carry out a simulation. This involves testing in an artificial way which is similar to the real situation.

Testing can also concern the appearance of your solution. Here, good testing will involve getting opinions. You might find it helpful to draw up a list of questions on a response sheet.

When you have completed your testing, your results may include the need to start designing again, to improve the solution further still.

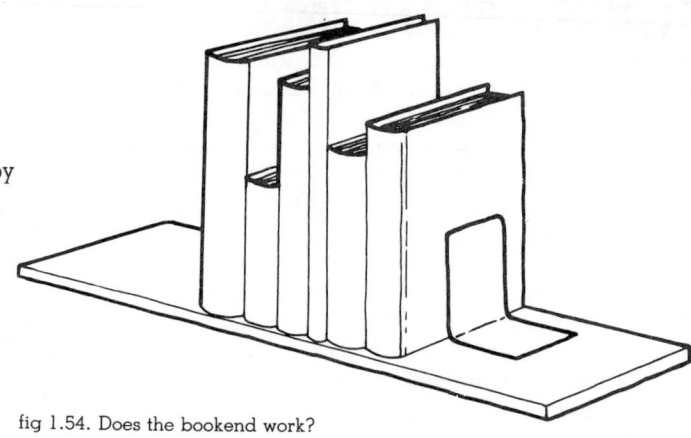

fig 1.54. Does the bookend work?

STAGE 9: EVALUATION

This final stage is concerned with assessing the whole process, from deciding upon the task, creating ideas, and leading to a fully realised design which has been tested. The evaluation should be critical in identifying faults in the process, lessons you have learnt for the future, and possible improvement to your design. Throughout an evaluation, it is wise to be constructive. The following questions may be helpful:

1. Did I use my time effectively?
2. What are the strengths and weaknesses in my design?
3. How can I improve my design?
4. Did I manage to overcome the problems which arose during the project?
5. Were the planning and working drawings adequate?
6. What are the views of others about the solutions?
7. How would I tackle it again doing the same or similar projects?

On completing the project it is clear that you are now ready to begin again, because there is no doubt the solution could be improved. This shows clearly that designing is a fully integrated process and this framework should assist you in solving your design problem.

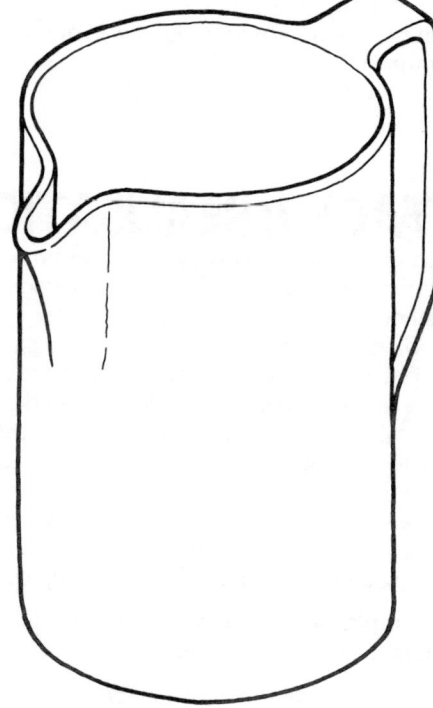

fig 1.55. How would you evaluate and improve this product?

DESIGN FOLDER LAYOUT

Examination boards are likely to suggest a layout for design folders and this should be followed. If no structure is provided, the layout opposite may help you organise your work so it can form a piece of effective communication.

Throughout the design folder, all pages should be named, numbered, and different sections identified.

The presentation is important, but rough sketches should be included to show your thoughts. Do not put research booklets or leaflets in without explaining their value to your design work. Do put letters written and received in the folder. If possible use photographs to explain stages of modelling and making which may be lost in a final solution.

1. Title page — your name, title of project, school and year.
2. Contents page.
3. Design brief.
4. Investigation — analysis, research data and specification.
5. Ideas and their evaluation.
6. Development.
7. Planning.
8. Working drawings.
9. Making — may be shown with sketches or photographs.
10. Testing.
11. Evaluation — including report diary if used.

BASIC GRAPHICS

Basic graphics refers to the graphic skills and techniques used in Technology. It involves drawing, using colour, displaying information and presenting ideas.

Drawing has many purposes, ranging from recording visual information to expressing ideas or feelings. In Technology you are concerned with drawings as a means of **communication**. You may want to communicate with yourself, clarify your own thoughts, try to visualise an idea or explain to someone else how something works or how it is made.

Colour is very important in terms of communication. It is used to help highlight ideas or to show graphically the type of material to be used.

You often need to display information or give instructions. This can be done graphically using drawings or diagrams. Details of electronic circuits can be shown schematically, methods of manufacture can be shown in simple stages.

Presentation is important in Technology. Ideas need to be well presented in order for them to be easily understood. Finished project work also requires a high standard of finish and presentation. Good work can often be spoilt by poor presentation.

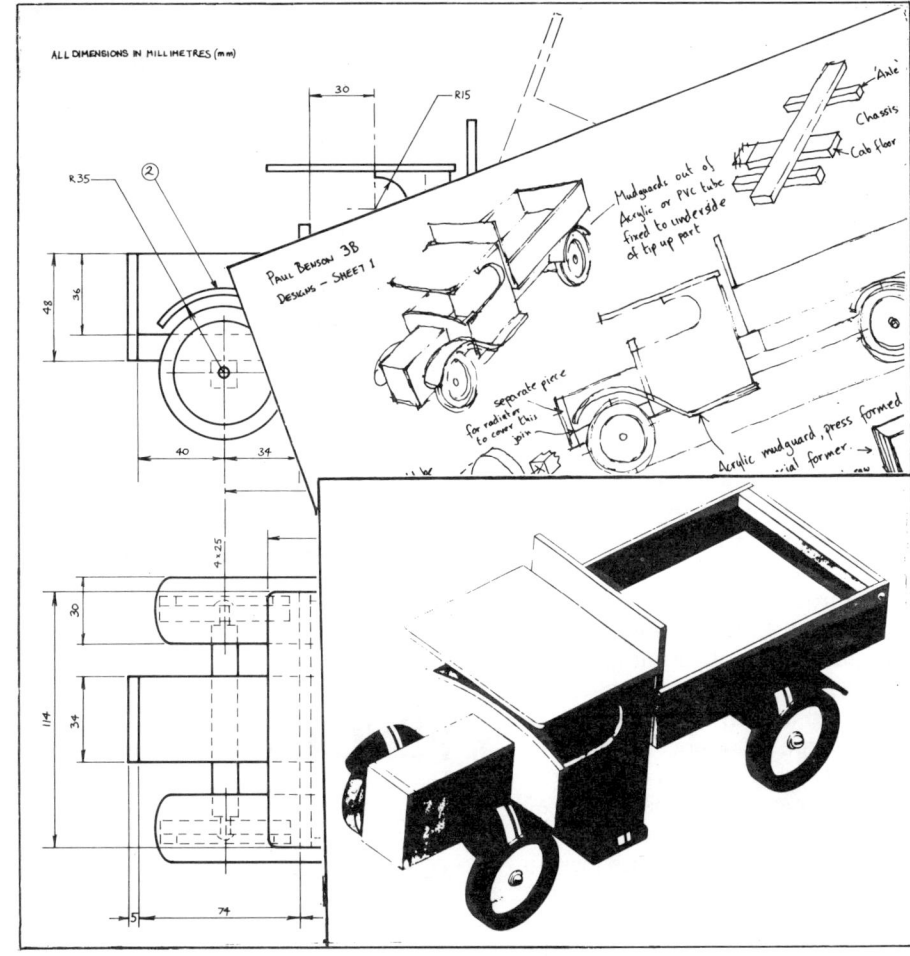

fig 2.1. Examples of graphics related to designing and making

IDENTIFICATION OR PROBLEM BRIEF ANALYSIS	Understanding the problem using freehand sketches and notes.
Precise written statement of what is required.	SPECIFICATION
RESEARCH	Recording information and data using sketches, notes, diagrams and graphs etc.
Sketch ideas using a variety of techniques.	IDEAS
EVALUATION OF IDEAS	Notes and sketches
Show how product is to be made using sketches exploded views etc.	DEVELOPMENT OF CHOSEN IDEA
WORKING DRAWING	Orthographic drawings, sections and presentation drawings.
PLANNING PROCEDURE REALISATION	Charts and diagrams to show material requirements and stages involved.
TESTING AND EVALUATION	Record testing and evaluation using sketches and notes.

fig 2.2. Graphics and the design process

MATERIALS AND EQUIPMENT

Graphic work in Technology does not require vast amounts of material and expensive equipment. A very high standard of work can be achieved with relatively inexpensive equipment.

PAPER

Drawings can be made on a variety of surfaces, but in Technology paper and card are normally used. There are many different types of paper available, ranging from expensive hand or mould made papers to machine made papers. Two types of paper commonly used in Technology are, **layout paper** and **cartridge paper**.

Layout paper is very thin, rather like white tracing paper. It is ideal for sketching because it is thin enough to trace through when you need to redraw, yet it is white enough to draw on with pencil, ink or felt-tipped pen. However, as it is rather fragile, it is not suited to final drawing work or presentation drawing.

Most of our drawings are done on cartridge paper which is heavier and not as fragile as layout paper. Cartridge paper can be used with pencil, ink and a little colour wash, but if large amounts of water colour or other forms of 'wet' colour are to be applied then it may be necessary to use a much heavier watercolour paper.

Layout paper and cartridge paper are both available in the standard A sizes. This system of paper sizing is very easy to remember. It ranges from A6 to A0, A6 being the smallest. A5 is twice as big as A6, A4 twice as big as A5 and so on. You are likely to use A4 or A3 for most of your graphic work.

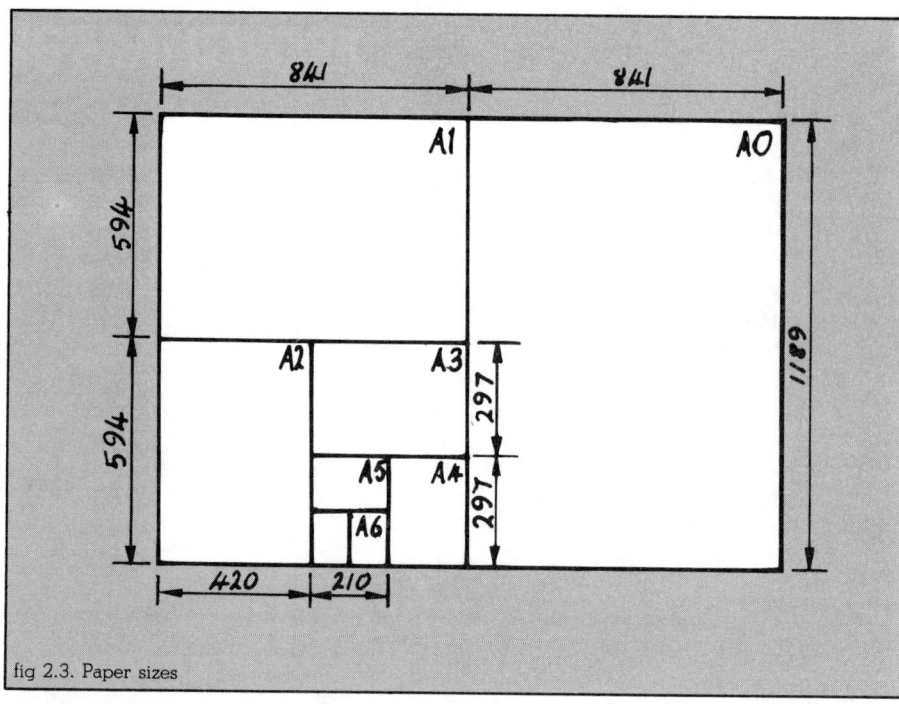

fig 2.3. Paper sizes

PENCILS

The most common piece of drawing equipment is the pencil. The traditional wood and graphite pencils are available in nineteen different grades, ranging from the very hard **9H** to the extremely soft **EE**. You will only need two pencils, a 2H and an HB. Use the 2H for construction lines and layout and the HB for sketching and the final lining in of the drawing.

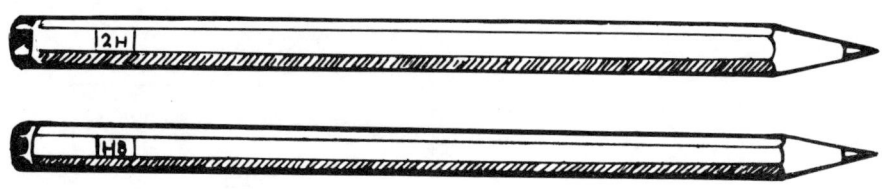

fig 2.4. Traditional pencils

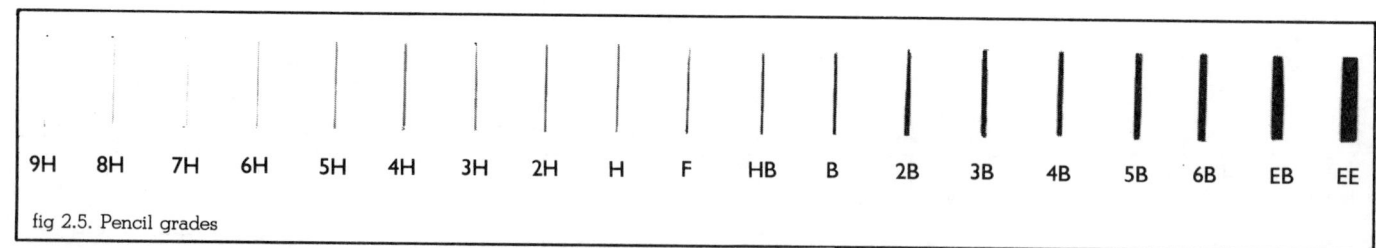

fig 2.5. Pencil grades

Some people prefer to use a **clutch pencil**. This has a plastic barrel rather like a 'biro', but inside there is a continuous length of graphite.

The **automatic fine lead pencil** is very similar to the clutch pencil except that it contains a polymer based 'lead' which may be less than half a millimetre in diameter in order to give a very fine, consistent line.

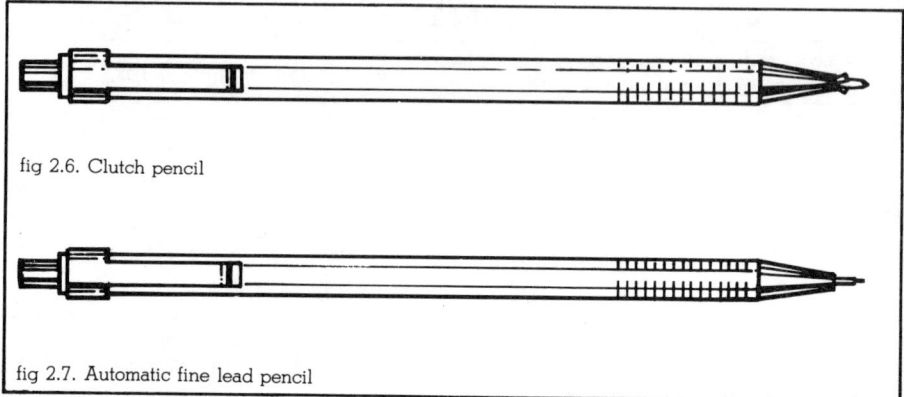

fig 2.6. Clutch pencil

fig 2.7. Automatic fine lead pencil

All pencils except the automatic line lead pencil need to be sharpened. There may be a desk mounted sharpener in your classroom or you may have a pocket sized sharpener of your own. Clutch pencils usually have a detachable sharpener on the end of the barrel. Many designers like to keep their pencils really sharp by occasionally rubbing the point on a piece of very fine glass paper. If you try this take care not to get the graphite dust on your drawing.

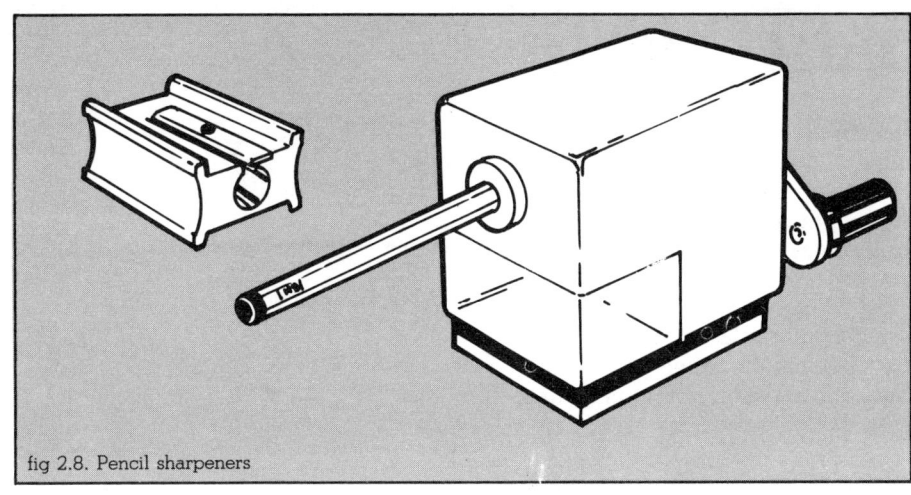
fig 2.8. Pencil sharpeners

PENS

Some designers prefer to use pens, especially if their work has to be copied or printed. There are some very fine line **fibre-tipped** pens available, but they do wear out in time and cannot usually be refilled. If you are doing a lot of drawing in ink it may be worth buying a **technical pen**.

Technical pens have a very fine, hollow nib which gives a consistent line. They are available in a variety of line widths, but the most common sizes are 0.35mm, 0.5mm and 0.7mm. Technical pens can also be used with a wide range of stencils for either lettering or symbols.

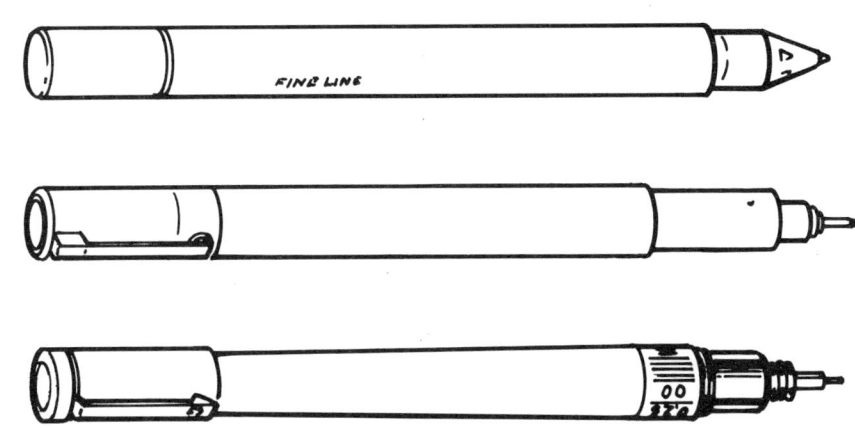
fig 2.9. Pens

fig 2.10. Technical pens

fig 2.11. Erasers

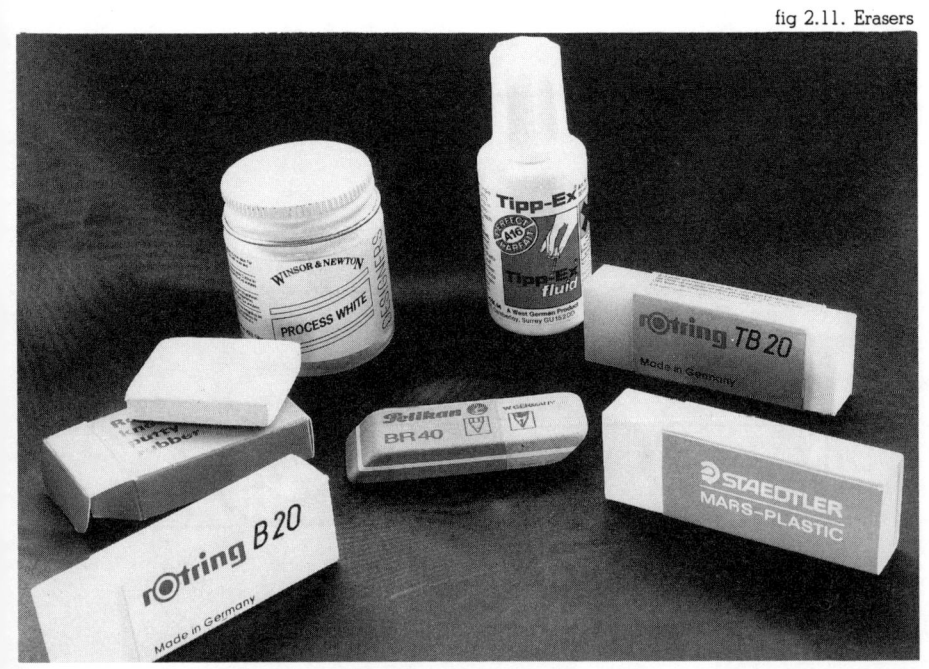

ERASERS

At some time you are bound to make mistakes. Pencil lines can be rubbed out in the normal way with a rubber, but you may find it less messy to use a **plastic eraser** instead of the traditional rubber. A **kneaded** or **putty rubber** is also useful for cleaning your paper after the drawing is finished, and for removing graphite dust which has been rubbed into your paper.

Mistakes in ink are much more difficult to remove. An ink rubber can sometimes be used, but be careful not to damage the surface of the paper. If a mistake cannot be removed with a rubber then you may need to cover it with either white ink or poster paint.

If you use correcting fluid to cover mistakes, make sure you read the instructions on the label carefully.

SKETCHING

Sketching is very important in this subject. It is extremely useful when you are getting your thoughts down on paper quickly. Ideas can be explained far more easily with sketches than with words.

Sketching is usually done **freehand**. Using a ruler or a straight edge takes too long, breaks up the flow of your ideas and prevents you from getting your thoughts down quickly.

START SKETCHING

If you have not sketched before, take a pencil and a piece of paper, and practice drawing a series of horizontal lines. Draw the lines about 50mm long and aim to keep them as straight as possible without using a ruler. Try to draw with your **arm** and not just by moving your wrist. Work quickly and freely.

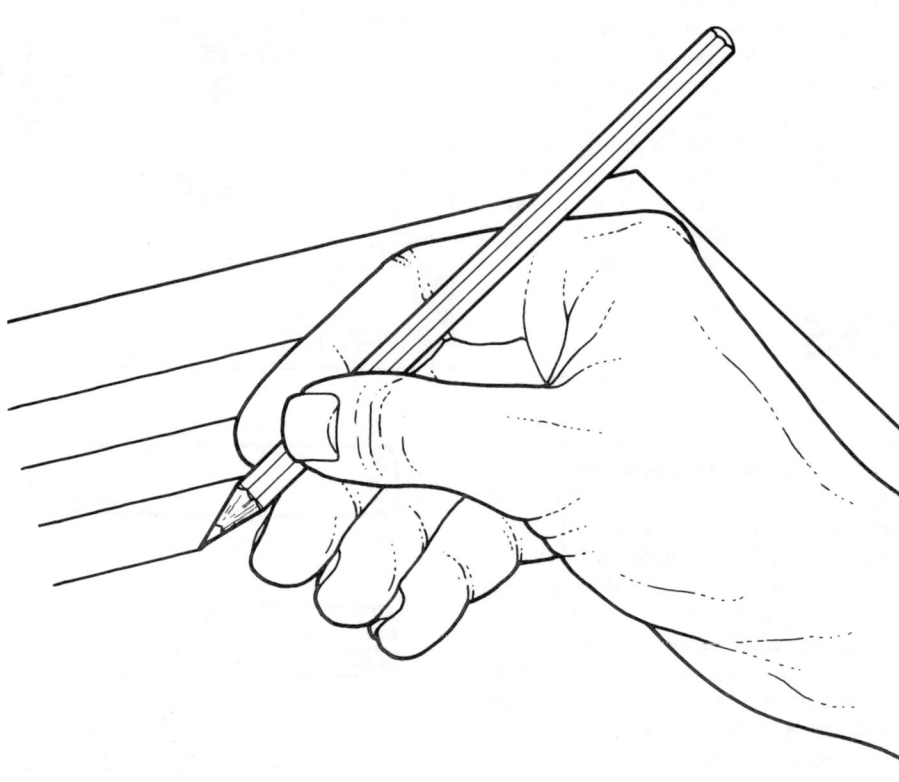

fig 2.12. Holding the pencil correctly

When you have drawn horizontal lines try doing the same with vertical lines. Work from the top of the page downwards and try to keep the lines as vertical as possible. Once you feel confident with vertical lines, you can practise drawing diagonal lines too. Work in both directions and try to keep your lines at approximately 45°. Don't forget to keep your pencil sharp!

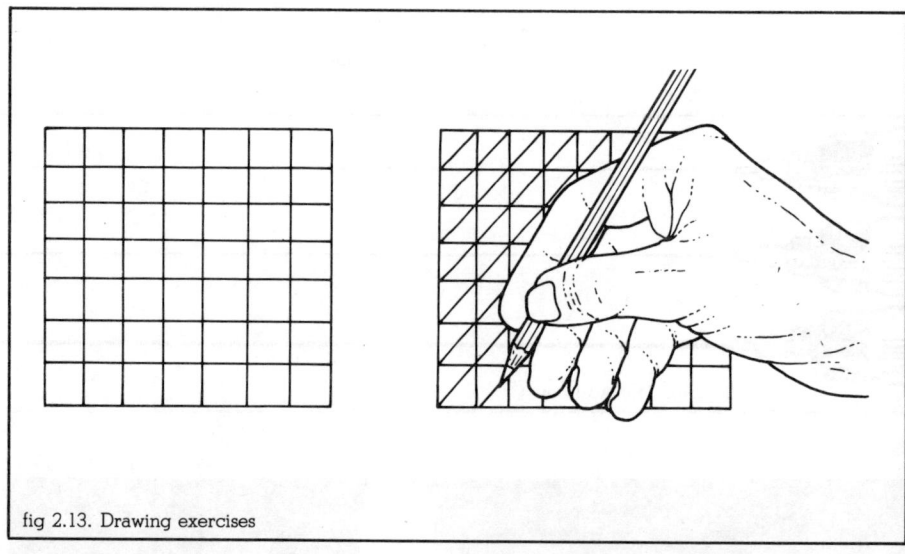

fig 2.13. Drawing exercises

When working freehand there are several traditional drawing techniques which are useful to help you produce a realistic three-dimensional drawing.

OBLIQUE DRAWING

A simple way of making three dimensional drawings is the method known as **oblique drawing**. This method is used when it is important to show the front view of an object. The oblique lines are usually drawn at 45° to the horizontal.

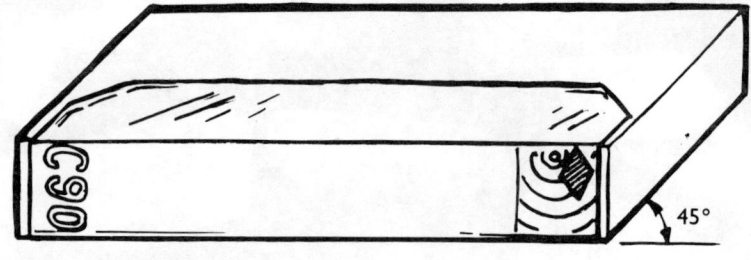

fig 2.14. Oblique

ISOMETRIC DRAWING

Isometric drawing is also useful when sketching freehand. It is best imagined as the object being turned until the horizontal lines appear to be at 30° to the horizontal. This method shows more of the top of the object.

Both these methods of drawing distort the view of the object slightly. If a more realistic view is required then it is best to draw it in **perspective**.

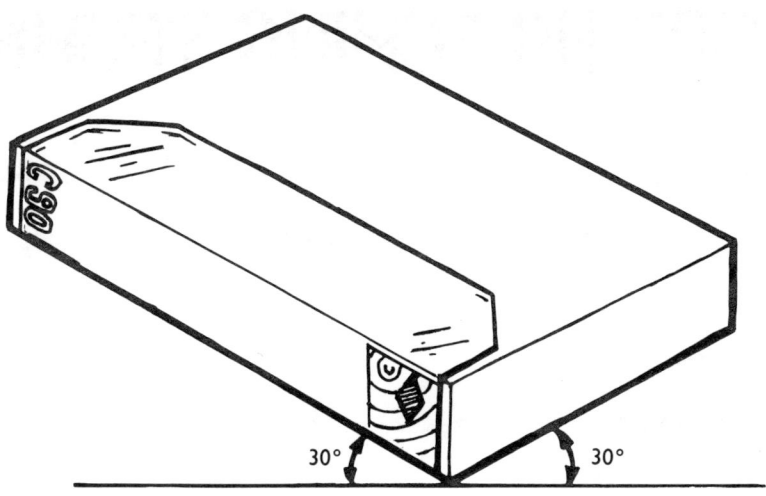

fig 2.15. Isometric

PERSPECTIVE DRAWING

Perspective drawing is based on the fact that lines appear to converge and meet at the **vanishing point**. You have probably seen this visual effect when looking down a railway line or a very straight road.

There are two types of perspective drawing: **single-point** and **two-point perspective**.

Single-point

In single-point perspective all horizontal lines converge and meet the one common vanishing point.

In the example shown, single-point perspective shows the front and one side of the object. If we want to show the top surface, then it is best to use two-point perspective.

fig 2.16. Single-point perspective

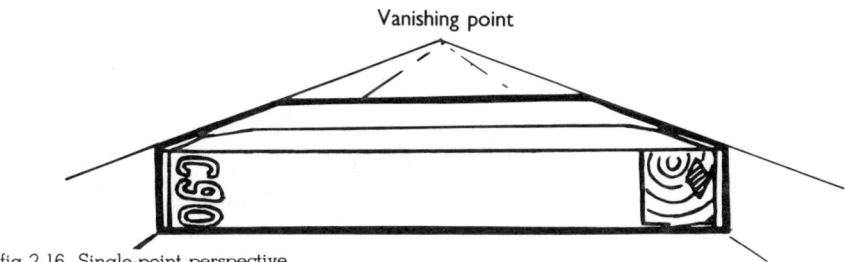

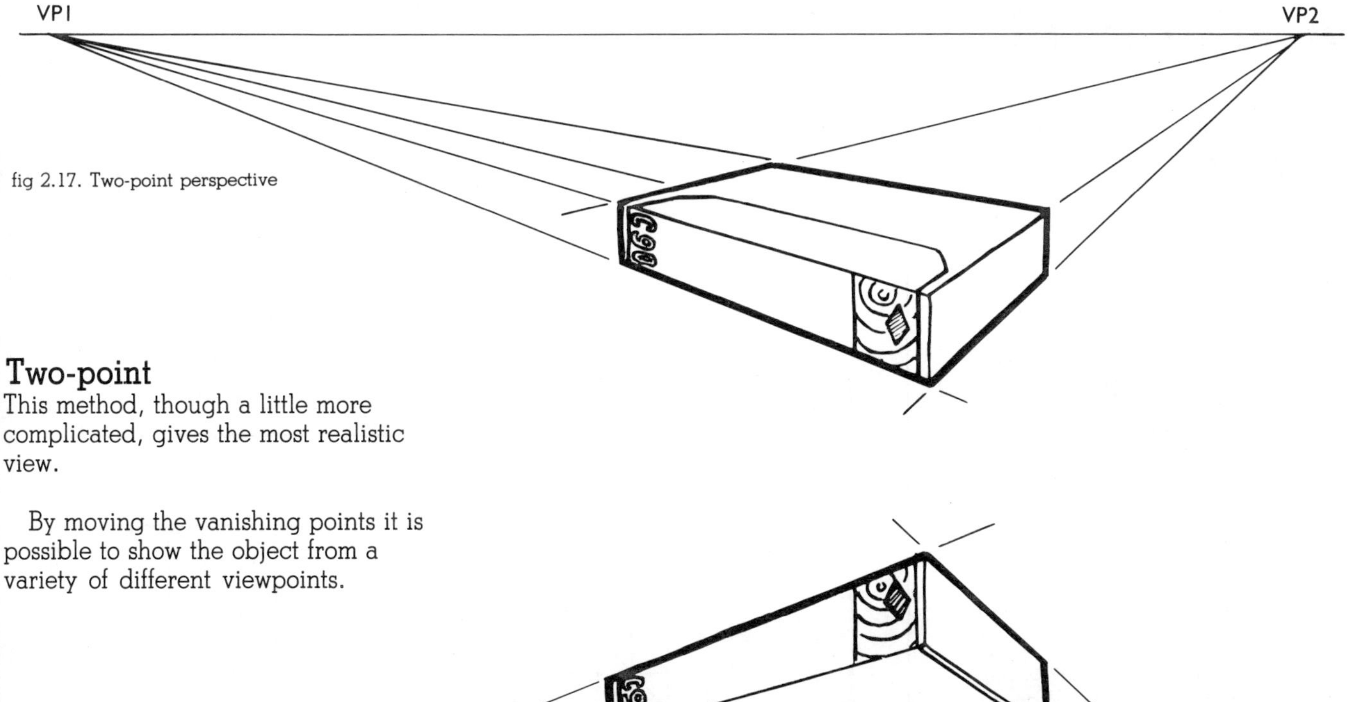

fig 2.17. Two-point perspective

Two-point

This method, though a little more complicated, gives the most realistic view.

By moving the vanishing points it is possible to show the object from a variety of different viewpoints.

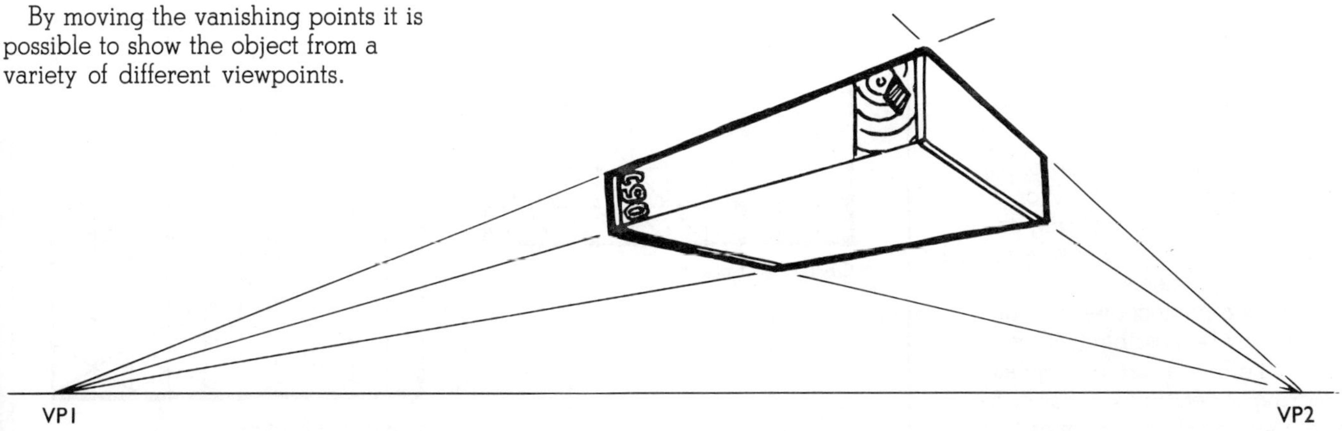

fig 2.18. Changing the viewpoint

MORE SKETCHING TECHNIQUES

VPI ◄──────────────────────────────► VP2

CRATING

A useful way of constructing a drawing or sketch is to draw a simple box or crate which will enclose the object you want to draw. This is known as **crating**. Lightly draw the crate first using any of the drawing methods shown so far and then draw the object within the crate.

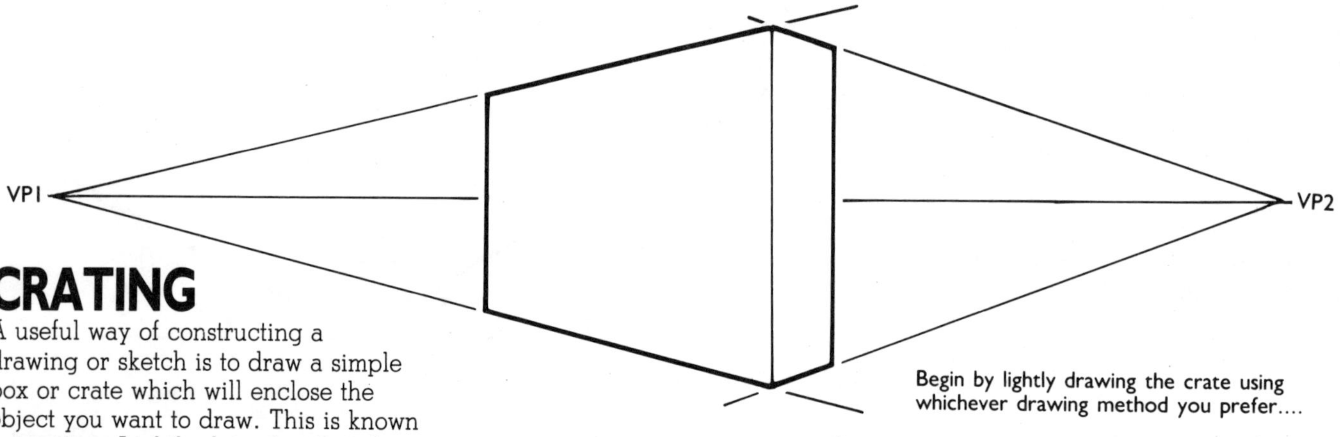

Begin by lightly drawing the crate using whichever drawing method you prefer....

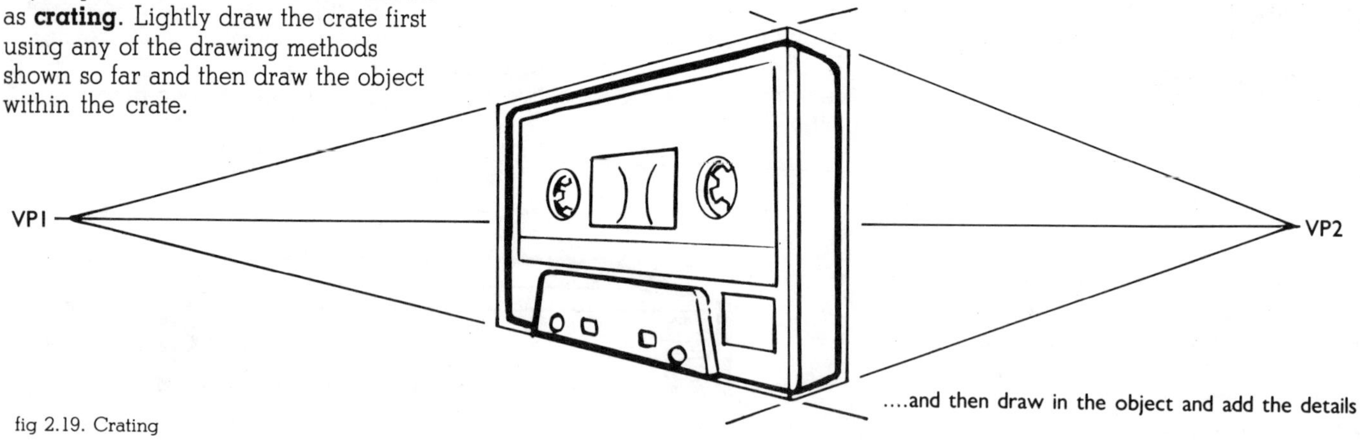

VPI ◄──────────────────────────────► VP2

....and then draw in the object and add the details

fig 2.19. Crating

BACKING SHEETS

Grids printed on to a backing sheet and used underneath your paper are very useful in helping you to learn to sketch. They work particularly well with layout paper and can be seen through most medium cartridge paper.

fig 2.20. Using a backing sheet

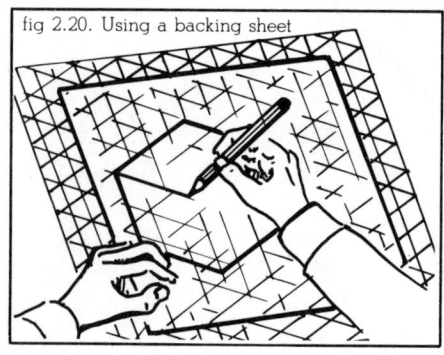

A backing sheet consisting of horizontal lines is also useful for making sure that your lettering or handwriting is level and evenly spaced on your design sheets.

If there is a photocopier in your school it may be possible (with your teacher's help) to produce a selection of backing sheets for your own use. When freehand drawing remember ... Do not use a ruler or a straight edge and try to work quickly and freely.

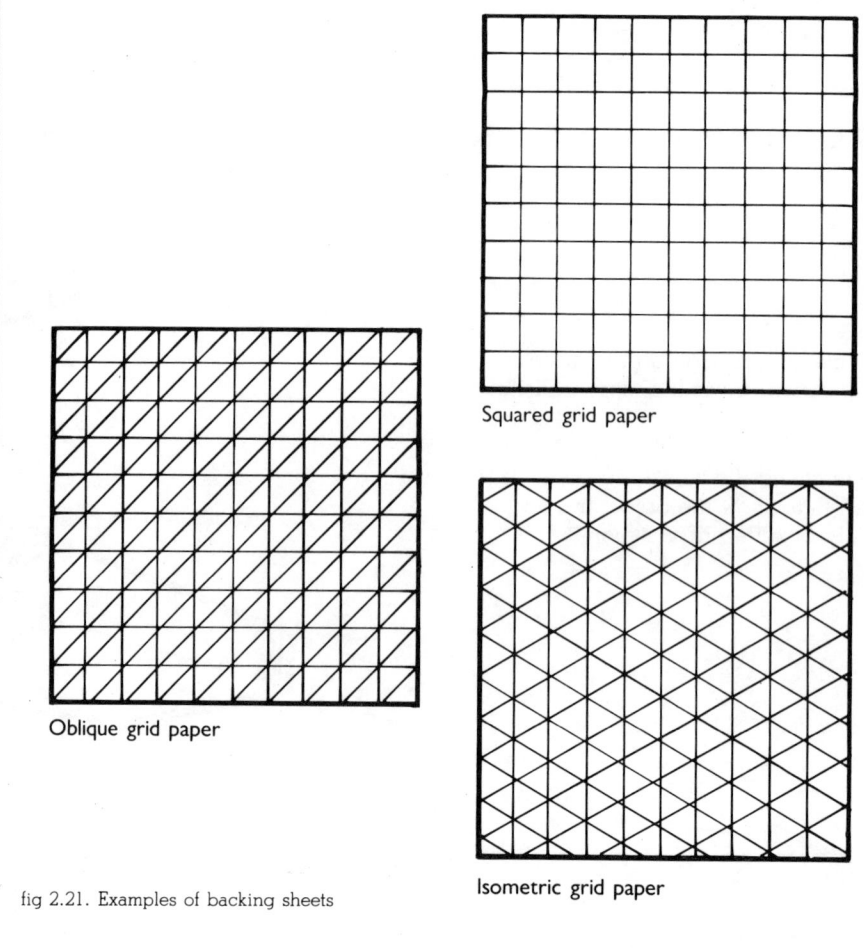

Squared grid paper

Oblique grid paper

Isometric grid paper

fig 2.21. Examples of backing sheets

ELLIPSES

Freehand sketching of objects made up from straight lines is simple, but when curves and circles are introduced it becomes a little more complicated.

As soon as circles are viewed from an angle they have to be drawn as **ellipses**. They no longer appear round, but oval or **elliptical**.

Sketching ellipses is not difficult. The method is illustrated in fig 2.23. This method of sketching ellipses will work whichever drawing method you use, as shown in fig 2.24.

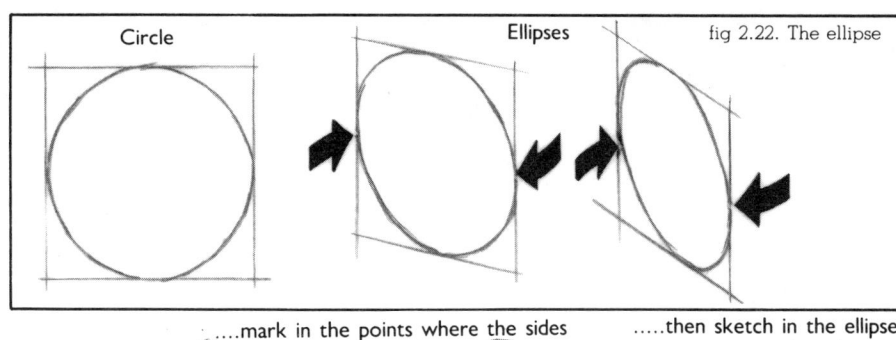

Circle   Ellipses   fig 2.22. The ellipse

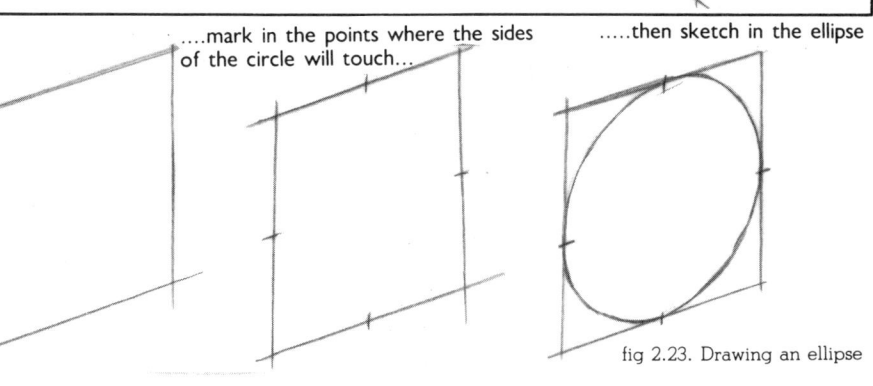

....mark in the points where the sides of the circle will touch...

.....then sketch in the ellipse

Begin by drawing a rectangle which will enclose the circle.....

fig 2.23. Drawing an ellipse

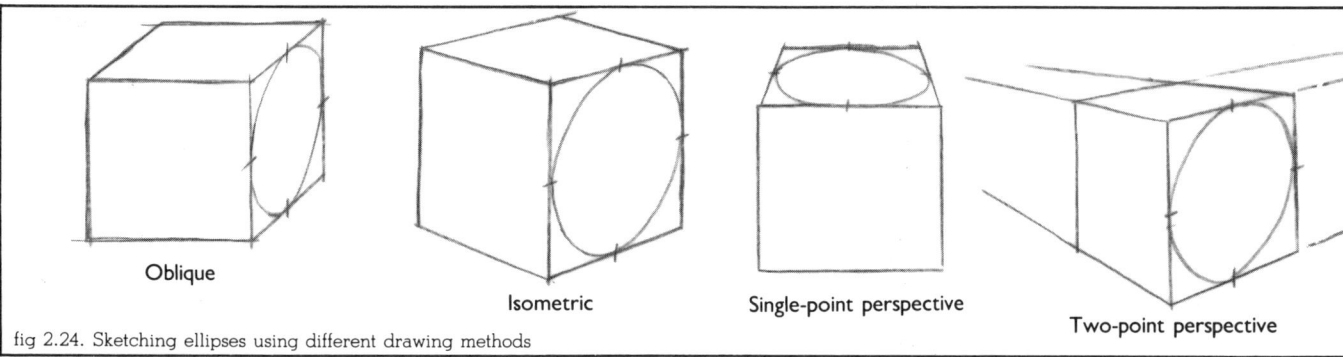

Oblique   Isometric   Single-point perspective   Two-point perspective

fig 2.24. Sketching ellipses using different drawing methods

SKETCHING CURVED SURFACES

Curved surfaces can be drawn quite easily using the same technique.

Begin as before, marking the points where the curved shape will touch the box and then lightly sketch the shape. When you are happy that the shape is correct go over it with your HB pencil to line it in.

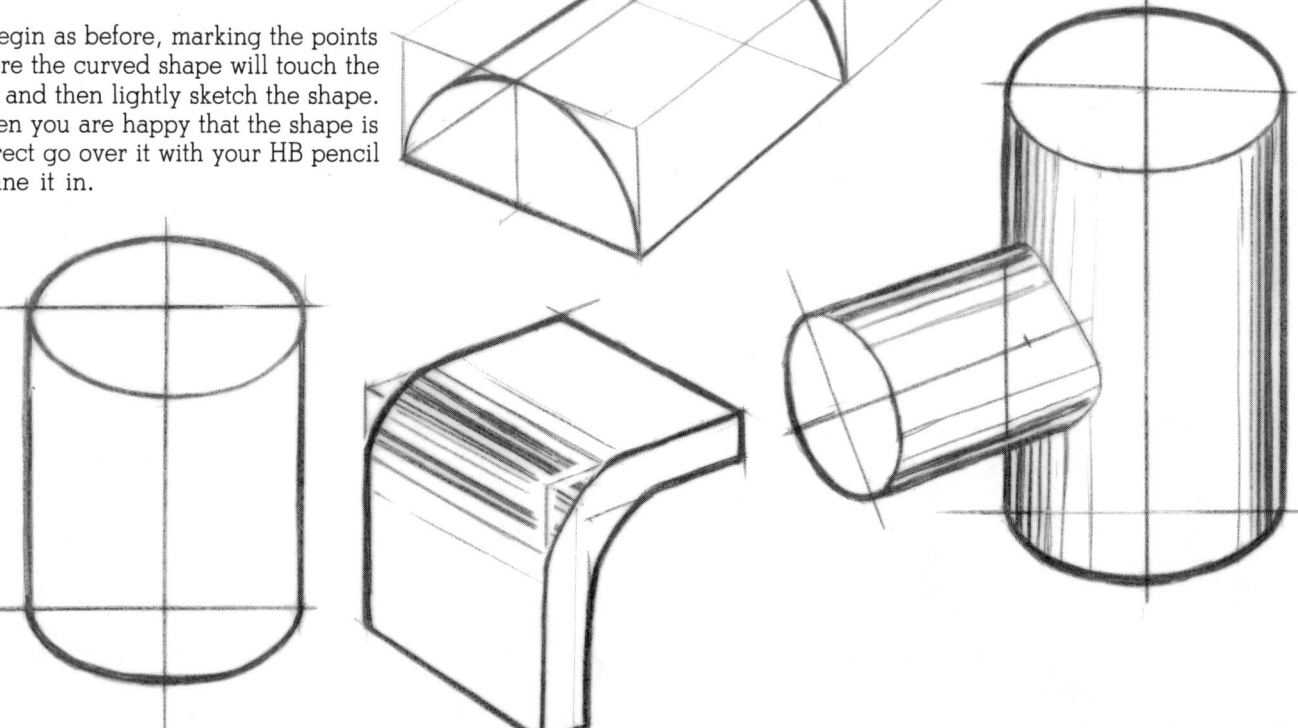

fig 2.25. Sketching curved surfaces

COLOUR

Colour is very important in presentation, it can improve a piece of work or it can ruin it. It is also important to consider the colour of your finished product while still at the design stage. An unsuitable colour could ruin a good design.

There is a great variety of materials available to use, but it is better to keep to those which you can use with confidence. Pencil crayons, felt-tipped pens and watercolours are probably the most suitable.

PENCIL CRAYONS

Pencil crayons are available in a wide range of colours and tones, and are simple and clean to use. By varying the pressure of the pencil on the paper it is possible to produce a good range of **tones** and graduated effects.

FELT-TIP PENS AND MARKERS

Using **felt-tip pens** requires a little more care. Some of the inexpensive types are often only available in a limited range of bright colours which are not really suitable for Technology work. However, the more expensive designer's graphic markers are excellent, especially when used in conjunction with pencil crayons. Graphic markers are available in a variety of subtle colours and tones. The grey colours are very useful for showing shadows and representing metal.

When using graphic markers, always replace the cap immediately after use as they can dry out very quickly. It is also important to use a sheet of scrap paper under your drawing as the ink tends to soak through. It is best to work from light to dark with markers, leaving the paper white to represent the lightest areas in the drawing and putting in the dark areas last.

WATERCOLOURS

Watercolour can be used as a wash on drawings, but it does require practice to apply well. It can only be used on a fairly heavy paper, and cannot be successfully applied to layout paper. Cartridge paper is suitable, but in order to prevent it from wrinkling, needs to be soaked in water, stuck down to a drawing board with gummed paper tape and allowed to dry. As it dries it will shrink and tighten, and will not then wrinkle when watercolour is applied.

When working with watercolour, work with your paper taped to the board, held at a slight angle. The paint can then be allowed to run down the paper and can be controlled with the brush. Any surplus paint can be lifted off the paper with either the brush or a piece of tissue. This technique may take a little practice to master.

It is essential when using watercolour to keep your water clean, otherwise the colour will appear muddy. Keep a separate jar of water for washing your brush.

fig 2.26. Colour materials

USING COLOUR

Colour can be used as a highlighter by applying a small area of colour around the idea or object drawn on the paper (fig 2.27). Alternatively, the object itself can be coloured.

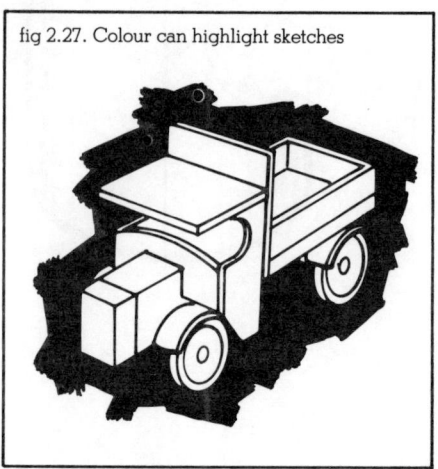

fig 2.27. Colour can highlight sketches

fig 2.28. Colour combinations can show texture

Another excellent use of colour is to use it to show the type of material to be used. A combination of watercolour or graphic marker and pencil crayons is very successful for this. Apply a wash or a background colour and then pick out the details such as woodgrain or reflections with pencil crayons. A white pencil can be very effective when used to show reflections or highlights on shiny surfaces.

The use of colour on a sheet of ideas can improve them enormously or ruin them if not properly applied. Take care with your choice of colour, do not use bright, striking colours such as purple or bright green and avoid using fluorescent markers. Keep to subtle, earthy colours such as browns, yellows or oranges and take care when using other stronger colours. The object is to make the ideas stand out, not the colour itself.

RENDERING

In Technology, rendering refers to the techniques used to make simple drawings look more realistic and interesting. It can be done quite simply using shading or texture in the form of lines or dots.

SHADING

Shading is usually done using a soft pencil to create an area of tone. By varying the pressure of the pencil it is possible to create a variety of tones ranging from light grey to dense black. It is not possible to shade successfully using a pen. Lines, dots and hatching are used instead. Darker tones can be created using dots or lines close together.

Another method is to use weighted lines to give the impression of light and dark. Weighted lines are simply heavier, darker lines. Fig 2.29 shows the use of this technique.

LIGHT AND SHADE

When shading, it is a good idea to think about the direction of the light falling on the object you are drawing. Apply the shading to the side furthest from the light. This technique adds depth to simple shapes and makes them look more interesting. Drawing the position of shadows will also add to the realism of a drawing.

First determine the position of the light source and estimate where the shadows will fall, then lightly draw in the outline of the shadow and fill in with any of the shading techniques.

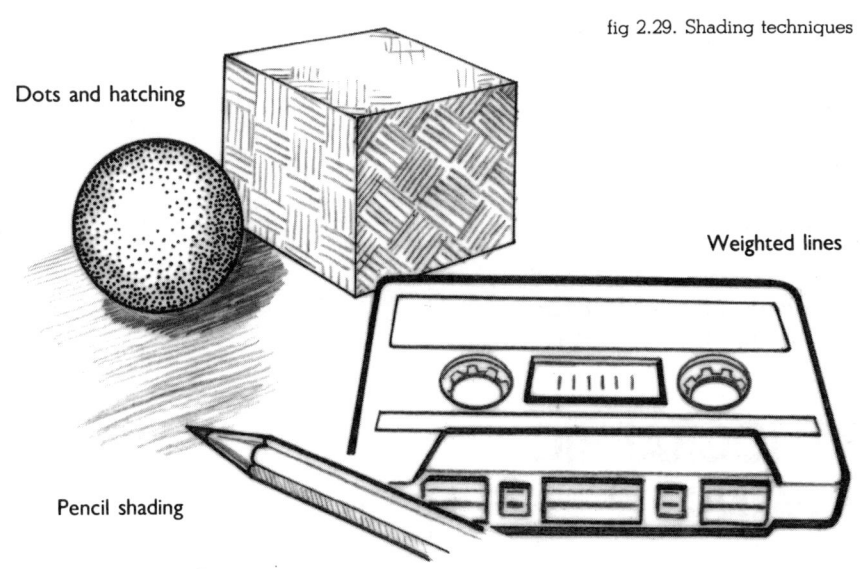

Dots and hatching

Weighted lines

Pencil shading

fig 2.29. Shading techniques

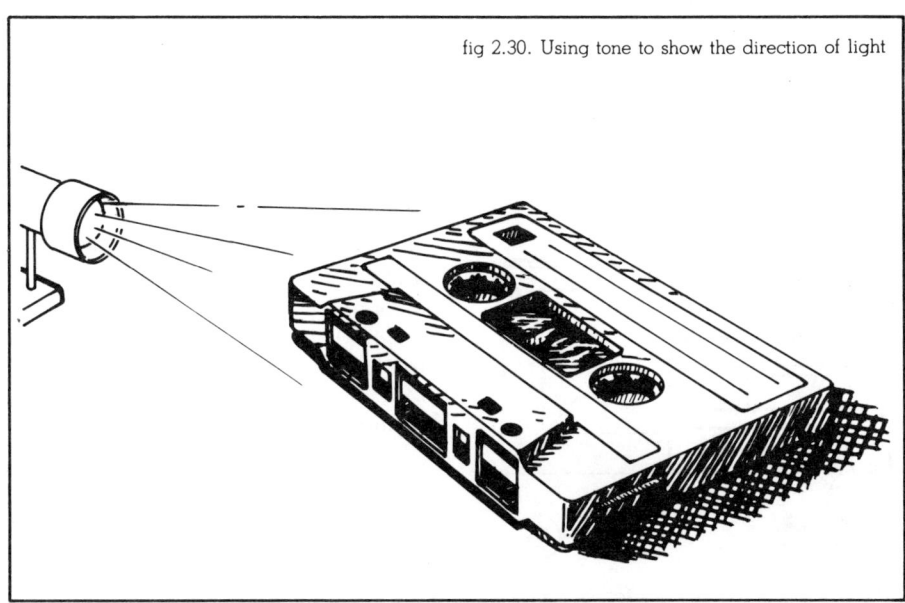

fig 2.30. Using tone to show the direction of light

fig 2.31. Using rendering to show materials

REPRESENTING MATERIALS

Tone, lines and dots can also be used to show the surface texture of an object and suggest the material that it is made from. For example, parallel lines can be used to suggest a shiny surface, while dots can be used to give the impression of a smooth, matt surface. Alternatively, dots can be used to show a stippled, textured surface such as foam rubber or plastic.

A variety of lines can be used to represent wood grain, while small circles positioned close together will look like a leather grained surface. Metal tubes or cylinders can be shown by using broad stripes of dark tone. This method is also used to show reflections in polished surfaces.

WORKING DRAWINGS

Working drawings contain the information needed to make the object you have designed. They need to convey details such as dimensions, materials to be used, construction details or assembly instructions.

Working drawings usually use orthographic projection, but they could also include other views in oblique or isometric projection. You will have to decide what is the best method for the object you have designed.

ORTHOGRAPHIC DRAWING

There are two types of orthographic drawings: **first angle projection** and **third angle proejction** (see fig 2.33). Both types show three separate views of the object: front view, end view, plan view. The difference between first and third angle projections is simply the way the views are arranged on the paper.

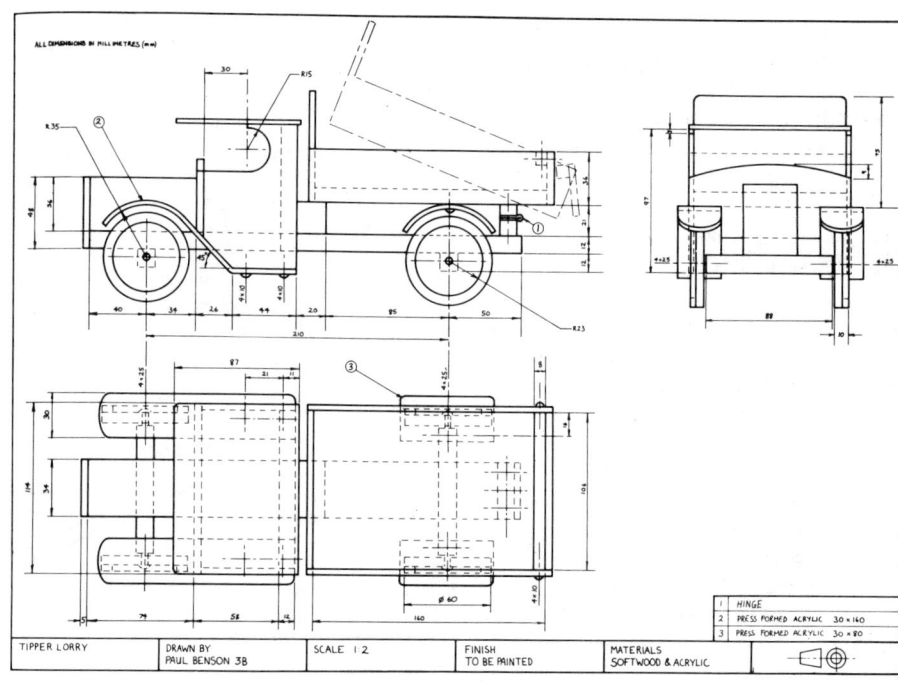

fig 2.32. An example of a working drawing

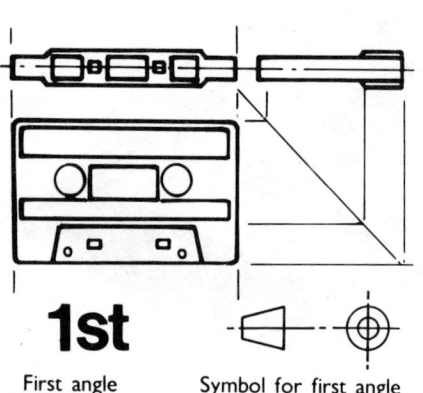

1st

First angle   Symbol for first angle

First angle is mainly used in Britain, but third angle, used in the USA and the continent, is becoming more popular. Third angle has the advantage that the adjoining views appear next to each other on the drawing.

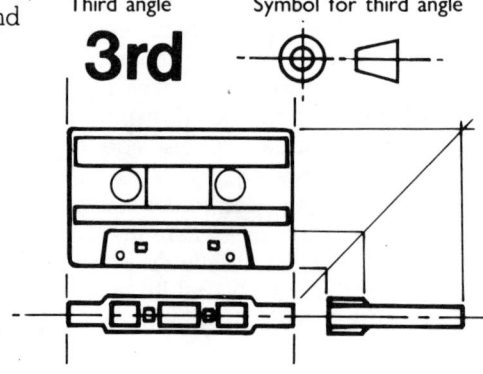

Third angle   Symbol for third angle

3rd

fig 2.33. First and third angle drawings

EQUIPMENT

In order to make accurate working drawings you will need several simple items of equipment. You will need a drawing board and a tee square or a drawing board fitted with parallel motion.

In addition to your HB and 2H pencils you will need two set squares, a 30°/60° square and a 45° square, a pair of compasses and either clips or tape to hold your paper on the drawing board.

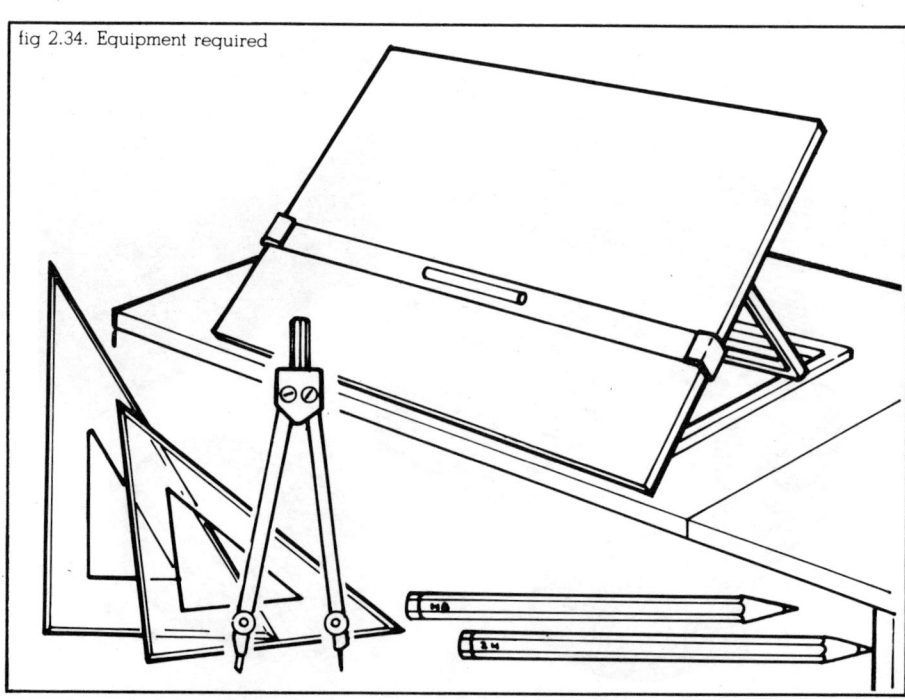

fig 2.34. Equipment required

DIMENSIONING

As working drawings are used to provide information for construction, they are usually drawn to scale and contain the necessary measurements or **dimensions**. In order that everyone can understand the dimensioning used, it is important that a standard set of rules is followed. These rules have been set out by the British Standards Institute (BSI) and are in a booklet known as *BS7308.* There should be a copy of *BS7308* in your school for you to look at.

The main points to consider when dimensioning drawings are as follows.

1. Do not allow dimensions to confuse the drawing. The dimensions should be spaced well away from the drawing and the lines should be lighter than those of the outline.

2. Dimensions should be read from the bottom or right hand side of the sheet.

3. Drawings should not be confused by too many dimensions. Each dimension should appear once only and it is not necessary to include dimensions which can be worked out by adding or subtracting others.

4. Arrowheads should be small, sharp and neat.

5. All dimensions should be shown in millimetres.

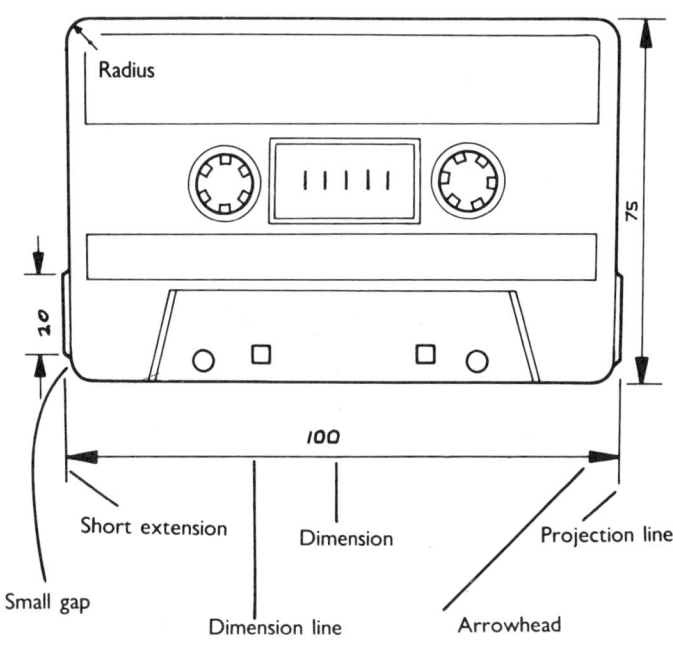

fig 2.35. Dimensioning

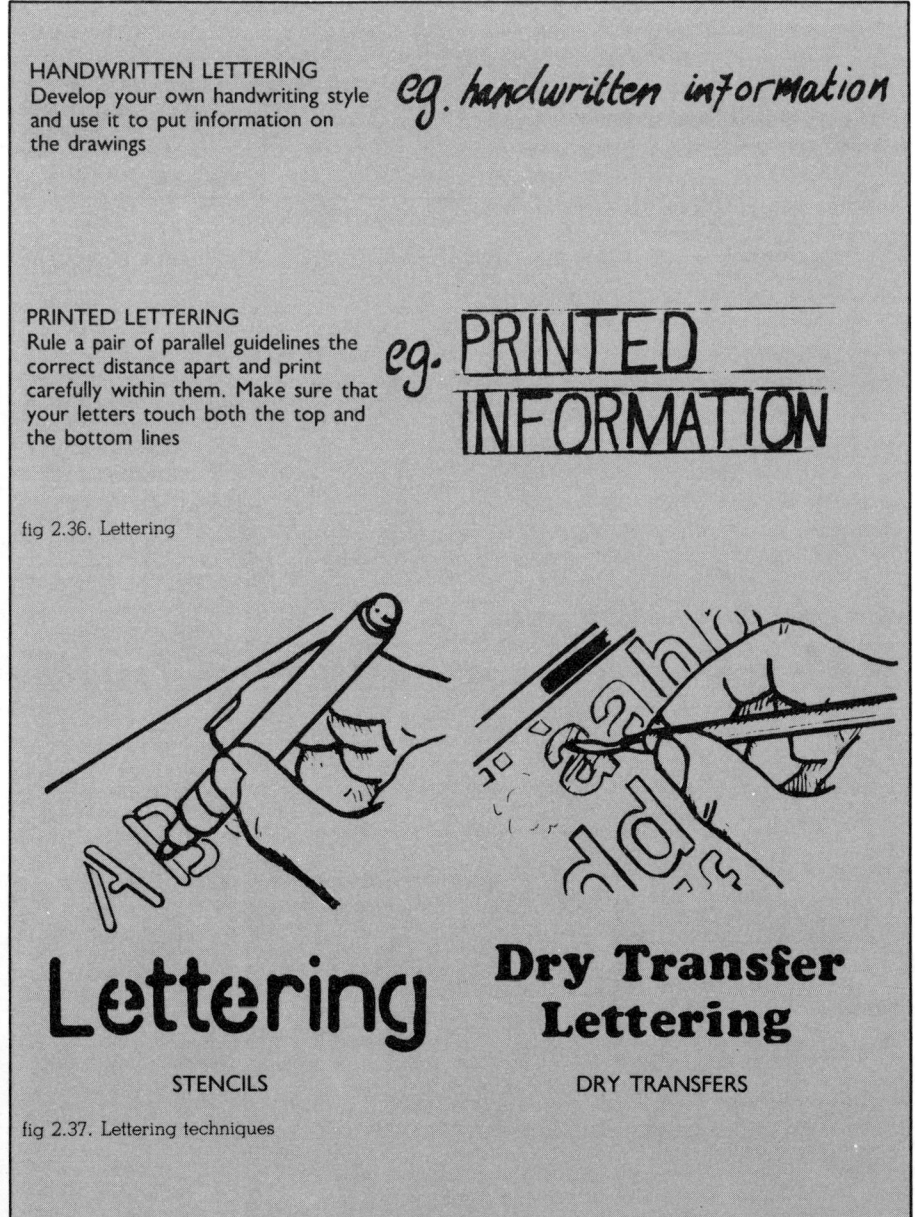

HANDWRITTEN LETTERING
Develop your own handwriting style and use it to put information on the drawings

eg. handwritten information

PRINTED LETTERING
Rule a pair of parallel guidelines the correct distance apart and print carefully within them. Make sure that your letters touch both the top and the bottom lines

eg. PRINTED INFORMATION

fig 2.36. Lettering

Lettering
STENCILS

Dry Transfer Lettering
DRY TRANSFERS

fig 2.37. Lettering techniques

LETTERING

You may need to include some written information on your working drawing. This can be done in handwriting, stencils or dry transfer lettering.

It is not difficult to develop your own handwriting style for notes and instructions. Titles and dimensions need to be printed in block capitals. The size of the letters depends on the importance of the information. General information is usually about 6mm high, while titles are 8mm high. Main headings can be up to 10mm high.

Plastic lettering stencils can be bought in a variety of different sizes and styles. Special stencils are available for use with draughting pens which correspond with the size of the nib.

Take care when using stencils to ensure that the lettering is level. Some stencils can be rested on a tee square or a parallel motion straight edge.

Dry transfer lettering gives very professional results and is available in a wide range of lettering styles. The main disadvantage is cost. It can work out rather expensive if used on every drawing.

SEQUENTIAL DRAWINGS

One way of planning how to make the object you have designed is to work out a planning procedure. This usually consists of a list of the materials and parts required and step-by-step details of construction. A planning procedure will help you to plan and organise your work from cutting up the material or obtaining the components, to the final assembly of the finished product.

This type of information is usually conveyed with the use of **sequential diagrams**. Instructions for self assembly furniture are often given in this way. In Technology it is a good method of clarifying your thoughts and making sure that you know exactly what you are doing.

Sequential diagrams can be drawn as simple block diagrams to show the various processes to be used, or they can use pictorial views when more detailed information is required. Written information can also be added in the form of titles and notes. This is known as **annotating. Exploded drawings** can also be useful at this stage to show how the various parts are assembled.

SCHEMATIC DIAGRAMS

In the case of project work involving electronics, details of circuit designs and other systems can be shown diagrammatically. Standard symbols or conventions are used to represent the components. This type of drawing is known as a **schematic diagram**.

The circuit diagram shown in fig 2.40 is an example of a schematic drawing. It is much easier and less time consuming to draw the internationally agreed (British Standard) symbols than it is to draw the actual components and connections. Full details of these symbols can be found in the BSI booklet *PP7303 'Electrical and Electronics Graphical Symbols For Schools And Colleges'*.

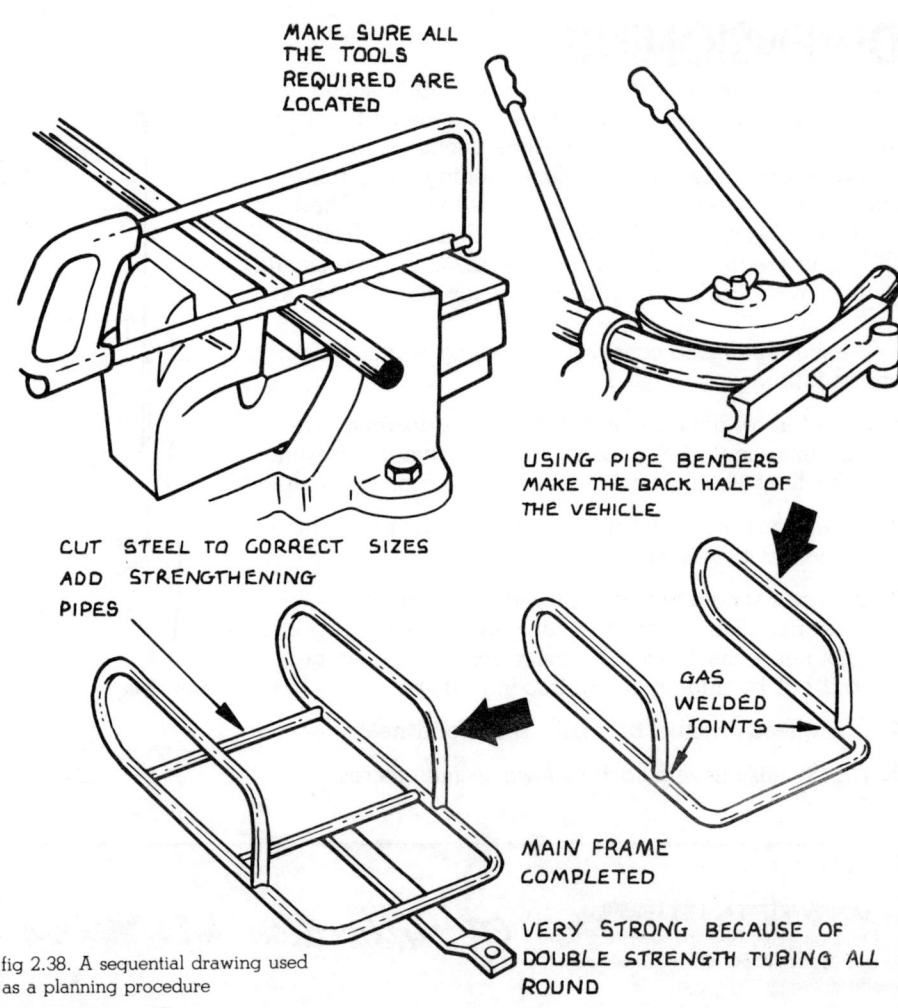

MAKE SURE ALL THE TOOLS REQUIRED ARE LOCATED

CUT STEEL TO CORRECT SIZES ADD STRENGTHENING PIPES

USING PIPE BENDERS MAKE THE BACK HALF OF THE VEHICLE

GAS WELDED JOINTS

MAIN FRAME COMPLETED

VERY STRONG BECAUSE OF DOUBLE STRENGTH TUBING ALL ROUND

fig 2.38. A sequential drawing used as a planning procedure

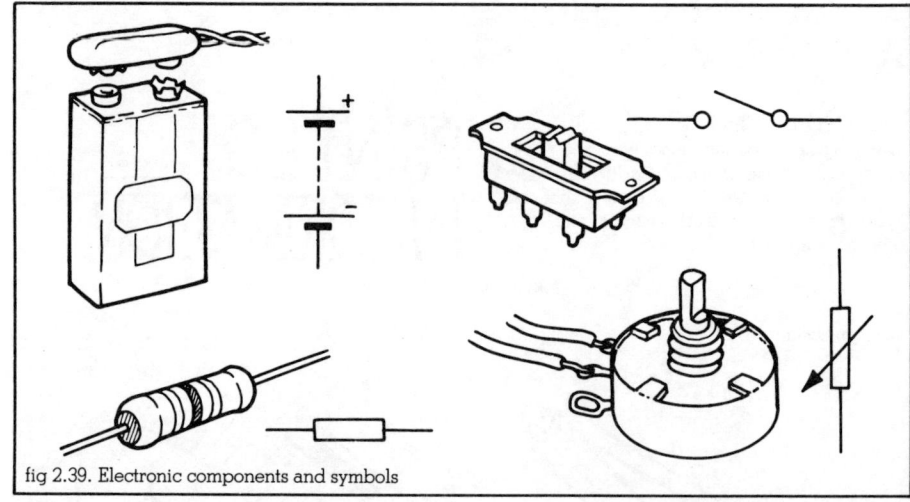

fig 2.39. Electronic components and symbols

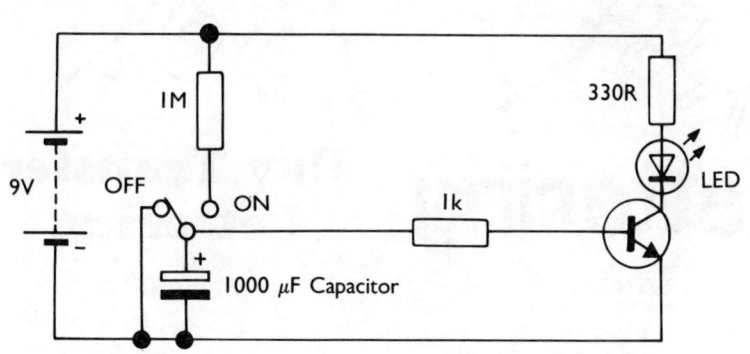

fig 2.40. A circuit diagram

TECHNOLOGY IN SOCIETY

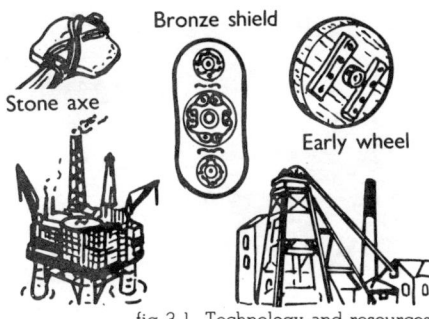

fig 3.1. Technology and resources

Stone axe
Bronze shield
Early wheel

Technology began when primitive people first fashioned pieces of stone into axes and other simple tools. They soon learned how to use fire for keeping warm, cooking and for smelting and using metals. The development of technology also led to the development of mathematical and analytical skills. The thinking processes developed, and the ability to solve problems improved.

The population of the world has increased so rapidly that it has been estimated that more people have lived on earth in the last 100 years than in the previous 10,000 years. The needs of such vast numbers of people are enormous, and increasing all the time.

Technology provides vital help to supply some of these needs. It helps to provide food, water and shelter, and provides the ability to build equipment which allows resources to be explored, located and extracted from the earth.

Medical technology helps to control the spread of disease and provides emergency surgery and life support systems. Many diseases which were fatal 100 years ago, can now be controlled with a simple injection. Antiseptics and anaesthetics have made surgery safer and more successful.

Technology has improved communications. It is now possible to speak to people across the other side of the world, using satellites to transmit your telephone conversation. Transport has improved; road, rail and air services link towns and cities and provide a practical transport network.

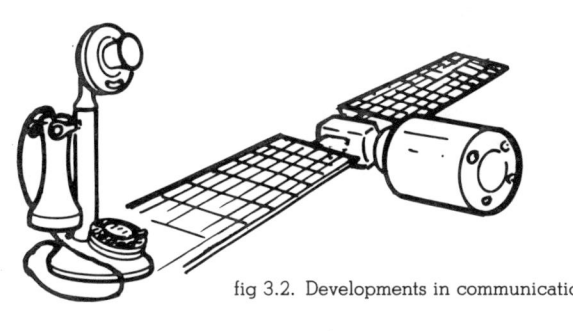

fig 3.2. Developments in communications

Technology has affected you probably more than any other generation. Many inventions and technological developments have affected the way we live, yet we take most of them for granted. Can you imagine life without radio, TV, washing machine or microwave cookers? Many of these examples of domestic technology have been made possible because of the developments which have taken place in electronics in recent years. It is now possible to put a miniature electronic circuit onto a tiny chip of silicon. We call these miniature circuits **integrated circuits**. By using integrated circuits, many pieces of equipment which were large because of the space required for the components, can be made smaller.

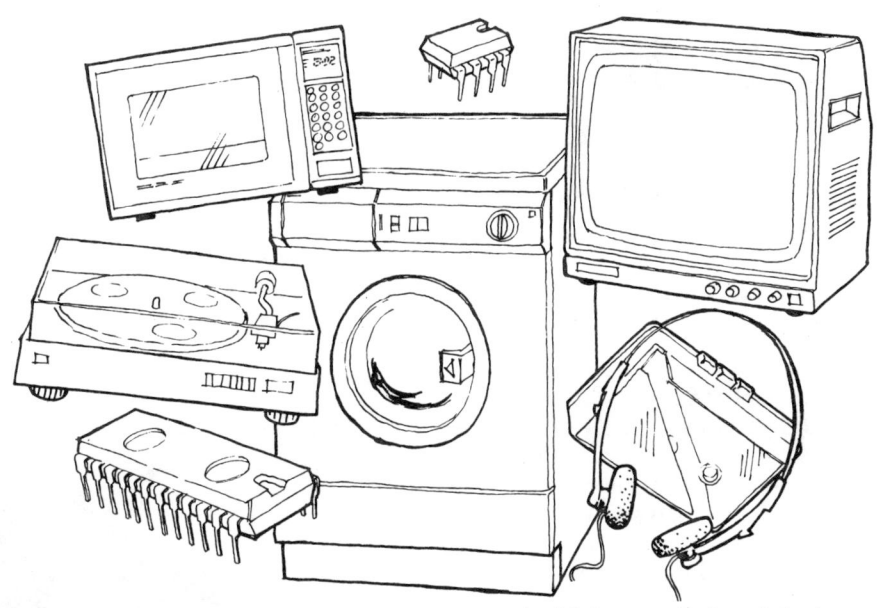

fig 3.3. Recent technological developments

fig 3.4. Robotic production line in industry

The integrated circuit and the microprocessor have been responsible for the development of more efficient, automated machines. Computer controlled machines can be used to produce large numbers of components quickly and accurately. Robotic production lines are used to build motor vehicles. A direct result of this has been to release workers from repetitive tasks and allow them to carry out those which are better done by people. Another effect has been to increase our leisure time. Many people already work less hours during the week than they did 30 years ago. It is very likely that this trend will continue and the working week for most people will be shortened even further.

TECHNOLOGY AT HOME

The impact of technology on society can be seen clearly in the home. The modern home now contains many examples of domestic technology ranging from labour saving devices to home entertainment.

For example, the appliances found in the kitchen show how technology has affected our everyday lives. Many people now have more freedom from the drudgery of domestic chores thanks to machines such as vacuum cleaners, washing machines and dishwashers. A reduced workload in the home has allowed more individuals, who were traditionally responsible for running the home, the opportunity to continue with their careers.

Washing clothes is perhaps easier now than ever before. The combination of powerful detergents and automatic washing machines means that wash day now consists of loading the dirty clothes into the machine, adding the required amount of detergent and switching on.

fig 3.5. Domestic technology in the kitchen

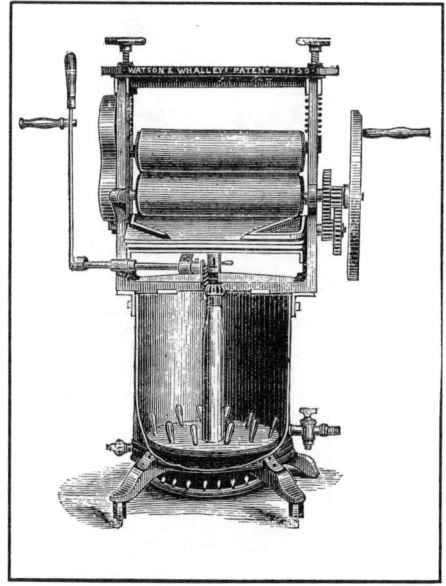

fig 3.6. Gas heated washing machine circa 1884

The washing machine is a fairly recent development. For centuries washing was done in a tub and the clothes were agitated with a three legged stick known as a dolly. During the latter half of the nineteenth century someone had the idea of putting the washing into a wooden box and turning it over with a handle. The illustration in fig 3.6 shows a gas heated washing machine of 1884. It had a handle to agitate the washing, as electric motors were not fitted until about 1914. Compare this with the machine shown in fig 3.7 which is microchip controlled and has a considerable number of program options available. Once loaded and switched on, this machine can wash while you are at work or operate on low cost energy while you are asleep.

fig 3.7. Microchip controlled washing machine

Most homes have a choice of entertainment available. Colour television provides a choice of network TV channels at present, but recent developments in communications will soon bring cable and satellite TV into our homes. Pre-recorded television programmes and films can be watched on video and pre-recorded music is available in a variety of formats. The micro computer can also be used as a form of entertainment and there is a huge variety of computer games available at present (fig 3.8).

Listening to music has come a long way since the turn of the century when recorded music was played on a wax cylinder. Fig 3.9 shows an Edison phonograph. A clockwork motor rotated the wax cylinder which vibrated a needle connected to a diaphragm. The sound created by the moving diaphragm was amplified by the large horn. Compare this to the compact disc player in fig 3.10 which uses a laser to decode digital information stored on a disc.

fig 3.8, 3.9, and 3.10 Entertainment technology – early phonograph, computer and hi-fi

TECHNOLOGY AT WORK

There is probably no better place to see the effects of technology than at work, in industry, offices and shops. The use of technology has changed many people's working lives.

Industry has always made use of technological developments to improve output and productivity. The developments which took place during the Industrial Revolution improved manufacturing skills and turned Britain into the first industrial nation. Fig 3.11 shows Henry Maudsley's first screw cutting lathe which was built around 1800. Lathes for shaping wood were known to the ancient Greeks and metalworking lathes were later developed for clock and watchmakers. Machines had also been invented which would produce metal screws, but Maudsley's design became the prototype for the modern lathe, one of the most widely used machine tools.

Recent developments in computer technology have made it possible to

fig 3.11. Maudsley's lathe

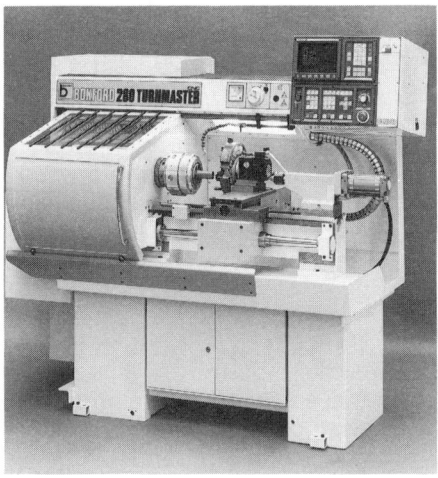

fig 3.12. A CNC lathe

fig 3.13. A word processor

control a lathe using a computer. This is known as computer numerical control or **CNC**, and is particularly useful when used in conjunction with computer aided design or **CAD** as it is known. The two systems are used together in **computer aided design and manufacture**. This eliminates much of the need for traditional drawings, as the item to be manufactured can be designed and drawn using the computer. The information can then be fed into the machines directly. This system is very useful when accurate, but repetitive work is required. Sophisticated CNC machines are used in industry yet they still have the same basic form as Maudsley's lathe.

CAD means that drawings can be stored on computer discs and updated or changed very easily without redrawing completely. Similar developments have taken place in the office. Many offices have their own computers which can deal with accounts, wages and other figures such as sales or productivity details. Typewriters are gradually being replaced by word processors which can be used to prepare, edit and print text. Fig 3.13 shows a typical word processor which can relieve the typist of much of the boring copying and redrafting of documents and letters. It can even be used to address envelopes.

Computers can be linked together using a telephone line or modem. This enables information to be sent or received virtually instantly and so there is no need to rely on the postal services. Who knows, in the next few years it may well be possible to find offices which do not use paper at all.

fig 3.14. Bar code on a film package

fig 3.15. Computer controlled checkout system

The retail industry has also been affected by the recent changes in technology. You will no doubt have noticed that most items you buy have a series of thin black vertical lines on them somewhere. This is known as the bar code. Fig 3.14 shows an example of a bar code on a film box. The code enables the computer to identify the product and its cost. It is 'read' by passing a light pen over it. When used in shops it can tell the till how much the item costs and will also deduct it from the stock list automatically. This makes stock control easier because the computer can be programmed to re-order automatically when stocks are low. Fig 3.15 shows a supermarket checkout system developed by IBM. The assistant simply passes the goods over a laser beam which automatically records the purchase and totals the bill.

These are just some of the developments which have affected the working lives of many people. They have all been made possible by developments in the field of **microelectronics**.

TRANSPORT

The first significant development in transport technology was the invention of the wheel, which appears to have originated about 3000 BC. Wheels which were found in Mesopotamian tombs were made from three planks of wood, held together with cross pieces. It is interesting to note that a natural knot hole was usually used as an axle hole. The timber around the knot is very hard and would resist wear as the wheel turned on a fixed axle. The invention of the wheel gave people the opportunity to transport themselves and their goods. It meant that they no longer had to be hunters or farmers, but could develop their manufacturing skills, specialise in their work and organise themselves into urban communities. Food and other goods were transported to them and trading took place.

fig 3.16. Early wheel

fig 3.17. The HS 125 and its ancestor, the 'Rocket'

Without transport the Industrial Revolution could not have taken place. The manufacturing industries needed raw materials and goods to be transported quickly and cheaply. **Canals** could transport heavy loads much more effectively than horseback and were used to carry coal from the mines to the factories. By the 1760s, coal was being carried by barge instead of horseback, cutting transport costs by half. The advent of the **railways** in the nineteenth century took much of the traffic away from the canals and by the middle of the century many of them were allowed to fall into disuse. The first public railway, the Stockton and Darlington, was opened in 1825. The railway engineer George Stephenson proved that steam haulage was a practical proposition and the railway became an ideal carrier of heavy loads. For the next 100 years or so it was to supply most of the transport needs of British industry.

The **steam engine** provided power and transport for industry, but it was the invention and application of the internal combustion engine which made private motor transport possible. The first successful **internal combustion engine** was really only a modified steam engine which used gas as a fuel, but in 1867 the firm of Otto and Langen began building their 'silent' engine. This was to be the first modern internal combustion engine and was to have an important influence on the American inventor **Henry Ford**.

To begin with motor cars were expensive to produce and not many people could afford them, but motor buses were designed and it was not long before public service vehicles gave people new travel opportunities. In America, Henry Ford saw the possibilities of motor transport for all and by producing cars on a **production line** he was able to mass produce them instead of making them one at a time. The Model T Ford became the world's first **mass produced** motor car with over 15 million built between 1908 and 1927. In England, Herbert Austin was designing low cost cars. The Austin Seven, a small car capable of 50 mph and 60 mpg sold for £125. Motoring soon became available to everybody. Freedom and mobility were no longer only for the wealthy.

fig 3.18. Model T Fords coming off the production line

fig 3.19. Developments in air travel

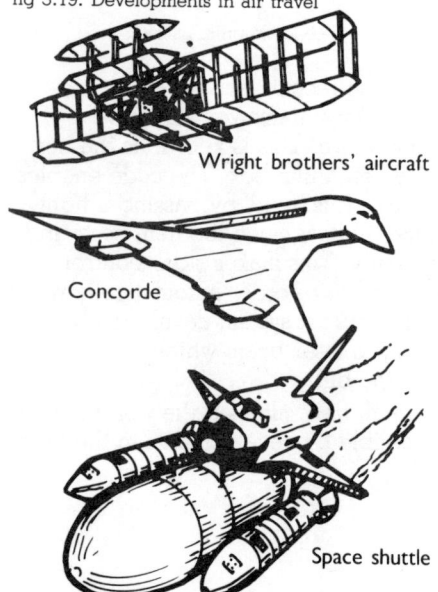

Wright brothers' aircraft

Concorde

Space shuttle

The internal combustion engine also made successful **powered flight** possible. It was light, and powerful enough to enable an aircraft to fly. In 1903, the Wright brothers made the first successful powered flight. Their aircraft was in the air for 12 seconds and travelled a distance of 120 feet. In 1909, Louis Bleriot flew across the English Channel and in 1927, Charles Lindbergh flew non-stop from Paris to New York in 33 hours and 39 minutes. Aircraft developed rapidly both in military and civil use and the invention of the **jet engine** during the Second World War had a considerable effect on air travel. The first high speed jet airliner appeared in 1949 and it is now possible for passengers to fly to America and other countries at supersonic speeds. The invention of the **liquid fuel rocket** made space flight possible and further developments led to satellites, manned spacecraft, moon landings and reusable space shuttle vehicles.

Technology has a considerable effect on transport. We only have to look at the modern motor car to see this. Design and construction, safety and comfort, economy and efficiency have all been developed and improved over the years. Transport is an industry in itself and provides work for many people. A modern technological society relies very heavily on its land, sea and air transport systems. They not only give people more freedom and mobility than ever before, but also provide goods and services that were not previously available.

COMMUNICATION

Communication ranges from verbal and written communication to highly sophisticated **telecommunication** systems. Technology has helped to develop communications and in turn, society has gained from these developments.

The invention of **printing** is usually credited to Gutenberg, but some of the earliest books were printed in China from wooden blocks. During the 15th century, printing in the West benefited from the developing paper technology, improved presses and Gutenberg's invention of movable metal type. These developments led to an increased demand for books and newspapers. Recently, the newspaper and printing industry has been affected by the increasing use of computer technology and desk-top publishing. Type can now be set using a computer and VDU screen and no longer needs to be set by hand. Fig 3.20 shows hand typesetting at the beginning of the seventeenth century.

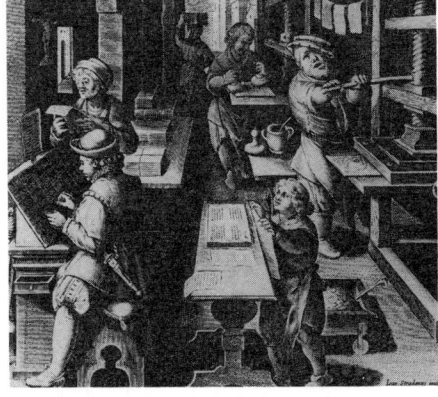

fig 3.20. Typesetting by hand circa 1600

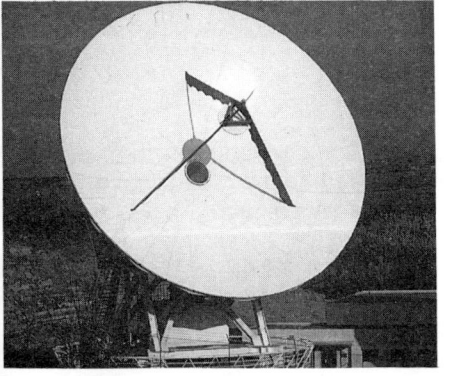

fig 3.21. Satellite dish

fig 3.22. Bell's telephone

TELECOMMUNICATIONS

Telecommunications refers to the transmission and reception of information using electricity and radio or light waves. There have been major developments in this area over the last hundred years or so.

The **telegraph system**, developed in the 1830s, was used to send messages by many of the railway companies. Samuel Morse devised a signalling system known as the **Morse Code** and by the middle of the nineteenth century commercial telegraph companies were in operation.

In 1876 Alexander Graham Bell demonstrated that people could talk to each other from a distance and two years later in 1878, the first telephone exchange opened and the development of the **telephone system** began. Continents were linked together by laying undersea cables, but in the last 25 years it has been possible to 'bounce' telephone conversations from one country to another using telecommunications satellites orbiting the earth. It is now possible to receive television programmes from other countries by satellite in the same way. The first telecommunications satellite 'Telstar', was launched in 1962 and completed each orbit of the earth in 160 minutes. Telstar received and transmitted telephone calls and television signals from America to Europe and vice versa. A recent development in telecommunications is the use of **fibre optics**. Telephone conversations or signals are converted into light using a laser and the beam is sent along a tiny optical fibre. This fibre is rather like a very thin tube and when the light reaches the other end of the tube it is converted back into a signal by a demodulator.

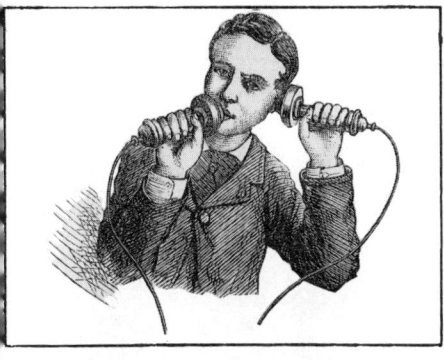

In less than one hundred years, the **radio** has developed into a major form or entertainment and communication. In 1896 Marconi invented a wireless telegraph system and a year later formed the Wireless Telegraph and Signal Company. This was the birth of the radio. Transmitting morse code was simple, but the sound of the human voice and music presented more of a problem. The impulses from the microphone needed to be combined with a more powerful carrier wave for transmission, then separated again and changed back to sound in the receiver. This was made possible with the development of the **thermionic valve** known as a triode valve. The potential of the radio was soon realised and it has developed as an important form of communication.

Similar developments have been made in **television**. Early black and white programmes were broadcast just before the start of the Second World War, but it was not until after the war that the full potential was exploited. Colour broadcasts began in the late 1960s and experiments with simultaneous stereo broadcasts on radio began shortly after. Recent developments have enabled viewers to receive a wide range of information from their TV sets in the form of text. This is known as Teletext or Ceefax and can be received using specially adapted TV sets. The service is offered free of charge and offers news coverage, weather forecasts, travel reports and a variety of other information.

fig 3.23. Early valve radio and modern transistor counterpart

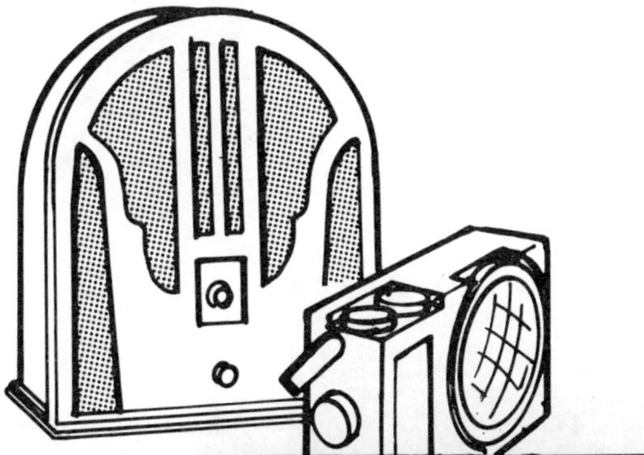

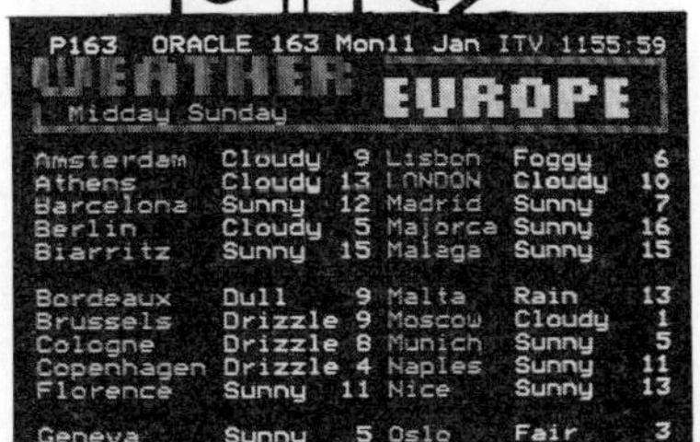

P163 ORACLE 163 Mon11 Jan ITV 1155:59

WEATHER EUROPE

Midday Sunday

Amsterdam	Cloudy	9	Lisbon	Foggy	6
Athens	Cloudy	13	LONDON	Cloudy	10
Barcelona	Sunny	12	Madrid	Sunny	7
Berlin	Cloudy	5	Majorca	Sunny	16
Biarritz	Sunny	15	Malaga	Sunny	15
Bordeaux	Dull	9	Malta	Rain	13
Brussels	Drizzle	9	Moscow	Cloudy	1
Cologne	Drizzle	8	Munich	Sunny	5
Copenhagen	Drizzle	4	Naples	Sunny	11
Florence	Sunny	11	Nice	Sunny	13
Geneva	Sunny	5	Oslo	Fair	3
Gibraltar	Bright	15	Paris	Cloudy	11
Innsbruck	Sunny	1	Rome	Sunny	13
Jersey	Cloudy	10	Venice	Sunny	6
Las Palmas	Cloudy	20	Vienna	Fog	-2

WINTER WOE? BOOK A HOLIDAY! 580/590

fig 3.24. Modern TV information screen

MEDICAL TECHNOLOGY

Within the last 200 years there have been many technological developments in medicine which have affected the lives of most people living in western society today. Up to the early 1800s there had been very few advancements since the time of the ancient Greeks. Life expectancy is now far greater than it was at the start of the nineteenth century and the number of children dying at birth or not reaching maturity is also considerably less.

Communication has affected medical science. Improved transport has meant that doctors can get to their patients quickly, but it has also enabled doctors to interchange their ideas and methods with each other more quickly than ever before. The telegraph and the telephone were key technological developments which allowed medical information to be passed on quickly from one doctor to another. The Flying Doctor Service, operating in remote parts of Australia, relies very much on the use of the radio and aircraft. Nearer home, many hospitals and groups of doctors operate flying squads. These are fully equipped vehicles which are ready day and night to deal with any accident or emergency.

fig 3.25. Australia's flying doctor service

fig 3.26. Early electrocardiograph

Since 1800, medical technology has benefited from the rapid growth of science and technology. Much of this has come from developments in the textile and engineering industries. Chemistry was important in the production of textiles and, during the mid-nineteenth century, a new branch of science concerned with the study of the human body's chemical system grew up. This became known as **biochemistry**.

Engineers began to develop new methods to produce precision instruments and specialised steels for industry. Soon these were available to doctors and surgeons. Engineers could also produce specialist equipment, such as microscopes, which proved to be invaluable for research and diagnosis.

Technology provides doctors with a number of pieces of equipment which help them to carry out tests and enable them to make accurate diagnoses. Much of this equipment owes its existence to the work done in physics, electricity and electronics in particular. The **electrocardiograph** was invented in 1903 and was based on the theory that the heart produces a tiny electrical current as it pumps blood around the body. Today, the electrocardiograph is a routine instrument for examining people with heart conditions.

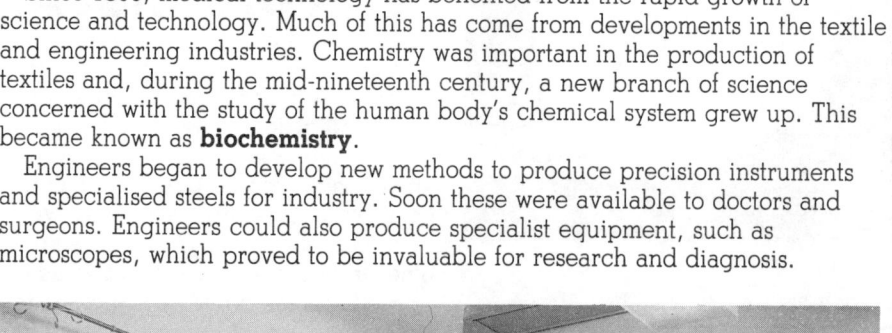

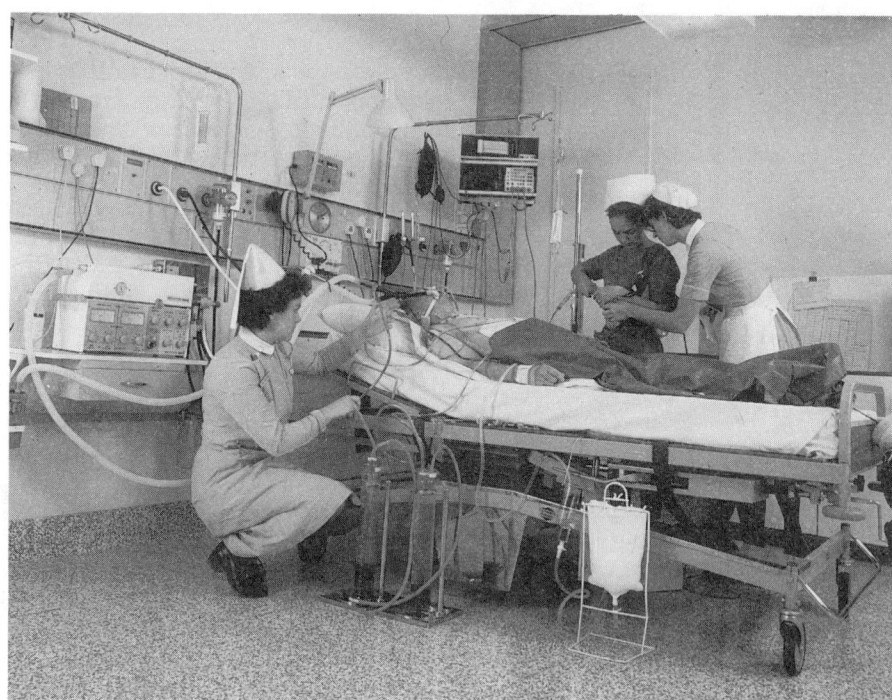
fig 3.27. Modern intensive care ward

Electronics and microelectronics have been used extensively in medical technology. Intensive care units rely on electronics to monitor the conditions and progress of patients. Fig 3.27 shows a modern intensive care unit. Compare the equipment with that shown in fig 3.26.

Many people rely on such equipment to enable them to live relatively normal lives. For example, a considerable number of people would die from kidney disease if it were not for dialysis machines which do the job of a normal kidney for them. People suffering from heart problems rely on electronics to make sure their hearts beat regularly. A small electronic device known as a pacemaker is fitted inside the patient's chest and generates electrical impulses which stimulate the heart. The idea for pacemakers was first suggested in 1862, but it was not until years later that a successful cardiac stimulator was built. Now with the use of microelectronics, it is possible to make them smaller than ever before.

fig 3.28. Pacemaker

MANAGING TECHNOLOGY

In looking at the effects of technology on society, it is easy to see only those things which are of benefit. Technology is associated with certain problems and must not be seen as the solution to all our problems or the cure for all ills. Page 30 looks at how technology has affected medical science. One outcome of this has been to increase life expectancy: people are, on average, living longer now than they did last century. This might not immediately appear to be a problem, but if we consider that many people will live longer in retirement than ever before, then it becomes an economic problem. How much extra will this cost the nation in terms of pensions and care for the elderly? This might appear to be a very selfish and uncaring attitude, but it is a direct result of developments in medical technology.

fig 3.29. Home for the elderly

fig 3.30. Unemployment in the 1930s

It is impossible to apply technology without changing the world in which we live. Technology brings about change which affects society and civilisation. Lifestyles change as use is made of the new discoveries. It is only human nature to resist change, especially that which will affect personal lifestyles. For example, in the past the introduction of new technology has caused riots and revolts. The Luddites rioted and destroyed new textile machinery at the beginning of the last century because they felt that their lifestyle and livelihood was threatened.

Fig 3.30 shows a dole queue in the 1930s. Unemployment has risen recently due to the effects of the world recession. Does it mean that the increasing use of technology will bring about a return to mass unemployment? If change is resisted then perhaps it does. Lifestyles must change if people are to come to terms with technology. The majority of people already work a shorter working week than they did 20 or 30 years ago and in most cases their standard of living is higher now than ever before.

In order to manage technology there needs to be a shift of emphasis in terms of employment and the deployment of the workforce. Many people are required to work in the technological industries to produce the technology and support the industry. Microelectronics, for example, is still a rapidly expanding area.

Where jobs are best done by machines, then workers should be occupied in tasks which are better suited to being done by people. More people will be required to work in the leisure and service industries, while others will be required to care for the increasing number of elderly people. Many traditional occupations will disappear and be replaced by new ones. Living with technology will, for most people, mean coming to terms with change.

fig 3.31. Modern leisure facilities

fig 3.32. Chemical disaster at Bhopal, 1986

The use of technology requires people to be responsible for their actions. Over the last hundred years, many nations have been irresponsible in dealing with pollution. Natural resources have been taken out of the earth, the landscape has been scarred with spoil heaps and waste products and toxic chemicals have been released into the air and the sea. In the 1950s many people in Japan died when unwanted mercury was dumped into the sea. The mercury contaminated the fish and so got into the food chain. More recently in India, hundreds of people were blinded when dangerous chemicals escaped from a nearby factory. For many years Scandinavian countries have suffered from the effects of smoke from factories being carried across the North Sea by the prevailing wind. The sulphur dioxide in the smoke dissolves in the moisture in the air and eventually falls as acid rain. The smoke can be cleaned considerably before it is released into the atmosphere, but to do this costs money. The use of nuclear energy emphasises the need for responsibility. Nuclear waste is harmful for thousands of years and needs to be disposed of carefully so that it cannot affect water supplies or harm the food chain.

It is clear that technology has much to offer the world, but it is also clear that control and management is required if we are to get the best from it. With the proper control and responsible attitudes, it could be the hope for the future and a way of life for coming generations.

EXERCISES

1. Talk to an older relative and compare your present lifestyle with theirs at your age.

2. What do you understand by the term 'mass production'?
   Can you name **three** products which are produced in this way?
   What are the advantages of mass production?

3. Technology has had an important effect on industry. Can you think of **three** other areas which have been affected by the use of technology?

4. Technology has made life easier for the housewife. Can you give **two** examples of labour saving domestic technology and explain how their introduction has affected the user?

5. Explain what is meant by the term 'paperless' office and describe how it is possible for this to become a reality.

6. Roads and motorways took much traffic away from the railways. Can you suggest why this happened?

7. Why was transport so important to industry at the time of the Industrial Revolution?

8. Can you think of **three** ways in which technology has affected modern motor car design?

9. What is meant by the term 'telecommunications'?

10. Explain what is meant by the term 'life expectancy'? Describe how developments in medical technology have affected it.

11. What is the function of the heart pacemaker?

12. What is an electrocardiograph used for?

13. Can you explain why it is necessary to exercise control over the use of technology?

14. What is meant by the term 'pollution'?

15. Can you think of **three** areas or parts of our environment which can easily be polluted?

16. What is meant by the term 'acid rain'?

17. Can you explain why unemployment has been linked to the increasing use of technology and suggest ways in which it can be avoided?

PRINCIPLES OF CONTROL

Control is an essential part of everyday life. Most activities require some sort of control. You need to control your hand to pick up a book or turn a page. You can control a TV or model car by remote control. Computers can control machines and robots to work with great accuracy and speed.

The way in which things are controlled can seem very complicated, but even complex control systems can be simple to understand if you use the right approach to them. This is called the **systems approach**. It involves looking at the main building blocks that go to make up a system. It is rather like using building blocks to build a model house. Each block links to others to make up the whole house. In a control system each control block links to others to produce the required system. The basic building blocks that go to make up a system are shown in fig 4.2.

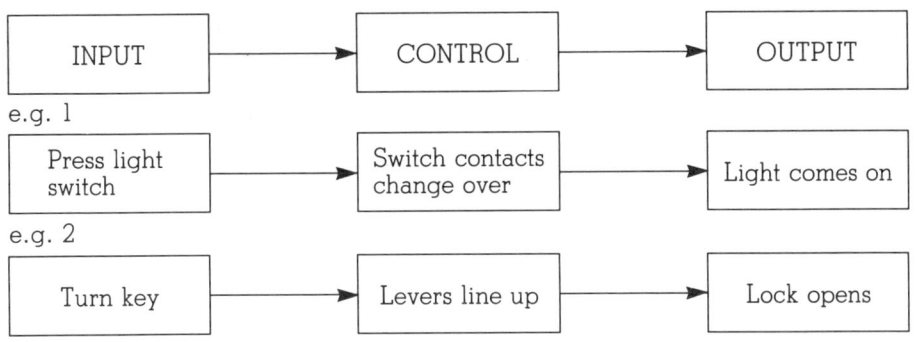

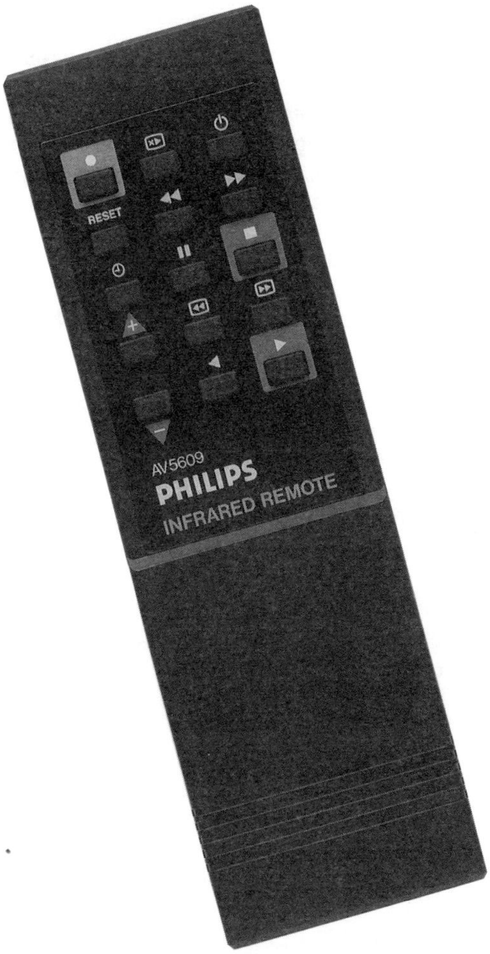

| INPUT | CONTROL | OUTPUT |

e.g. 1

| Press light switch | Switch contacts change over | Light comes on |

e.g. 2

| Turn key | Levers line up | Lock opens |

fig 4.2. Basic building blocks of a control system

As you can see from these examples, the systems approach can be used to explain any control situation without getting involved in the detail of how it is done. In industry, it is used to identify the various parts of a problem before starting to work out the detail. You can use it in a similar way to design solutions to control problems.

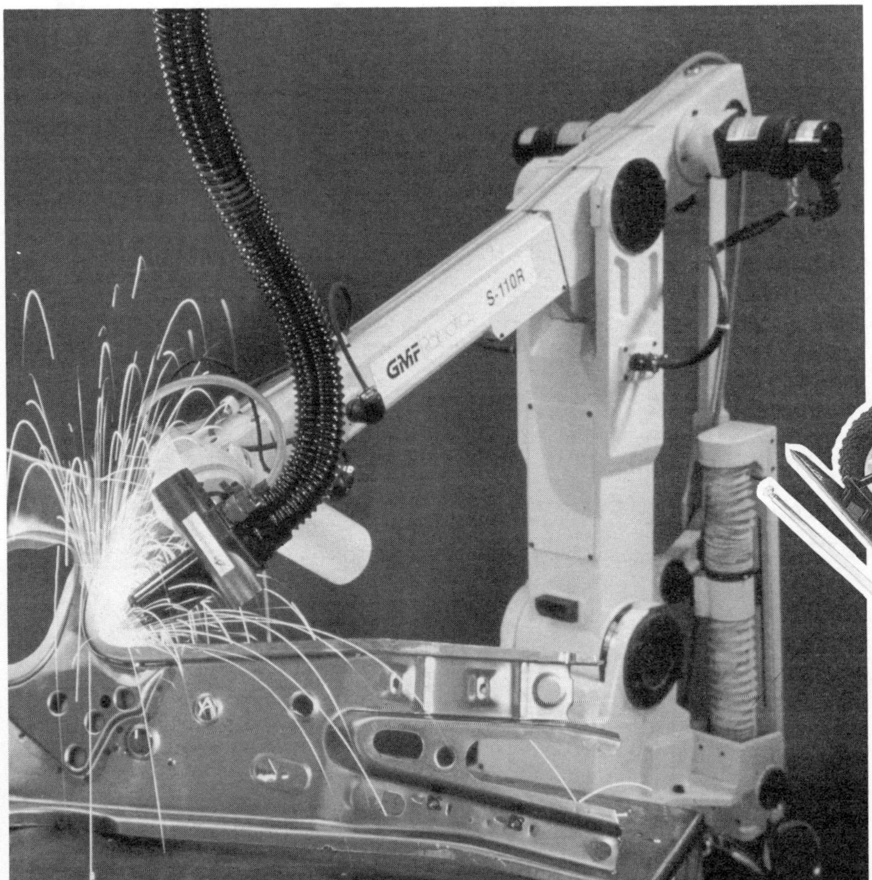

fig 4.1 Examples of control systems

SYSTEMS APPROACH

As already mentioned, the basic building blocks of any control system are **input**, **control** and **output**. Each of these may be in many different forms, depending on the system required.

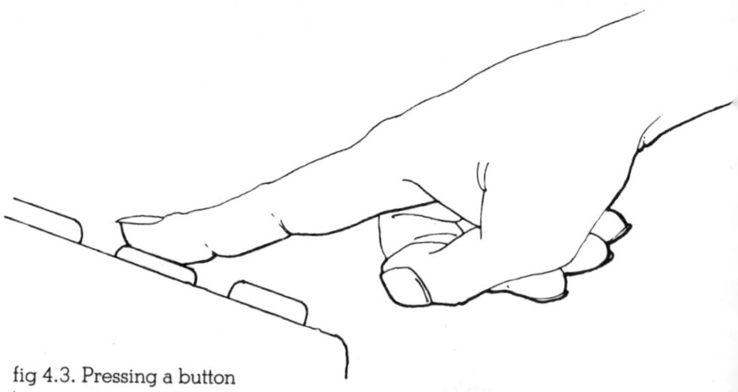

Inputs

The input may be some kind of **movement**, for example a finger pressing a button, a motor moving a lever or a component passing a sensor. Alternatively, it may be some kind of **change in the environment**, such as a change in temperature, a change in light levels or a change in moisture content.

Controls

The control may be **changing the size of the input**, for example, increasing the volume of a sound (an amplifier changes a small sound into a louder one) or decreasing the size of a movement (fine adjustment of binoculars). It could also involve **changing one kind of input into a different kind of output**, for example changing a movement into a noise (triggering off an alarm) (fig 4.4), or a temperature change into a movement (closing a greenhouse window when the air temperature falls), or one kind of movement into another (turning a handle to close a vice).

Outputs

Outputs may be **movements**, for example turning a wheel (using an electric motor) (fig 4.5), positioning a robot arm (using stepper motors) or raising a weight (using a pulley system). Outputs may also be **sounds** such as music (from a speaker) or an alarm signal (from an electric bell or siren), or they may be **lights**, for example a visual alarm signal (flashing lights on an ambulance) or visual display (LED to show when something is switched on, disco lights which flash in time to the music.)

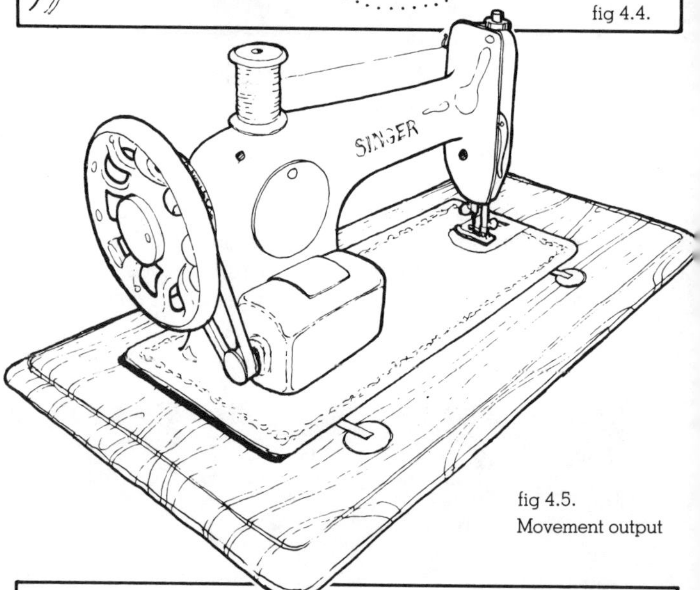

There are two basic types of system, **open loop systems** and **closed loop systems**.

OPEN LOOP SYSTEMS

Open loop systems are the simplest systems, being made up of a number of blocks connected together in a **linear** way. This means that each block leads directly to the next without any kind of checking. Fig 4.6a shows the **block diagram** of an open loop system.

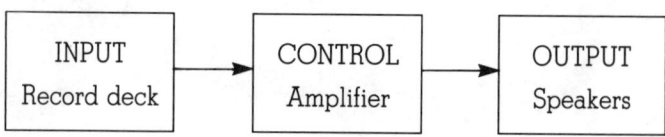

INPUT		CONTROL		OUTPUT
Record deck	→	Amplifier	→	Speakers

fig 4.6a. Block diagram of an open loop system

In this system the small electrical signals from the record deck are changed by the amplifier, into larger electrical signals which can be heard as music through the speaker.

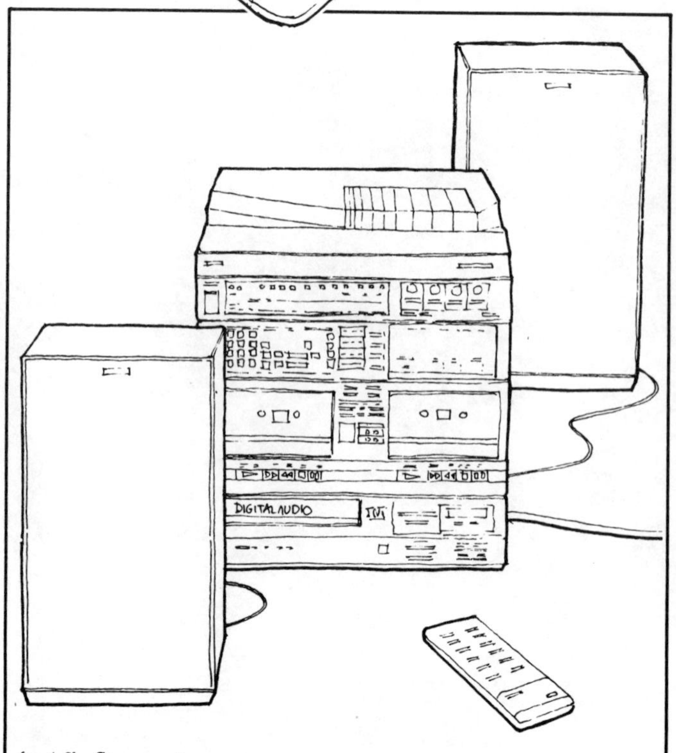

fig 4.3. Pressing a button

fig 4.4.

fig 4.5.
Movement output

fig 4.6b. Stereo system

CLOSED LOOP SYSTEM

A closed loop system not only carries out an action, but also has some way of checking to see if that action has been carried out correctly. A system which has some means of checking is said to have **feedback**. Fig 4.7a shows a block diagram for a control system with feedback.

INPUT Record deck	CONTROL Amplifier	OUTPUT Speakers	CHECKING Is volume correct?

FEEDBACK

fig 4.7a. Block diagram with feedback

In this simple example the feedback is provided by the person using the system. Many everyday examples of feedback also involve people; for example, getting the brightness of a TV picture right, getting the temperature of a shower right, or controlling a cooker ring so that the vegetables are boiling without boiling over.

CONTROL SYSTEMS

More sophisticated systems check themselves. The central heating system in a house is an example of this type of system. Fig 4.8a shows the block diagram of a central heating system. It is controlled by the thermostat which senses the temperature in the house.

TEMP. SENSOR	CONTROL UNIT	BOILER SWITCH	BOILER/PUMP RADIATORS

FEEDBACK

fig 4.8a. Block diagram of a central heating system

The thermostat is checking the temperature all the time. If the temperature falls below the set level the boiler is switched on, and when the temperature reaches the set level again the boiler is switched off.

To keep a central heating system switched on all the time would be quite expensive, so in most houses it is normally switched off at night. Therefore, the programmer usually includes a timer, so that the boiler can be controlled to only come on at certain times. It will only react to temperature readings from the thermostat during the times when it is programmed to be on. Thus the boiler will only come on if it is the right time **and** the temperature is below the set level. Fig 4.9 shows this as a block diagram.

TIMER	CONTROL UNIT	BOILER SWITCH	BOILER/PUMP RADIATORS
TEMP. SENSOR			

FEEDBACK

fig 4.9. Block diagram of a central heating system with timer

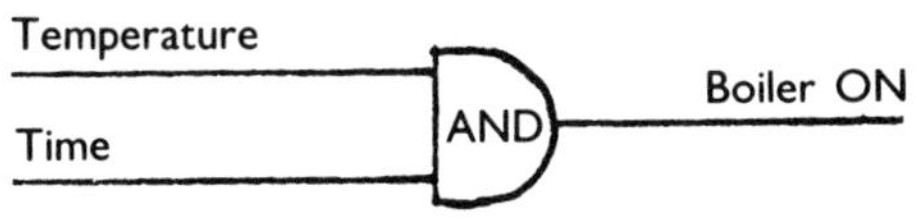

The control unit can be shown as a **logic** diagram (fig 4.10). Since this system needs both the temperature sensor and the timer to send signals to switch on the boiler, it is called an AND system. The AND logic can also be shown as a truth table (fig 4.11). This shows all the possible states of the inputs and the state of the output in each case.

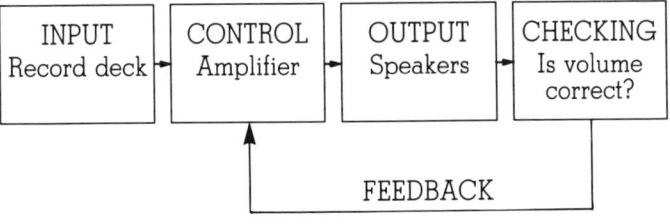

fig 4.7b. Adjusting the volume

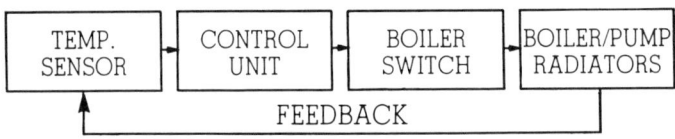

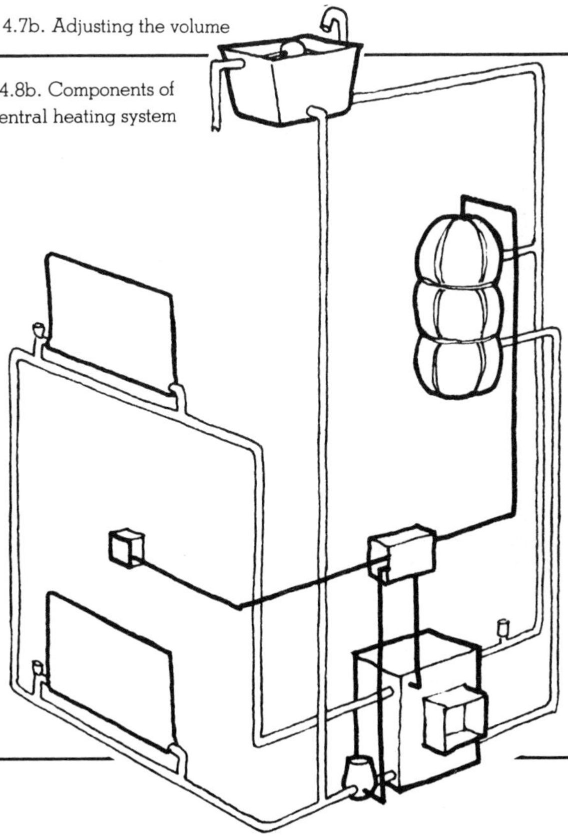

fig 4.8b. Components of a central heating system

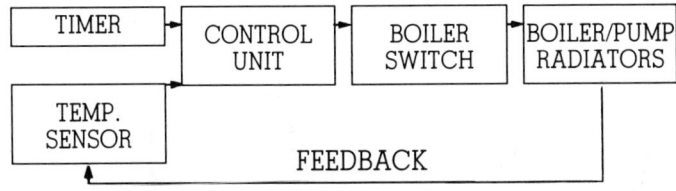

fig 4.10. Logic diagram

TEMPERATURE	TIMER	BOILER
off	off	off
off	on	off
on	off	off
on	on	on

fig 4.11. AND truth table

CONTROL SYSTEMS IN THE HOME

There are a number of control systems found in a modern home. Most of them are electronic systems which use 'microchips'. They include calculators, watches, computers, videos and washing machines (fig 4.12). Electronic control systems are the most common for several reasons: they are small, cheap, readily available and easy to power from mains or batteries.

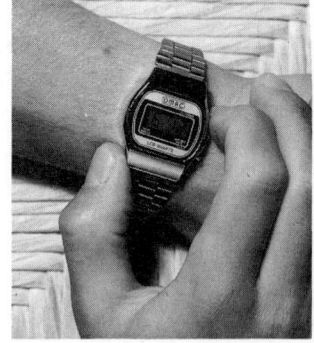

fig 4.12. Control systems in the home

WASHING MACHINE

The control system of a modern automatic washing machine has to take into account several different factors. These are door position, water level, water temperature, wash and spin times and drum speeds. Most of them are decided when you select which washing program to use.

Fig 4.13 shows a block diagram of the system. You can see that this is quite a complex closed loop system using feedback to keep a check on water level, water temperature and drum speeds.

How does it check these things?

Door position

The machine will not start any program unless the door is fully closed and locked. When the door is closed it completes an electrical circuit which heats up a **heat sensitive pellet**. This expands as it gets hot, pushing a mechanical lock into place and closing a switch (fig 4.14). The switch signals the control unit that the door is closed and locked. Only when it has received this signal will the control unit start the wash program.

Water level

When a wash program first starts it has to open the valves which allow the water in. There are usually two of these valves, one for hot water and one for cold (fig 4.15). Each must be controlled separately depending on the water temperature needed for that program. The valves are **solenoid operated**, i.e. they are opened and closed electrically.

The rising water level is checked by the water level sensor. This is a **pressure sensor** (fig 4.16). The pressure of the air in the plastic tube rises as it is compressed by the rising water. The pressure sensor keeps the control unit informed as to the pressure reached and the control unit uses the information to decide when to close the water inlet valves.

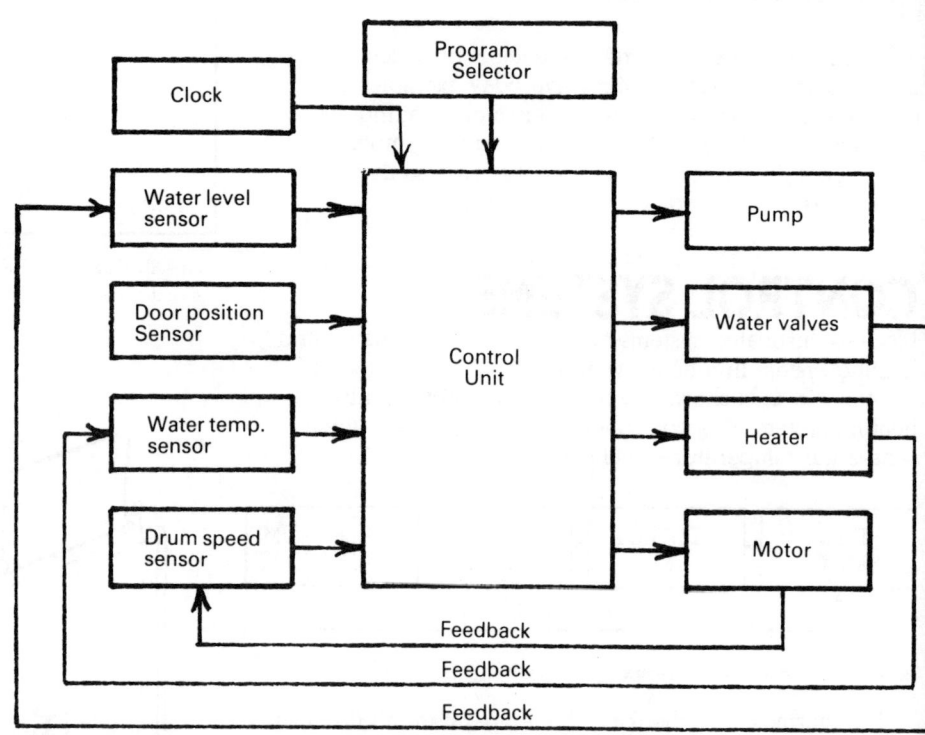

fig 4.13. Block diagram of washing machine control systems

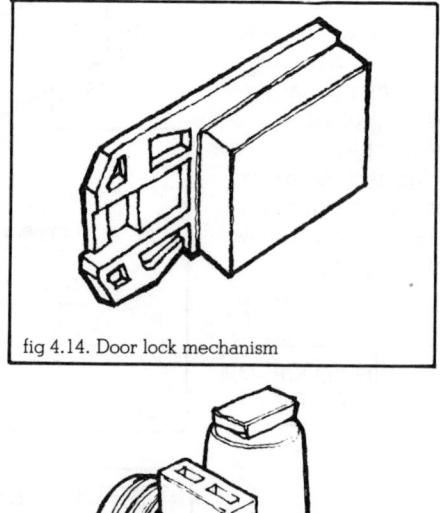

fig 4.14. Door lock mechanism

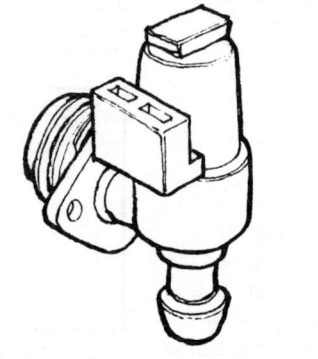

fig 4.15. Solenoid operated valve

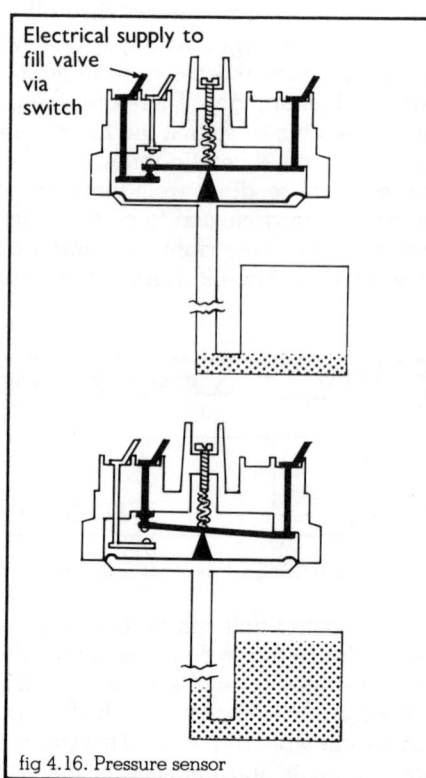

fig 4.16. Pressure sensor

Water temperature

The temperature sensor, a type of **thermometer** which fits inside the washer drum (fig 4.17), measures the water temperature and signals it to the control unit. The control unit compares it with the temperature needed for the program being used. If the water temperature is too low, the control unit will switch on the **heater**. The temperature sensor continues to check the temperature and keep the control unit informed. Once the correct temperature is reached, the control unit switched off the heater and moves on to the next stage of the program.

Clock

The control unit includes a memory which tells it how long each stage of a program should last. The times may be different for each program. The **electronic clock** built into the control unit keeps the memory of the control unit informed so that each stage of each program is timed correctly.

Drum speed

During the washing and spinning cycles of the program, the drum has to spin at various speeds. Most machines use three different speeds: 53 rpm, for washing; 83 rpm, for distributing the load before spinning; 1000 rpm, for spinning.

The control unit signals the motor to produce these speeds. The motor starts up slowly, then gradually increases speed. The speed sensor, a **tachogenerator**, keeps the control unit informed as to the speed that has been reached. The control unit uses the information to control the power to the motor and so controls the speed of the drum at all times.

Pump

The pump is switched on by the control unit to pump the water out of the drum during spinning. It is a rotary pump driven by a separate motor (fig 4.19).

Control unit

The control unit (fig 4.20) is the heart of this system. It receives and sends out signals which control all the activities of the machine. It is also capable of diagnosing faults which may occur, stopping the program and informing the service engineer what is wrong. It is a small dedicated computer which, like other computers, uses the language of **logic**.

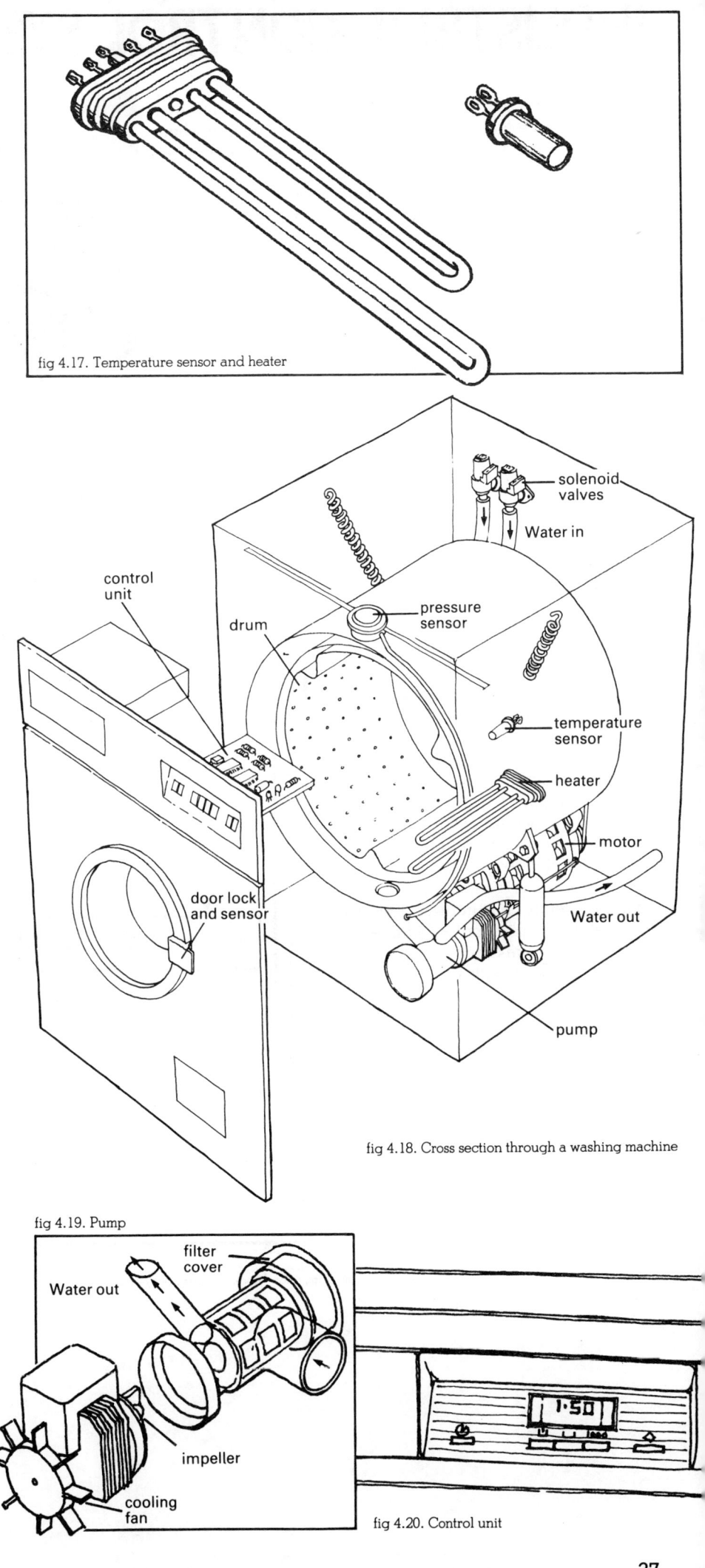

fig 4.17. Temperature sensor and heater

fig 4.18. Cross section through a washing machine

fig 4.19. Pump

fig 4.20. Control unit

INDUSTRIAL CONTROL SYSTEMS

Control systems of one sort or another are vital to industry. They make sure that operations are carried out as and when required, for example, when a valve should be opened or a furnace switched on. Most are electronic systems because they are reliable, versatile and efficient. Efficient control systems help to reduce the cost of making a product by making good use of materials and time. They not only control when and how things happen, but also give information about what is happening in the system. While appreciating the importance of electronic control systems, you should not forget that they do not do the work. The work is normally done by pneumatics, hydraulics or electric motors.

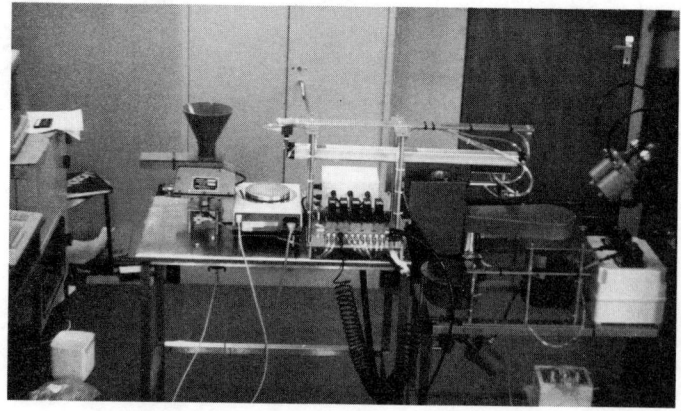

fig 4.21. Automatic sample preparation unit

AUTOMATIC SAMPLE PREPARATION UNIT

The automatic sample preparation unit, designed for the chemical industry (fig 4.21), is an example of an industrial control system. Its job is to take a quantity of powder and mix it with liquid before passing it on to a chemical analyser. This sounds like a very simple job, but in reality it is very difficult to do accurately. The quantities of powder and liquid have to be controlled very accurately if you want to get reliable results from the analyser. Before this unit was designed the operation took about 90 minutes, using the unit it takes about 10 minutes and the results are much more accurate. The whole unit is controlled and checked by a computer. Fig 4.22 shows a block diagram of the system.

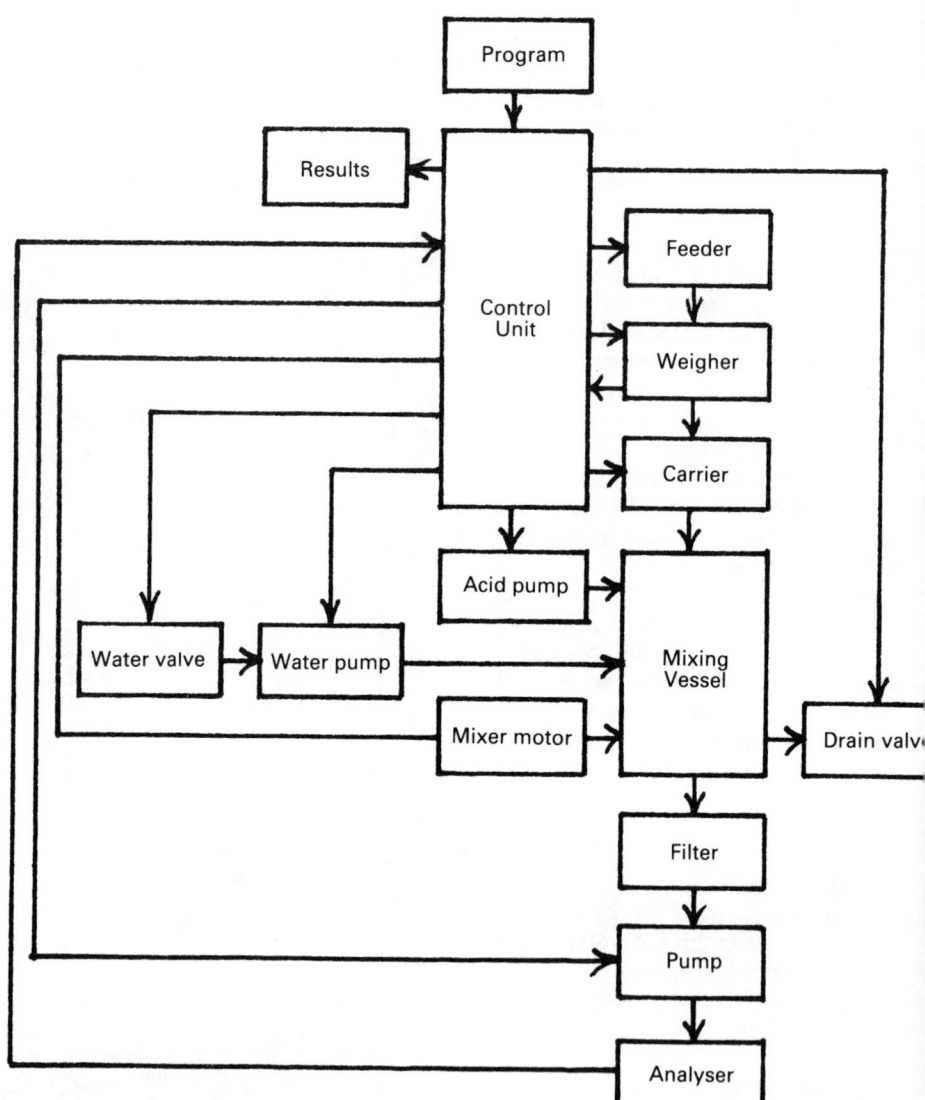

fig 4.22. Block diagram of system

Weigher

The weigher (Fig 4.23) is one of the most important parts of the system. It must weigh very exact quantities, to the nearest one hundredth of a gramme. It begins by weighing the empty 'cup'. Powder is then put into the cup until it is within 90% of the weight needed. The last 10% is put in very slowly so that it can be stopped at any time to get the amount exactly right.

fig 4.23. Weigher

Carrier

The cup is then picked up by a pneumatic arm (fig 4.24) and taken to the mixer. Once there it is tipped into the mixer with a 'quick flick' action to make sure powder does not stick to the cup. Even after this, nothing is left to chance. The cup is taken back to the weigher and reweighed. By comparing its weight before and after carrying the powder the computer knows exactly how much powder went into the mixer.

fig 4.24. Carrier

Mixer

A precise amount of water goes into the mixer at the same time as the powder (fig 4.25). If required, acid may also be added. It is important that the water goes in before the acid, otherwise it would get very hot. The flow of these liquids is controlled by pneumatically operated valves and pumps (fig 4.26); the pumps allow precise amounts of liquid to be added. The mixer motor, controlled by the computer, starts the mixer motor slowly and gradually takes it up to several thousand rpm.

After mixing, the sample is ready for analysing. It is pumped out through a filter to the analyser. Once empty the mixer is cleaned out by water ready for the next sample. The analyser results are displayed and printed out by the control unit.

fig 4.25. Mixer

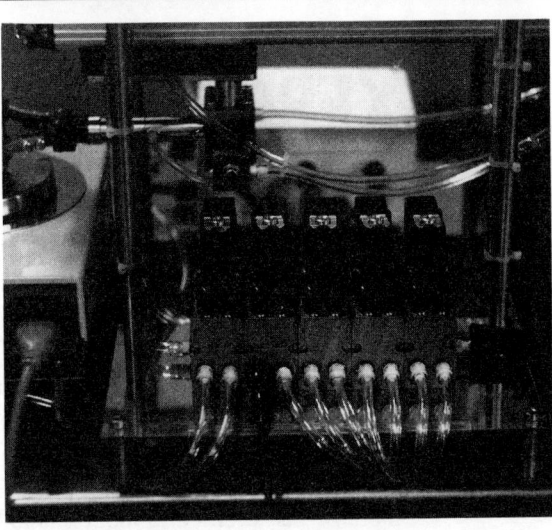

fig 4.26. Pneumatic valves

THE SYSTEMS APPROACH IN USE

Problem
An electronic system which will count packets of biscuits as they pass, one at a time, along a production line is required.

Basic block diagram
This shows the basic parts of the system and gives some detail about what each part has to do (fig 4.27).

INPUT	CONTROL	OUTPUT
Must sense when a packet passes.	Must switch on the output when the input detects a packet.	Must count how many times it has been switched on.

fig 4.27. Basic block diagram

Detailed block diagram
This identifies the components needed for each part of the system (fig 4.28).

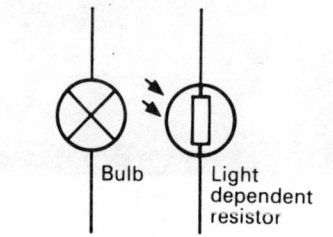

INPUT
The packets passing along the production line could be sensed in a number of ways (e.g. sound, light, weight). This designer decided to use a light beam, shining across the production line. As a packet passes it will break the beam.

Bulb | Light dependent resistor

fig 4.28a. Input circuit

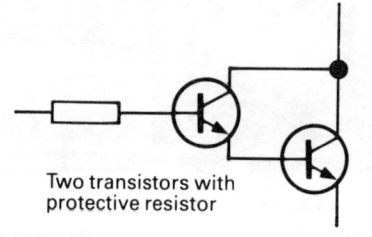

CONTROL
This part of the system must switch on the output every time the input senses a packet. The designer decided to use two transistors, connected so as to give a reliable switching action.

Two transistors with protective resistor

fig 4.28b. Control circuit

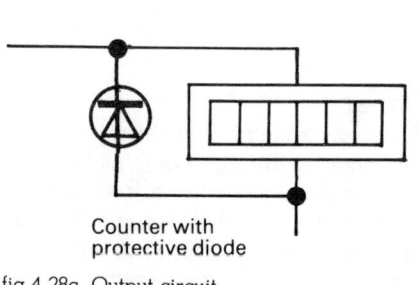

OUTPUT
Every time the control switches on, the output must add one to its count. The designer decided to use an electromechanical counter which counts each time it is switched on.

Counter with protective diode

fig 4.28c. Output circuit

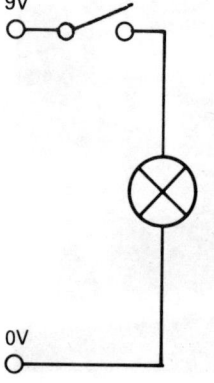

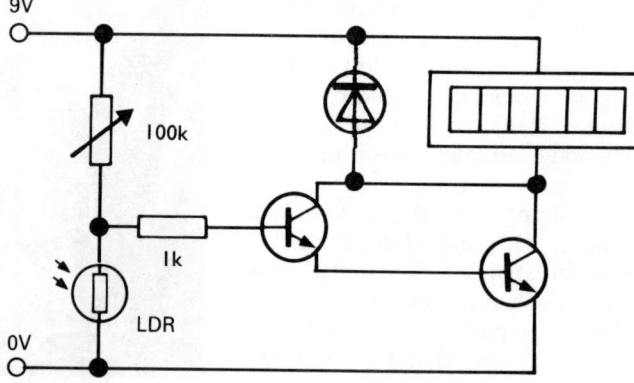

fig 4.29. Circuit diagram

Circuit diagram
Fig 4.29 shows exactly how the blocks are connected together to complete the system. The three basic blocks can still be seen in the circuit. Before making the final circuit the designer may model it to make sure it works.

Final solution
Working from the circuit diagram, a printed circuit board is designed and the circuit built. A case which will protect the circuit and fix all the parts of the system firmly in place must also be designed. Once fixed in position the system can be tested and adjusted so that it will work accurately.

EXERCISES

1. Draw a block diagram showing how sound going into a microphone is amplified and played through speakers.

2. Give three reasons why electronic control systems are used widely in industry.

3. A system to open or close a greenhouse window automatically according to the temperature in the greenhouse is needed. Use the systems approach to design a solution to the problem.

4. Use the systems approach to design a mechanical system which will change the rotary input motion of an electric motor into an oscillating output motion. The oscillating motion is to be used to operate a fuel pump.

MATERIALS AND THEIR APPLICATION

When something is built, the builder has to use his experience to select the most appropriate material from which it is to be made. The choice of materials available is increasing all the time as new materials with special properties are being developed.

The materials available to early man were restricted to natural materials such as wood, bone, animal skins, clay and rocks. He used these to build tools and shelter for his family.

The materials at your disposal include many metals, woods and plastics. The electronic materials now available are many and varied. The choice of material will depend on what properties are required in the specific situation. The use of materials specifically designed for modelling electrical circuits or plumbing systems may be part of the development stage of a project and the material used could well be different to that chosen for the final product.

The materials described on pages 42 and 43 are only a few of those that you could use in your project work. Further information about materials or processes is available in the Collins CDT Design and Realisation book. This chapter is intended to act as a guide to aid your choice of materials in your project work.

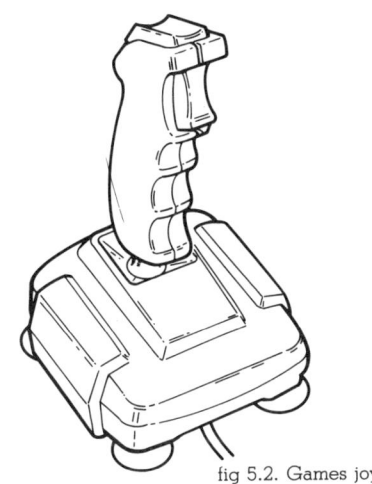
fig 5.2. Games joystick

fig 5.1. Electricity pylon

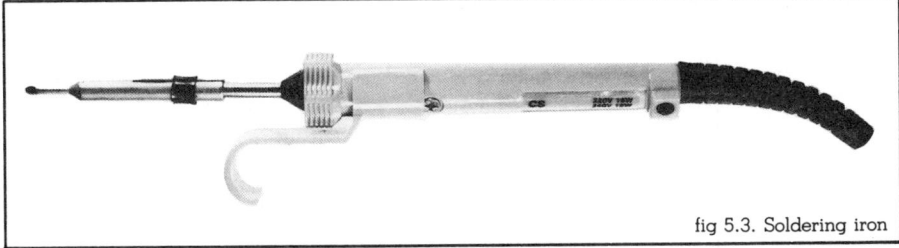
fig 5.3. Soldering iron

fig 5.4. Weight training bench

fig 5.5. Suspension bridge showing use of natural materials

fig 5.6. Car dashboard

Look at the pictures on this page and try to understand why the various materials have been chosen to do the job required.

CHOOSING MATERIALS

Everything that you make, must be made from some sort of material. The choice of material is very important. In order to make this choice you need to know the properties of the material.

1. Does the material need to be light or heavy?
2. How strong must it be?
3. Does it need to be flexible or stiff?
4. Is colour important?
5. What kind of finish is required?
6. Is cost very important?

Once these and other questions have been answered, you will be able to choose a suitable material.

METALS

Material	Colour/MP	Composition	Properties	Uses	Common forms
Aluminium	Light grey 650°C	Pure metal	Light, soft, ductile, highly conductive of heat and electricity	Aircraft, boats, engine components, saucepans foil for packaging, drink cans	Bar, rod, sheet, tube, angle
Duralumin	Light grey 650°C	Aluminium 4% copper 1% manganese	As aluminium and will age harden and machine well	Aircraft and vehicle parts Where strength and lightness are required	Bar, rod, sheet, tube, angle
Aluminium LM4	Light grey 650°C	Aluminium 3% copper 5% silicon	Casting alloy, good fluidity, easily machined Improved hardness and corrosion resistance	General purpose casting alloy	Ingots
Copper	Red 1100°C	Pure metal	Very malleable, ductile and tough High heat and electrical conductivity Corrosion resistant Easily soft and hard soldered	Wires especially electrical cables and conductors Water pipes, soldering iron bits, welding nozzles Printed circuit board conductor	Bar, rod, tube, sheet
Brass	Yellow 980°C	65% copper 35% zinc	Very corrosion resistant Casts well, easily machined Good conductor Easily soft and hard soldered Can be work hardened	Castings, forgings Switch contacts Decorative metalworking	Bar, sheet, rod ingot, tube, angle
Mild steel	Grey 1400°C	Iron + 0.15% to 0.3% carbon	High strength, ductile, tough, fairly malleable Cannot be hardened and tempered Can be case hardened Low cost	General purpose construction work Nuts, bolts, washers and screws Tubular furniture	Bar, sheet, rod wire, tube, girders, angle iron, nuts and bolts
High carbon steel	Grey 1400°C	Iron + 0.7% to 1.4% carbon	Hardest of the carbon steels, but less ductile and malleable Can be hardened and tempered	Cutting tools such as drills, files, plane and chisel blades Wood and plastic cutting saws	Small bars, rod and strip
Tin plate	Shiny grey 230°C	Thin sheet steel coated with pure tin	Mild steel core gives strength and ductility Pure tin coating makes it resist corrosion and easy to soft solder	Tin cans, light metal work Small boxes	Sheet only
Soft solder	Dull grey 250°C	60% tin 40% lead	Soft, low melting point Joins to other metals easily	Jointing copper, tin plate, and all electrical soldering	Wire or bar

PLASTICS

Name	Chemical name	Properties	Uses	Common forms
Low density polythene	Low density polyethylene	Wide range of colours Good electrical insulator Good chemical resistance Flexible soft Service temperature 60°C	Squeezy bottles, toys TV aerial lead insulation Plastic sacks and sheets	Powders, granules, sheet, film
High density polythene	High density polyethylene	Wide range of colours Stiff and hard Can be sterilised Fades in light unless stabilised Easily injected and blown	Buckets, bowls and other household wares Milk crates, boxes, barrels, some machine parts	Powder, granules, sheet, film
Rigid PVC	Rigid polychloroethane	Wide range of colours Tough at room temperature Can be used outdoors Stiff and hard Good for fabricating	Plumbing fittings, pipes and gutters Curtain rails Roofing and constructional sheets	Powder, granules, extrusions, sheet
Expanded polystyrene	Expanded polyphenylethane	White Light, good insulator Absorbs shocks crumbles easily Burns with a poisonous gas	Heat and sound insulation Packaging Bouyancy in canoes and small boats	Sheet and beads
Acrylic	Polymethyl methacrylate	Stiff, hard, clear or opaque Many colours available Very durable, can be polished easily Good electrical insulator Ten times more impact resistant than glass Safe with food Can be bent and formed easily at about 160°C	Signs for shops Aircraft canopies Double glazing Baths, furniture, caravan windows	Sheet, rod and tube
Nylon	Polyamide	Hard, tough, wear resistant, self lubricating High melting point	Bearings, gear wheels, clothing, packaging	Powder, chips, rod, tube, sheet
Polyester resin	Polyester resin	Stiff, hard, brittle, much stronger when laminated with glass or carbon fibres	Boats, canoes, chairs, encapsulating electronics	Liquid and pastes
Epoxy resin	Epoxy resin	Good insulator, good adhesive	Two part glues for metal Aerolite glue for wood	Powder, pastes

WOODS

Common name	Colour	Properties	Uses
Redwood	Cream to pale brown	Soft, easily worked, straight grained, finishes well, low cost	Most used softwood, all inside joinery in houses Low cost constructional work
Mahogany	Reddish brown	Easily worked if straight grained Takes a good finish, strong, medium weight	Furniture, veneers and plywood
Balsa	Light cream	Very light in weight Easily cut and glued Takes paint poorly	Light structures, modelling, prototype case building
Birch plywood	Light cream	Strong, easily painted, knot free usually in thinner sheets	Covering box frames, large flat surfaces
Marine plywood	Reddish brown	Usually mahogany based Strong, waterproof and expensive	Boats, exterior use for high quality structures
Chipboard	Cream speckled	Low cost sheet material can be veneer or melamine faced Interior use only	Flooring, interior furniture, wardrobes, knock down furniture
Blockboard	Cream or reddish	Very strong constructional sheets	Strong box making Can be waterproofed for exterior use

JOINING MATERIALS

Type of fastening	Uses
Round wire nail	Used for general construction, the heads cannot be punched below the surface. Good to hold thin sheet to a frame, head sticks out, but does not pull through easily.
Oval wire nail	Used for better quality work, the head can be driven below the surface and the hole filled prior to painting.
Panel pin	Used for fixing thin sheet to small frames and for small lap joints, such as in a small box.
Countersunk screw	Most common general purpose screw. They are used to join wood, fit hinges, and brackets. The head fits flush with the surface.
Round head screw	Used to attach fittings that are made from thin metal where countersinking could make the joint lose strength.
Corrugated nail	Used to join frames together at butt joints.
Bolt	Used to fasten pieces of metal together where they must be held rigidly. The plain portion of the shank resists wear and movement in the fastening hole.
Set or machine screw	Used to hold metals together where there is to be no movement. Used in machine assembly. Different head shapes are available.
Self tapping screws	Used to fasten thin sheets together or fittings to thin sheets.
Rivets	Used to fix metals together permanently. Can be countersunk headed or round headed. The round headed rivets are used on thin sheets or for decoration.
Pop rivet	Used to fasten sheet metals together. Can be fastened from one side, i.e, fastening to tube, or to inaccessible places.
Grub screw	Used to fasten components to a rotating shaft. The head can disappear into the metal.

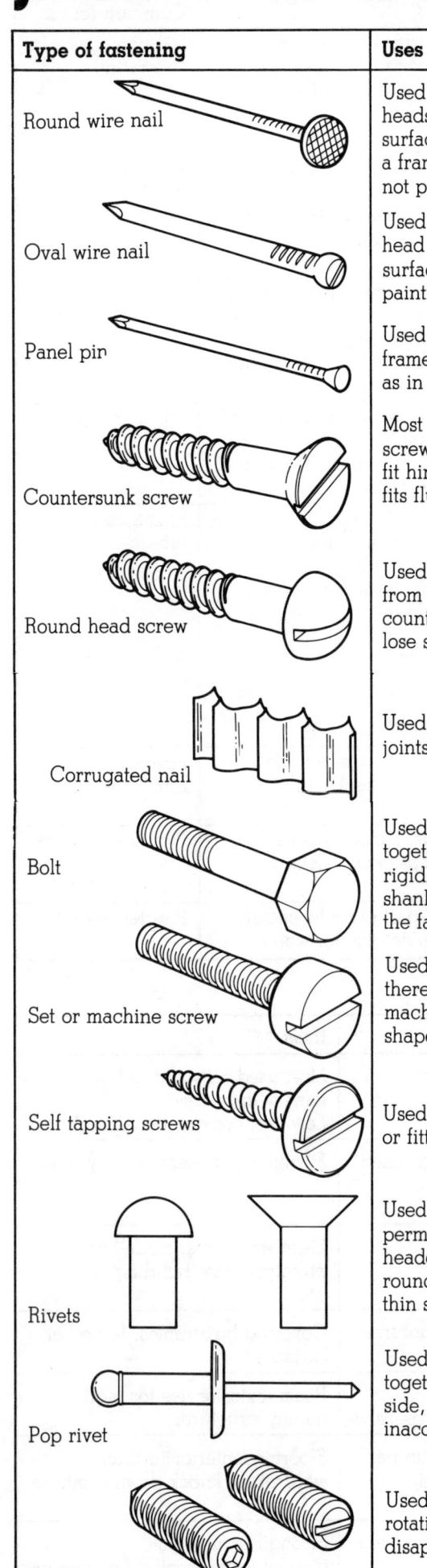

BOX JOINTS

Many projects will be housed in a box structure. Here is a selection of possible alternatives.

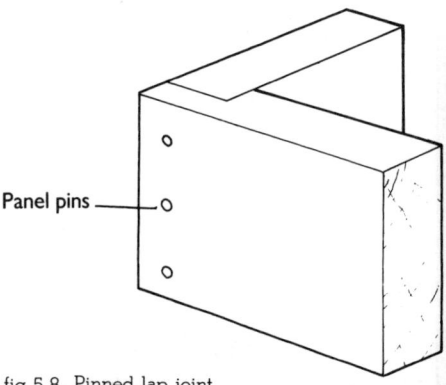

Panel pins

fig 5.8. Pinned lap joint

Other possibilities in wood are:
Comb joint
Dovetail joint
Butt joint

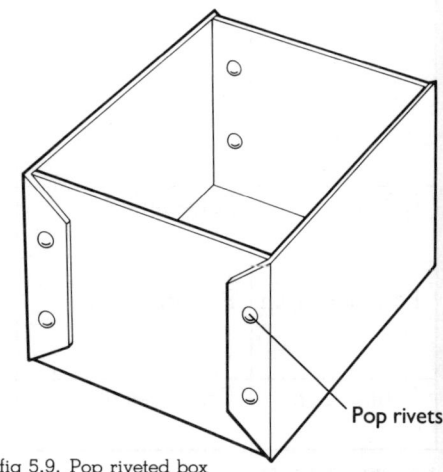

Pop rivets

fig 5.9. Pop riveted box

Other possibilities in metal are:
Butt joint
Lap joint soldered
Lap joint riveted
Lap joint screwed

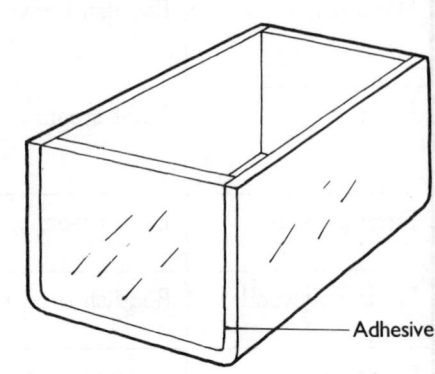

Adhesive

fig 5.10. Glued acrylic box

Other possibilities in acrylic are:
Butt joint glued
Butt joint screwed
Lap joint glued

FRAME JOINTS

Wood is a structural material that can be used in many different ways, either in a frame type structure such as a climbing frame or a chair, or in a box type structure such as a container for an electronic circuit or a cabinet for a weather station. The methods of joining the wood are numerous and varied.

Metals are materials of superior strength. They are used in many different situations. The methods used to join them are as varied as their different uses. These pages are meant more for reference than for detailed information, more information can be found in the Collins CDT Design and Realisation book.

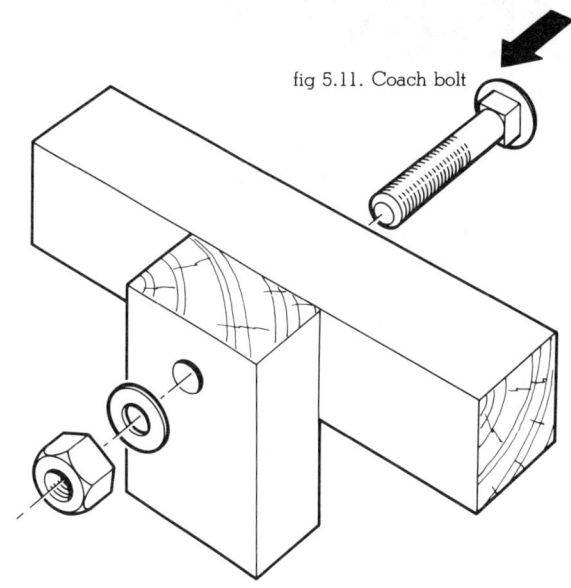

fig 5.11. Coach bolt

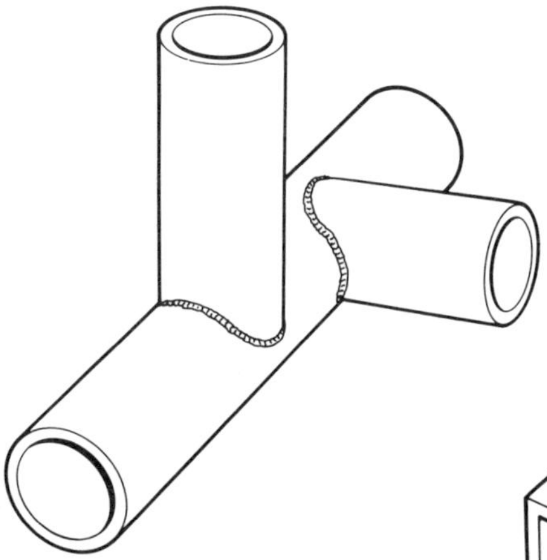

fig 5.12. Bronze welding or brazing

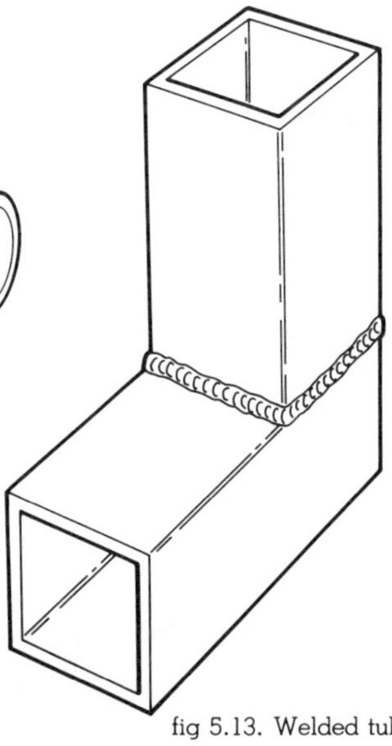

fig 5.13. Welded tube

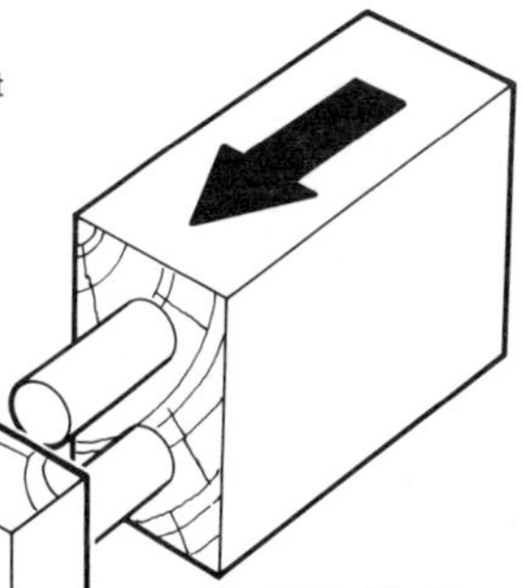

fig 5.14. Dowelled joint

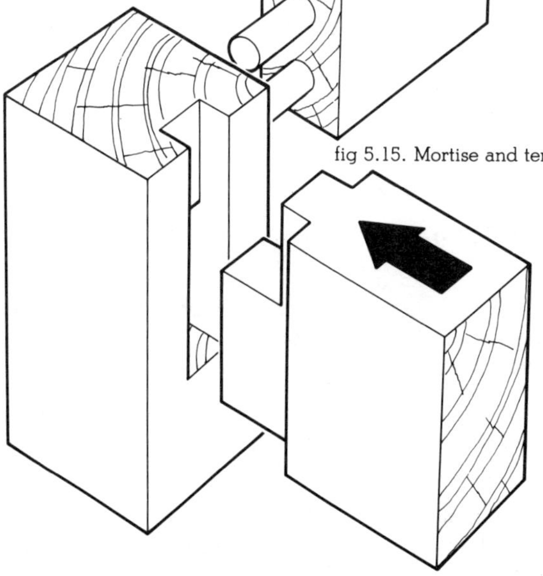

fig 5.15. Mortise and tenon joint

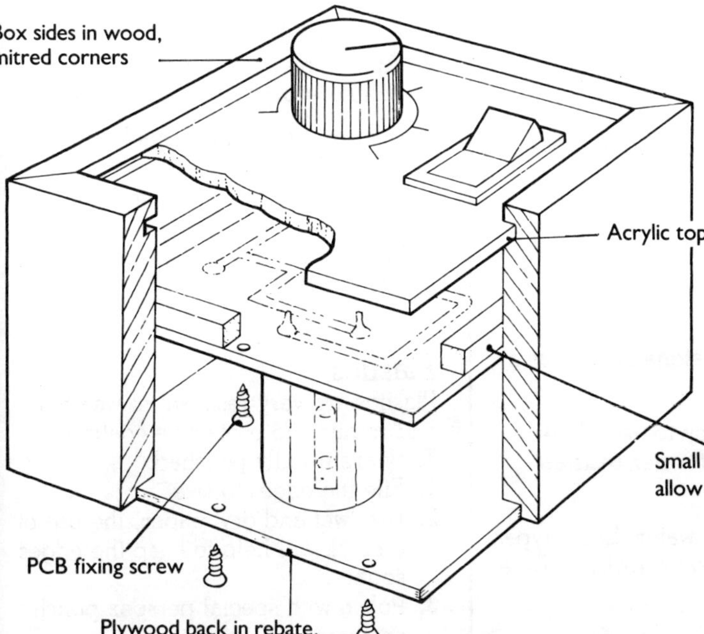

Box sides in wood, mitred corners

Acrylic top in groove

Small batten glued to side to allow screw fixing of PCB

PCB fixing screw

Plywood back in rebate, fixed with screws

fig 5.16. Typical box construction

The cutaway box construction in fig 5.16 shows how the display panel, PCB board and back cover can be assembled. Note that the display panel is permanently held once the box is completed, but the back and PCB board can be removed for battery replacement and repair. This kind of access requires careful consideration when constructing box housing.

ADHESIVES

Adhesives, or glues, are designed to bond materials together. As there are many different types of material to be bonded, a wide range of adhesives has been developed.

The strength of a glued joint depends on three things:
1. the area to be bonded;
2. the strength of the glue when set;
3. the bond between material and glue.

To get a strong glued joint:
1. the area to be glued should be as large as possible;
2. the correct glue should be used; and
3. the surfaces to be glued should be as clean as possible.

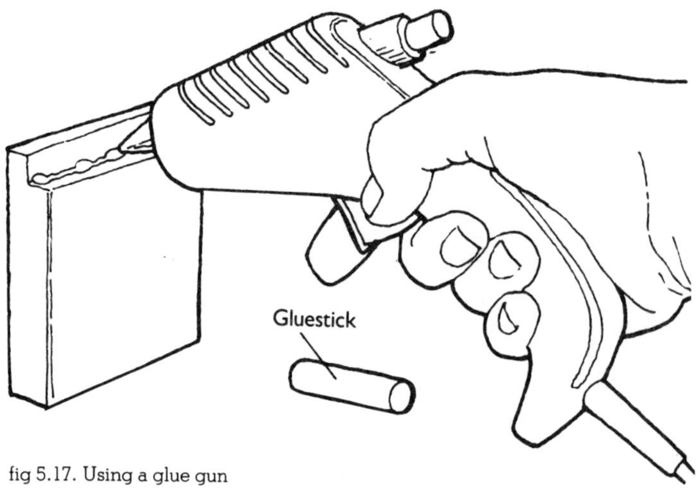

Glue gun

This heats up a solid stick of glue and melts it. When the trigger is pressed, glue comes out of the nozzle. The glue sets very quickly as it cools. The glue can be used to bond almost any material and is particularly effective on small areas.

Gluestick

fig 5.17. Using a glue gun

FINISHES

Wood

Finishes are applied to wood to preserve the timber from moisture and to enhance the decorative effect of the grain.

1. Plane the wood smooth.
2. Use glass paper of varying grades to produce a fine surface. Rub along the grain only.
3. The timber can be stained to enhance the colour.
4. A clear varnish is applied.
5. Rub down the surface gently.
6. Apply a further coat of varnish

Paint in one of many colours can be applied instead of varnish.
1. Plane and glass paper smooth
2. Paint on a primer coat to make the paint stick to the wood properly.
3. An undercoat is then applied dried and rubbed down.
4. A gloss coat is applied to give a fine smooth and colourful surface.

Metals

Steel for external use must be protected from the damp to prevent rusting.
1. Smooth the surface with a file then emery cloth.
2. Apply one coat of a special cellulose paint such as Hammerite.

Copper and brass are usually simply polished. For external use a coat of spray lacquer will keep the shine longer.

Aluminium can be polished or painted as steel.

Plastics

Plastics are very resistant to weather and require no protective coating. They are usually polished.
1. File the edges to size.
2. Use 'wet and dry' paper, the use of a block will help to keep the edges square.
3. Polish with special perspex polish or Brasso on a cloth.

SELECTING THE CORRECT GLUE	
Acrylic (interior) Tensol cement No. 6	Ready mixed adhesive, use tape to mask unjointed areas is suggested.
Acrylic (exterior) Tensol cement No. 7	A two part mix for exterior use.
Expanded polystyrene PVA	A white ready mixed glue.
Fabrics Fabric glue e.g. Copydex	A white rubbery glue, especially made for glueing fabrics.
Metal Epoxy resin glue e.g. Araldite Rapid	A two part mix that will stick almost anything. Slow to set fully.
Plastic laminates Contact adhesive e.g. Evo Stick	A thick brown rubbery glue that is spread on both surfaces and allowed to dry before placing together.
Plastics to other materials Epoxy resin glue e.g. Araldite Rapid	A two part mix that will stick almost anything. Slow to set fully.
PVC Tensol No. 53 Gloy PVC repair.	Can be readily glued. Acetone used as a glue solvent.
Rigid polystyrene Polystyrene glue.	Model glue is commonly used. Acetone is used as a glue solvent.
Timber (interior) PVA Polyvinyl Acetate e.g. Resin W	A ready mixed adhesive that is easy to use. Also useful for paper and card. It is not water resistant.
Timber (exterior) Synthetic resin e.g. Cascamite	A white powder that is mixed with water. Some types have a separate hardener. It is a water resistant glue.

FORMING PLASTIC

Thermoplastic materials such as acrylic can be shaped by forming when they are hot. Using a plug and yoke former the hot material can be shaped by press forming. It is possible to press acrylic of up to 6mm in thickness using this method. Thinner acrylic and high density polystyrene can be shaped by vacuum forming.

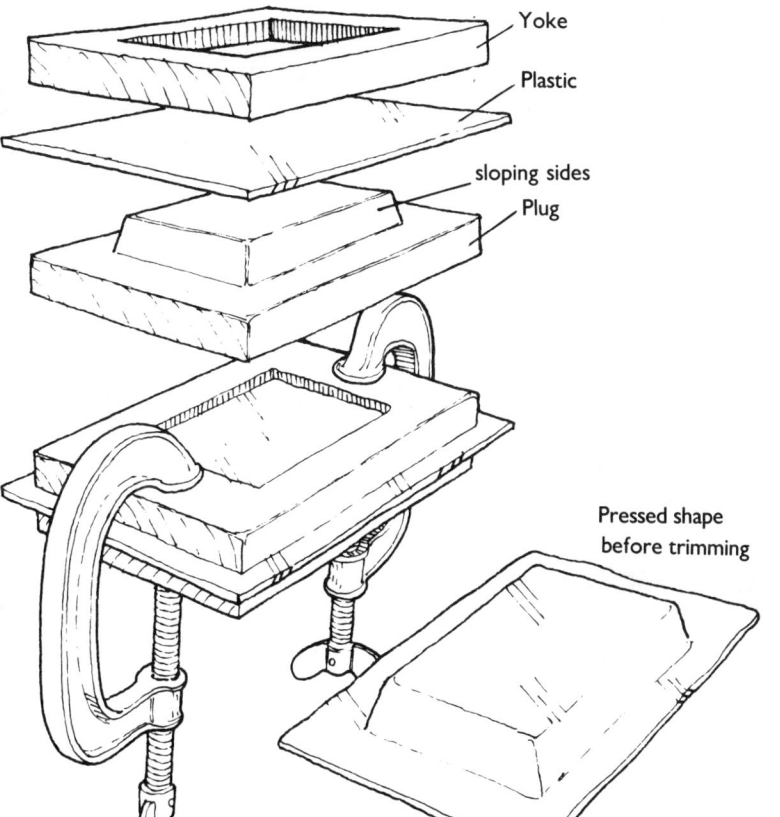

Yoke

Plastic

sloping sides

Plug

Pressed shape before trimming

fig 5.18. Two part former

PRESS FORMING

Press forming is a way of forming 'dish' shapes. A two-part former is needed. The process is as follows.

1. Place the whole piece of acrylic in an oven set at 160°C and heat it until it goes floppy (in 10 to 15 minutes).
2. Wearing protective gloves, remove the acrylic from the oven and quickly place it between the two parts of the former.
3. Quickly press the two parts together, trapping and shaping the acrylic. The former and acrylic can be pressed using a vice, cramps or a press.
4. When the work has cooled, release it from the former.
5. The edges will need to be trimmed.

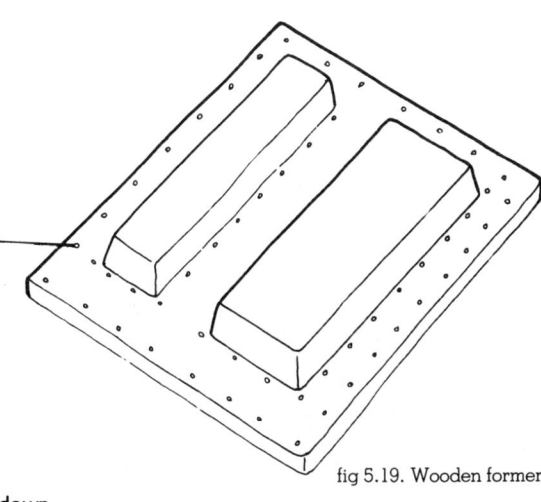

Air holes

fig 5.19. Wooden former

VACUUM FORMING

Vacuum forming is another way of forming thermoplastic sheets. It sucks out the air from under a heated plastic sheet, so pulling it down onto a former. To vacuum form successfully you should proceed as follows:

1. Make a former for the shape you need. The sides of shapes on the former should slope slightly.
2. Place the former on the vacuum forming machine.
3. Clamp the thermoplastic sheet onto the machine and heat it until it goes soft.
4. Lower the hot sheet onto the former and flick the switch to suck the air out from underneath.
5. Allow the work to cool and then remove the formed sheet from the machine.

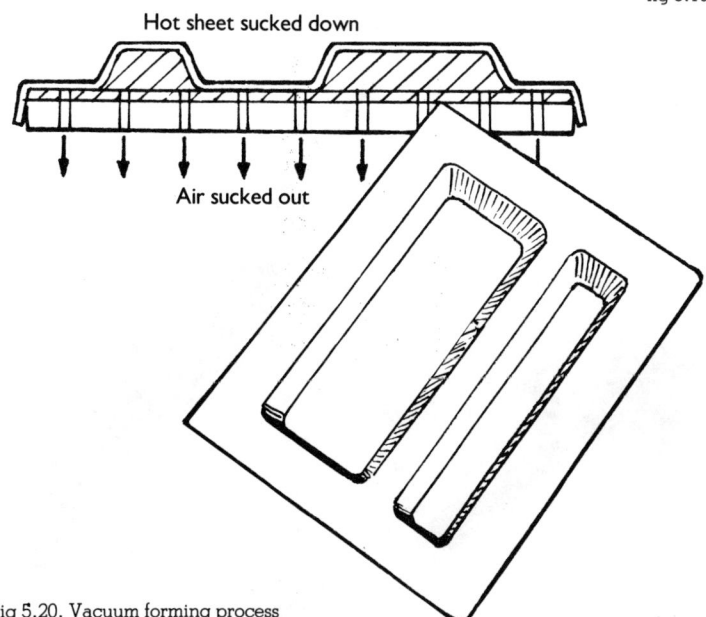

Hot sheet sucked down

Air sucked out

fig 5.20. Vacuum forming process

MODELLING

Materials used for modelling ideas must be quick and easy to use. The choice of material will depend on the type of project or part of project being modelled.

MECCANO

Meccano can be used to model structures or mechanisms. The metal strips are strong and easily joined with the supplied 5/32" BSW nuts and bolts or with the alternative M4 sizes. Motors and gears can be easily fitted.

FISCHERTECHNIK

Fischertechnik construction kits use a system of building blocks which can be fastened together by lugs at the end of each block. These lugs fit into slots in the ends or sides of other blocks. This system is difficult to use, but produces very strong, mechanised structures. Motors and gears can be used to give models mobility. The grippers (fig 5.22) and the BBC buggy (fig 5.23) are both made from this type of kit.

fig 5.21. Meccano buggy

fig 5.22. Simple robotic grippers

fig 5.23. BBC buggy

fig 5.24. LEGO gear mechanism

LEGO TECHNIC

The use of LEGO bricks as a creative building system for children is well known. LEGO has also developed a technical building system with gears, pneumatic cylinders and motors, called LEGO Technic, which can be used to model many different technological situations. The 'bricks' are easily assembled, but suffer from the fact that they easily come apart under relatively small forces.

CARD AND PLASTIC SHEET

These materials can be cut and joined easily to form flat faced structures. Their use in building 3D boxes and cases should be part of a development stage in designing. The use of bifurcated rivets enables jointed mechanisms to be modelled.

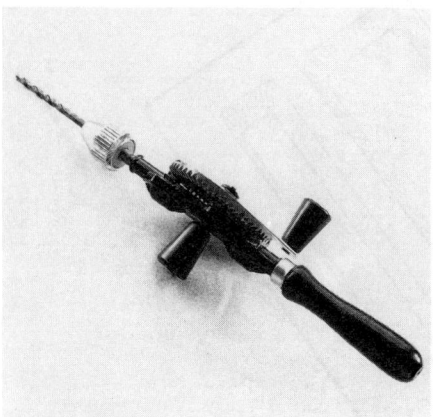

fig 5.25. Manufactured hand drill

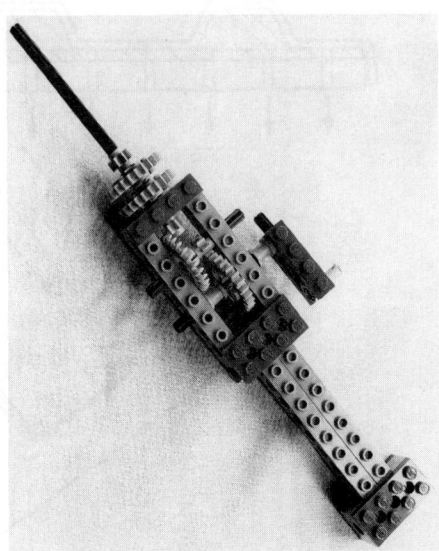

fig 5.26. LEGO drill mechanism

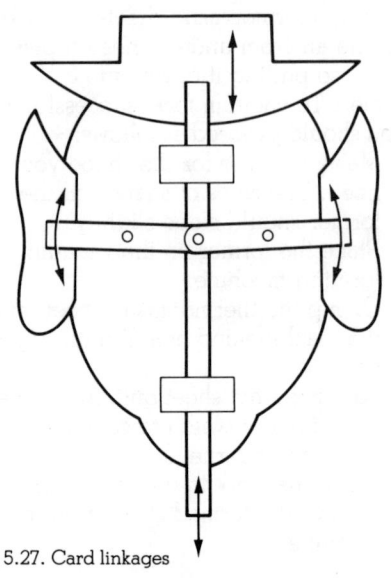

fig 5.27. Card linkages

LOCKTRONICS

This is a system that uses component carriers that plug into a mother board. It is possible to build electronic circuits very quickly. The building of complex circuits with ICs is very difficult. All circuits that have been designed need to be modelled in some way before permanent circuits are built on printed circuit board or strip board.

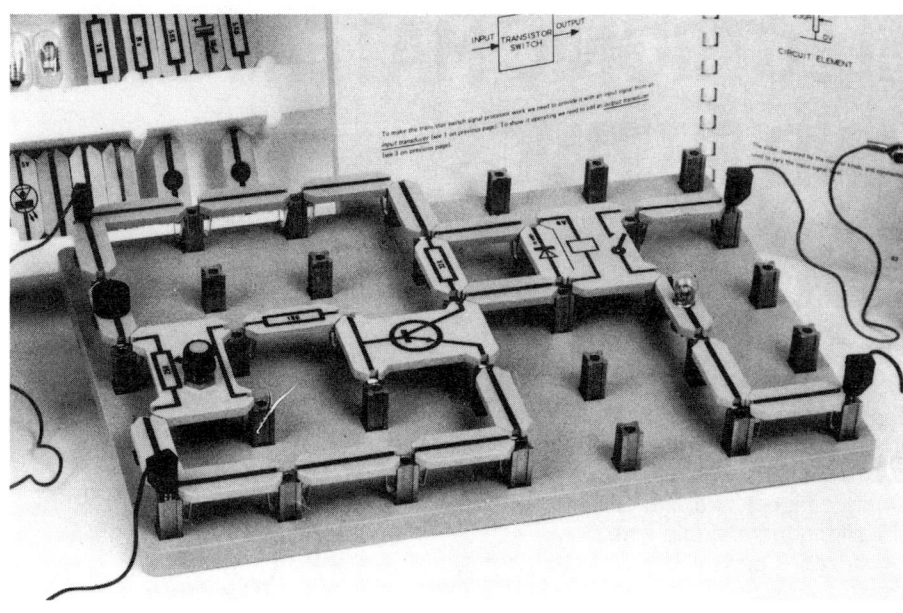

fig 5.28. Locktronics board

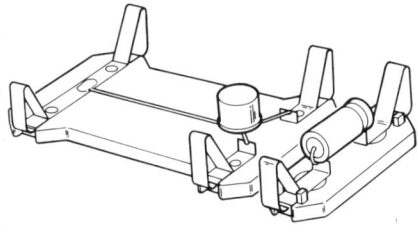

fig 5.29.
Locktronics units showing underside and components

BREADBOARDS

These modelling devices are made by various manufacturers and are all slightly different. The system is based on components being plugged into a board. The interconnections are made with single strand wire bared at the ends. You may find it difficult to use the boards at first, and may even make a few wrong connections. With practice, though, you will find it easier. You will find the boards very useful. They can be used with ICs and can model very complex circuits if built with care.

fig 5.30. Breadboard modelling

SYSTEMS ELECTRONICS KITS

These building blocks are built on small PCBs that can be connected together. They are designed to be used as building blocks, with each block having a small circuit that can be linked with others to model a complete system. It is not possible to add your own ICs into the system without building extra PCBs. It is a very quick and easy method of developing circuits from standard building blocks.

fig 5.31. Alpha systems kit

ELECTRONIC MATERIALS AND FITTINGS

Sockets
Sockets and plugs are used to connect single wires of a temporary nature. Uses include connections to power supplies and test equipment (fig 5.32). The wire is connected by soldering.

Crimp connectors
Crimp connectors are another method of terminating a single wire. Connections may be disconnected, but are seldom separated. They are used for connections to spade terminals on switches, and are also often used for connections in vehicle wiring systems (fig 5.33).

Ribbon cable
Ribbon cable is a flat strip of many conductors laid side by side. It can be of a single colour with a stripe on the first wire, or can be rainbow coloured.

IDC plugs and sockets
Insulation displacement connectors (IDCs) are used to connect ribbon cable. The connector has tiny blades that cut through the insulation to make contact with each of the many wire cores. It is a very quick method of connecting up to 50 wires in a single squeeze of a vice. They are usually used to connect many wires to a PCB by means of a plug and socket. The BBC micro uses them to connect the printers and disc drives (fig 5.34).

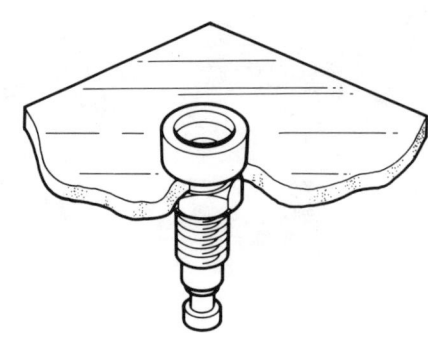

fig 5.32. 4mm socket

Male bullet Butts Female bullet

Blade Shrouded receptacle Receptacle

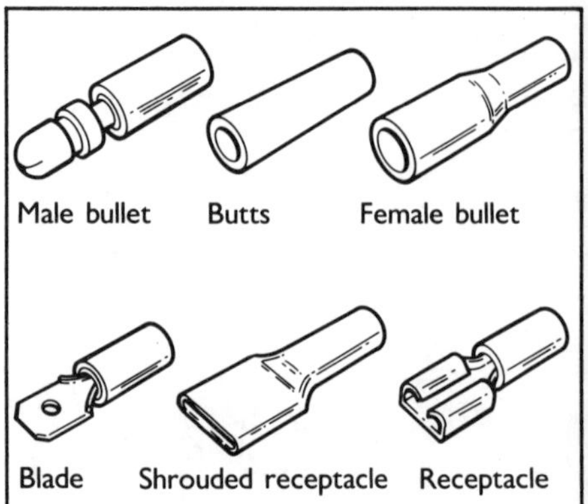

fig 5.33. Crimp connectors

fig 5.34. IDC connectors

DIL connectors
Dual in line (DIL) connectors are used to make permanent connections to a PCB. The pin spacing is 0.1 inch with a central gap of 0.3 inches (fig 5.36). Fig 5.37 shows the IDC connection on the top of the DIL spacing.

DIL sockets
Dual in line sockets are used to connect integrated circuits to a PCB. Soldering the chips directly can damage them by overheating. If they are soldered directly into a PCB they are impossible to remove if they fail in use (fig 5.38).

fig 5.36. Pin spacing

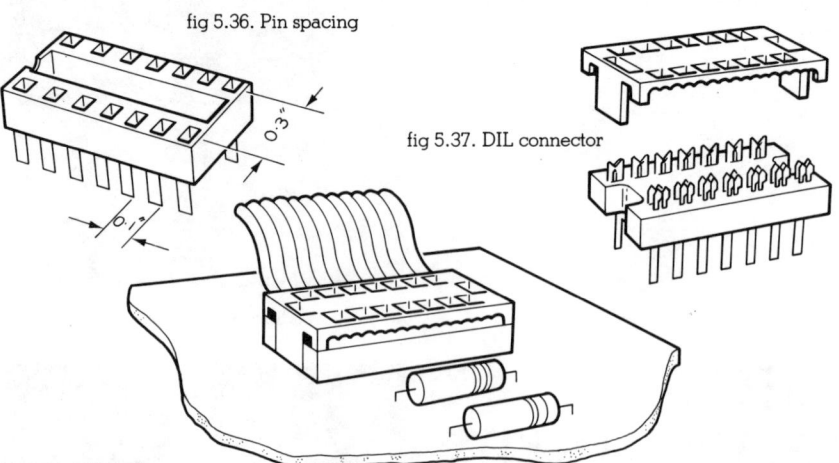

fig 5.37. DIL connector

fig 5.38. DIL socket

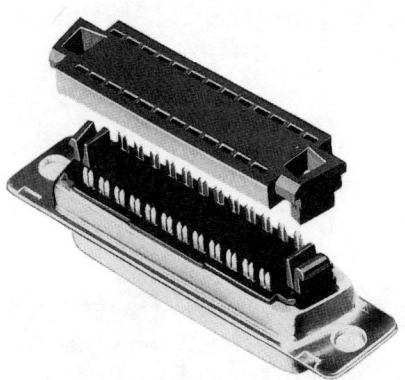

fig 5.35. 'D' connectors

'D' connectors
'D' connectors are used to fasten a number of wires into one durable socket and plug. They can be mounted into the box of an electronic project giving a strong easy connector. The wires are soldered to the pins singly. The plugs should be mounted in a plastic plug housing to cover the bared wires from the user (fig 5.35).

fig 5.39. ZIF socket

ZIF socket
Zero insertion force sockets are used for connecting integrated circuit chips to PCBs. The chips that usually require these expensive sockets are ROMS and EPROMS, as part of a computer system. They allow the easy insertion and removal at the turn of the locking lever (fig 5.39).

WHAT IS ENERGY?

Energy is all around you, though it cannot be seen or touched. The food we eat gives us energy to walk, run, eat and to keep warm when it is cold. To make something move we must charge it with energy, for example we give a golf ball energy by hitting it. Cars are fuelled with power or energy in the form of petrol. Can you think of anything that does not need energy in one form or another?

When energy is used we say that **work** is being done. If a car is being driven along the road, work is being done. If it is driven further, more work has been done. In the present day and age enormous amounts of energy are used in the industrial world. Fig 6.4 shows how energy is used in the UK. Without the ability to harness this energy, we would rely on the same sources as early man.

fig 6.1 100 metre race at finish

fig 6.3. A rocket at take off

fig 6.2. A volcano erupting

WHERE DOES ENERGY COME FROM?

All energy ultimately comes from the sun: the food we eat, the petrol used in cars, the hydrogen in rocket fuel. Some energy, such as coal and oil, was stored many years ago, other energy is still reaching the earth today in the form of sunlight. If we could capture only one hundred thousandth of the solar energy falling on the earth we would have more than sufficient power for all our needs.

Energy sources can be classified in two forms, capital and income.

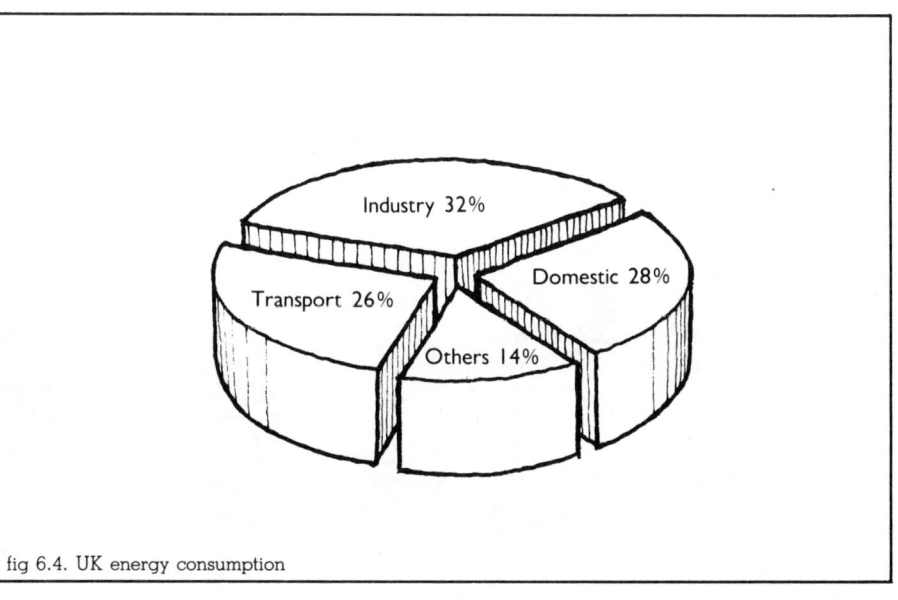

fig 6.4. UK energy consumption

(Industry 32%, Domestic 28%, Transport 26%, Others 14%)

CAPITAL ENERGY SOURCES

Capital energy sources are sources of energy that, once used, cannot be replaced. This energy store was created over many millions of years. The fuels are in a very concentrated form and are easily converted into readily useable energy. For example, a lump of coal could have taken 10 million years to form, but only takes a few minutes to burn on a fire.

FOSSIL FUELS

Coal

Coal was formed by the decomposition of large plants which existed 350 million years ago. When they died they formed a large mass of vegetable material, in a similar manner to a compost heap built by a gardener.

These plants grew in a swamp environment which slowly dried out to become solid land. This was, in turn, covered by sand and rock washed down from the mountains to the valley bottom.

The pressure of the layers of sand over many millions of years compressed the material, squeezing out the moisture to form a rock-like material. Later earth movements have caused some of these layers to become folded and buried deep below the present day surface. Extracting these seams requires deep mining and high costs. Other coal seams are very close to the present surface and can be extracted by quarrying in open cast coal mines, which provides cheaper coal.

At the present rate of world use there should, with known resources, be sufficient coal to cover our needs

fig 6.5. Sea oil production platform

for the next five or six hundred years. Coal is one of the main source materials for the plastics industry. The use of coal for burning may well not be the best use of this important resource.

Gas

Gas is found in conjunction with oil, trapped at the top of the pocket of oil. This resource needs very little cleaning before it is ready to be piped to the house or power station to be burnt.

Oil

Oil was formed in the shallow seas around the land masses. The sea bed was covered with the remains of plants and animals that lived in the water, and as the land was steadily washed into the sea these remains became covered with sandy material. Under the pressure of the accumulating sand and rock, these sea bed remains were converted to oil and gas. Unfortunately these gasses and oils are squeezed and bubble through the overlying rock and escape. However, sometimes the oil and gas cannot escape because the overlying rock is not porous, and if this rock has been twisted to form a dome-like structure the oil and gas becomes trapped. It is these pockets of oil and gas that are drilled into to supply our needs.

The earliest oil wells were drilled on land. It is now recognised that oil pockets can be under both the sea bed and dry land. The cost of extracting oil from the sea bed is much higher in exploration and recovery costs than land based operations.

At present usage rates the world's oil reserves will last less than 50 years. Some authorities on the subject expect the oil to run out in the next 30 years, thus a convenient source of energy could disappear in the very near future. Can you imagine cars without oil and petrol? Oil is another major source of material for the plastics industry.

fig 6.6. Open cast coal mine

Peat

Peat is a resource that is burnt on open fires in moorland areas, and also in power stations in Ireland where it is found in large deposits. Peat is formed in areas of water on mountains. Mosses grow around the edges of the lakes and slowly cover the whole surface. After many hundreds of years the sphagnum moss has rotted and sunk into the lake, while new moss has been growing on the surface. Eventually the lake becomes a peat bog, a very wet, rich, dark, sticky sludge. The bog is then drained and the peat cut and dried ready for burning.

fig 6.7. Traditional cutting and stacking of peat

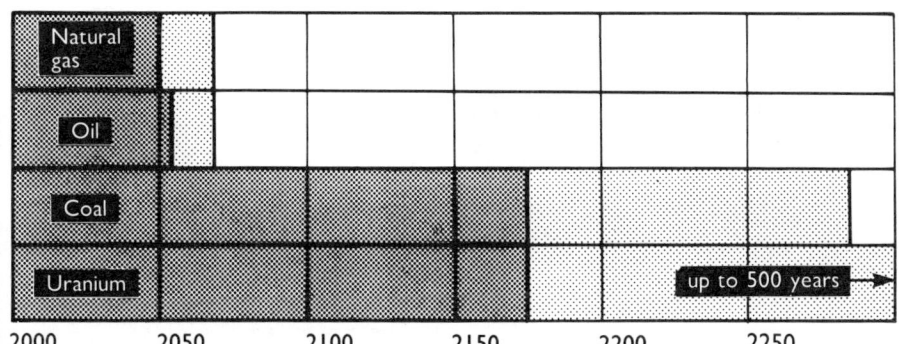

	2000	2050	2100	2150	2200	2250
Natural gas						
Oil						
Coal						
Uranium				up to 500 years →		

fig 6.8. Estimated lifetimes of fossil fuels based on maximum (5.7%) and minumum (1.9%) rates of consumption

fig 6.9. 'Sizewell B' nuclear power station

fig 6.10. Hot springs

NON-FOSSIL SOURCES

Nuclear fuel

Uranium is the source that is used to produce heat energy in **nuclear fission reactors**. The energy is released by splitting the atoms of the isotope uranium U-235. Uranium is a very concentrated source of energy, one kg can produce the same amount of energy as one million kg of coal. A problem with this type of fuel is that the waste produced is very harmful and remains so for many hundreds of years. It must be disposed of in a very safe manner. Many people are very much against this material as an energy source. However, in the near future there will be no gas or oil left, so nuclear energy will become more of a proposition despite its drawbacks and the effect of long-term storage of waste.

Geothermal energy

When it was formed many millions of years ago, the earth was a liquid. It cooled to form a crust upon which we live and is still cooling today. The centre of the earth, many miles below the surface, is still very hot. However, this heat is closer to the surface in some parts of the earth's crust. These places are associated with volcanic activity. We have all seen pictures of volcanic eruptions with their very hot lava flows.

By drilling deep down into the earth's crust we can reach rocks that are much warmer than those on the surface. Pumping water down these holes and then extracting through another hole produces hot water. In some areas of the world hot springs are common, they are a natural occurrence of water heated deep in the earth's crust and forced out under pressure.

Geothermal energy is sometimes classified as a renewable resource (see Income energy sources on pages 54–55), since the heat energy of the earth's core is so vast.

INCOME ENERGY SOURCES

Income energy sources are sources of energy that are being replaced at a faster rate than they can be used. All income energy comes from the sun, and will last as long as the sun. We can extract this energy by **direct** and **indirect methods**. The sun can provide energy directly by means of solar panels, solar furnaces and solar cells. We can also extract the sun's energy indirectly through photosynthesis, wind, waves, tides and water.

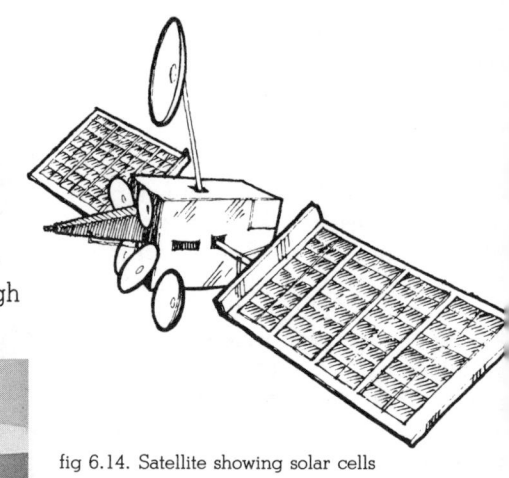

Solar panels

Where the sun is hot enough it can be used to heat water directly. Many houses use this type of heating for their domestic water. The houses have **solar panels** containing water on the roof, facing south to catch the rays of the sun. The problem with solar energy is that most of it is available in summer when there is the least need for heating. Heating our houses takes about 25% of all the energy requirements of Britain. About half the hot water required to heat a house could come from 4 or 5 m² of solar panel. This could be a very important way of saving energy.

fig 6.11. House with a solar roof panel

fig 6.14. Satellite showing solar cells

Solar cells

Photovoltaic cells, commonly called **solar cells**, can convert sunlight directly into electricity. Orbiting satellites and spacecraft use this method of producing power. At present these cells are still very expensive and are not used extensively to produce electricity. Studies are currently in progress in America to investigate the possibility of covering areas of the Arizona desert with solar cells. However, using present technology an area of 20,000 km² of solar cells would be needed to produce sufficient power for the USA, at a very high cost indeed.

Solar furnaces

In countries where the source of sunlight is very strong, concave mirrors are used to concentrate the sun's rays and produce very high temperatures. Most people have seen how focusing the sun's rays using a magnifying glass can burn paper. A **solar furnace** in the Pyrenees has been used to produce clean heat to melt metals for research purposes. Nomadic tribes in the hot deserts have used solar heating for cooking for many years.

Photosynthesis

The conversion of sunlight and carbon dioxide to sugar and starch in green plants by the process of **photosynthesis** is a means of capturing solar energy. This energy is then eaten by man and other animals to provide us with vital energy to live.

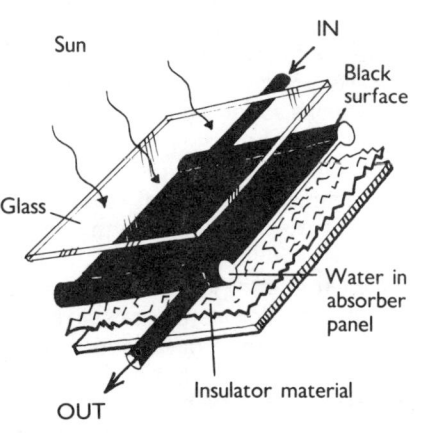

fig 6.12. Solar panel

fig 6.13. Solar furnace in the Pyrenees

fig 6.15 A wind farm

fig 6.16. Storm lashing coastal defences

Wind energy

The sun heats the earth much more at the equator than it does at the poles. This causes the air at the equator to rise, and the cold air at the poles moves to takes its place. (This is a very simplified version of events, in practice it is rather more complex.) This movement is called wind and has produced power through the ages. The windmills of Lincolnshire and Holland are well known examples of this type of energy capture. Man has used the wind to produce movement, the sailing clipper ships were excellent wind converters. At present the possibility of using wind power on large oil tankers as a means of cutting the cost of transporting the oil is being examined.

Large, modern windmills are used to produce electricity in remote areas. Around 2000 windmills are needed to produce power equivalent to that of a normal power station. Areas of shallow water around the coast, such as the Wash, have been suggested as sites for clusters of such windmills. Wind farms are now being developed in the Pennines and Scotland on exposed hill sites (fig 6.15).

Wave energy

Waves are produced by the action of wind blowing over the surface of the water. The force of these waves can clearly be seen during a storm as they batter the sea defences; however their power is not so easily extracted. A series of floating rafts can be used to change this wave energy into mechanical energy and then into electricity.

The sea around the Scottish Islands on the north west coast of Britain is a very good site for harnessing wave energy. This still has to be developed into a practical energy source.

fig 6.17. Scale model of wave energy converter

fig 6.18. Tidal barrage at Rance, N. France

Tidal energy

The tides are due to the gravitational pull of the sun and the moon. As the earth rotates, the effect of the tidal forces are also ever changing. The tides change twice a day, with two high and two low tides each day. The level of the water can rise as much as 15m in the Bristol Channel, though the national average is nearer 5m. It is not this vertical rising and falling of the water that is used as a source of energy, but the filling and emptying of large basins of water. The movement of water into and out of these basins can be used to provide much more power.

There is a scheme in operation at Rance in northern France driving turbines and generators as the tide fills and empties the estuary (fig 6.18).

Hydroelectric power

Some sea water is evaporated by the sun and the resulting water vapour is carried on the wind until it is forced to rise by mountains. At this point it cools, changes back to water and falls as rain. The water then flows back to the sea to start the cycle all over again. The returning water flows along rivers and streams, and can be used to drive turbines, producing **hydroelectric power**. Water is also stored in reservoirs where the energy can be extracted when it is needed. Large areas of the Nile have been dammed at Aswan to provide much needed power (fig 6.19). However, the need to drown large areas of land, often fertile valley floors, can be detrimental to the environment.

Income energy and conservation

The use of these types of solar energy sources will take on much more significance when, in the very near future, supplies of oil and gas become scarce and coal is being used up even quicker. Conservation of energy will become a much more important aspect of life.

fig 6.19. Hydroelectric power dam at Aswan, Egypt

FORMS OF ENERGY

There are many different types of energy. It can exist in a variety of forms, and though it cannot finally be used up, it can change from one form to another.

Mechanical energy

Mechanical energy exists in two forms, **movement energy** and **stored energy**. A cricket ball travelling at speed along the ground towards the boundary has movement energy. It is slowly losing this energy as it rolls along the ground until when it has no more energy, it stops. The energy has been transferred to the grass through bending the blades.

Stored energy is able to do work when released. A clock spring is wound up and stores energy which is released slowly to drive the clock. A book held in the air has stored energy, it can move if unsupported. The water in a dam stores energy, it can be released if the sluice is opened, rushing out with movement energy.

Sound energy is a type of mechanical energy. The vibrations in the air are detected by the eardrums.

Chemical energy

Energy is stored in the chemical bonds within a substance. A chemical reaction can give out large quantities of energy, in the form of heat for example. Burning is a chemical reaction. Energy is released from coal, oil and gas by oxidation (burning). Chemical energy is involved in the digestion of food, which gives us energy to move and to keep warm. A battery is a chemical reactor, it gives out electrical energy.

Electrical energy

Electrical energy is a very convenient form of energy. You cannot see electrical energy, only its effects on various devices. It can be used to make movement with a motor, heat from an electric fire, light from a filament of wire, and is very easily transported along wires and cables. Care should be taken when using electrical energy as it can also kill very easily.

Heat energy

When an object is heated its individual molecules vibrate at a higher speed. If this vibration is at the frequency of light we are able to see the object become red or even white hot. A light filament is a hot strip of wire giving out light vibrations. If water is heated to over 100°C, it will change to steam, and if in an enclosed tin will blow off the lid with the released energy. The lid then has movement energy!

Nuclear energy

Nuclear energy is the energy stored in the atoms of uranium. It is released in the form of heat energy by the splitting or fusing of the atoms in a nuclear reactor.

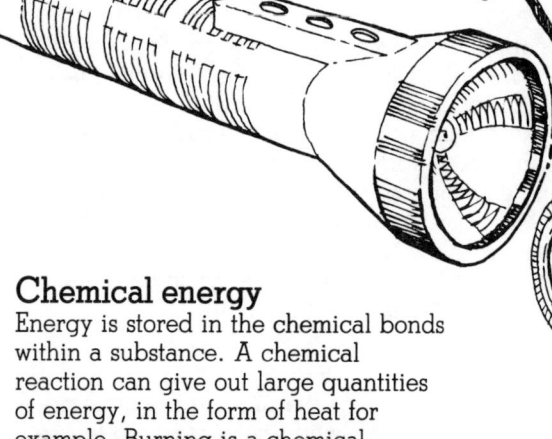

fig 6.20. Forms of energy

Electromagnetic radiation

Solar energy reaches earth by means of radiation. It can cross even a vacuum. There are many forms of radiation, they differ only in the frequency of the waves. Common forms include X-rays, light, infra-red, ultra violet and radio waves.

CONVERSION OF ENERGY

All energy comes, ultimately, from the sun. To make energy easily available to do work, it needs to be stored and converted. Some forms of energy are easily stored until we wish to use them. Some types of energy are much easier to transport than others. There are very grave problems involved with the transport of uranium.

Power stations

Power stations are large installations that convert stored energy from coal, oil or atoms into electrical energy.

Petrol engines

Petrol engines convert chemical energy in petrol into heat energy in the engine then into movement energy to provide transport.

Turbines

Turbines convert movement energy into electrical energy.

Electrical motors

Electrical motors convert electrical energy into movement energy.

Electric kettles

Electric kettles convert electrical energy into heat energy to provide hot water.

MEASUREMENTS INVOLVING ENERGY

Mechanical energy is measured by the amount of work done or that may be done. First, what is work? Work is the **force** used times the **distance** moved:

$$\text{WORK} = \text{FORCE} \times \text{DISTANCE}$$

Distance is a unit of measurement you have all used, but what is a force? It is a push or a pull: wind exerts a force on the sail of a boat, gravity is a force, making objects fall. Isaac Newton is famous for explaining the gravitational pull of the planets and their effect on apples! The unit of force, a **newton** (N) is named after him. A force must have a direction as well as a **magnitude** (size). A medium sized apple resting in your hand exerts a force of about 1 N, due to gravity and the mass of the apple.

Problem

Elizabeth has had a sack of potatoes delivered to the back door. She now has to move the sack from the back door to a shelf in the pantry. Calculate the total work she will be required to do. The resistance to dragging the sack across the floor is 500 N, the force required to lift it is 870 N, the shelf is 0·7 m high, and the door is 4 m from the shelf.

We can calculate the amount of energy used to move the sack from the back door to the pantry from the force required to drag the sack along the floor and the distance travelled to the pantry:

work done = force × distance moved
(dragging)

$$= F \times s \text{ (Nm)}$$
$$= 500 \times 4$$
$$= 2000 \text{ Nm}$$

If the sack of potatoes has also to be lifted onto the shelf, the total energy expended will include the work done in lifting the sack as well as the work done in dragging it across the floor:

work done = force × height
(lifting)

$$= F \times h \text{ (Nm)}$$
$$= 870 \times 0·7$$
$$= 609 \text{ Nm}$$

total work done = dragging work + lifting work
$$= 2000 + 609$$
$$= 2609 \text{ Nm}$$

The units are newton metres (Nm) or **joules** (J), 1 J = 1 Nm.

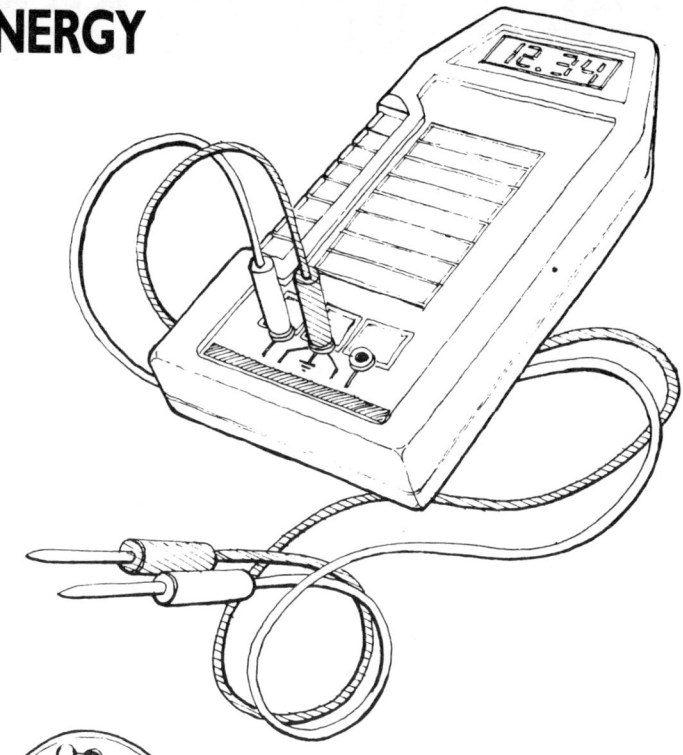

All forms of energy are measured in joules. An oil tanker carrying 4500 litres of oil contains about 160 thousand million J of energy. One thousand million J is called a gigajoule, thus the oil can be said to be 160 gigajoules of energy. A lorry carrying four and a half tonnes of coal is transporting about 120 gigajoules of energy. On a smaller scale an HP2 battery in a torch supplies about 150 J of energy into a pocket torch. A car petrol tank containing 45 litres (10 gallons) of petrol contains as much energy as 12 million HP2 batteries!

Think of the chain of energy conversions involved in producing a cup of tea.

1. Radiant energy from the sun.
2. Chemical energy in the plants converting the radiation to sugars and starches.
3. Mechanical energy to compress the plants into coal.
4. Mechanical energy required to extract the coal.
5. Mechanical energy required to convey the coal to the power station.
6. Chemical energy to heat energy to electrical energy in the power station.
7. Electrical energy to heat energy to boil the water.
8. Mechanical energy to pour the water over the tea bag.
9. Mechanical energy to pour into the cup.
10. Have we any energy left to drink the tea!

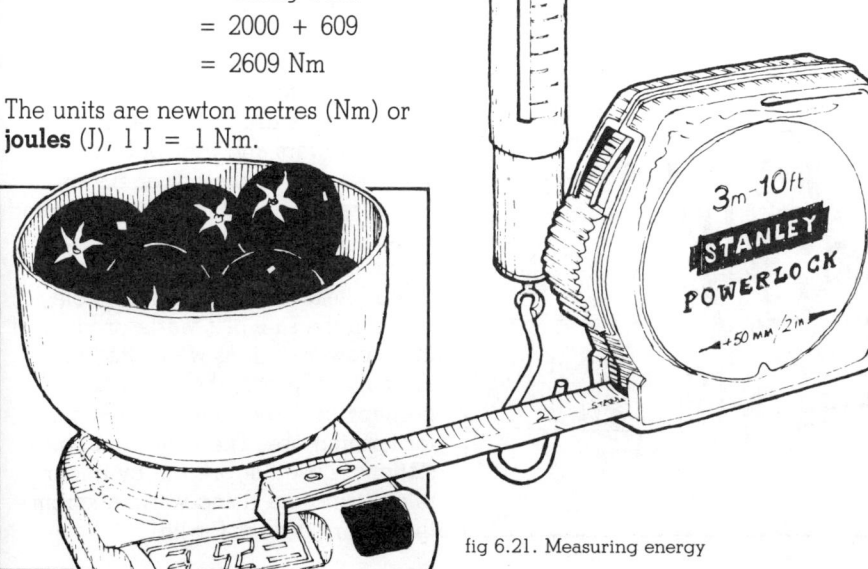

fig 6.21. Measuring energy

CONSERVATION OF ENERGY

Energy is needed to mine metal bearing ores and transport them. Heat energy is needed to melt the ores and heat the metals before they are rolled. Mechanical energy is needed to cut metals and machine them.

Energy is needed to grow crops, especially as more fertilisers are now used. Energy is required to transport, cook and package crops.

Energy costs money and is being used up at an alarming rate. The capital energy resources will not last very long, so we must take care to make the best use of them, without waste. Although there is a colossal amount of energy available from the sun directly, it is difficult to harness using known technology.

ENERGY IN INDUSTRY

Inbuilt obsolescence

It takes almost seven million J, or 0·2 litres of petrol, to make an aluminium drinks can; four million J to make a plastic squash bottle; five and a half million J to make a glass milk bottle. A cardboard box to hold these items will use another ten million J. Only one of these items is re-usable, the other three items are discarded and more energy will be needed to replace them. Are we being short-sighted when we waste this valuable energy?

To save energy every article made should have a suitable lifetime.

Would much extra energy be required to make a car body last longer? Probably not. Unfortunately if cars lasted longer fewer cars would be built, and the workers in car factories would probably not agree with this loss of output and jobs.

Saving materials

By careful engineering it would be possible to reduce the amount of materials used in making a product. For example, material can be saved by **injection moulding** and **casting**, rather than **machining** from the solid. However, we must still have enough material to make the article safe in use.

Reclaiming materials

If we simply bury all our waste materials, we will have to provide new materials (e.g. ores, oils) for every product we use. The reclamation of materials must be a priority in saving energy and resources, though some processes involved in reclamation are high in cost.

Some local authorities now have special factories to process household rubbish, reclaiming metals, glass, plastics and paper, and using the incinerator to provide heating for council property. Surely this is much better than throwing it all away!

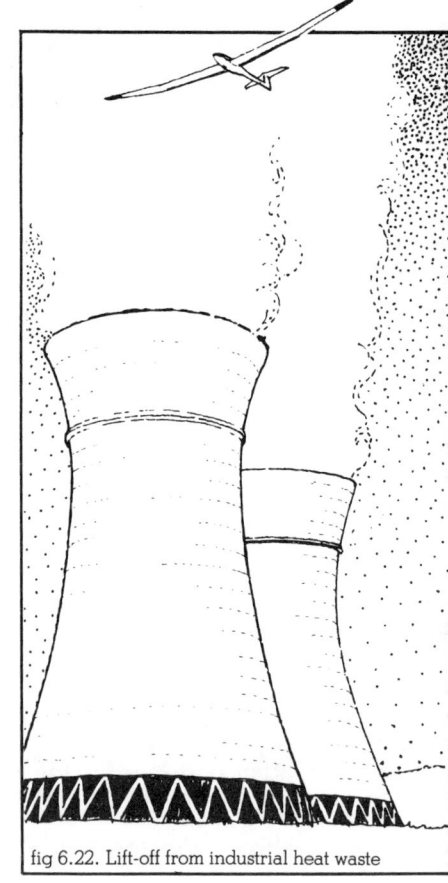

fig 6.22. Lift-off from industrial heat waste

Waste heat

The iron and steel industry uses about 20% of the total energy used in Britain. The efficiency of the blast furnace and the conversion of iron into steel have both been improved, but if these two operations are not carried out on the same site, the metal must be cooled and then reheated at the second site, which is a great waste of heat energy. Steel must be red hot for rolling. If the rolling mill is next to the furnaces and converters, the metal can be kept hot and will not need to be reheated, saving more energy.

Large, new integrated steel works have been built to make use of the saving in energy, therefore, cost.

Power stations produce much heat waste in the form of hot steam and water. This is usually cooled in large cooling towers and the cold water re-used in the power station. If this water was used to heat houses and nearby factories, the waste heat could be sold, representing a saving in overall heat energy and a possible profit in the sale of a waste product. If these power stations were sited in agricultural areas, the heat could, perhaps, be used to heat glasshouses for farming use. The heat from power stations is used at present by glider pilots using the rising warm air from the cooling towers for lift!

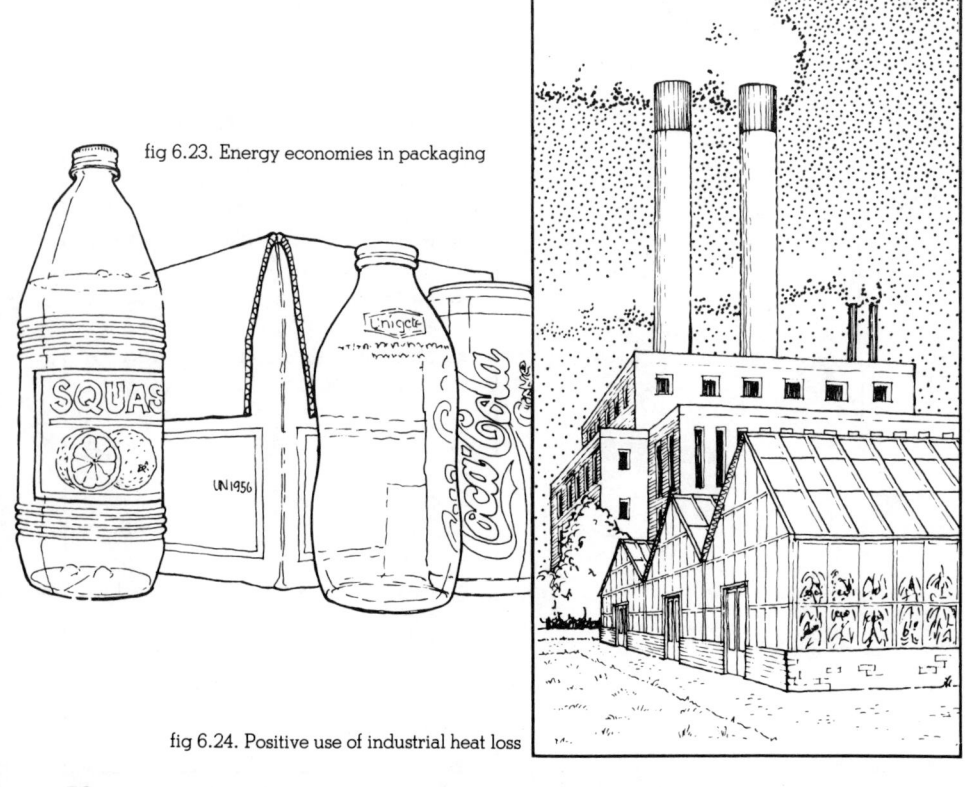

fig 6.23. Energy economies in packaging

fig 6.24. Positive use of industrial heat loss

ENERGY IN THE HOME

About 30% of the energy used in Britain is used to heat our homes and places of work.

Why do we need heating? Stone age man had no heating other than food. He did have insulation, in the form of clothing, furs from the animals he ate as food. Insulation is still the solution to economical use of heat in our homes. Fig 6.25 shows the relative energy needs of different types of housing. The energy required to heat a terrace house is almost half that required to heat a detached house, even less is required to heat a middle floor flat. Perhaps we should all live in middle floor flats!

With a little thought and insulation the detached house could require only as much heating as a terrace house.

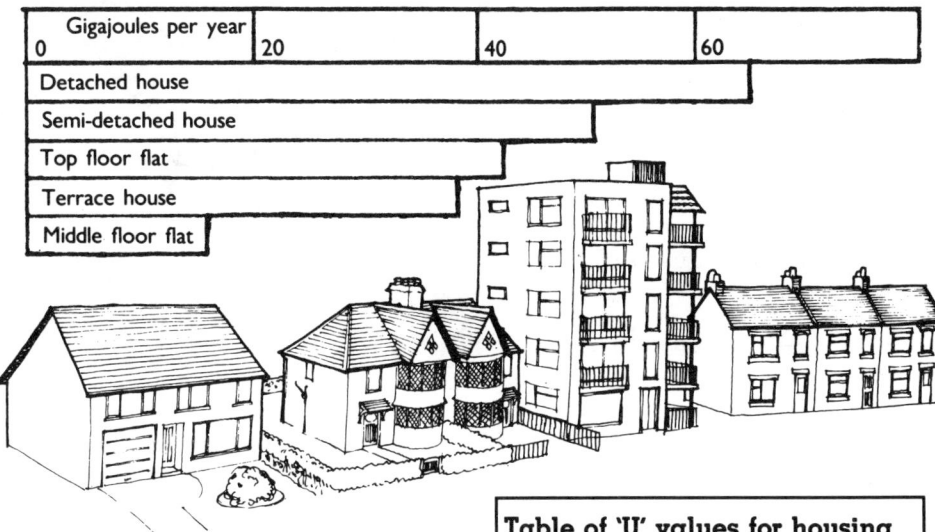

Gigajoules per year			
0	20	40	60
Detached house			
Semi-detached house			
Top floor flat			
Terrace house			
Middle floor flat			

fig 6.25. Energy consumption of different types of housing

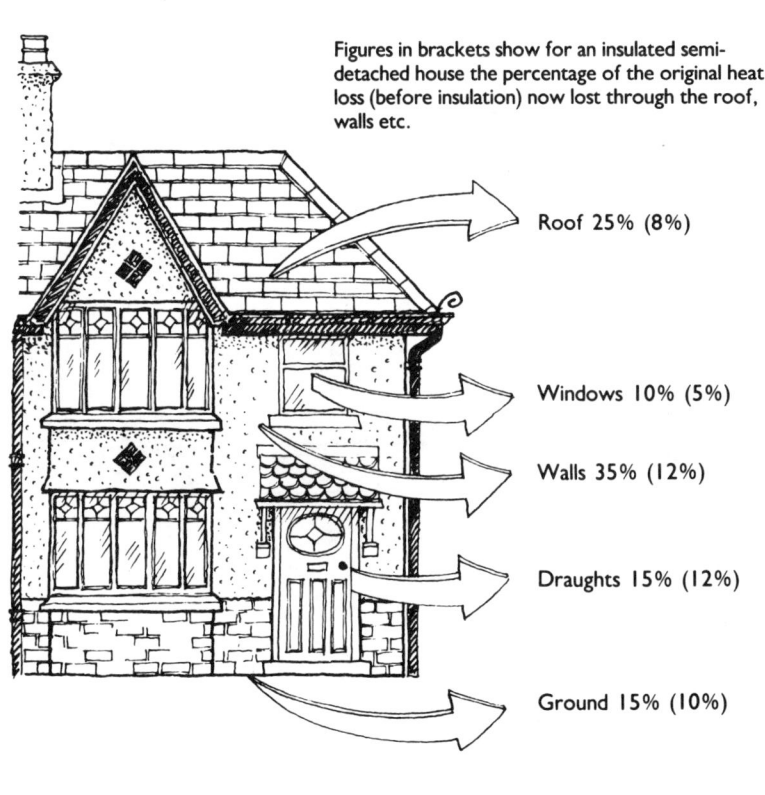

Figures in brackets show for an insulated semi-detached house the percentage of the original heat loss (before insulation) now lost through the roof, walls etc.

Roof 25% (8%)

Windows 10% (5%)

Walls 35% (12%)

Draughts 15% (12%)

Ground 15% (10%)

fig 6.26. Percentage heat loss in uninsulated semi-detached house

Table of 'U' values for housing constructions

(Lower number = slower heat loss rate)

Corrugated asbestos	7.9
Tiles on battens	2.0
Tiles + 80 mm fibreglass	0.4
Solid brick 120 mm	3.6
Solid brick 240 mm	2.7
Cavity wall 280 mm	1.5
Cavity with insulation	0.5
Single glazing	5.6
Double glazing	2.8
Concrete on ground	1.1
Concrete and wood block	0.9
Timber on joists	2.3

Insulators

Materials that do not allow heat to travel through them are called **insulators**. Materials that allow heat to pass through easily are called **conductors**. This is the same terminology that is used when describing electrical conductors and insulators (see 'U' table above).

Good insulators include air, glass fibre, wood, wool, cork, polyurethane foam, polystyrene foam and straw board.

All metals are good conductors.

If you place your hand on a material that is at normal air temperature and it feels cold, then you can be sure that it is conducting away your body heat. It is a conductor. If it feels warm, then it is likely to be an insulator.

If we clothe the walls, roofs and floors of our houses in an insulating material we will conserve the heat inside.

The cost of installation of these insulators must be balanced against the cost of heating our homes. Many people feel that the cost to the world in saved energy may be worth much to future generations.

Heat can travel in one of three ways: **conduction**, **radiation** and **convection**. It can also be transferred if the matter in which it is contained is physically removed, for example warm air being removed from a building by ventilation.

Materials that are used in buildings all allow heat to escape through them, but some are much better than others. The 'U' value is a measure of this. Heat loss through the building materials can be calculated using the formula:

$$\text{Rate of loss} = A \times t \times \text{'U'} \quad (\text{watts, W})$$

where A = area of the material (m^2), t = temperature difference, inside to outside (°C), 'U' = 'U' value ($W/m^2/°C$).

This will give the rate of heat loss not the quantity. The quantity can only be calculated if we know for how long the heat is escaping.

EXERCISES

1. Energy is classified as capital or income energy. Explain what you understand to be the difference between income and capital energy.

2. (a) List the main sources of capital energy.
(b) List the main sources of income energy.

3. It has been suggested that income energy should be used in preference to capital energy. Why is this? Explain your answer with examples.

4. How can wind energy be changed into electrical energy? List all the changes in energy states that occur in this process.

5. Tidal barriers are built to capture large amounts of water at suitable estuary sites. This can produce large amounts of energy from turbines as the water moves into and out of the basin. What are the arguments against barrages? Why should we think carefully before building them around our coast-line?

6. Photosynthesis is the means by which plants capture solar energy. How can this energy be released to provide various other forms of energy?

7. List the chain of energy conversions involved in cooking and eating an egg for breakfast.

8. Give examples of a change from electrical energy to some other form of energy in your home or school.

9. Explain how energy could be saved in your home.

10. How could the waste heat from power stations be used more profitably?

11. Make a list of disposable goods that are available for our use.

12. The race to extract energy can prove to be harmful to the environment. Does this always happen? Use a local example to explain how it did or did not harm its local environment.

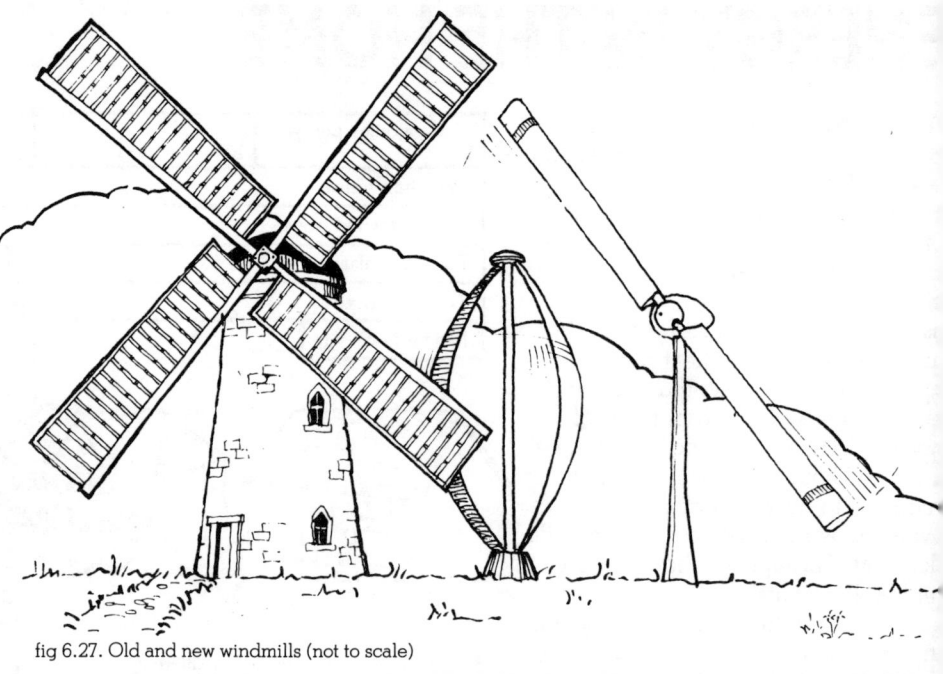

fig 6.27. Old and new windmills (not to scale)

fig 6.28.

fig 6.29.

ELECTRONICS

As you look around your school and home you will be aware of devices that use electricity. If you were to turn off the electricity in a house, very few things would work. However, this would not have been the case 100 years ago. We have come to accept that electrical energy is available almost everywhere. We have electrical energy available in our houses for heating, cooking, washing, lighting and many other uses. We have portable electrical energy in the form of batteries for cassette players, torches, watches and electronic games.

Although electronic devices are often small and complicated, they are made from only a few different types of component. Computers are very complicated devices, but are made almost entirely from millions of tiny transistor switches that can only be turned on or off.

The development of the electronic watch gives us an example of the speed of the changes in this technology. In July 1969 Neil Armstrong walked on the moon. Despite the high technology that this required, he would not have been able to describe a digital watch, they had not been developed. The dials in the spacecraft were not digital, but analogue, as are most speedometers in cars. Watches and clocks were mechanical devices driven by springs.

In less than 20 years we have progressed to digital watches which cost only a few pounds, and are even given away with petrol! We have watches with calculators, with voice output and even with a computer terminal. This is the age of electronics. Who can tell what we will be wearing on our wrists in another 20 years time?

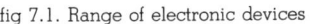

fig 7.1. Range of electronic devices

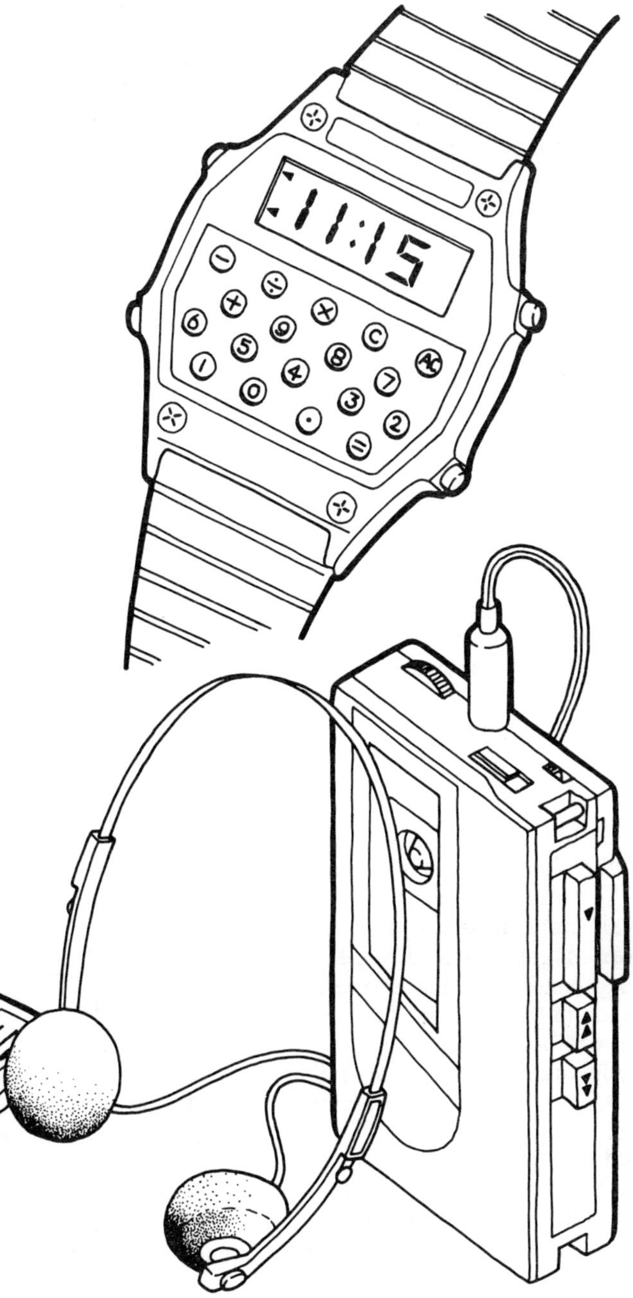

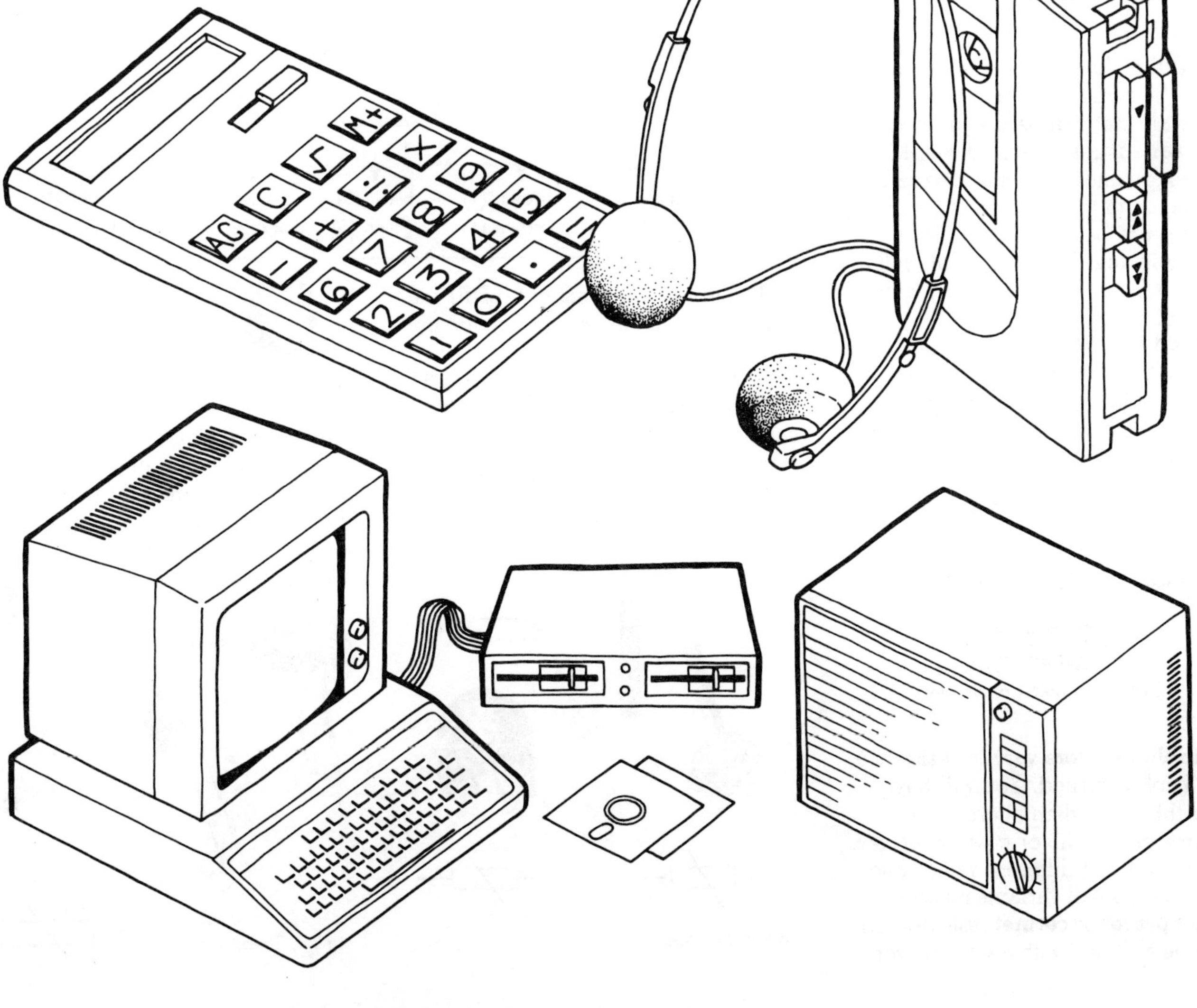

BASIC COMPONENTS

Electronic devices, as we have said, can be very simple or very complicated. However, they are all built from a few very basic components which can be put together in an infinite number of ways to produce many varied solutions to many varied problems. It is the use of these components in a properly organised manner that is the basis of technological design.

Battery

The power for most of your electronic devices will come from a **battery**. Batteries provide a safe source of electrical energy. The chemicals inside the container will provide 1.5 volts of electrical energy. However, you do not need to understand the complexities of the chemical generation of electricity to be able to use a battery.

You will need more than 1.5 volts to power your circuits. The most common batteries for project work are the PP3 and the PP9, both of which are 9 volt batteries.

Resistors

These components will restrict the flow of electricity. The unit of resistance is called the **ohm**, the symbol for which is Ω. The resistor is made from materials that can be constructed to give various resistance values. Metals are good conductors and have a low resistance to the passage of electricity. Plastics are bad conductors of electricity and have very high values of resistance. There are many different types of resistor.

Fixed resistors are cheap and simple to use. The carbon film, fixed value resistor is the most common one in use. The values vary from a few ohms to 10M ohms (10 million ohms). They are always marked with a colour code, which consists of coloured bands to indicate the value (see fig 7.3).

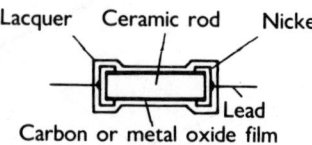

Variable resistors vary in resistance as a knob is turned. You will have used this type when altering the volume on a radio or cassette player. They can be of different values and sizes. Very small variable resistors called **preset** or **cermet** resistors can only be adjusted with a screwdriver.

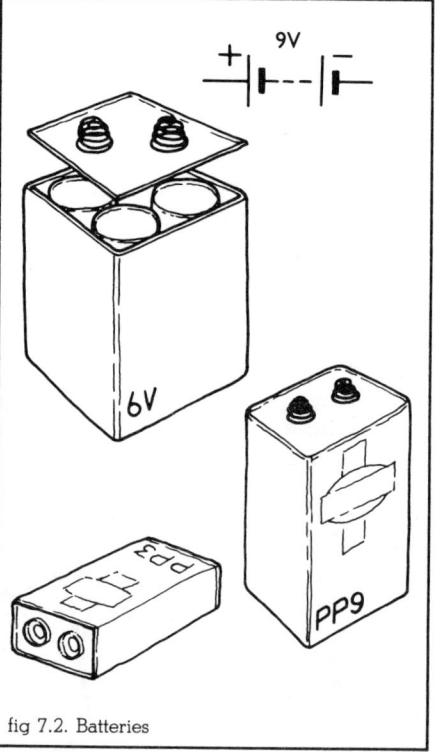

fig 7.2. Batteries

Wires

The various components of an electronic circuit are connected together by wires, made with a copper core and a polyvinyl chloride (PVC) covering. The copper is a good conductor of electricity, while the PVC is a good insulator. Wires or cables can be made with a **solid core** or with **stranded cores.** Stranded wire is used where movement is required in the connection, such as test leads or wiring to mains plugs. Solid core wire is cheaper, but does not allow much movement before the wire snaps.

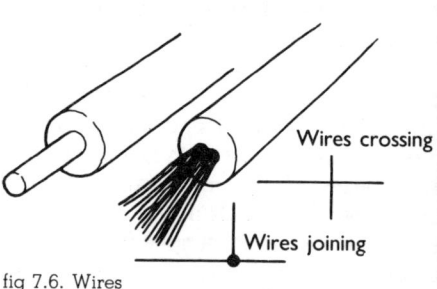

fig 7.6. Wires

fig 7.3. Resistance colour code

Resistance Colour Code

Number	Colour
0	black
1	brown
2	red
3	orange
4	yellow
5	green
6	blue
7	violet
8	grey
9	white

1st number — 2nd number — number of 0s — tolerance
gold ± 5%
silver ± 10%

brown (1)   black (0)   brown (1) = 100Ω or **100R**
brown (1)   green (5)   orange (3) = 15000Ω or **15k**

Light dependent resistors (LDR) are devices that act as resistors that can change in value. The change is brought about by the amount of light that is allowed to fall on the sensitive cadmium sulphide surface. The resistance can change in value from 10M ohms in the dark to 1k ohms in daylight. This makes them suitable for detecting light levels (fig 7.5).

fig 7.5. Light dependent resistor

A thermistor or thermal resistor is a device whose resistance can be changed as its temperature is changed. This makes thermistors suitable for use in temperature detection circuits (fig 7.7).

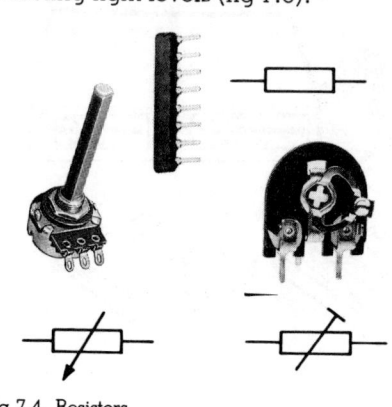

fig 7.4. Resistors

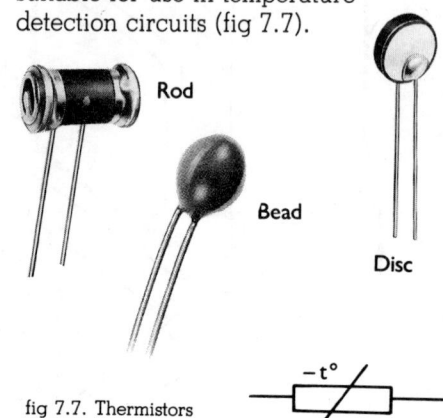

Rod

Bead

Disc

fig 7.7. Thermistors

Capacitors

The most commonly encountered component after the resistor, is the capacitor. A capacitor is a component that can store electrical charge. It consists of two metal plates separated by an insulator called a **dielectric**. The plates are very close to each other, but electricity cannot pass across the dielectric (fig 7.8).

When the capacitor is connected to a voltage supply, a current will flow while the capacitor collects its charge, rather like a bucket filling with water. When the **capacitor** is full the flow stops. The capacitance of a capacitor, C, measures its charge-storing ability. The units of capacitance are **farads** (F). These are very large units and most capacitors you will use will be measured in **microfarads** (µF). One microfarad is equal to one millionth of a farad (see page 215 for an explanation of SI units).

Electrolytic capacitors must be connected the correct way round if they are not to be damaged.

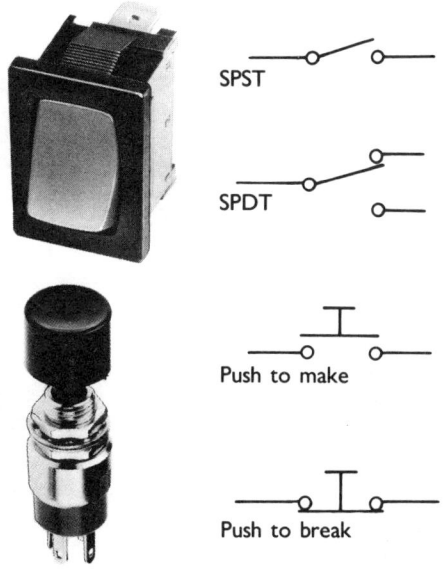

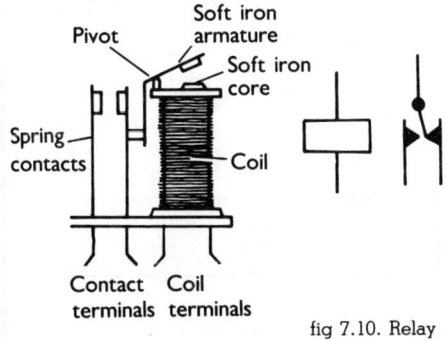

fig 7.9. Switch types

Relays

Relays are devices that are used as interfaces to allow a small voltage and current to switch to a large voltage and current. Household mains voltage can be switched with a battery operated relay.

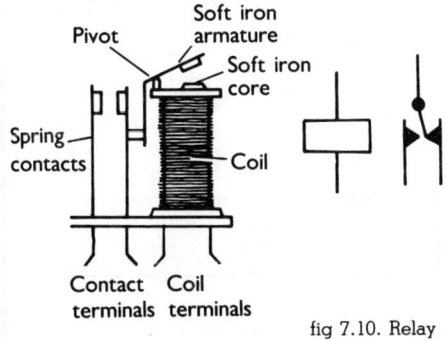

fig 7.10. Relay

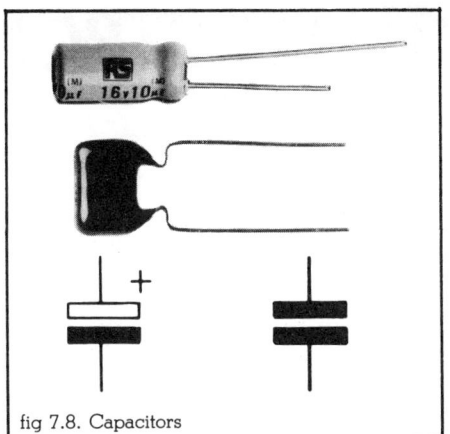

fig 7.8. Capacitors

Light Emitting Diodes (LEDs)

LEDs are the most common optical devices in electronics. They have a coloured top that gives off light when current flows through it. The output level is lower than a bulb, but power consumption is very low. Most electronic devices use LEDs (e.g to indicate volume). Most LEDs are rated at 20mA with a voltage drop of 2V.

Switches

Switches are not electronic components, but mechanical devices that can make (connect) or break (disconnect) a circuit. Switches have different numbers of **poles** and **throws**. The poles (P) are the number of separate circuits that the switch makes or breaks at the same time and the throws (T) are the number of positions to which each pole can be switched.

Push-button switches can have a spring return or can stay down until pushed again. They are usually single pole type.

Slide switches are single pole or double pole change over type.

Toggle switches are often to be seen on equipment as on-off switches.

Keyboard switches are available singly or as complete keyboard units.

Rotary switches are used when many poles are required.

Speakers

Speakers are devices that convert electrical energy into mechanical energy (movement) in a paper cone, producing sound waves in the air.

fig 7.11. Speaker

Diode

Diodes are semiconductor devices that allow the current to flow in one direction only.

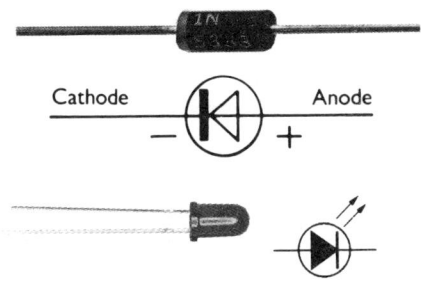

fig 7.12 Diode (top, middle) and LED (bottom)

Transistors

Transistors are semiconductor devices made from two diodes. They have three connecting wires and form the basis of most electronic circuits that involve electronic switching. Transistors may be large or small, all have their uses. One problem is choosing the most suitable one for the circuit you are building.

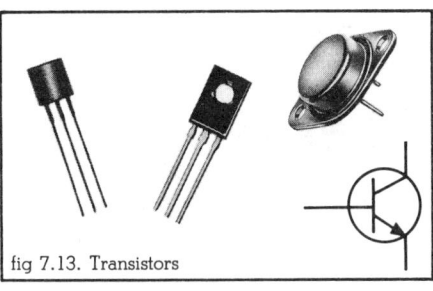

fig 7.13. Transistors

Integrated circuits

Integrated circuits are small electronic circuits that can have many different components on a small slice of silicon in a single unit. They are the central part of all modern circuit designing. A standard package is used, in standard sizes.

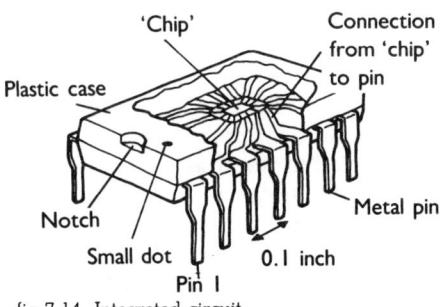

fig 7.14. Integrated circuit

fig 7.15. Inside a computer

DESIGNING WITH ELECTRONICS

You have seen some of the components that can be used to make electronic circuits. Your aim is to be able to use these components linked in different ways to solve various problems. This is designing. In order to do this you must have a fundamental knowledge of how electrical circuits and component groups work.

A simple circuit

A circuit is, as the name suggests, a circular path or route. For example, a motor racing circuit is a circular route, with a starting point and a finishing point linked by roads. In the case of an electronic circuit the start and finish are at the power source, in the example (fig 7.16), a battery, and a switch are linked by electrical pathways.

If the circuit in fig 7.16 is broken by the opening of the switch, the path is not complete and the lamp is not lit. Close the switch, the circuit is complete and the lamp is now lit. A poor connection is often the reason for a circuit not working. In the same example, if the wire to the bulb holder has not had the plastic insulation stripped before connecting, the circuit is incomplete because the plastic is an insulator and prevents the passage of electricity. The circuit must be made from conducting components.

The circuit you wish to produce must be drawn in a way that other designers will understand. This means using a simplified method: wires are drawn as vertical or horizontal lines and the different components have their own symbols (7.17). This is the language of the electronic designer. The circuit diagram is often drawn as in incomplete circuit. It is recognised that the two rails are connected to the same power source, so though it is always there, it is left out of the diagram for clarity. Instead the two rails are marked with their voltages (fig 7.18).

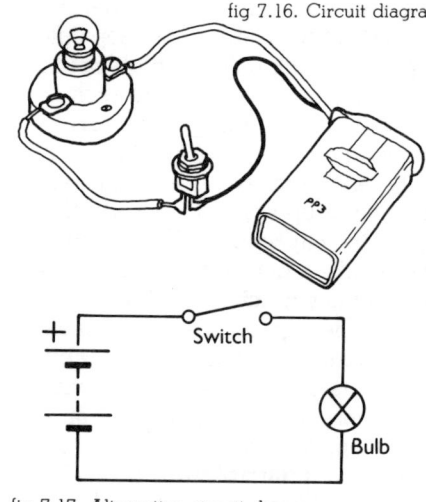

fig 7.16. Circuit diagram

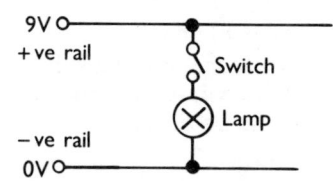

fig 7.17. Alternative circuit diagram

fig 7.18. Alternative circuit diagram

POWER SOURCES

Every circuit must have a source of electrical energy. The one you are most likely to use is a battery or a power pack running from the mains supply acting as a battery substitute.

A single dry cell

You do not need to know how a dry cell works as part of this course, but details can be found in textbooks on electricity. For our purposes, the chemicals in the cell produce an electrical energy supply of 1.5 volts.

Cells can be connected in **series**, that is positive to negative, to produce a higher voltage (fig 7.21). A PP3 battery has six cells each producing 1.5 volts, making a total of 9 volts.

Alternatively, cells can be connected in **parallel**, that is positive to positive, negative to negative (fig 7.22). The total voltage produced is still 1.5 volts, but the battery will last much longer than a single cell.

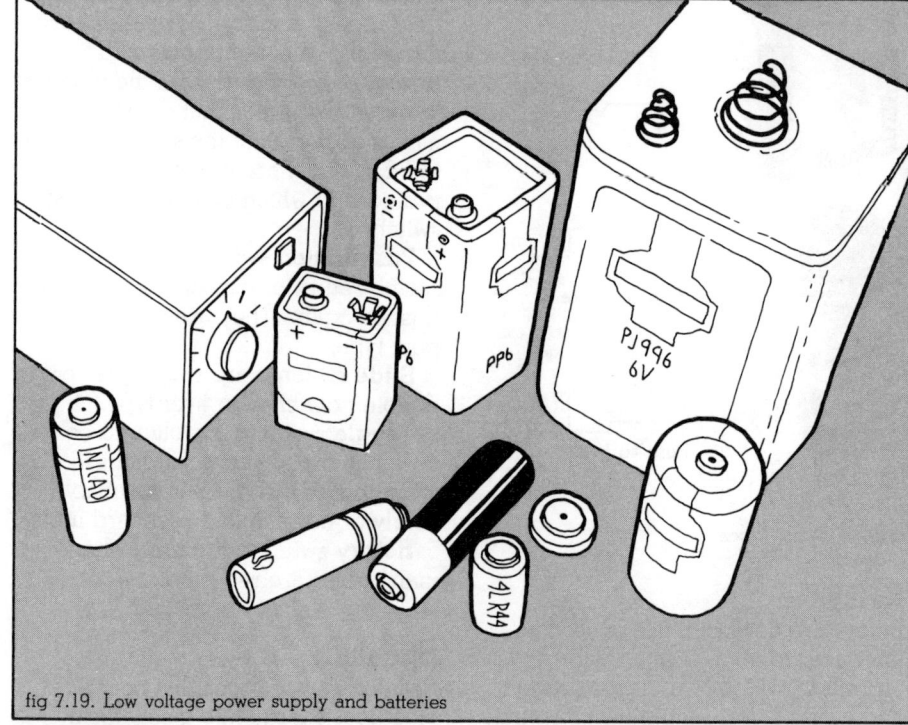

fig 7.19. Low voltage power supply and batteries

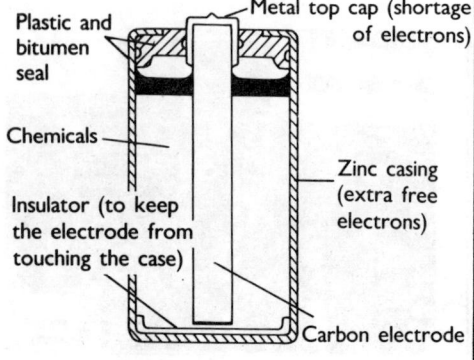

Plastic and bitumen seal

Metal top cap (shortage of electrons)

Chemicals

Insulator (to keep the electrode from touching the case)

Zinc casing (extra free electrons)

Carbon electrode

fig 7.20. A single dry cell

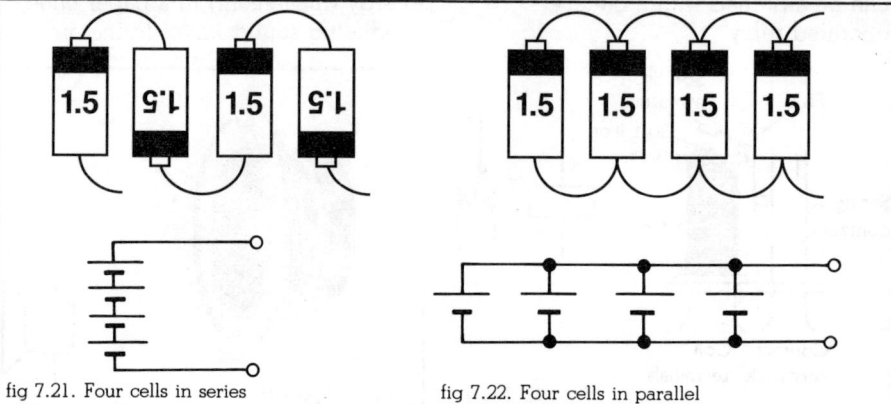

fig 7.21. Four cells in series

fig 7.22. Four cells in parallel

When the battery is connected in a circuit, electrons will flow. The amount of flow, the **current**, is measured in **amps** (A). The **voltage** can be seen as being similar to the pressure of water in a high tank connected by pipes as in fig 7.23. The amount of water flowing in the pipe is the current and the restriction in the pipe is a **resistor**. This analogy to a plumbing system helps us to visualise the electrical flow in a circuit.

If the tank is positioned higher the pressure is greater and the water flows faster in the pipe. The higher the voltage the more power there is in the circuit to drive the electrons. If there is a larger restriction in the pipe, the flow of water is smaller. The larger the resistor in the circuit the smaller the current.

The relationship between the voltage, the current and the resistance in an element of the circuit is always the same. This was discovered by Georg Simon Ohm in 1827, and is called **Ohm's Law**.

Ohm's Law states that the current, **I**, flowing through an element, is directly proportional to the voltage drop, **V**, across it at a constant temperature. It is usually written in the form:

$$V = IR$$

It is possible to calculate any one of the three if the other two are known, by rewriting the equation as either:

$$I = V/R \qquad \text{or} \qquad R = V/I$$

Given a voltage, it is possible to find a suitable resistor to give a specified current value.

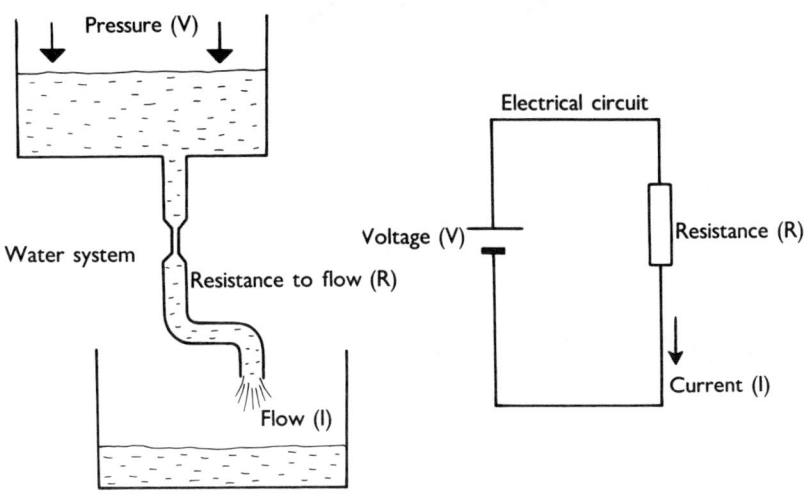

fig 7.23. A plumbing analogy

Design problem

A **light emitting diode** (LED) is a device that gives off light when an electrical current passes through it. It has limitations, first the current must flow in the correct direction as for any diode, and second the current must not exceed 20mA or the LED will be damaged.
What resistor must I use to restrict the current to 20mA if I am using a 9 volt battery?

Using Ohms Law to find the resistance:

$$\mathbf{R} = \mathbf{V/I}$$
$$= \mathbf{9V\text{-}2V/20mA}$$
$$= \mathbf{7/0.020}$$
$$= \mathbf{350\ ohms\ (\Omega)}$$

The **nearest standard value** above this is 390R.

fig 7.24.

I would, therefore, use a 390R resistor in a circuit with an LED and a 9 volt supply.

SYSTEMS ELECTRONICS

Designing with electronics is made much easier if we use a **systems approach**. The problem to be solved is broken down into blocks of components, which are then designed separately using basic components or special 'chips'.

Each circuit will have at least three main blocks, an **input**, a **control** and an **output**. These can be shown as a block diagram like fig 7.25. The input part of the circuit senses and reacts to specific changes, for example, changes in the temperature, moisture level or light level of the surrounding environment. The control part of the circuit reacts to changes in the input and then controls (changes) the output accordingly. The output is switched on and off by the control part of the circuit.

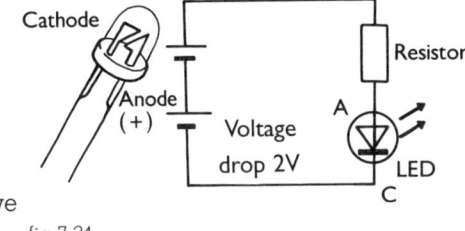

fig 7.25. Block diagram for systems electronics

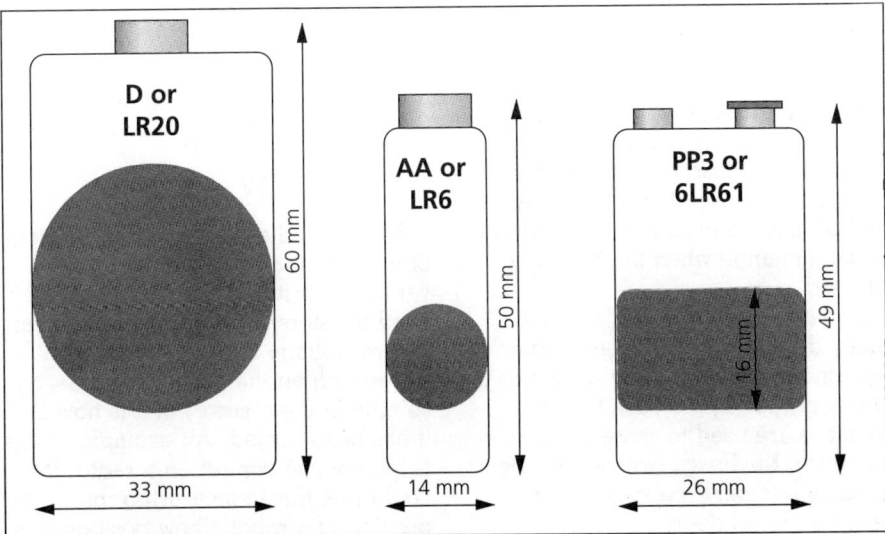

fig 7.26 Common battery sizes

SENSING CIRCUITS

In order to build circuits that will control things, we must be able to sense the environment around the project. We must change the surrounding variations of light, heat and moisture into electronic signals that can be processed by the control element of the system.

SWITCH SENSOR

Movement of a switch by a person can be sensed and turned into an electrical signal.

The circuit diagram in fig 7.28 shows a power supply to the switch (SW1), with one resistor. Let us see what happens, not how it happens, but what happens, when the input is changed.

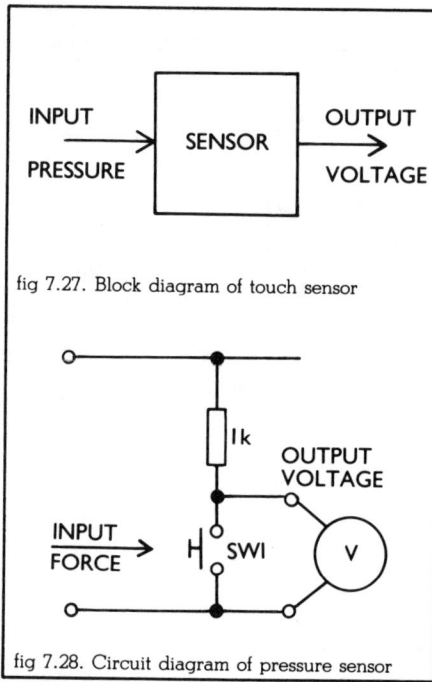

fig 7.27. Block diagram of touch sensor

fig 7.28. Circuit diagram of pressure sensor

First, with no input force the switch is open, and the voltage output is 9 volts. Second, with an input force at the switch, the switch is closed, and the voltage output is 0 volts.

In order to see this we must be able to see the output. A **multimeter** is one means of registering a voltage or **potential difference**. A multimeter set to a suitable voltage range with the positive (red) lead connected to the output of the circuit and the negative (black) lead to the 0V line will show a voltage change when the button is pressed.

The switch used could be one of many different types, depending on the amount of force to be sensed or the shape of the object. Push button switches are used to sense finger pressure. Keyboard switches can be used as bumper type switches or to sense a closed door.

POTENTIAL DIVIDER

A potential divider is not a sensing circuit, but a basis on which other sensing circuits are built (figs 7.29 and 7.30).

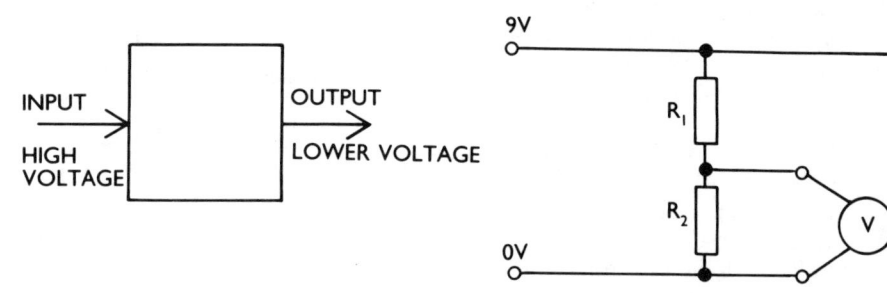

fig 7.29. Block diagram of potential divider

fig 7.30. Circuit diagram of potential divider

In fig 7.31 three examples of resistor values are shown. In the first example the resistors are of the same value, and the voltage output will always be half the supply voltage, thus here it is 4.5 volts. With R2 larger than R1 the output is over half the supply voltage. With R2 smaller than R1 the output voltage is lower than half the supply voltage.

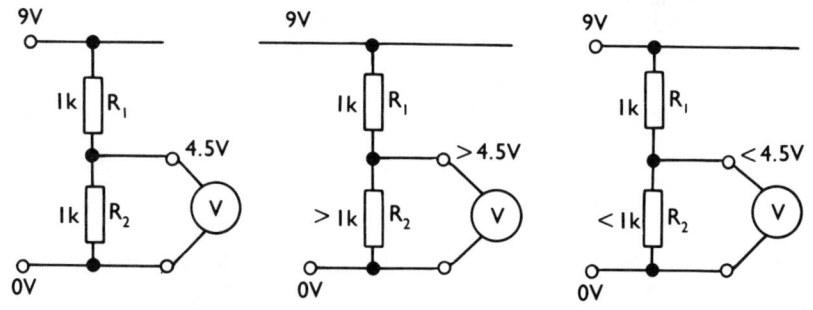

fig 7.31. Three examples of resistor values controlling output voltages

The voltage in the circuit passes through the two resistors and back to the battery. Part of the voltage is across one resistor and part across the other. By adjusting the resistor values you can have any output voltage between 0 and 9 volts.

The actual values of the output voltage can be calculated from the formula:

$$\text{Voltage out} = \frac{R2}{R1 + R2} \times \text{supply voltage}$$

For example, to find the output voltage in fig 7.32:

$$\text{Voltage out} = \frac{R2}{R1 + R2} \times \text{supply voltage}$$

$$= \frac{1k}{8k + 1k} \times 9V$$

$$= 1V$$

Figure 7.33 shows a special version of the potential divider which uses a variable resistor in place of the two fixed resistors. The output voltage can be any voltage between 0 and 9V, depending on the position of the spindle of the resistor, that is how far it has been turned. An example of this is the volume control on a radio. We could link this to measuring the position of a robot elbow position.

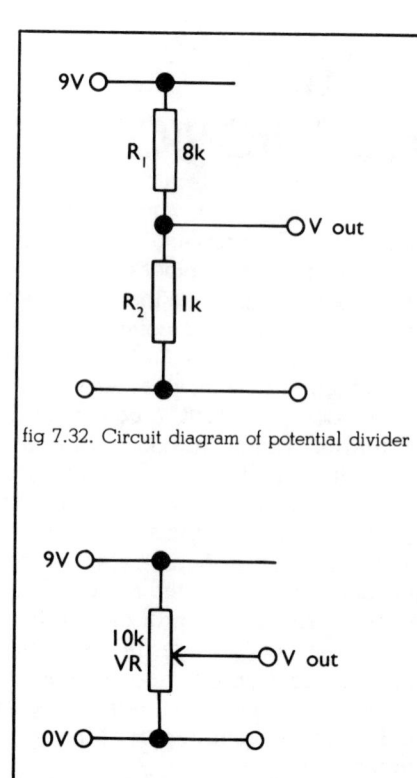

fig 7.32. Circuit diagram of potential divider

fig 7.33. Potential divider with variable resistor

LIGHT SENSOR

A **light dependent resistor** (LDR) is a device that changes its resistance value as the level of light falling on the window changes. There is a grid-like layer of cadmium sulphide behind the clear window.

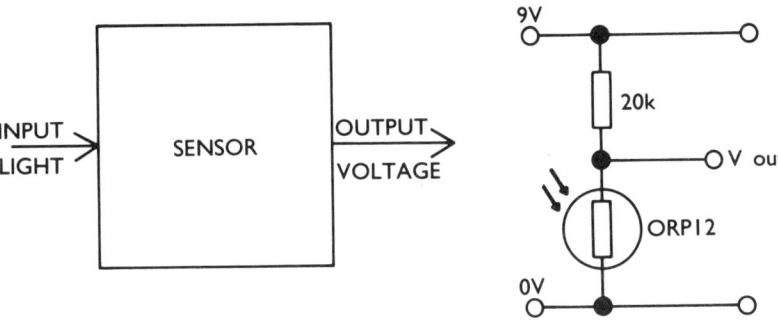

fig 7.34. Block diagram of light sensor

fig 7.35. Light sensor circuit

The specifications for an LDR, code name ORP12, give details of the device. In daylight the resistance is about 1k ohm, changing to about 10M ohms in the dark, with the graduations between these two extremes.

The values of the output voltages can be worked out using the same equation as used in the potential divider:

$$\text{Daylight V out} = \frac{R2}{R1 + R2} \times 9V = \frac{1k}{21k} \times 9 = 0.4V$$

$$\text{Darkness V out} = \frac{R2}{R1 + R2} \times 9V \quad \frac{10000k}{10020k} \times 9 = 8.9V$$

From daylight to darkness, the output voltage can vary from almost 0.4 to almost 9 volts. Intermediate light levels will give a voltage between these limits. Therefore, by changing a light level into a voltage related to that light level, we can sense light.

TEMPERATURE SENSOR

A **thermistor** is a device whose resistance drops as the temperature increases. It can be used in a similar manner to the light sensor circuit to sense temperature, the output voltage increasing as the temperature drops.

If the resistor and the thermistor are interchanges in the circuit, i.e. the thermistor is placed in the top position, the output voltage will now increase as the temperature increases.

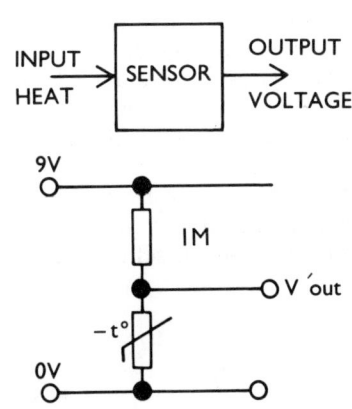

fig 7.38. Temperature sensor

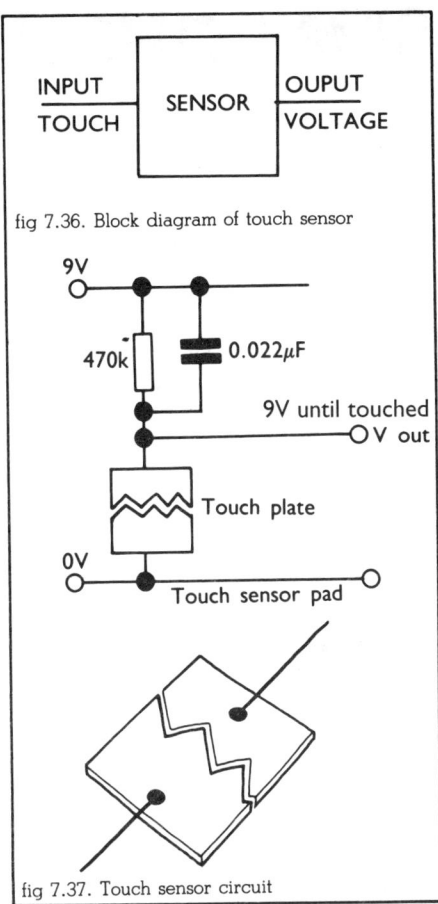

fig 7.36. Block diagram of touch sensor

fig 7.37. Touch sensor circuit

TOUCH SENSOR

The touch of a finger bridging between two copper strips on a PCB connected to a resistor and a capacitor will change a voltage input to a circuit (fig 7.37).

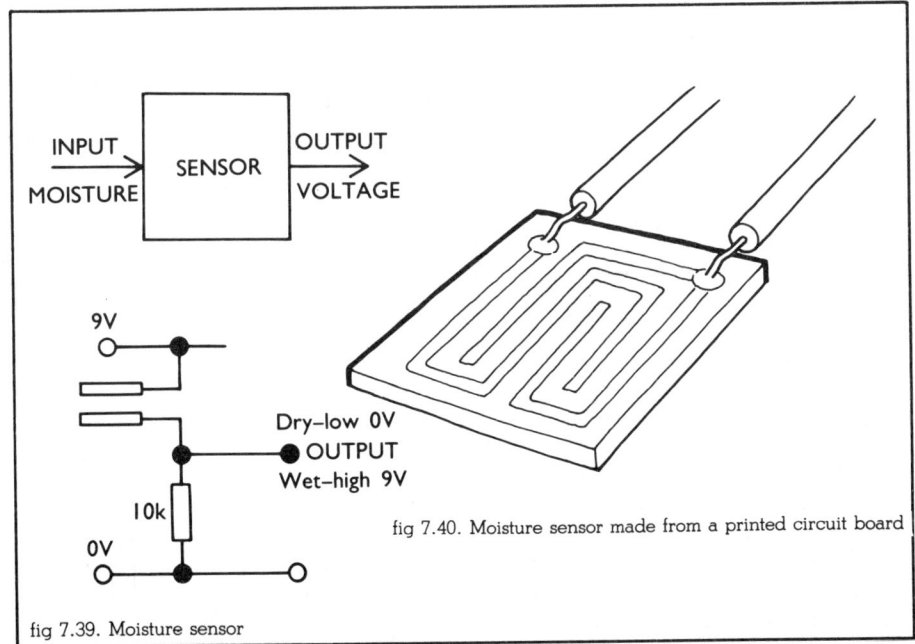

fig 7.39. Moisture sensor

fig 7.40. Moisture sensor made from a printed circuit board

MOISTURE SENSOR

You may wish to use a moisture sensor to test for the moisture content in a plant pot, for example, or the amount of moisture produced from perspiration (this is the principle behind the lie detector). The sensor relies on the fact that water is a conductor and its presence in varying amounts alters the resistance. For example, wet soil has a lower resistance than dry soil. The sensor is a probe consisting of two conducting rods which are inserted into the material to be tested. The rods are not touching but close to each other.

SEMICONDUCTORS

A semiconductor is a material specially made from silicon which, depending on how it is used , is either a conductor or an insulator. It is not necessary at this stage to know how it works in detail, but it is important to know how to use these very important materials, without which no micro electronics would exist. It is the basic material used to manufacture all **'silicon chips'** used in all computers, digital watches, electronic games and many more items.

DIODE

The simplest form of semiconductor device is called a **diode**. This device is used as a one-way component, the electrical current will only pass in one direction.

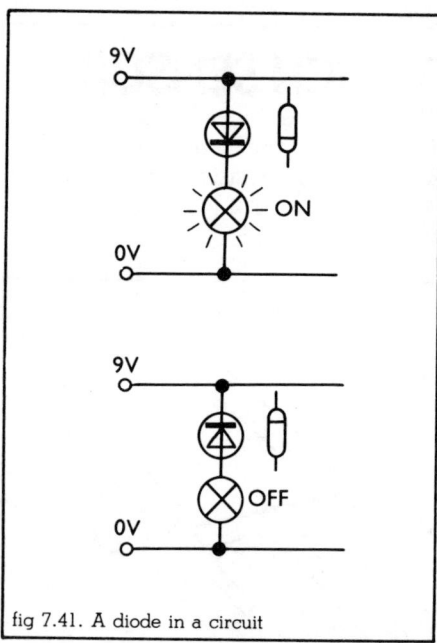

fig 7.41. A diode in a circuit

Light emitting diode (LED)

There is a special type of diode available which gives off light like a bulb when the current flows in the correct direction. The device can only stand a small amount of current before being damaged. The maximum current varies with different types, but is typically 20mA. The example on page 65 explains how to protect the LED when in use in a circuit. An LED can be used as an indicator in place of a light bulb, but it cannot replace a bulb in a torch as the amount of light given out is too small.

TRANSISTOR

The development of the transistor in the mid 1950s was the beginning of the electronic age. This device can be very small, many thousands can be made on a single piece of silicon of 1 mm². A single transistor has three connections as shown in fig 7.43.

The transistor will work as a conductor or as an insulator in a manner similar to a press switch. The input to the switch is a finger pressing the button, the input to a transistor is a voltage at the base. If the transistor is connected as in fig 7.44a the lamp is not lit, it is equivalent to the open switch circuit fig 7.44b. However, if the transistor is connected as in fig 7.45a with the base connected to the 9V rail, then the lamp will light, because there is a voltage input. Fig 7.45b shows the equivalent closed switch circuit.

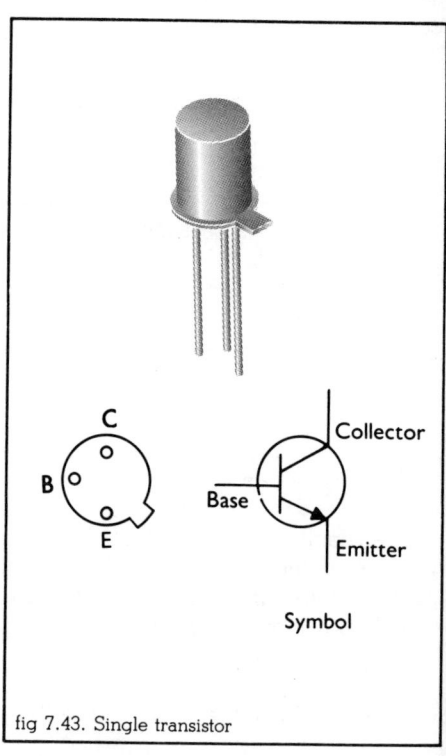

fig 7.43. Single transistor

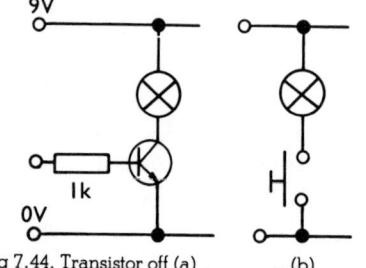

fig 7.44. Transistor off (a)   (b)

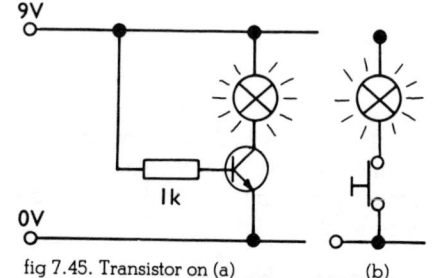

fig 7.45. Transistor on (a)   (b)

The voltage to the transistor **base** is fed through a resistor, called a **current limiting resistor**, which restricts the amount of current reaching it. Too much current will damage the transistor very easily. In the above example, the voltage fed to the base was either 0V when the lamp was off, or 9V when the lamp was on. However, it does not require the full 9V to switch this transistor on, any voltage over 0.6V will do just as well. This voltage of **0.6V** is the **threshold voltage** at which the transistor changes from an insulator to a conductor.

To summarise, if the voltage input at the base is over 0.6V, the transistor switches on and allows a current to flow from the collector to the emitter.

Transistor specifications

Transistors can be made in different forms and sizes to suit various applications. Looking at a list of transistors in an electronics catalogue will give you an idea of the many different types that are available. The catalogue will also list various properties and specifications for each of the transistors. In order to choose the type of transistor needed for a specific application, you need to understand these specifications.

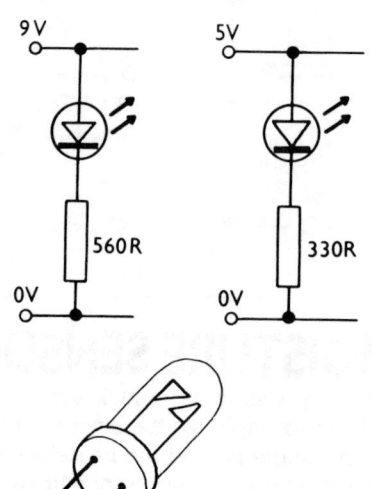

fig 7.42. Light emitting diode

Transistor gain

Transistor gain is one of the properties we should understand. The current flowing into the base of the transistor (I_{BE}) is much smaller than the current flowing from the collector to the emitter (I_{CE}).

The **gain** of the transistor is the **ratio of I_{CE} to I_{BE}**, and it will vary from type to type. A small transistor called **BC108** has a gain (**hfe**) of 200–800 (the gain is not very precise on low cost types). Where a very small current can turn on a large current, the input signal is said to have been **amplified** to become a larger output current.

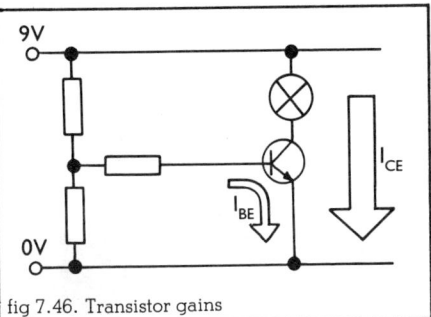

fig 7.46. Transistor gains

Transistor power

The power of a transistor is related to the size of the collector current that can pass before the transistor becomes very hot and stops working. The current varies from a few milliamps to many amps. The transistor type 2N3055 will allow a current of 15A at the collector, while the BC108 will only allow 100mA at the collector.

Transistor selection

Figure 7.47 lists a few different types of transistor. They all work in exactly the same manner, but have different applications. It is not possible to build a transistor with both a high gain and a high power current. Each device is a compromise depending on requirements.

Type	Gain	Ic max
BC108	200–800	100mA
TZX300	50–300	500mA
BFY51	40	1A
TIP31	10–40	3A
2N3055	20–70	15A

fig 7.47. Transistor types

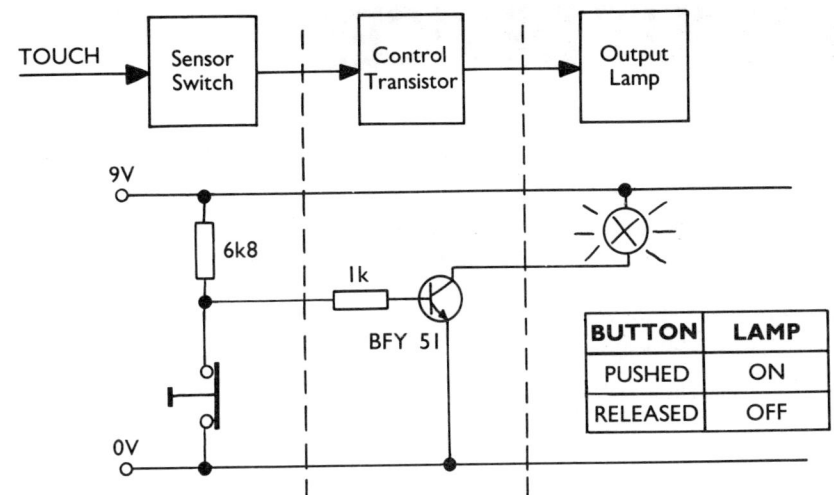

fig 7.48. Complete control circuit

DESIGNING WITH TRANSISTORS

Linking a sensor, a control and an output as shown in fig 7.48, gives a simple circuit that will detect pressure on a switch and show this by lighting a lamp. When the switch is pushed the voltage at the base changes from 0V to 9V. The transistor control switches from **OFF** to **ON** and the lamp indicates the state of the output. The light remains off until the button is pressed.

The following circuits all have sensors, a control and a lamp output. You should now be able to follow their workings.

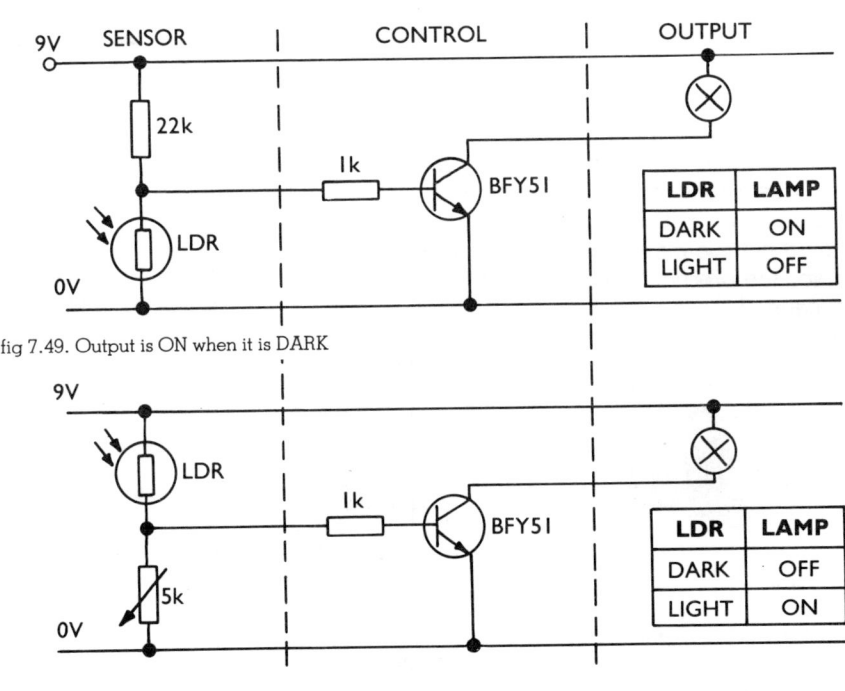

fig 7.49. Output is ON when it is DARK

fig 7.50. Output is ON when it is LIGHT

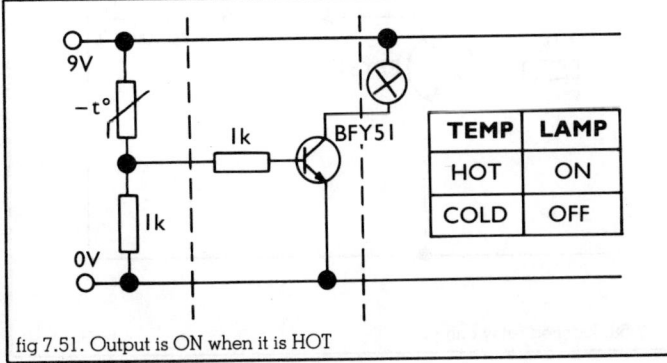

fig 7.51. Output is ON when it is HOT

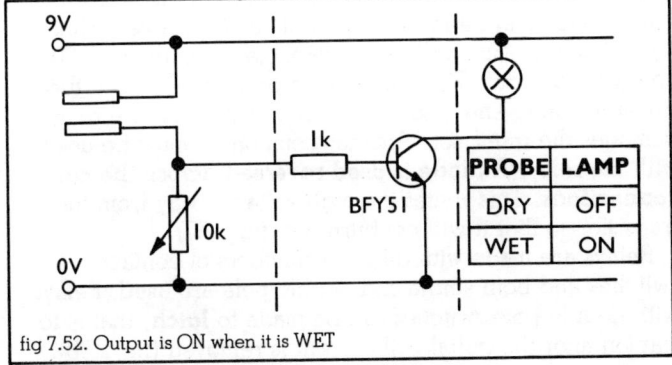

fig 7.52. Output is ON when it is WET

OUTPUT DEVICES

In a way, output devices are the opposite of sensors. They take an electrical signal and give out light, heat, sound or movement. The circuits on the previous page all have an output through a bulb. The current passing through the bulb heats the wire to white hot, thereby giving off a strong white light. The problem with bulbs and batteries is that they use a fairly large current and quickly flatten the battery. If the bulb is only an indicator to show that something has happened, then the lower light level and lower current demands of an LED and its resistor could be used (fig 7.53).

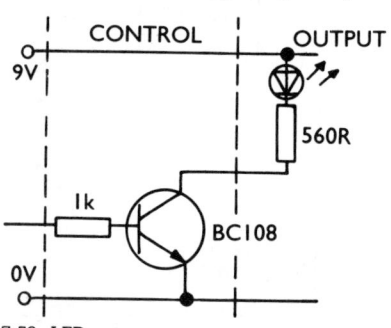

fig 7.53. LED output

Sound output

The transistor can switch on a small buzzer which will give a sound output. It is used as a direct replacement for the bulb in the circuit, but must be connected with the correct polarity, i.e. the red lead connected to the 9V rail.

Relay

A relay is a device that can be used to separate two circuits. The control circuit energises the coil and switches the contacts which are completely isolated electrically from the coil. This means that this output device can be used to switch a mains circuit, though from a safety point of view, you should not use mains voltages in your projects. The higher voltage presents dangers, 240V will kill you, a 9V battery cannot!

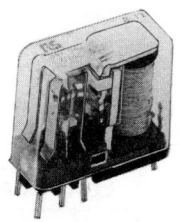

fig 7.56. Relay

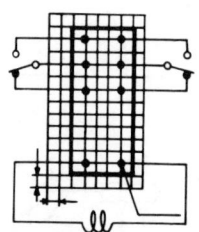

Pitch/Underside view

When the relay is **energised**, the current in the coil creates a magnetic field and this pulls down the lever to operate the contacts of the switch. When the relay is **de-energised** the contacts return to the original position and the coil loses its current. However, when losing its magnetism, the magnet induces a voltage in the coil wire. This voltage can be in the region of 200V for a very short time. The transistor will not be capable of conducting 200V and will burn out. Thus, we have a problem. We can turn a transistor on to energise a relay, but turning off the relay damages the transistor. Another component must be used with the relay. **A diode is used reversed across the coil connections.** This routes the high voltage away from the transistor so that it will not burn out (fig 7.57).

Relays are made with different numbers of contact switches and both single and double pole are used. Relays with double pole switches can be made to **latch**, that is to stay on after the initial coil current is removed (fig 7.58).

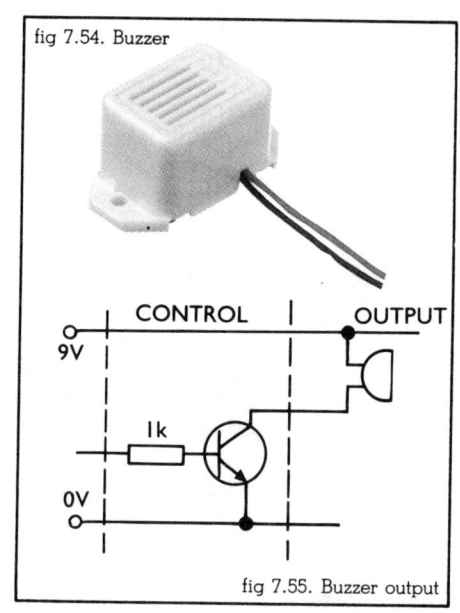

fig 7.54. Buzzer

fig 7.55. Buzzer output

Solenoid

A solenoid is a device for converting electrical energy into mechanical energy. The current in the coil causes a magenetic field to pull the soft iron core into its hollow centre. This straight line motion can be used to lock doors and operate switches and valves. Solenoids are especially useful for connecting electronic control circuits to pneumatic valves (fig 7.60). As in the case of the relay, which is a special adaptation of the solenoid, a diode is required if the solenoid is switched by a transistor.

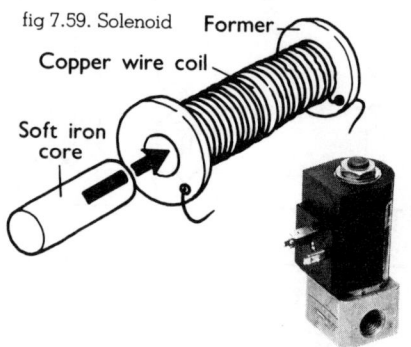

fig 7.59. Solenoid

fig 7.60. Pneumatic solenoid

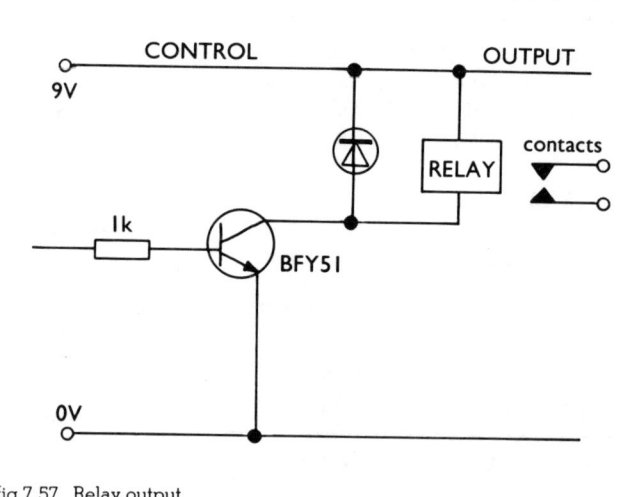

fig 7.57. Relay output

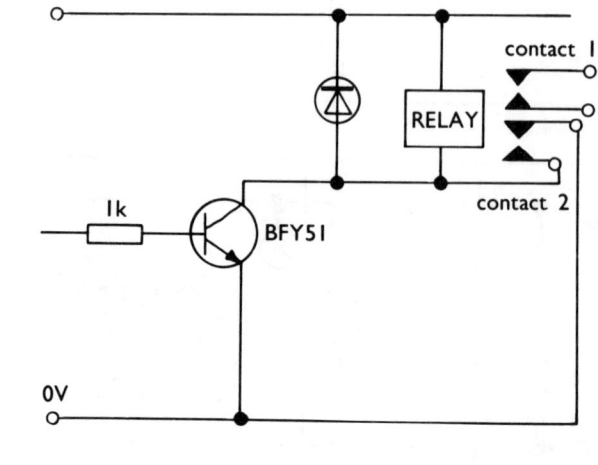

fig 7.58. Latched relay output

Thyristor

A **thyristor** or **silicon-controlled rectifier (SCR)** is a semiconductor device that can be used to latch an output. It has three connections, two of which, the **cathode** and the **anode**, are connected as a normal diode. The third, the **gate,** is connected to the control voltage. If there is a voltage at the gate, the thyristor will conduct and remain conducting even if the voltage at the gate is then removed. It can only be turned off by switching off the supply to the anode (fig 7.61.)

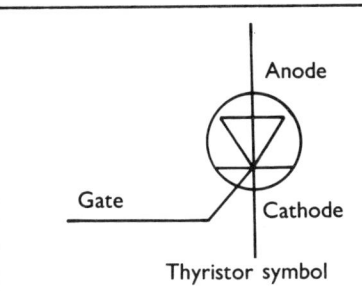

fig 7.61. A thyristor in a circuit

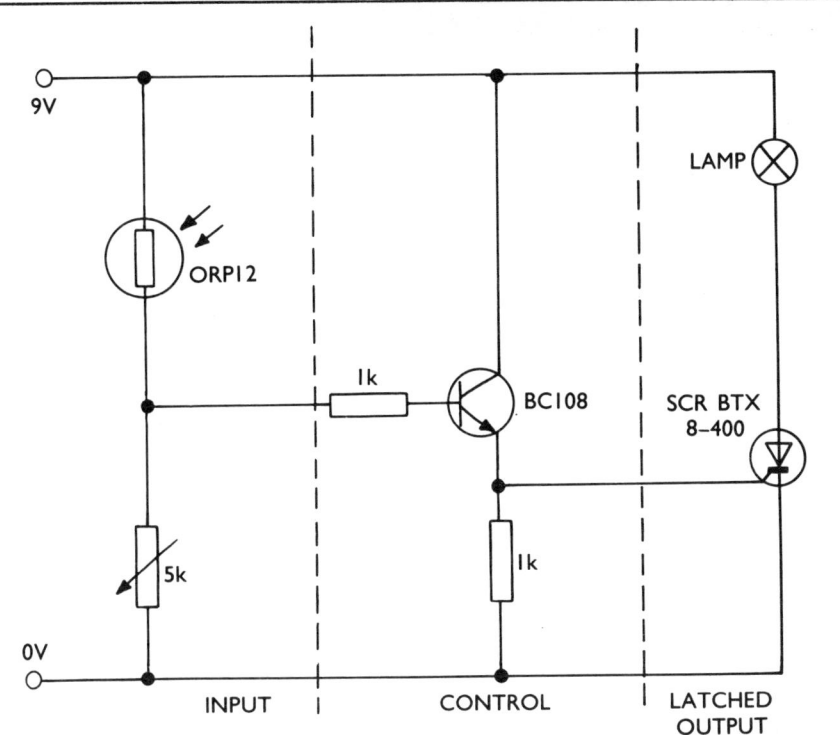

Electric motor

An electric motor is a device for converting electrical energy to mechanical energy, providing rotary movement. Small electrical motors run from 6 or 9V, while larger, more powerful motors such as starter motors or wiper motors in cars, run from 12V.

The d.c. motor is shown in fig 7.62. It has a permanent magnet providing a stationary magnetic field (**stator**) and electromagnets rotating within this field (**rotor**). The rotor consists of several, often three, wire coils wound round an iron core and connected to a **commutator**. When an electric current passes through the rotor coil, the coil rotates and the current is delivered through brushes at the commutator.

When the motor is first switched on, current is applied to coil number one, creating a magnet with a north and south pole. The outer north pole is attracted to the permanent magnet south pole and movement occurs. The contacts on the commutator then change to coil number two, which becomes a magnet and is attracted to the permanent magnet, while coil number one is switched off. This changing magnetisation of the coils produces movement, in small motors it is usually at around 2000 revolutions per minute (rpm).

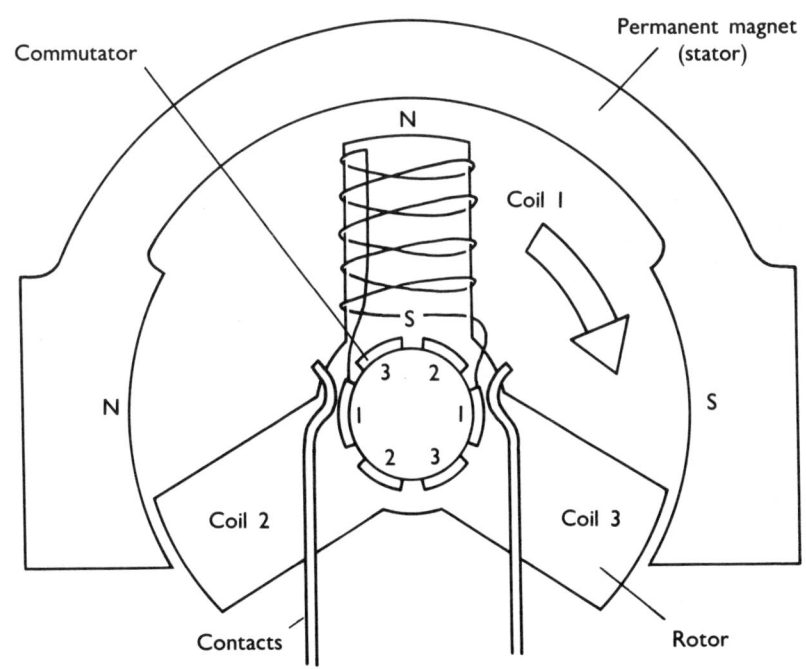

fig 7.62. Dc motor

fig 7.63. A small electric motor

fig 7.64. An electromechanical counter

Electromechanical counter

An electromechanical counter is a specialised type of solenoid. The movement of the solenoid advances a numeric display, enabling the device to count how many times it is switched on or off. It would be useful for counting people entering a room if a sensor could sense people passing by.

SPECIAL TRANSISTORS

As we have seen on page 69, transistors can have either a high gain or a high output current, but not both. If two transistors are used together, one with a high gain and one with high power, we can produce a control circuit that has both properties. It is called a **darlington pair.**

Figure 7.65 shows a typical circuit using a darlington pair. The output current through the power transistor BFY51 is dependent upon that transistor's power rating. The gain of the pair is found by multiplying the gain of the first transistor by that of the second, which in this case gives a gain of 8000. We now have the equivalent of a single transistor with a gain of 8000 and an output current of 1 amp, and the circuit is now much more sensitive to changes in the light level. The sensor can detect an object, such as a hand, one metre away passing between it and a light source, such as a window.

The **trigger voltage** has also changed to twice (two transistors) the original 0.6 V, i.e. **1.2V** at the base input. Altering the variable resistor will set this for the light level you wish to detect. Special, single package darlington transistors looking outwardly like a single transistor, but having a very high gain and a high output current capability are available.

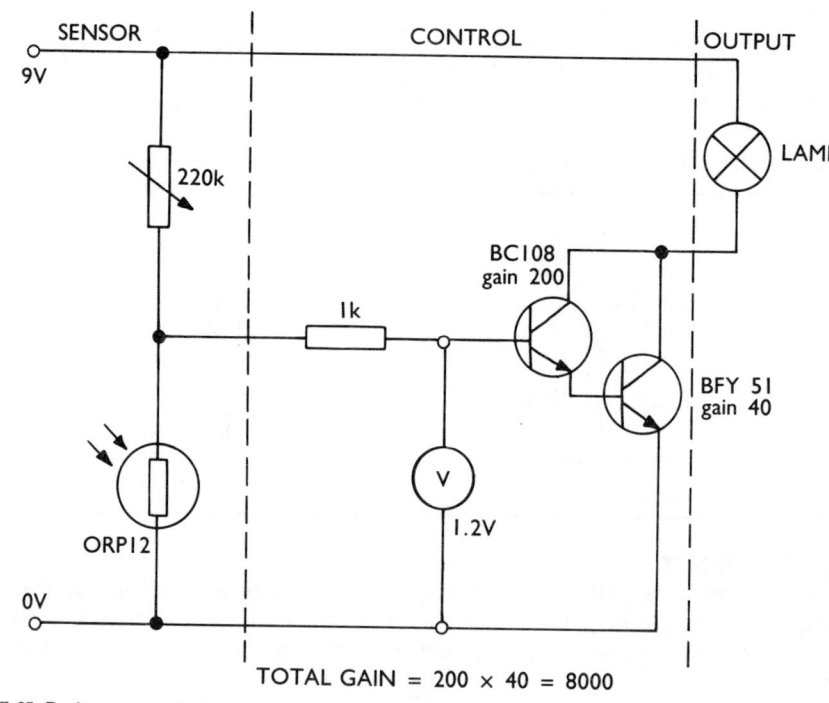

fig 7.65. Darlington pair light sensor circuit

CAPACITORS

A capacitor stores electric charge. It consists of two metal plates separated by an insulator called the dielectric. The charge stored depends on the size of the capacitor and the voltage applied to it. The unit of capacitance, the farad, is very large and most capacitors you will use are likely to be measured in microfarads (μF), (nanofarads (nF) or picofarads (pF) (see fig 7.26 on page 65 for SI prefixes).

Figure 7.66 shows a capacitor in a circuit. When switch one (SW1) is closed, a current flows into the capacitor until it is full. If R is a resistor with a high resistance the current flows slowly, and if it is a resistor of low value the current flows quickly, rather like the tap in fig 7.67. The amount of water flowing into the bucket is altered depending on whether the tap is fully open or almost closed.

Unlike voltage outputs we have seen so far, the output here is not a fixed voltage, but changes as the capacitor charges up. To start with the voltmeter will show 0V, increasing to 9V when the capacitor is fully charged. The time taken to charge will again depend on the size of the resistor. A graph can be plotted of the voltage output against the time taken (see fig 7.68).

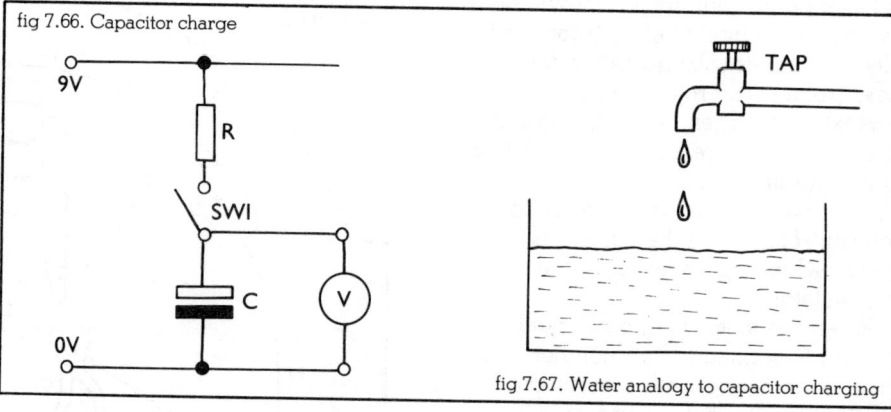

fig 7.66. Capacitor charge

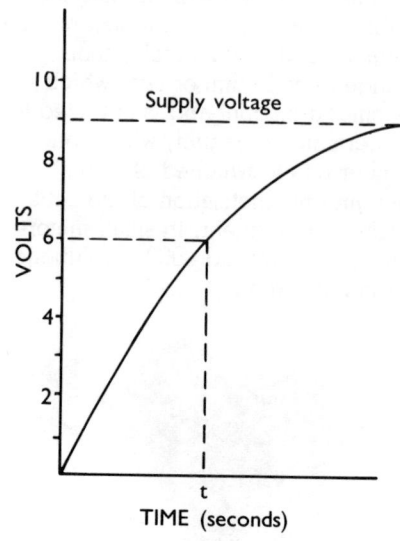

fig 7.67. Water analogy to capacitor charging

We can see that the first part of the graph is almost a straight line, which then curves and tails off as it takes longer and longer to fill the remaining space in the capacitor. The time taken to reach two thirds of the supply voltage is called the **time constant** and can be obtained from the following formula:

Time constant
$$t \text{ (sec)} = C \text{ (farads)} \times R \text{ (ohms)}$$
or
$$t \text{ (sec)} = C \text{ (}\mu\text{F)} \times R \text{ (M ohms)}$$

The time taken to charge depends on the resistor size and also, as we might expect, on the capacitor size. From this we can see that it is possible to alter the charging rate as we require by changing the size of the resistor and/or the capacitor to suitable values.

fig 7.68. Capacitor charge

TIME DELAY CIRCUIT

The capacitor and resistor can be used as a sensor (fig 7.69) to sense the passing of time. The voltage level required to trigger a transistor is 0.6V. If when the circuit is first switched on the capacitor is empty, it will begin to fill with charge and a current will flow into the capacitor through the resistor R, with the voltage at A increasing from 0V. When the voltage reaches 0.6V the transistor will switch on. By altering the sizes of the capacitor and the resistor we can arrange for there to be a delay before the transistor is switched on. The circuit is an adjustable time measuring device, that can measure from hundredths of a second to a few minutes, though the longer periods are often not measured very accurately.

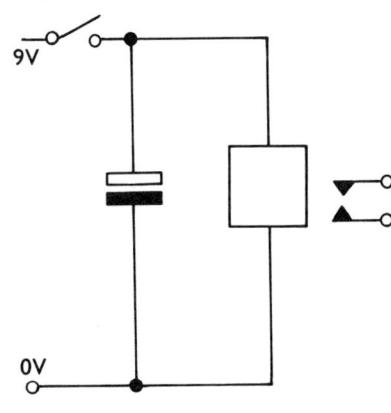

fig 7.73. A relay circuit

Other capacitor uses

The relay is energised as normal but when switched off holds on the relay for a short time. There is a delay after switching off (fig 7.73).

The capacitor charges up, and the relay is energised until the capacitor is fully charged, then the current stops. The relay energises for a short time only (fig 7.74).

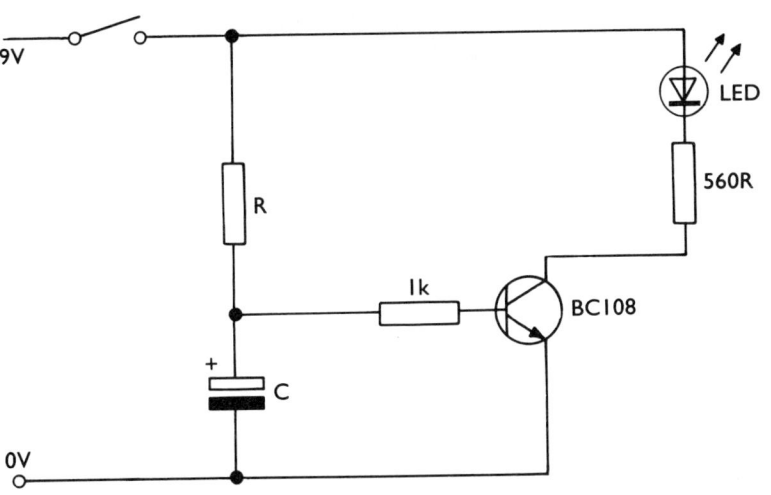

fig 7.69. Capacitor and resistor used as a sensor

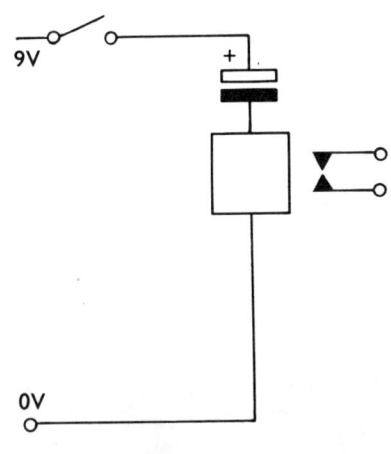

fig 7.74. A relay circuit

Discharging a capacitor

In fig 7.70 a capacitor is connected to and charged through the resistor, R_{IN}, and the voltage at the output is shown to rise slowly on the voltmeter, V, to a maximum of 9V. If when charged the switch is changed to connect to R_{OUT}, the capacitor will empty, slowly if the resistor is large, quickly if the resistor is small or there is no resistor. The reading on the voltmeter will decrease from 9 to 0V.

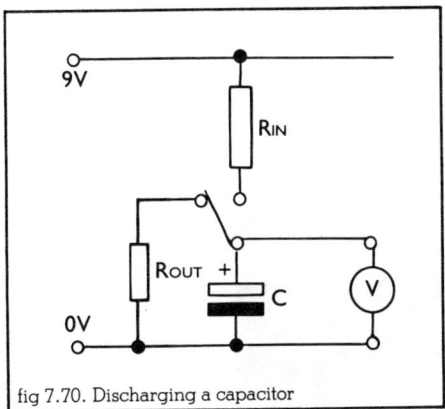

fig 7.70. Discharging a capacitor

fig 7.71. Capacitor discharge

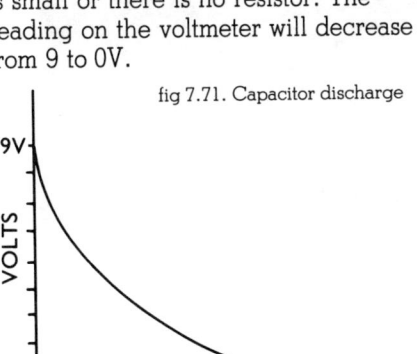

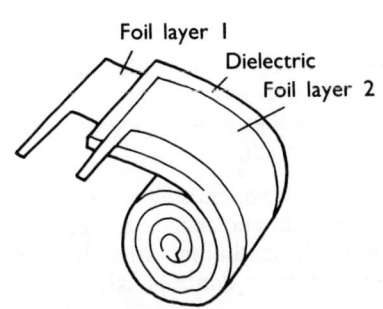

fig 7.72. Capacitor construction

fig 7.75. Comparative values of capacitors

pF	nF	μF
1000	1.0	0.001
1500	1.5	0.0015
2200	2.2	0.0022
4700	4.7	0.0047
6800	6.8	0.0068
10000	10	0.01
22000	22	0.022
47000	47	0.047
100000	100	0.1
220000	220	0.22
470000	470	0.47

INTEGRATED CIRCUITS

In 1950, Ferranti produced the world's first commercial computer, the Mark 1. The Mark 1 was 5m long, 3m high and 1m deep. It needed to be kept in an air conditioned room because it produced so much heat from its valves. It cost £1 million at present day values and was very unreliable. In 1977, the same company produced the first European microprocessor, the F100. The F100 was 5mm square, 0.5mm thick, 100 times more powerful than the Mark 1 and cost £50. In 1987, a similar processor costs £1.50 and is very reliable. A complete computer and memory can now be built into a credit card. The miniaturisation is due to the development of **integrated circuits (ICs)** or **'silicon chips'**.

In 1955, the transistor was beginning to replace the thermionic valve. It worked in a similar manner to the valve, but was much smaller. It could be made on a small piece of silicon or germanium. Once it was realised that it was possible to put two, then three transistors on to one piece of silicon, the race was on, microelectronics was born. The first working circuit on a chip was produced in 1958. The first ICs were used in defence equipment and space vehicles, but by 1963 were appearing in commercial products, such as hearing aids, sewing machine controls, calculators and computers. Now many thousands of components, such as transistors, diodes, resistors and capacitors can be built onto the silicon chip.

The chip is usually packaged in black plastic in a **dual in line (DIL)** form (fig 7.77), the connections from the circuit being connected to legs, making it look rather like a plastic insect.

ICs are now the basis of all electronic designing, supported by a few other basic components. All ICs look very similar, many have the same number of legs or connections. Micro chips are specially produced for control in the form of microprocessors, but there are others that have different uses. In this chapter we will investigate the uses of two of the chips that are available. Inspecting the semiconductor section of an electronics catalogue will show you how many and how varied these devices are.

fig 7.76. Integrated circuit chip

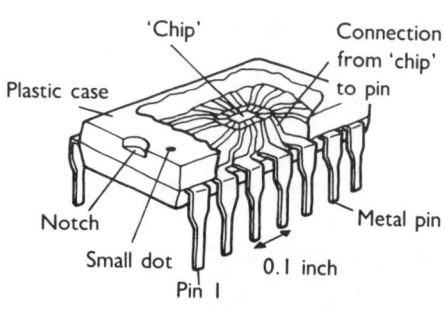

fig 7.77. Cutaway view of a chip

fig 7.78. Silicon chip

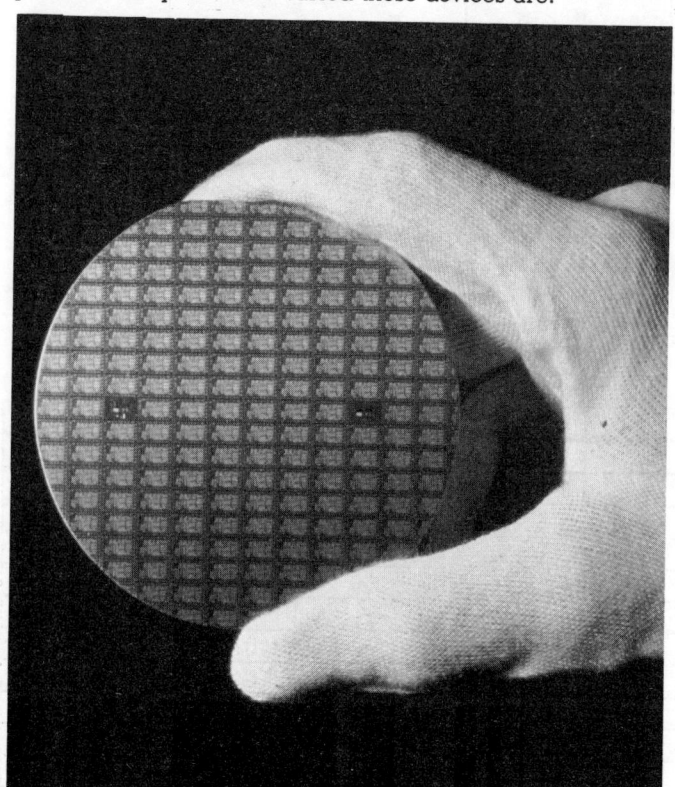

fig 7.79. Slice of silicon with chips visible prior to cutting

DEFINITIONS

Monostable: a system that has one stable state only. It can be made to change, but it always returns to the same stable state. A push button switch always springs back out when the pressure is released.

Bistable: a system that has two stable states. It can change between the two and can remain in either state. A toggle switch can be on or off.

Astable: a system that has no stable state, it changes from one state to the other all the time. A pendulum swings from one side to the other continuously.

fig 7.80.

555 TIMER

A 555 timer is a fairly simple chip that can be used in different ways as a control element in a circuit. It can be used as a **monostable** or an **astable** timer (see the table on page 74 for definitions). The chip is produced in an 8 pin DIL package (fig 7.81), it contains 25 transistors, 2 diodes and 16 resistors. When using it in a circuit you must treat it as a single component with required inputs and outputs. You do not need to know how all the internal components work. For more information there are entire books written on this chip alone.

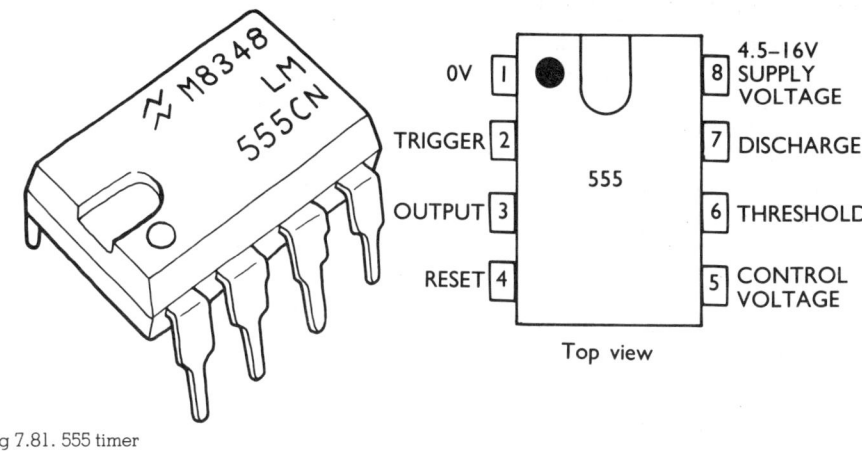

fig 7.81. 555 timer

Monostable timer

A one stable state timer, the monostable timer always returns to the original state (output 0V) after a certain time, set by two extra components, one resistor and one capacitor. You have already seen that these two components together can produce a rising voltage, it is this rising voltage that controls the time delay to be set.

Figure 7.82 shows the block diagram and fig 7.83 the circuit diagram for a monostable timer. The circuit diagram shows the pin numbers in the positions to suit the diagram, not the position on the actual chip.

Operation of the monostable timer:

1. Output 0V, the LED is off.
2. The timer is started when the button is pressed to provide a pulse start.
3. The output is now switched to 9V, the LED is lit.
4. The capacitor starts to charge.
5. When the voltage at the capacitor (pins 6 and 7) reaches two thirds of the supply voltage, here 6V, the output returns to the original 0V, its stable state.
6. The LED goes out again.

There are other possibilities for input and output for this timer. Figures 7.84–7.89 show a few examples.

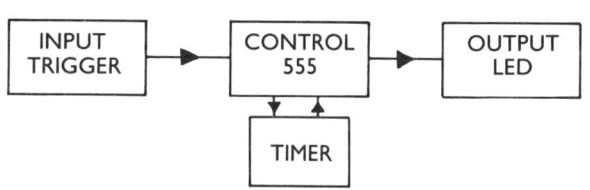

fig 7.82. Block diagram for a monostable timer

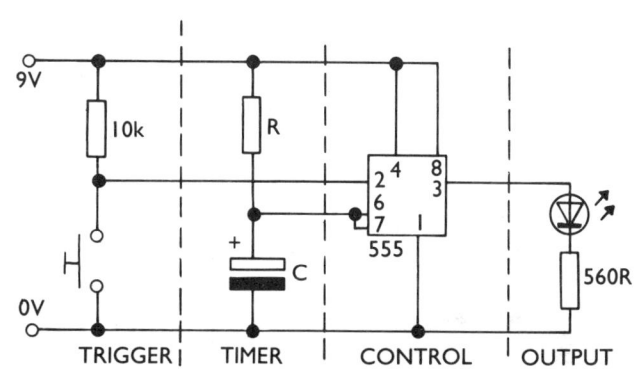

Frequency of the above monostable timer
t = 1.1RC usually calculated as t = RC
Units: R in MΩ
C in μF
t in seconds

fig 7.83. Circuit diagram for a monostable timer

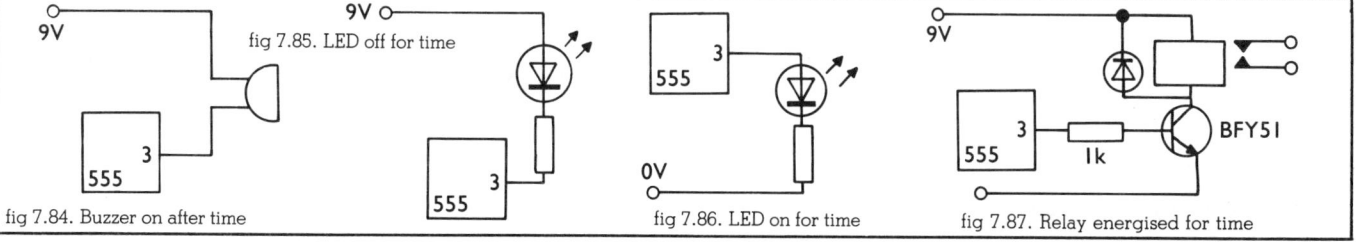

fig 7.84. Buzzer on after time

fig 7.85. LED off for time

fig 7.86. LED on for time

fig 7.87. Relay energised for time

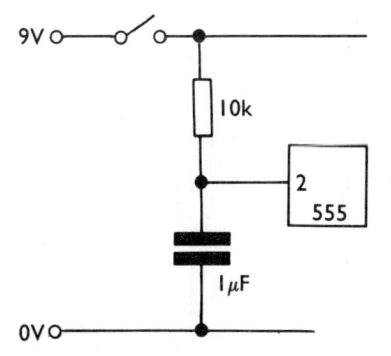

fig 7.88. Automatic start input

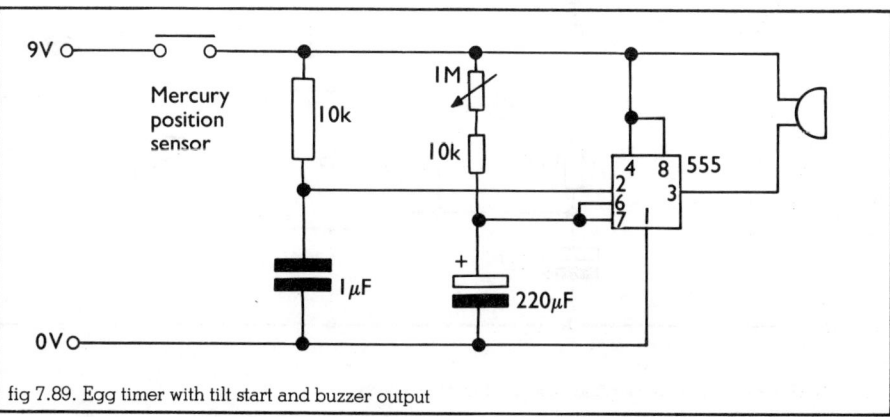

fig 7.89. Egg timer with tilt start and buzzer output

ASTABLE TIMER

An astable ouput is never still, in this case it will switch between two voltages, 0V and 9V, continually. A flashing light on an alarm is operated from an astable output. The device will also drive a speaker to produce a continuous tone if the rate of change is fairly fast.

The basic change to the circuit of the monostable timer, is that the trigger input, pin 2, is now connected to pin 6, the threshold, producing an automatic restart to the timing cycle. With this circuit it is possible to change the voltage output one million times a second (1MHz) or at a rate of many minutes per cycle. Thus, there is a very wide range of possible frequencies.

The **frequency** is the number of changes per second. It is measured in **hertz (Hz)**, one repetition each second is a frequency of 1Hz. The musical note middle C has a frequency of 264Hz.

fig 7.90. PCB with a 555 chip

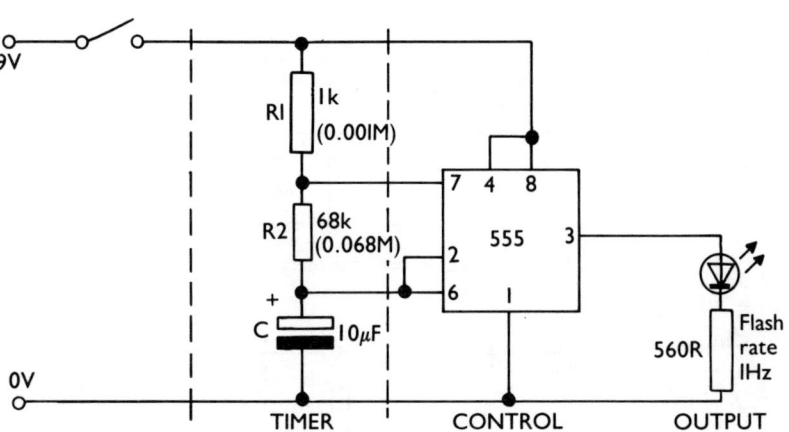

fig 7.91. Astable timer

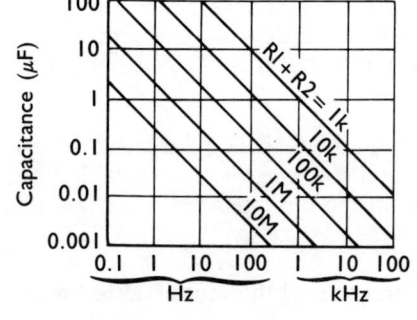

fig 7.92. Alternative way of calculating frequency of an astable timer

Frequency of an astable timer

$$f = \frac{1.44}{(R1 + 2R2)C}$$

$$f = \frac{1.44}{(0.01 + (2 \times 0.068))10}$$

$$f = \frac{1.44}{(0.01 + 0.136)10}$$

$$f = \frac{1.44}{1.37}$$

$$f = approx\ 1HZ$$

Units   R in MΩ

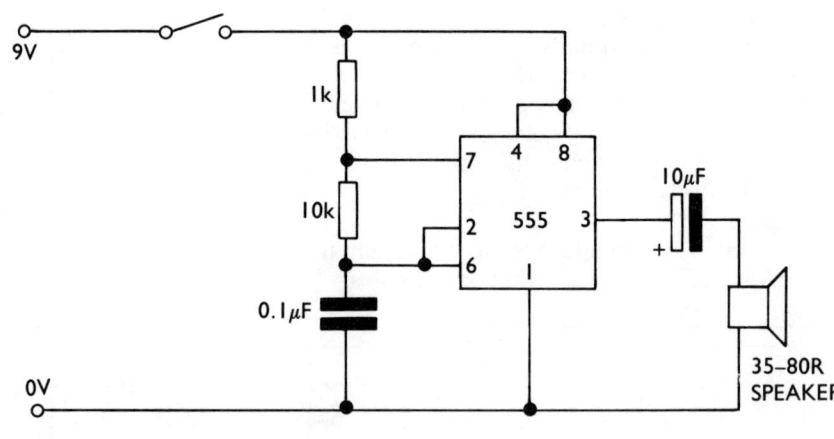

fig 7.93. Tone generator circuit

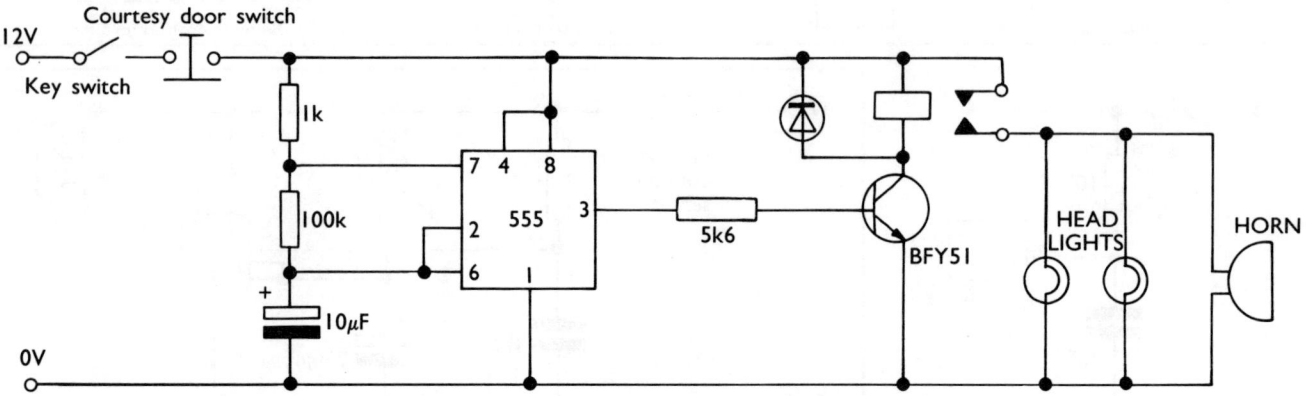

fig 7.94. Car alarm, output circuit to flash headlights and horn

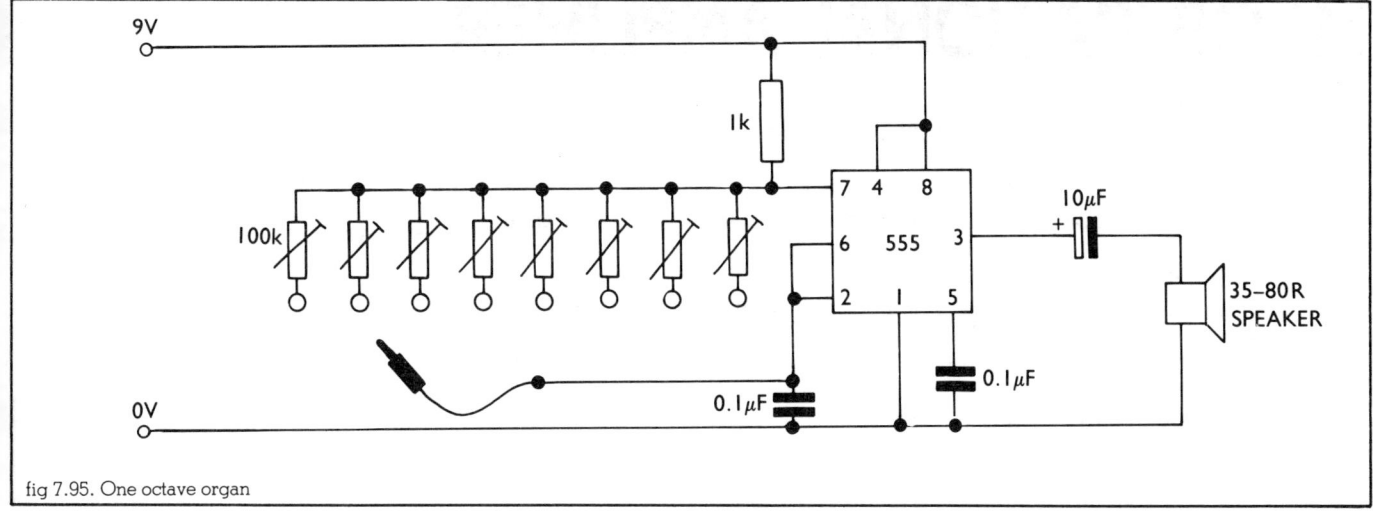

fig 7.95. One octave organ

TIMER VARIATIONS

It is possible to combine two astable circuits. If the first astable is set at a frequency of 1Hz and this then switches on a second astable with a frequency of 500Hz, a pulsed tone is produced through an output loudspeaker. If the output of the first astable is fed into pin 5 of the second astable, it is possible to produce a siren type noise. With experimentation, it is possible to produce many varied sounds from either one or two astable circuits. With an LDR in place of R2, the tone is varied by variations in light, it can be played like an instrument with the shadow from your hand! Connect a thermistor in place of the LDR and you could produce an audible thermometer.

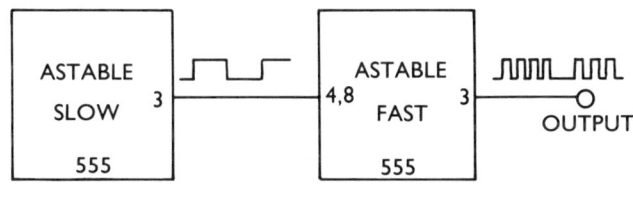

fig 7.96. Pulsed tone, block diagram

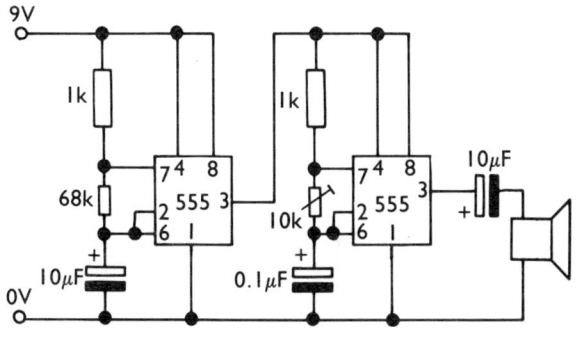

fig 7.97. Pulsed tone circuit

fig 7.98. Siren block diagram

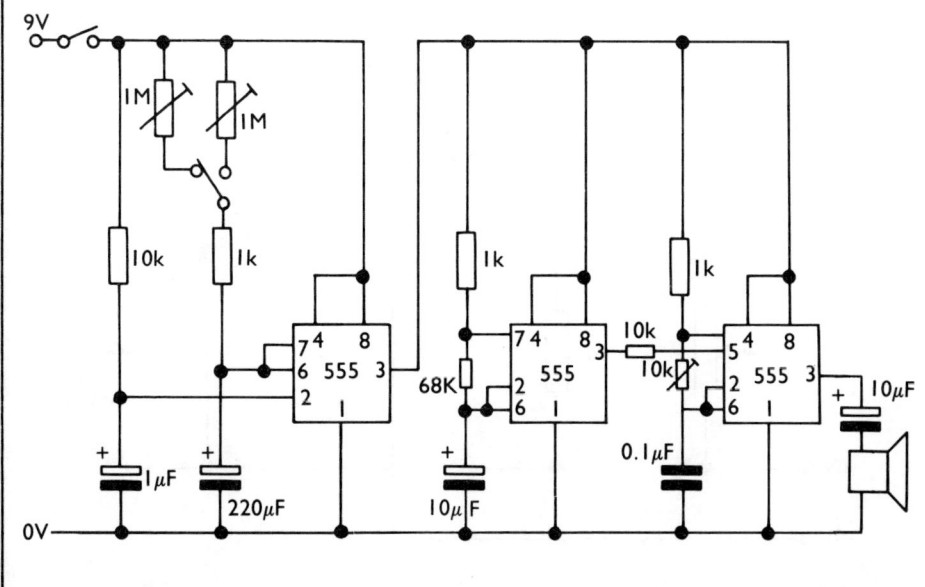

fig 7.99. Two position timer with siren output

Timer project

1. There must be two delay times, 3 and 4 mins.
2. Output to be heard from the next room.
3. Power from a 9V battery.
4. Boxed in a suitable manner for a kitchen.

You should now be able to understand this circuit for the above specifications. Break it into smaller blocks that you can understand. Draw a block diagram of the circuit.

741 OPERATIONAL AMPLIFIER

The 741 operational amplifier is another integrated circuit that can be used in a number of different ways. The 741 is a complex circuit on a silicon chip, consisting of 20 transistors, 11 resistors and one capacitor. It is packaged as an eight pin DIL chip, and apart from its code number, is identical in appearance to the 555.

The **operational amplifier**, or **op-amp**, has a gain of 100,000. It has two inputs, one, signified by a minus sign, is called the **inverting input**, and the other, signified by a plus sign, is called the **non-inverting input.** The symbol for an op-amp is a triangle as shown in fig 7.100.

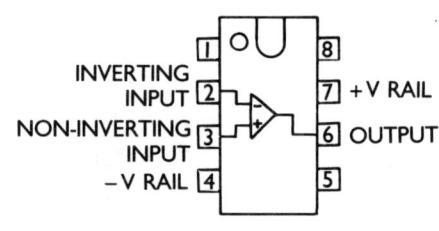

fig 7.100. 741 pin connections

The power supply requirements are rather different to any circuits we have seen so far. A split power supply is required. Two 9V batteries can give this, connected as shown in fig 7.101. The three supply rails are drawn as three lines on the circuit and labelled +9V, 0V and −9V.

The plus and minus signs in the triangle must not be confused with the positive and negative power supply voltages. These signs are always written on the inside of the triangle with the pin numbers around the outside. Often, for clarity reasons, the power connections are not shown on the circuit diagram, but they are always there, the chip will not work without them. The op-amp can work in many different modes, though it is only shown here as a differential amplifier or a comparator.

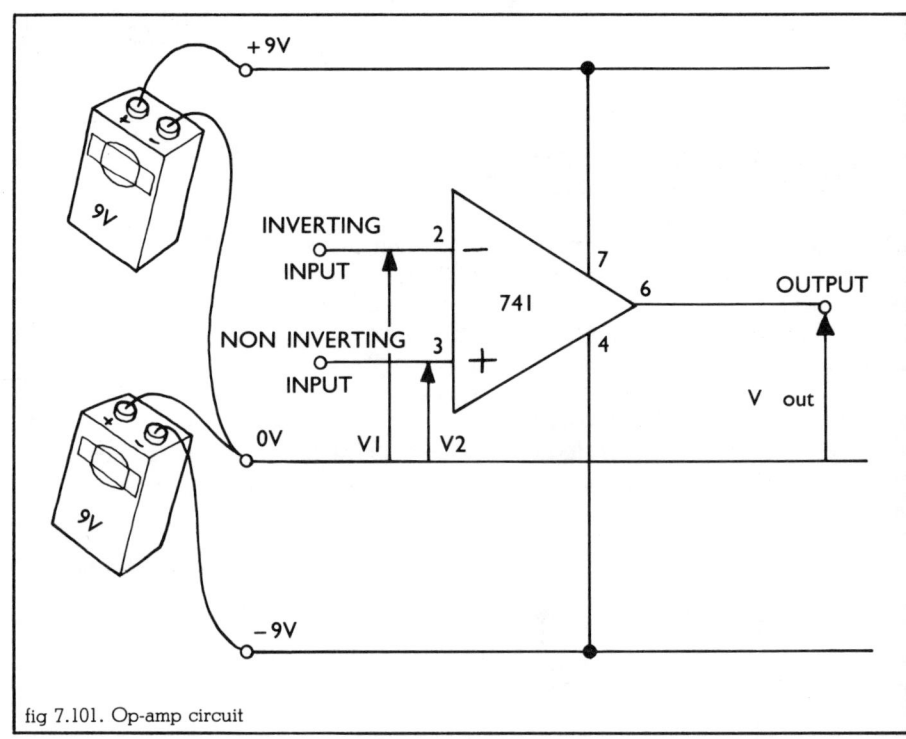

fig 7.101. Op-amp circuit

COMPARATOR

In the differential amplifier mode, the 741 will **amplify the difference in voltage between the two inputs.** The output will be +7V, 0V or −7V. In practice, as the gain is 100,000 to 1, the difference need only be 70 μV to swing the output fully one way or the other. It will be almost impossible to obtain an output of 0V. The output will only reach to within 2V of the supply voltages due to the internal make up of the op-amp.

+ input voltage	− input voltage	output voltage
large	small	+7V
small	large	−7V
equal	equal	0V

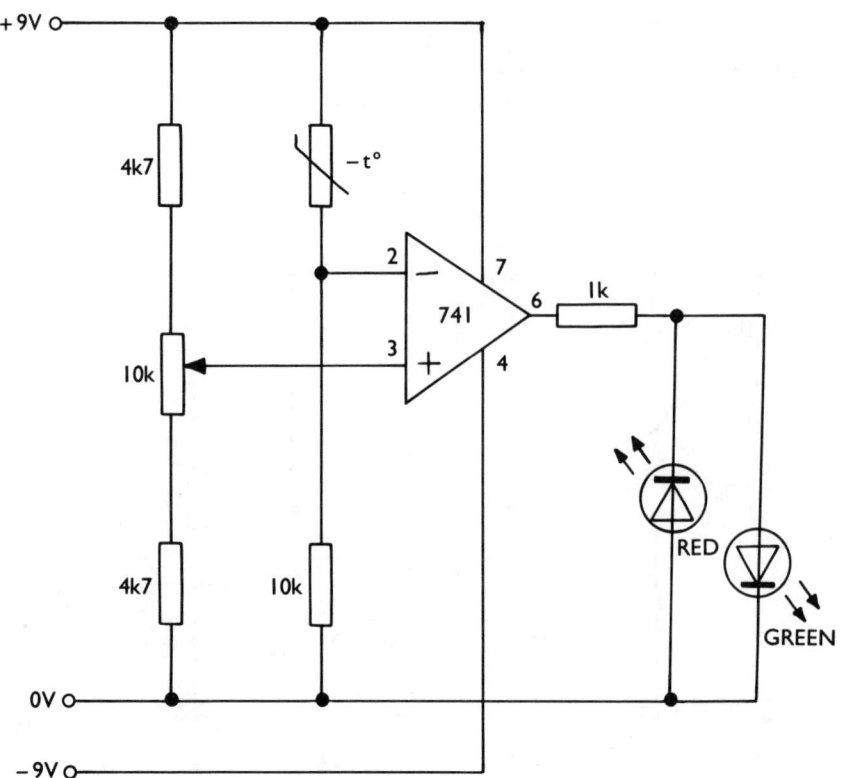

fig 7.102. Hot or cold detector

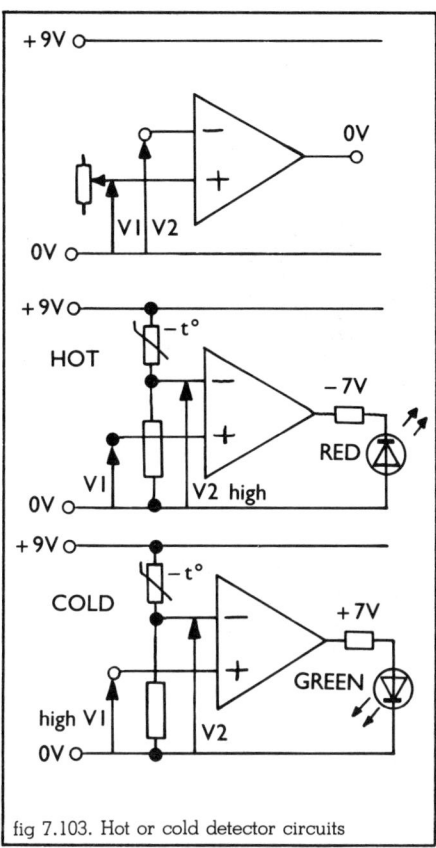

fig 7.103. Hot or cold detector circuits

Circuit explanation: hot or cold detector

With the thermistor at the setting temperature, the variable resistor is adjusted to balance the voltages at the two inputs. When V1 = V2 the output will be 0V. We have already said that this is almost impossible to achieve as the adjustment is very sensitive. This will set the trigger level for the temperature at which the device will operate. To get the thermistor at the setting temperature it can be placed in a beaker of water with a thermometer. The temperature of the water is then heated or cooled to the correct level and the variable resistor adjusted, thus giving an accurate temperature setting.

The thermistor works as a potential divider (see page 66). When the sensor heats up its resistance will fall causing V2 to increase. Voltage V1 was set previously and should not be touched, so because the voltage at the inverting input is larger, the output voltage will be negative, here −7V. As the −7V output is connected to the cathode of the red LED, this will light up, showing that the sensor is above the set temperature.

When the sensor is colder than the set level, the voltage V2 will decrease below V1. Thus, the non-inverting input will have the larger voltage, and the output will be positive, here +7V. The cathode of the red LED is now connected to a positive voltage and will not light. However, the green LED is connected up in reverse, and it will light, showing that the sensor is registering a temperature below the set level.

Thermostatically controlled fish tank heater

How the circuit works: when the thermistor is hot, that is above the set temperature, the output is −7V and the transistor is switched off. When it is below the set temperature the output is +7V and the transistor is switched on, energizing the relay which in turn controls the 24V heating element.

The output current of the 741 is only **10mA** as a **sink** or **source**. This will light an LED, just, but will not operate directly a relay, motor, lamp or a heater. For any of these to be operated there must be a **power transistor** requiring less than 10mA at the base.

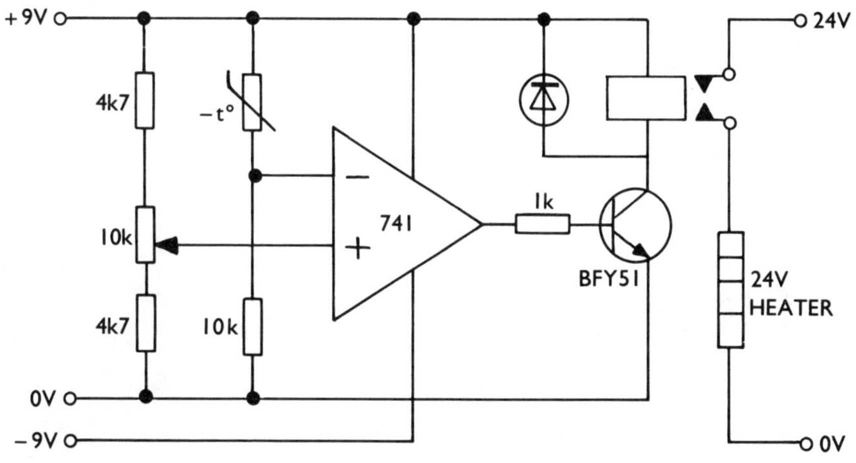

NEGATIVE FEEDBACK

Using the op amp as a comparator is useful, but it simply amplifies any difference in the input voltage to produce the maximum output. This is known as open loop gain. By using negative feedback you can gain greater control of the op amp and amplify the signal by specific amounts, as in audio amplifiers.

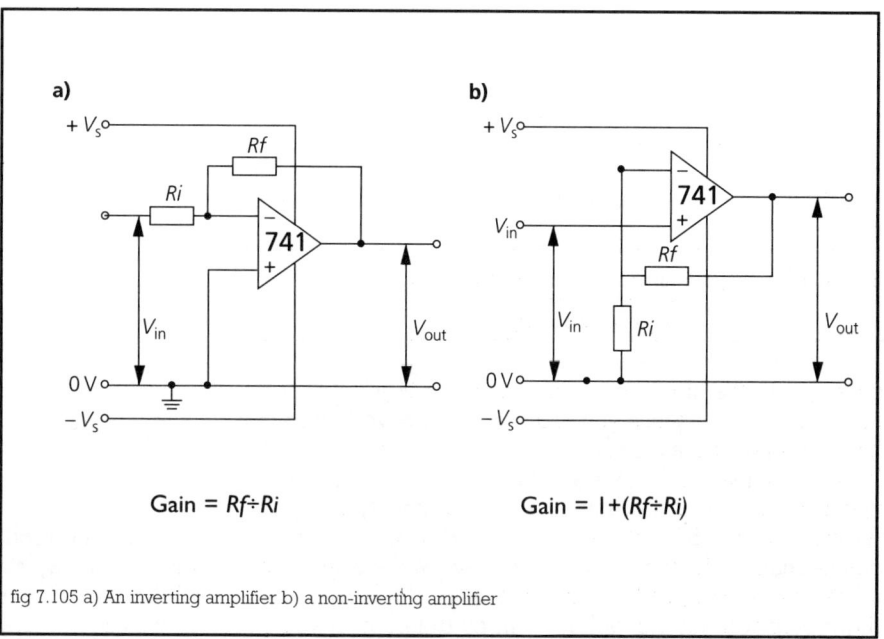

fig 7.104. Thermostatically controlled fish tank heater

a)

b)

$$\text{Gain} = Rf \div Ri$$

$$\text{Gain} = 1 + (Rf \div Ri)$$

fig 7.105 a) An inverting amplifier b) a non-inverting amplifier

ALTERNATING CURRENT

All the voltages studied so far have been steady values. A 9V d.c. (**direct current**) is available from batteries such as the PP3 and the PP9. Voltages from the mains supply are slightly different. These voltages are produced in power stations many miles from the user. If the supply is sent in the form of waves of voltage it loses less power in the cable it has been sent through. The voltage can start very high (e.g. 25kV), and be stepped down as required in sub-stations around the country. If a voltage is sent as d.c. over long distances, very little is left at the end of a 100-mile wire! The mains power is supplied as **alternating current** (a.c.). The voltage changes or alternates from positive to negative 50 times a second (50Hz). This voltage can be shown on the screen of an oscilloscope (fig 7.106) as a series of waves (see fig 7.107).

fig 7.108. Rectifiers

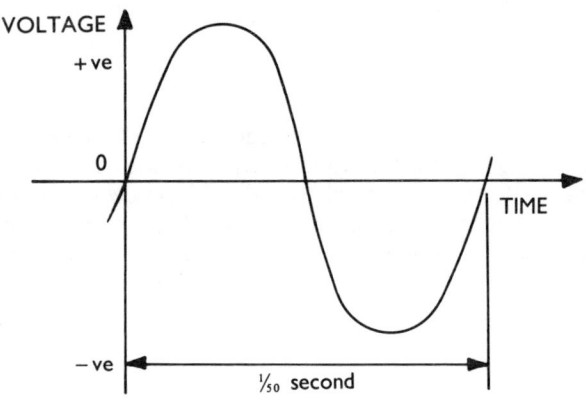

fig 7.107. Ac wave form

You cannot use the electricity straight from the mains socket for your electronic projects. They will not work, and you may be killed! Mains voltage is 240V a.c. You seldom need a voltage of more than 30V, and for most projects 9V is sufficient.

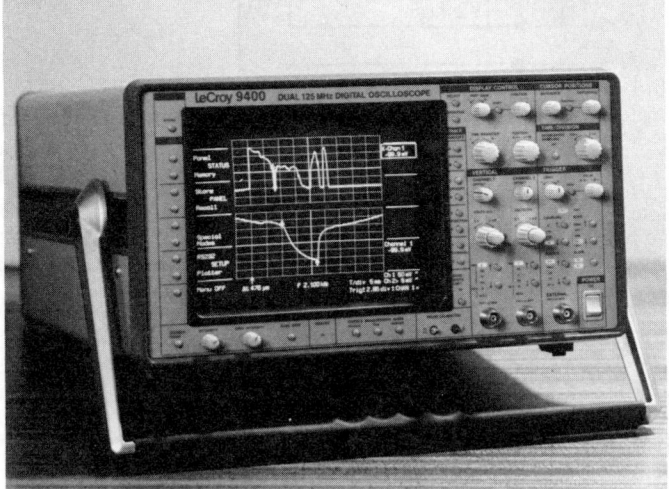

fig 7.106. Cathode ray oscilloscope (CRO)

fig 7.109. Transformer

TRANSFORMER

A transformer is a device that will change a.c.voltages into different a.c. voltages. It consists of a soft iron core, usually made from sheets of iron cut and made into a thick lamination, with two coils of copper wire around it.

As the current is passed through the **primary coil** in an alternating form, it creates a magnetic field in the core which changes with the changing voltage. The **secondary coil** is wound around the same core, but is insulated electrically from the first. The changing magnetic field in the core produces a changing electrical current in the secondary coil and a.c. current is produced. This device is used to change the usually high voltage in the primary coil into a lower voltage in the secondary coil. The ratio of the voltages in the coils is related to the number of turns in each coil:

Secondary voltage = turns in primary coil × primary voltage / turns in secondary coil

You can easily produce a much lower voltage than mains voltage, with the aid of a transformer but it is still no use to you in projects, because you require a steady voltage, not a changing one. If a motor is connected to this lowered (say 9V) a.c. voltage it will have an effect, but not as we may expect. It will run forward and backwards changing direction every 1/50th of a second (50Hz), because the voltage changes from positive to negative at the same rate, and the motor will just buzz and not move at all. However, a light bulb with an a.c. voltage will be different. If you

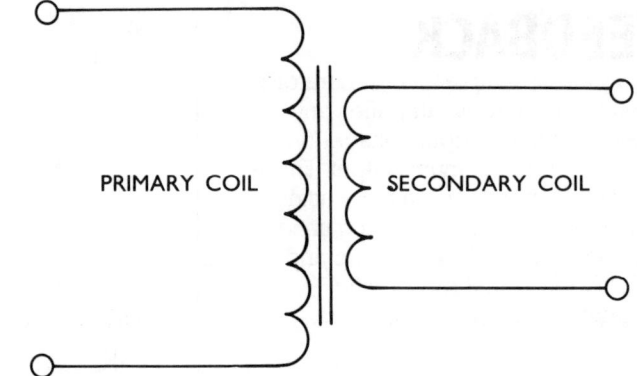

fig 7.110. Transformer symbol

PRIMARY COIL   SECONDARY COIL

change the direction of current flow the bulb will light which ever way round it is connected. Thus, an a.c. voltage will light a bulb properly. Indeed light bulbs in all our homes use a.c. voltage!

A rectifier

A diode will pass current in one direction only. The result of this in an a.c. circuit is that half of the wave is allowed to pass through and the passage of the other half of the wave form is blocked: it will look like fig 7.111. The diode is acting as a rectifier to produce a series of positive pulses, and a motor will run on this supply. The drawback to this is that half of the power available is wasted by blocking.

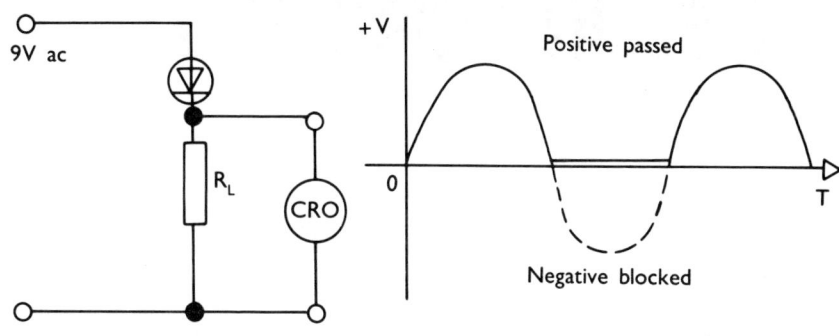

fig 7.111. Half wave rectifier

Full wave rectifier

A full wave rectifier can be made from four diodes built up into what is called a **bridge rectifier** (fig 7.112). This circuit will make use of all the power of the a.c. current. It is more usual to use a bridge rectifier as a complete chip rather than to build it from separate components.

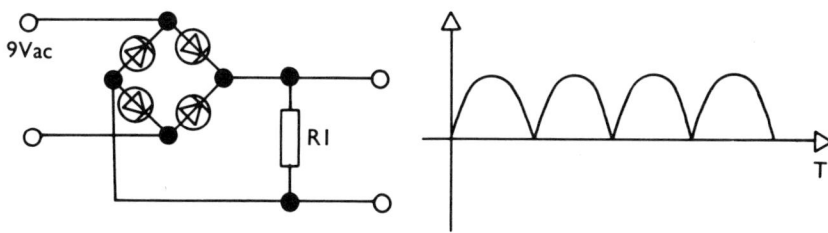

fig 7.112. Full wave rectifier

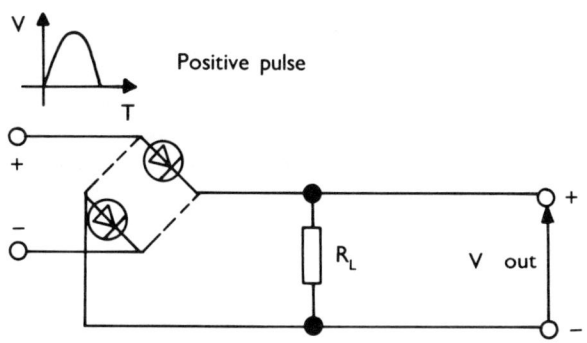

fig 7.113. Positive pulse

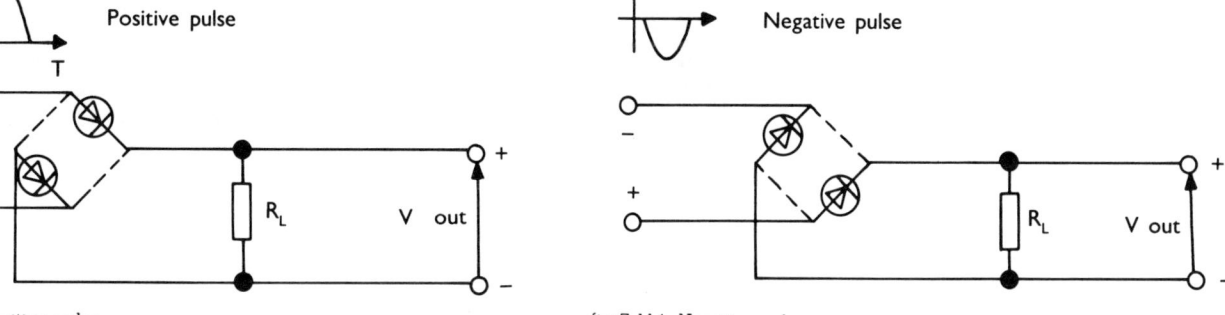

fig 7.114. Negative pulse

Smoothing circuits

The supply from a rectifier is still a pulsed supply and, as such, is not of any use to operate the transistor circuits shown so far. You need a steady supply to simulate a battery source. If a capacitor is incorporated into the circuit the bumpy supply can be smoothed to an almost steady voltage. The capacitor charges up quickly as the pulse of voltage is applied, then discharges slowly to hold up the voltage as the pulse drops away before the next pulse charges it up again. The wave form will now look like fig 7.115.

Simple smoothed power supply

Connecting the three components (transformer, full wave rectifier and smoothing capacitor) together, you can build up a simple d.c. voltage supply to run from the mains. It is possible to stabilise this output to achieve a very steady and accurate voltage, and this must be done if the supply is to be used to drive certain logic circuits. Special ICs are available to produce very accurate 5, 12 or 15V regulated supplies. It is not anticipated that students build power supplies for use in their projects. **Power supplies** or **'lab packs'** should be available for use in place of batteries for some applications.

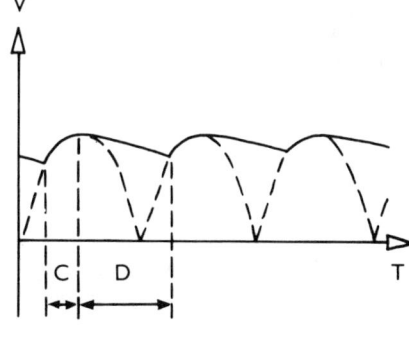

C–charging period
D–discharging period

fig 7.115. Smoothed output

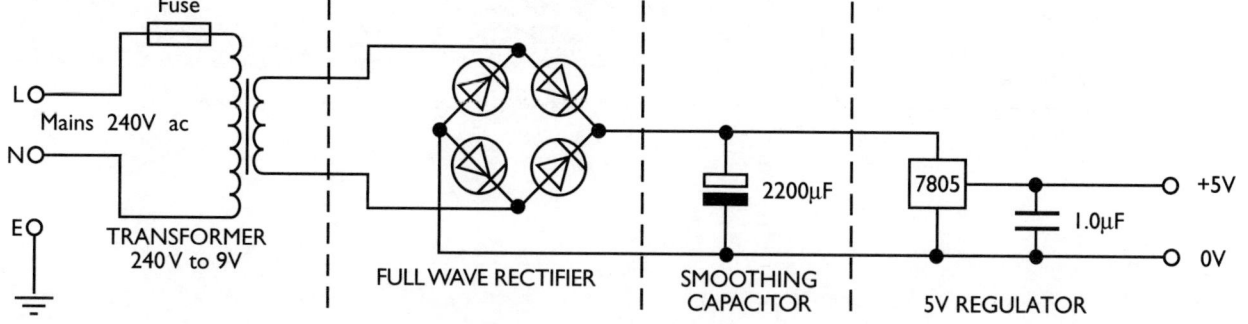

fig 7.116. Simple smoothed power supply

MODELLING CIRCUITS

Modelling techniques are needed to provide a quick method of testing electrical circuits and modifying circuit designs. It is rather shortsighted to spend many hours building a circuit into its final form before the circuit design has been checked to make sure it fits its specifications.

The method of modelling will vary depending on the availability of equipment and the complexity of the circuit. A very simple method for basic circuits is to use a piece of wood with screws and screw cups to provide a means of gripping the wires and components (see fig 7.117).

Components can be fastened into a series of units which clip or plug together. This method is very useful for developing circuits from a systems approach (see fig 7.118).

A single component version of this style board is very useful. The board is made from moulded plastic with square posts attached. The components on carriers are clipped in between the posts, which have holes suitable for 4mm plugs (fig 7.119). The advantage of this method is that the circuit drawing looks very similar to the circuit being built and is easily followed. However, it is not suitable for integrated circuits with many connections.

Prototype boards (fig 7.119) are available to suit the needs of IC modelling, though the circuit drawing will not look at all like the actual circuit. With care, prototype boards can be used to build very complicated circuits. The components themselves are used with connecting single core wires. The boards are available in different sizes, but they usually connect together into larger areas.

Computers are being used increasingly to model circuits mathematically. This is not an area that has much use in schools, but is used in an industrial design situation.

> The important thing to remember is not what type of modelling equipment is used, but why you model and build prototype circuits.

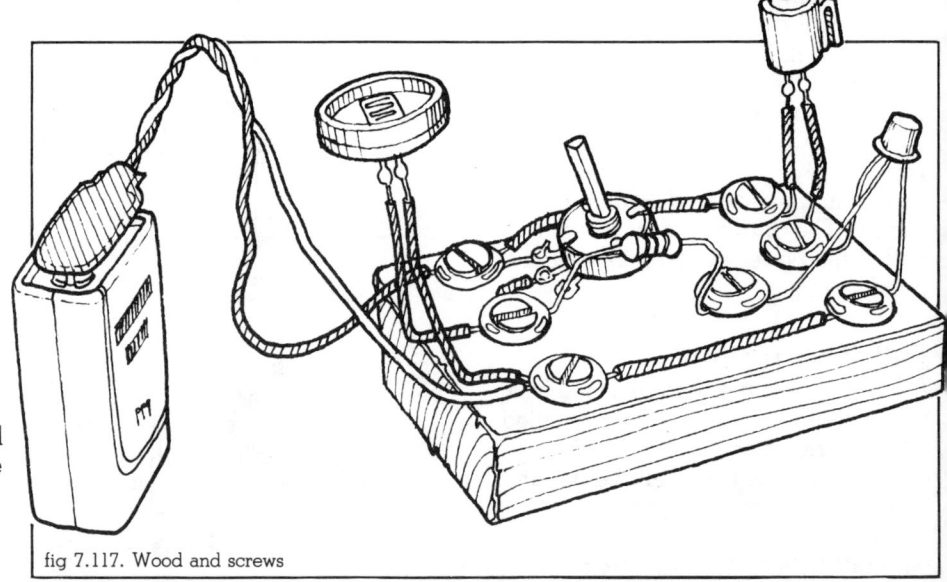

fig 7.117. Wood and screws

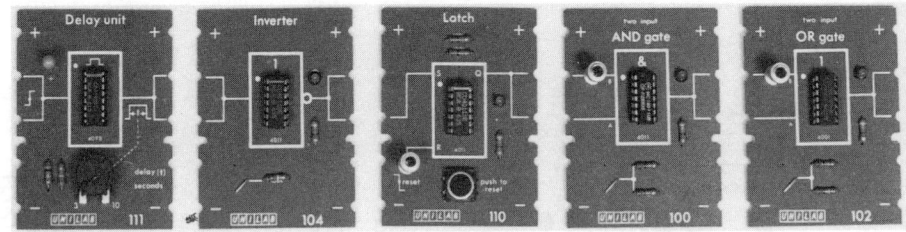

fig 7.118. Alpha type boards

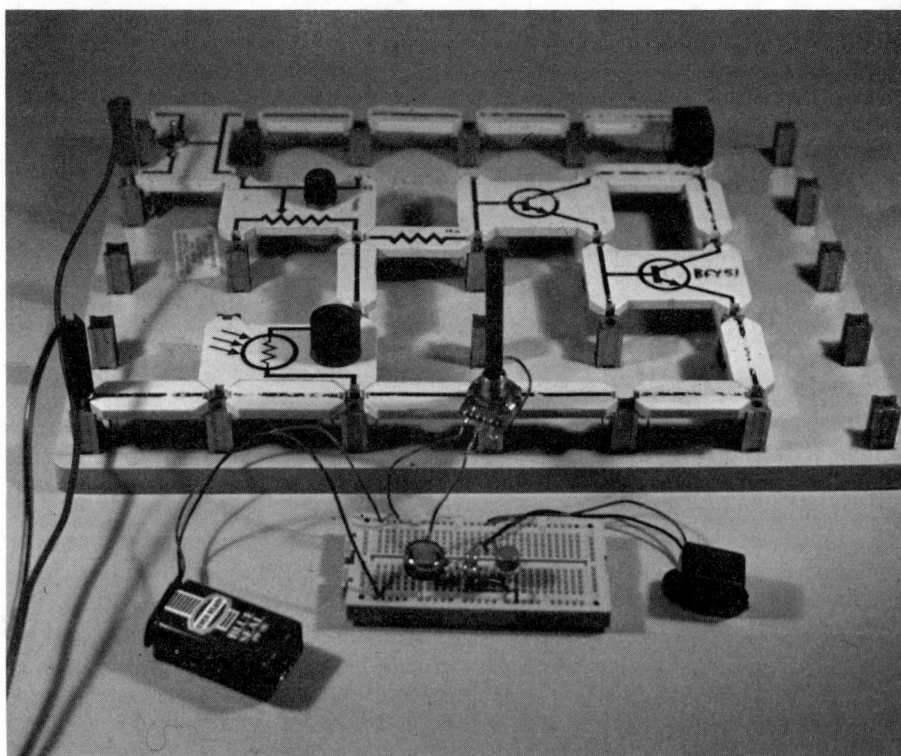

fig 7.119. Locktronics system and prototype board modelling the same circuit

CONSTRUCTIONAL TECHNIQUES

Soldering

Soldering is a permanent method of connecting components in an electrical circuit. Solder is a metal, an alloy of tin (60%) and lead (40%), that melts at around 200°C. The solder is used in the form of a wire that contains a core of **flux.** The flux melts with the solder and helps to stop the joint oxidising while it is being heated and to make the solder 'run' into the joint properly.

The technique of soldering is easily mastered, but simple rules should be followed for success.

1. All components must be clean and grease free.
2. Heat both parts of the joint to 200°C with a hot soldering iron.
3. Apply the solder, allowing it to run into the joint.
4. Allow it to cool without movement of the joint.
5. Use a heat sink when soldering diodes and transistors.
6. Use chip holders with ICs.

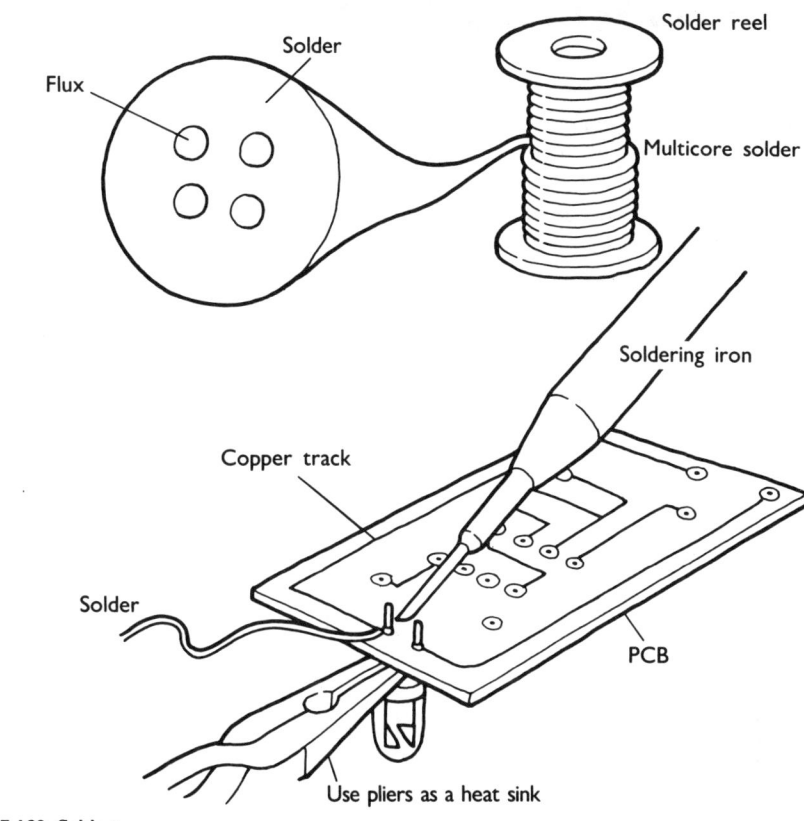

fig 7.120. Soldering

Many circuit problems are due to bad soldering or **'dry'** joints. As all permanent methods of construction will involve soldering it is very important to be capable of making good soldered joints.

Copper tape

The copper tape method of modelling provides a semi-permanent method of building a circuit (fig 7.122). Self-adhesive copper tape is used for all the tracks. The components are soldered into place simply by placing the leg directly onto the surface of the copper track (a very simple form of surface mounting). Special connecting pads are available so that components such as 8-pin dual in line ICs can be used in the circuit. This method is suitable for less complex circuits.

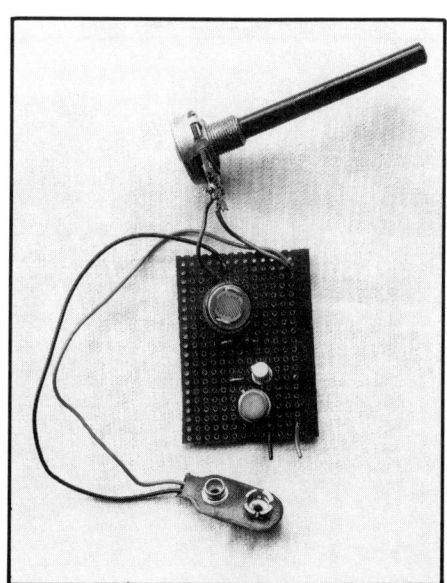

fig 7.121. Strip board

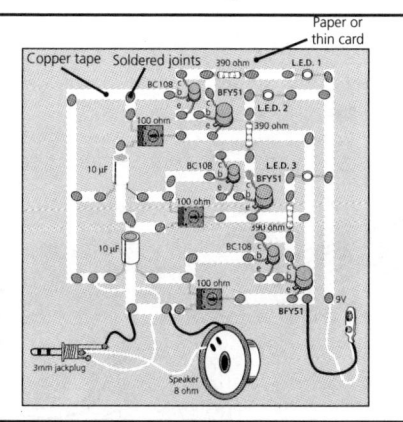

fig 7.122. 'Tracktronics', a copper tape circuit

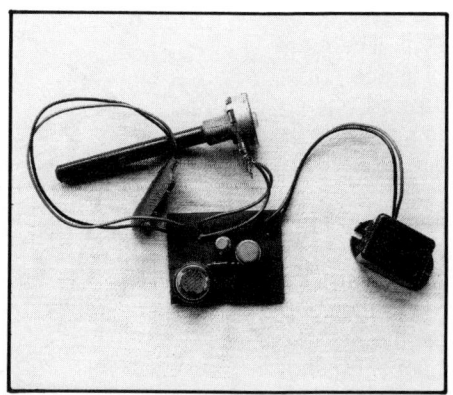

fig 7.123. Circuit board

Strip board

Strip board looks very similar to matrix board on the top side, but underneath it has strips of copper connecting the lines of holes. The circuit is built using the strips to connect the components. Breaks in the strip can be made with a small drill or a strip cutter turned by hand. Circuits shown in electronic magazines often use this method of construction, which is suitable for most circuit building. The components will not be in the position in which they were drawn on the circuit diagram. They must be re-arranged to suit the connecting strips, taking great care to connect the right strip (fig 7.121).

Printed circuit boards (PCB)

Printed circuit boards are made from a material similar to strip board or from **glass reinforced plastic** (GRP). The board does not have any holes and can be copper clad on one or both sides. PCBs are now used very widely in electrical devices. The components are held in a very compact and robust manner suitable for both mass production and limited production project work. Circuits built on these boards can be very complex, as used in microprocessor applications, or relatively simple as used in school project work (fig. 7.123).

PRINTED CIRCUIT BOARD CONSTRUCTION

Producing a circuit board as outlined below is not a very complicated operation. It involves lots of simple operations linked together as a production sequence. There are two main methods, using **direct acid resists** or using **light sensitive resists**. With a little care in using light sensitive resists you can produce PCBs of a very professional standard.

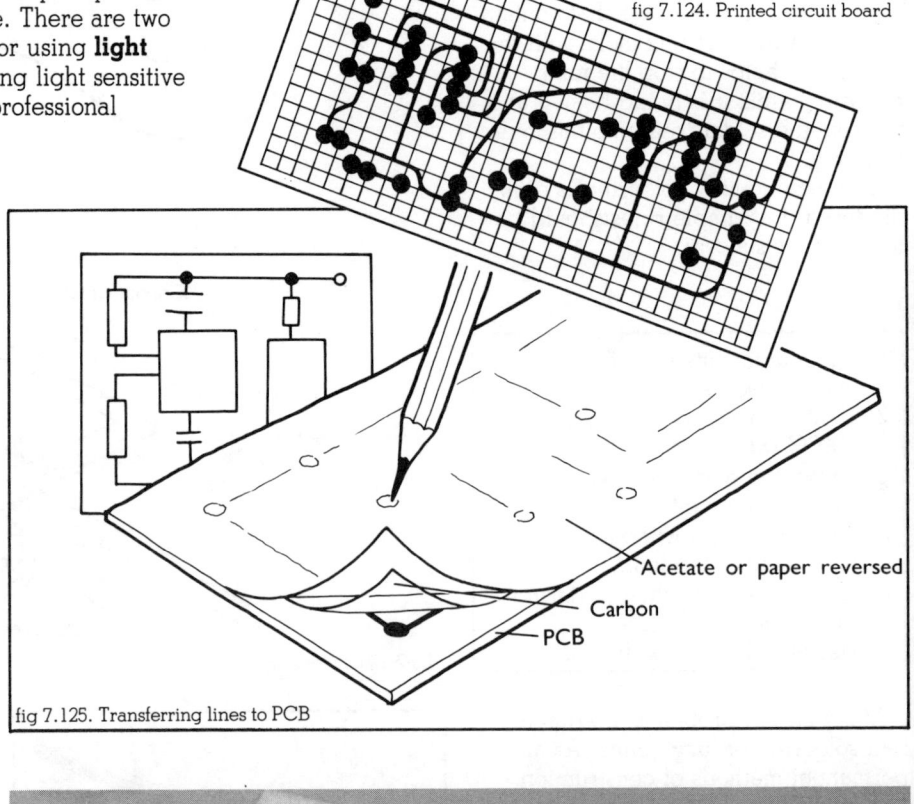

fig 7.124. Printed circuit board

Stage 1
Work from the circuit diagram to produce a circuit with no crossover wires. You must end up with the same electrical connections.

Stage 2
Arrange the components, chips, resistors, capacitors, diodes and transistors neatly, preferably in rows. These components can be used to make bridges, running connections under them.

Acetate or paper reversed
Carbon
PCB

fig 7.125. Transferring lines to PCB

Stage 3
Find the size of board needed, giving due consideration to the packaging of the project. Neatness and tidiness is important at all stages.

Stage 4
Draw out the board full size with all components marked and all connections laid out very clearly. This layout must be **drawn looking down on the components**, not from underneath. This can be done on gridded paper or acetate sheet with a 0.1 inch matrix. It can take quite a while to get the layout just right, and you should not be afraid to rub things out and try again if it goes wrong.

Stage 5
Check the circuit layout very thoroughly. Time spent here is very worthwhile.

fig 7.126. Materials used for transferring lines to PCB

Stage 6
Transfer the artwork to the copper on the PCB. There are two methods, direct drawing or photo transference. The direct method is only used on simple circuits and when a single circuit is required. The photographic method is used on all quality work and as a batch production technique.

A. Direct method: clean the copper with fine caborundum paper or wire wool. Turn the artwork over and, with a carbon sheet placed underneath, go over all the lines and pads with a pencil so they will be transferred to the copper side of the PCB. Remove the sheets to reveal the lines and apply dry transfer pads to all hole and component fixings. Then use either dry transfer lines, tape or an etch resist pen to draw in the joining lines.

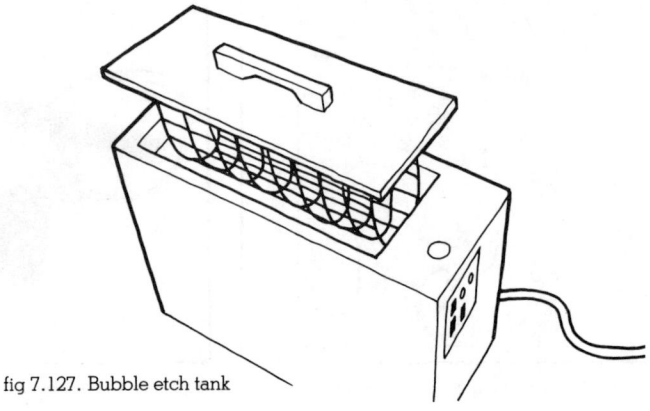

fig 7.127. Bubble etch tank

B. Photographic method: produce a clear acetate sheet with all the artwork on it. It is best to use the matt side. Dry transfers or indian ink pens can be used direct onto the clear sheet. Alternatively, a computer drawing or designing program can be used to produce the artwork onto a printer which can then be photocopied onto the acetate, or by plotter direct on to the acetate sheet or even the PCB itself.

The photo mask is then placed, as drawn, top side up onto the glass screen of the exposure unit (light is from below (fig 7.129). All writing to be on the finished PCB is applied to the reverse of this mask. Cut the PCB to size, remove the light protective layer and place the copper side down onto the mask. Close the box and expose for 2 to 6 minutes, depending on the boards being used, their age and the clarity of the mask. Trial and error will produce a time suitable for your materials.

Develop in a solution of **sodium hydroxide** for about 2 minutes or until all the blue has gone, leaving clear lines and pads. Gentle agitation helps here. The advantage of this method is that the mask can be reused very quickly for making many boards. It also produces boards of a very high standard of accuracy and definition.

Stage 7

The next stage is **etching.** The board is held in the cage of the bubble etch tank, immersed in **ferric chloride** solution and the agitator is switched on. Etching can be done in trays, but using an etch tank is much quicker and it is far safer not to have open dishes of acid in the workshop.

Stage 8

When etching is complete, remove the board and wash it thoroughly to remove all traces of acid. **Tongs** must be used when handling the acid coated boards.

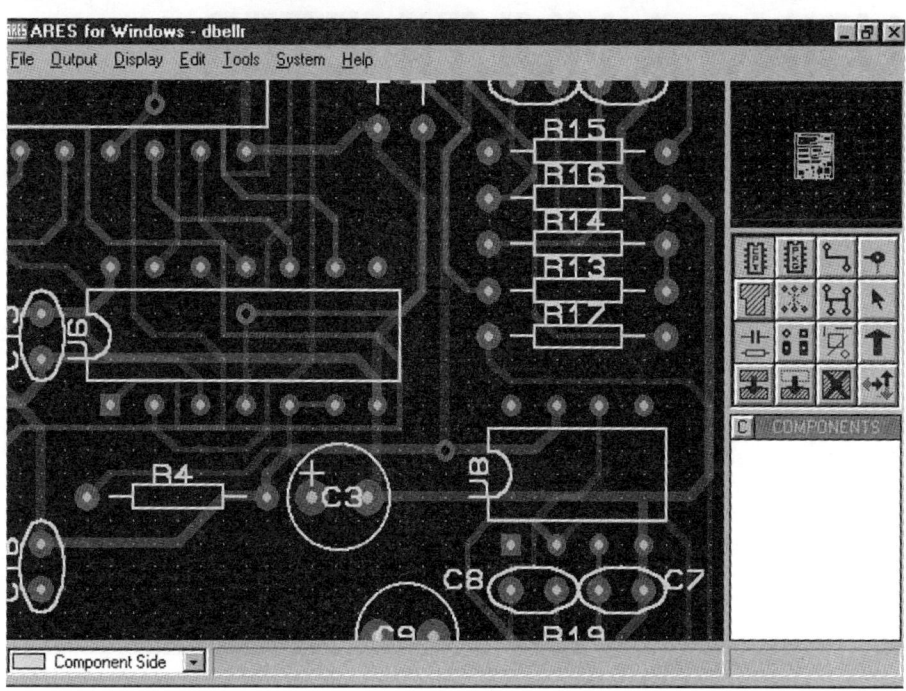
fig 7.128. A computer screenshot showing part of a PCB design

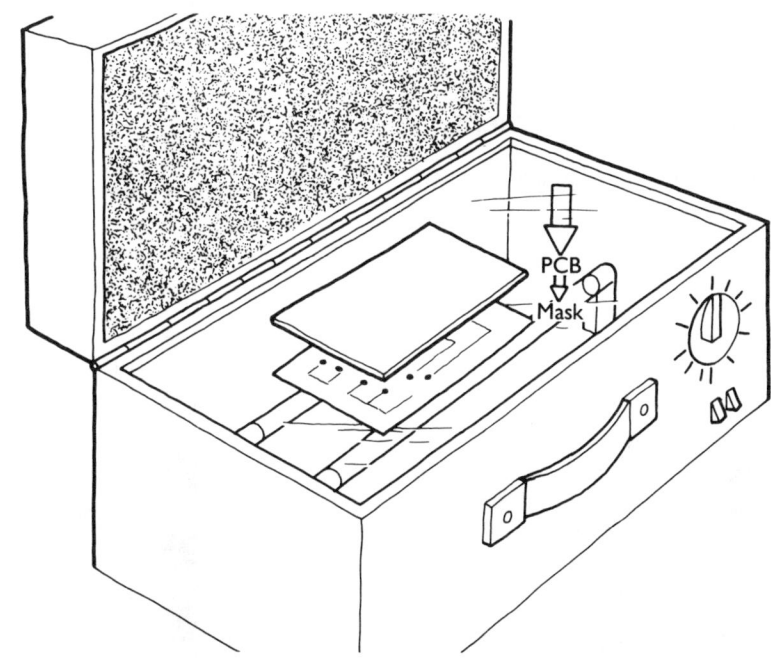
fig 7.129. Ultraviolet light box

fig 7.130. PCB built out of box

fig 7.131. PCB in box with battery

Stage 9

Drill all the holes using a 1mm drill as standard. Some very close pads will require 0.6mm or 0.8mm drills, and some large connectors will require 1.5mm drills. Using a drill stand makes the drills last longer and will help you to drill acurately.

Stage 10

Clean the PCB with wire wool or fine carborundum paper.

Stage 11

Solder in all components. Use holders for all ICs and a heatsink with all diodes and transistors. **The components should be on the top of the PCB, and the tracks and pads underneath.**

USING A MULTIMETER

The digital **multimeter** is simple to use and easy to read the output values. All multimeters will measure voltage, both a.c. and d.c., current and resistance. Some will also measure capacitance and transistor gain.

D.c. voltage measurement
Setting the meter:
1. Place black lead in black common socket.
2. Place red lead in red V socket.
3. Select d.c. not a.c.
4. Select voltage range required.
5. Touch the leads to the points to be measured and read off the value.

When taking voltage readings it is only necessary to **touch two points** with the probes in parallel with the circuit. The reading gives the voltage difference between the two points (fig 7.132).

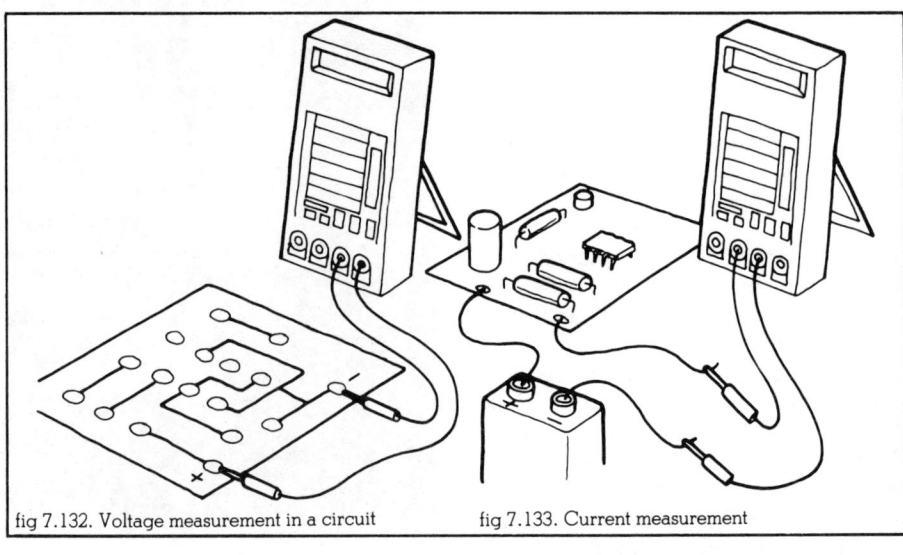

fig 7.132. Voltage measurement in a circuit   fig 7.133. Current measurement

D.c. current measurement
Setting the meter:
1. Place black lead in black common socket.
2. Place red lead in red mA socket.
3. Select d.c. not a.c.
4. Select amps.
5. Select current range required.
6. Connect the probes to the wire in which the current is to be measured. Correct polarity MUST be observed.

When taking current measurements it is necessary to **break the circuit** and connect the meter in series with the wire in which the current is to be measured (fig 7.133).

Resistance measurement
Setting the meter:
1. Place black lead in black common socket.
2. Place red lead in red ohms socket.
3. Select ohms.
4. Select ohms range.
5. Connect to the two ends of the component whose resistance you are measuring.

When measuring the resistance of a component it must be **disconnected from any circuit** otherwise the circuit will affect the reading (fig 7.134).

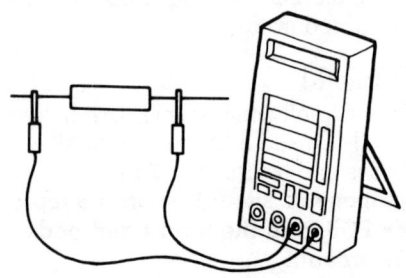

fig 7.134. Resistance measurement of a component

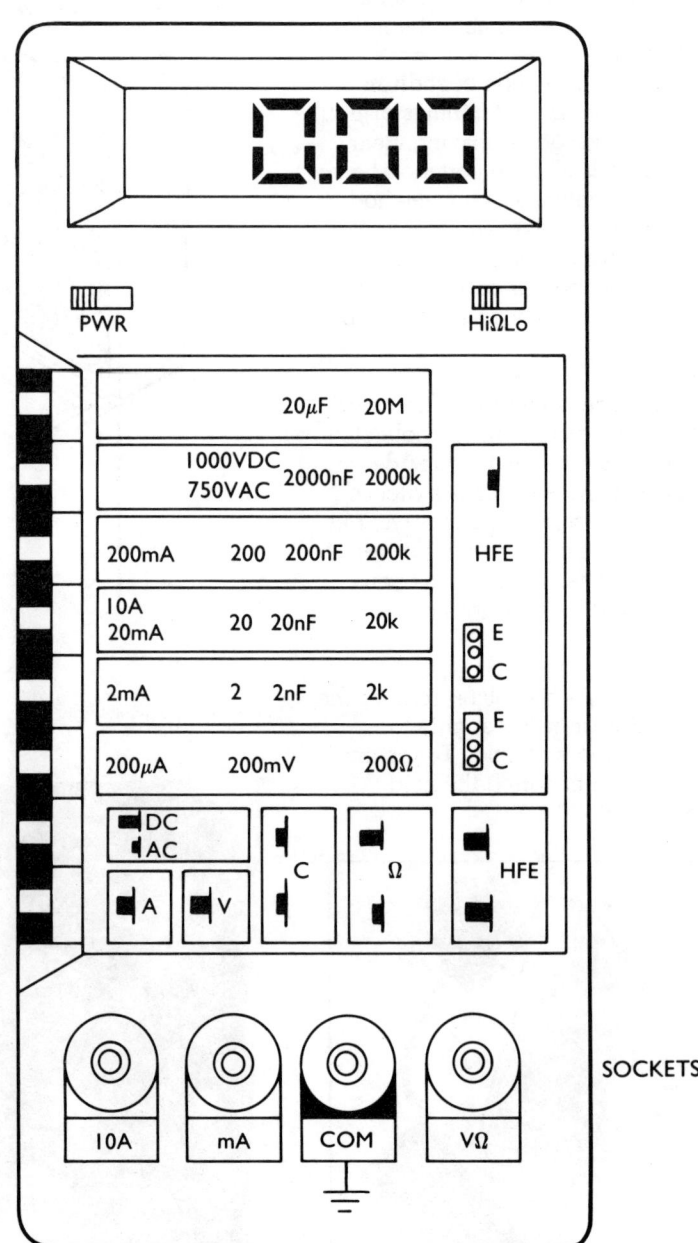

fig 7.135. Digital multimeter

USING MAINS SUPPLY

We have had electricity in our homes for many years and have learnt to use it in many different ways. It can be used for lighting, heating, cooling; for televisions, radios, music centres and computers. Electricity is now a part of our lives, and without it most devices in the home will not work. The electricity from the 'mains' has a voltage of 240V and is of alternating current type. The supply comes into the house through a meter which then registers the amount used through a **consumer unit**. The power is routed throughout the house to various socket outlets or direct to lighting circuits.

The electricity enters the system as two wires, the **live (L)** and the **neutral (N)**. A third wire, the **earth (E)** is connected to the top connection on all socket power points and is connected to a water pipe or an earth connection on the supply cable. This third wire is to prevent electrical shock if a fault develops in the cable or appliance connected to it.

A fuse

A fuse is a short length of thin wire that will melt when a stated current is passed through it. It is used as a safety device to stop a circuit from being over loaded. The fuse in a plug is of a cartridge type and must be replaced if **'blown'**. The correct rating of fuse must be used in each plug, the value can be found from the literature supplied with the appliance. Electrical appliances are rated by power, i.e. watts. For example, a light bulb may be 60W or 150W, an electric fire could be 3kW, a kettle 2kW. The power rating is calculated by multiplying the voltage by the current:

$$P = I \times V$$
$$P \text{ (watts)} = I \text{ (amps)} \times V \text{ (volts)}$$

To find the correct amp fuse to use, the current passing through the appliance is required. This is calculated by dividing the power by the voltage.

A 3kW fire: I = 3000/240 = 12.5A
A 2kW kettle: I = 2000/240 = 8.3A
A 150W bulb: I = 150/240 = 0.6A

From these calculations we then select either a 3, 5 or 13 amp fuse. The lower the value the safer is the appliance. For the fire and the kettle we would use a 13 amp fuse, and for the bulb, a 3 amp fuse.

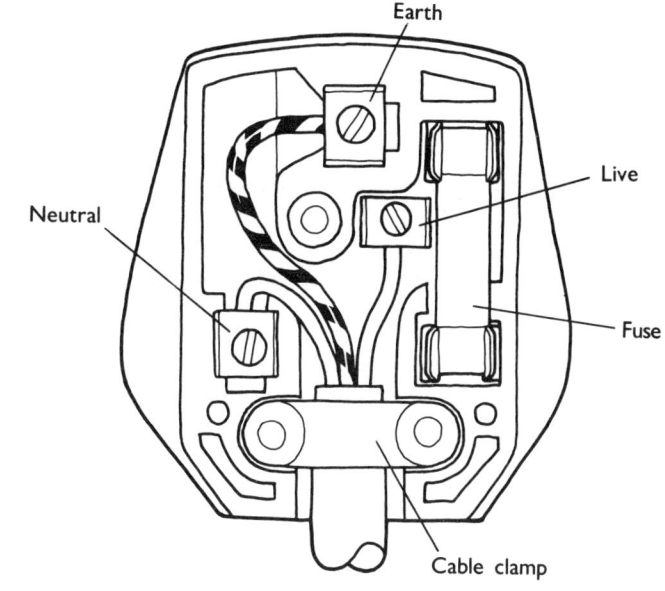

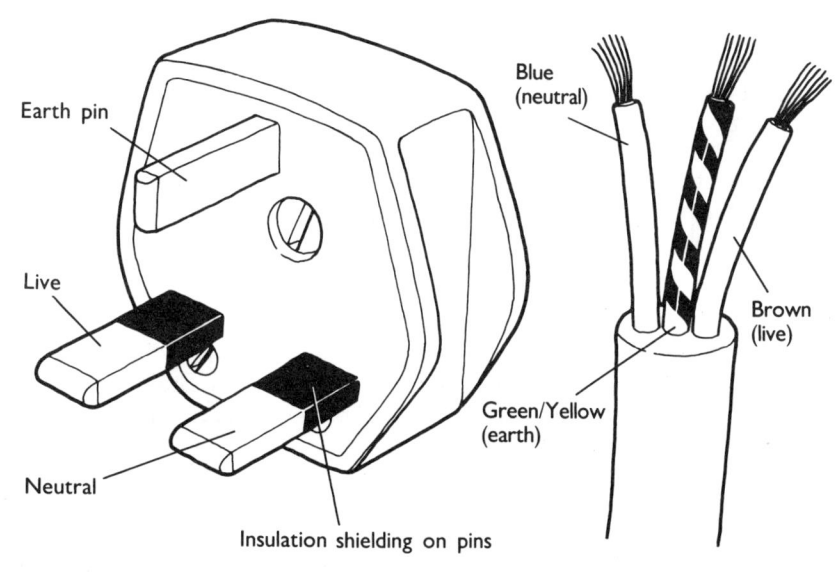

fig 7.137. Mains plug

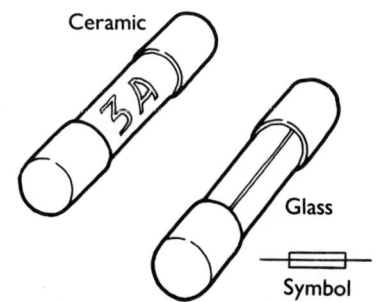

fig 7.136. Fuses

Electrical safety: some do's and do nots

Do use shielded plugs.

Do switch off the socket before inserting or removing a plug.

Do switch off the appliance before inserting or removing a plug.

Do select and use the correct fuse.

Do unplug a kettle when filling with water.

Do replace all cut, frayed or damaged cables immediately.

Do take care using mains electricity.

Do not touch mains plugs and sockets with wet hands.

Do not remove any casing from appliances when connected to the socket.

Do not connect to the supply with the case removed.

Do not 'test' by probing with a screwdriver.

Do not be tempted to 'play' with mains voltages, it could be fatal.

EXERCISES

1. Make a list of ten electrical articles in use in your home or school.

2. Make a list of ten electronic articles in use in your home or school.

3. The coloured bands on a resistor are orange, orange, red and silver. What is its resistance value?

4. You have been given a resistor, but unfortunately the coloured bands have been rubbed off. Explain how you could find its resistance value.

5. Explain the differences between solid core and stranded core wires. Give three examples of where the different types could be used.

6. I wish to use a light emitting diode as an indicator in my circuit. The voltage being used is 5V. What resistor will I require to limit the current in the LED to 20mA?

7. Draw a block diagram of the operation of a torch. Explain what happens in the three blocks, input, control and output.

8. A sensor to sense light levels can be built using an LDR. Draw two circuit diagrams to show how this can sense (a) falling and (b) rising light levels to give the same output. How do these two different circuits work?

9. A sensor is to be mounted on the front of a buggy to recognize when it has bumped into an object. What component could we use for the sensor? It must be able to recognise quite a small force.

10. Ian's project is to build an egg timer. He wishes to use a touch switch to start the timing, and would like to incorporate some form of logo into the switch to make it more interesting. Design a logo suitable for an egg timer that may be used as a touch switch. It could either be built from a PCB or a metal foil.

11. A sensor is required to count the number of people entering through a door to visit an exhibition. What sort of sensor could be used? Draw a circuit to use this sensor with a transistor control and output through an electromechanical counter.

12. What component would be suitable to use to interface an electronic sensing circuit with a pneumatic cylinder as the output? Design a circuit that could sense temperature and open a greenhouse window pneumatically when the temperature became too high.

13. A model lighthouse that simulates a flashing light is being built. Design a circuit that could be used to give a flash rate of 2 seconds on and 2 seconds off. To develop this further, you decide that the flash rate should be changed. It will now flash at the same rate for 20 seconds (5 flashes), not flash for the next 20 seconds, and then repeat this cycle. How should the circuit be modified to achieve this?

14. John is building a project to fit indicators to his bicycle. He has a 6V battery and four 4.5V bulbs. Draw a circuit that could be used. What type of switch is required?

15. Draw a block diagram for a burglar alarm. There should be some sort of latch and a delay to allow the user entry and exit when setting. Explain what sensors are to be used.

16. An electrically operated lock is being built. The lock mechanism will use a solenoid bolt. Design an operating system that will not need a key to operate, but will still be reasonably secure.

DIGITAL ELECTRONICS

Digital electronics is concerned with electrical systems made up of a series of switches. These switches are, in themselves, very simple; they are either on or off, nothing else will do. The complication arises in the way that they are switched, or what the input requirements are for the switch to operate.

A knitted Fair Isle sweater is a very complicated and involved piece of textile engineering. The patterns and colours are very involved. However, its basic construction consists of very simple stitches used in slightly different ways. When hundreds of these stitches are connected together they form a complicated knitted garment.

A computer is a very complicated digital circuit. It is made up from hundreds of components that are each very simple switching units connected together to form the full complicated switching unit called a computer.

Digital electronics are used in many items around us. The digital watch and calculator were two of the first widely used digital devices. Today an increasing number of electrical and mechanical devices have become digitized. The mechanical switching unit on a washing machine has now been replaced by a very small, reliable, low cost digital electrical control unit.

Other uses of digital circuits involve control units, controlling units for all manner of machines from production lines to sewing machines, digital compact disc players, clocks, timers, counters, meters, games, toys and a host of other items.

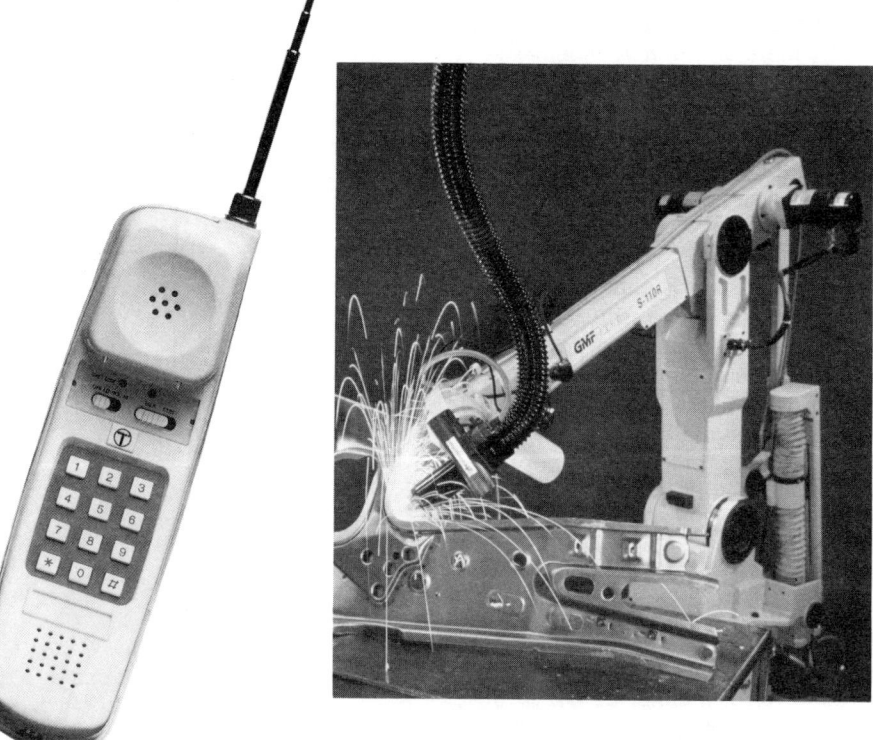

Fig 8.1. Examples of digital electronics

LOGIC

Digital circuits are built from simple digital switches, called **gates**. If you examine fig 8.2a, from your knowledge of electronics you should understand that with the switch, S, open as drawn, the output is connected to the 0 volt rail and is, therefore, at a voltage of 0V. When the switch is pressed the output is connected to both the 9 volt rail through the switch, and to the 0 volt rail through the resistor. The connection to the 9 volt rail is not through a resistive component and so the value of the output will be 9V. The output through the switching unit is either 0V or 9V.

Fig 8.2b shows two switches connected in series in a switching circuit. The output is connected to the 0 volt rail through a resistor and will have a voltage of 0V with the switches open. If S1 is pressed the output remains at 0V and the circuit is still broken at S2. If S2 only is pressed the result is the same, the output is still 0V.

The output only changes to 9V if S1 **AND** S2 are pressed.

Fig 8.2c shows two switches in parallel. Here, if either S1 **OR** S2 are pressed the output changes to 9V.

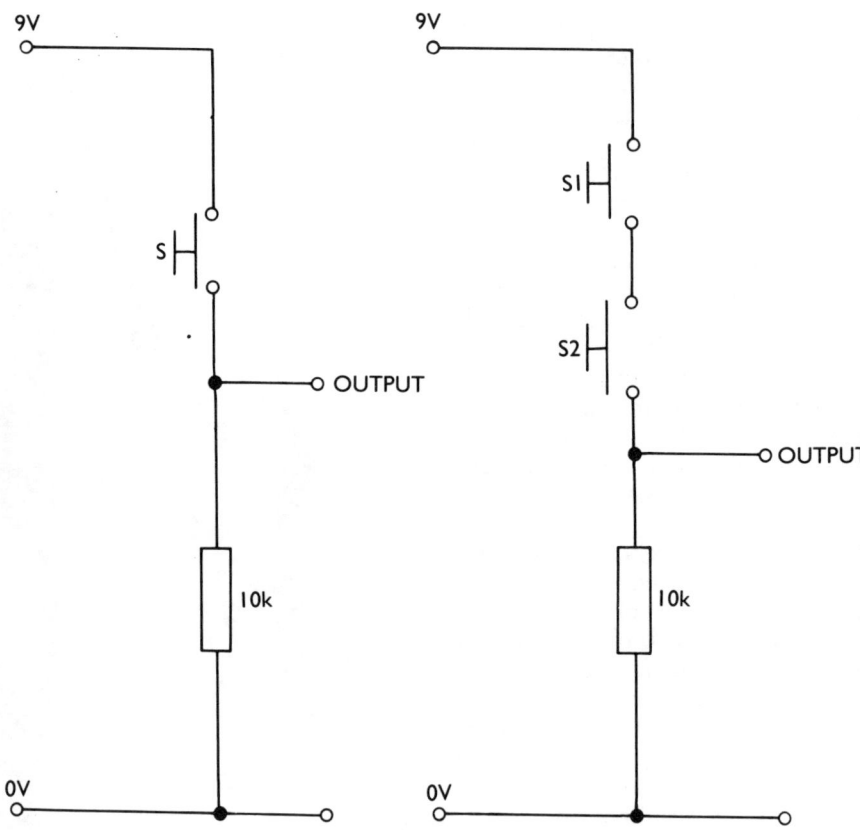

fig 8.2a. Single switch digital circuit   fig 8.2b. Double switch digital circuit

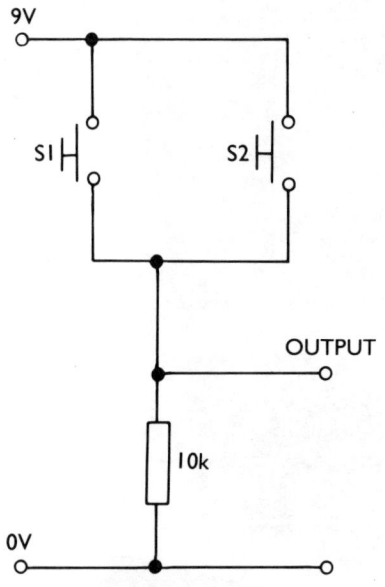

fig 8.2c. Digital circuit with two switches in parallel

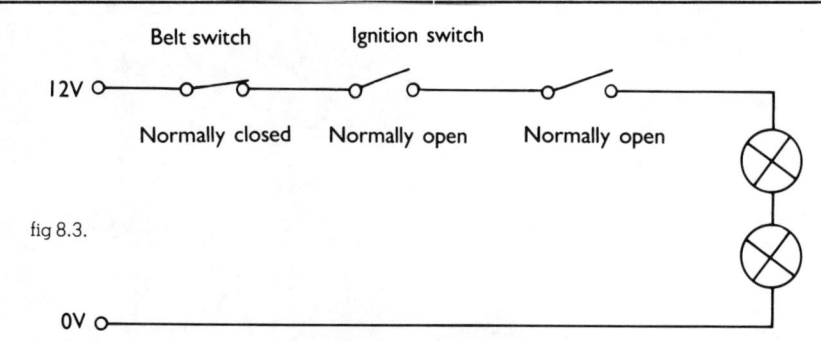

fig 8.3.

An example of this type of switching could be a seat belt warning sign, lighting up if the engine is on and a person is sitting on the seat. Switches could be incorporated under a seat to sense a seated person and included in the ignition circuit. This circuit would use the 12 volts available from a car battery.

An extra switch that must be opened when the belt is fastened can be used to sense when to turn off the warning.

Describing the circuit we would say that the light will be on IF the engine is on AND the seat is occupied AND the belt is NOT fastened.

AND
OR
NOT

This type of statement is said to be **logical**. Logic statements are the basis of digital systems, not all of which are electrical or electronic.

'If the watering can is tilted **AND** it contains water, it will water the plants.' This is a logical system, but there is no electrical connection. The following statement is also logical, but not electronic. 'If I start at 8.00 **AND** run very fast **OR** catch a bus, I will arrive at school on time and will **NOT** be in trouble.'

The three words **AND**, **OR** and **NOT** can all be represented by switches or logical decisions of one sort or another.

TRUTH TABLES

All possible alternatives of input to the gates can be charted in a table. This is called a truth table. The resulting output is given alongside the alternative inputs. The numbers 1 or 0 are used, here to signify on input either pressed (1) or not pressed (0) and as output either 9V (1) or 0V (0).

AND GATE

Fig 8.4 shows an AND switch or gate. If A **AND** B are pressed then Q is 9V. Because these switches or gates can be operated by pneumatic, electrical or mechanical systems, a symbol representng a general logic gate is used to simplify matters. The logical operation of pneumatic and electrical AND gates is identical, though the internal components are different.

OR GATE

Fig 8.5 is an OR gate or switch. If A **OR** B are pressed then Q is 9V.

NOT GATE

Fig 8.6 is a NOT gate. If A is **NOT** pressed then Q is 9V, or if A is pressed then Q is **NOT** 9V.

NAND GATE and NOR GATE

Fig 8.6b shows a **NAND** gate. This is made by connecting a **NOT** gate with an **AND** gate (NotAND). A **NOR** gate, similarly, is a **NOT** gate with an **OR** gate connected to it.

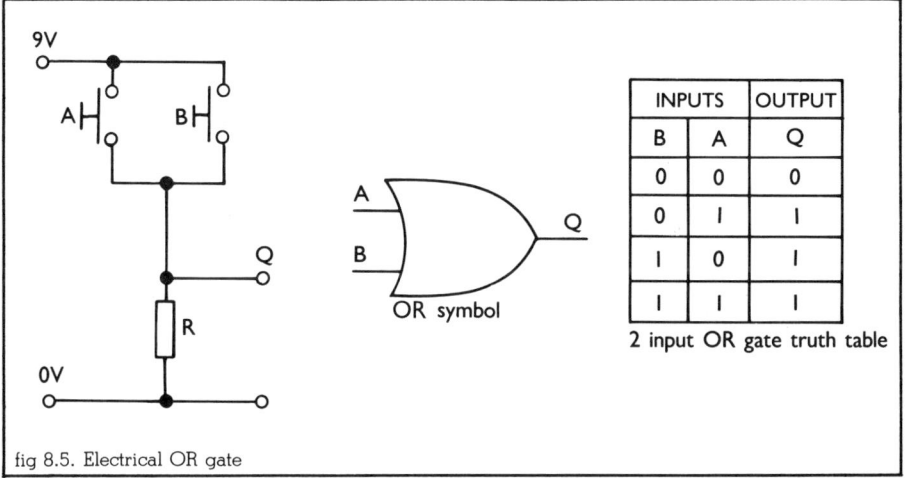

INPUTS		OUTPUT
B	A	Q
0	0	0
0	1	0
1	0	0
1	1	1

2 input AND gate truth table

fig 8.4. Electrical AND gate

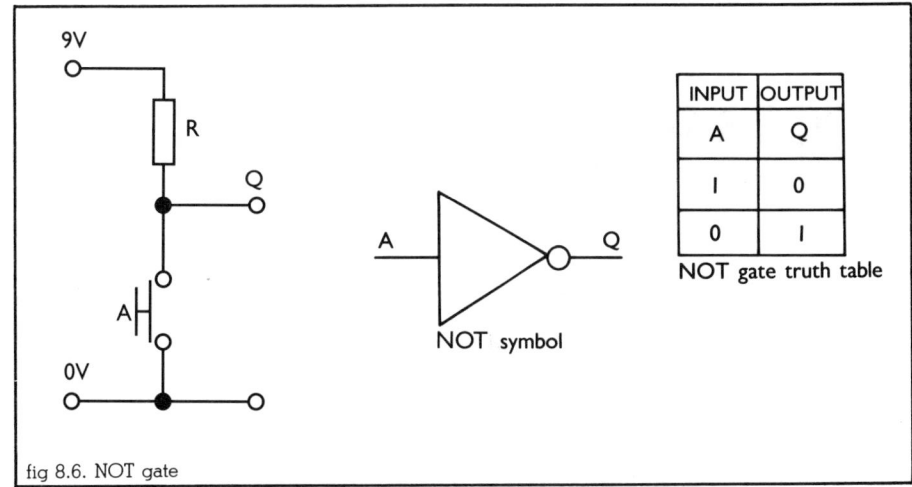

INPUTS		OUTPUT
B	A	Q
0	0	0
0	1	1
1	0	1
1	1	1

2 input OR gate truth table

fig 8.5. Electrical OR gate

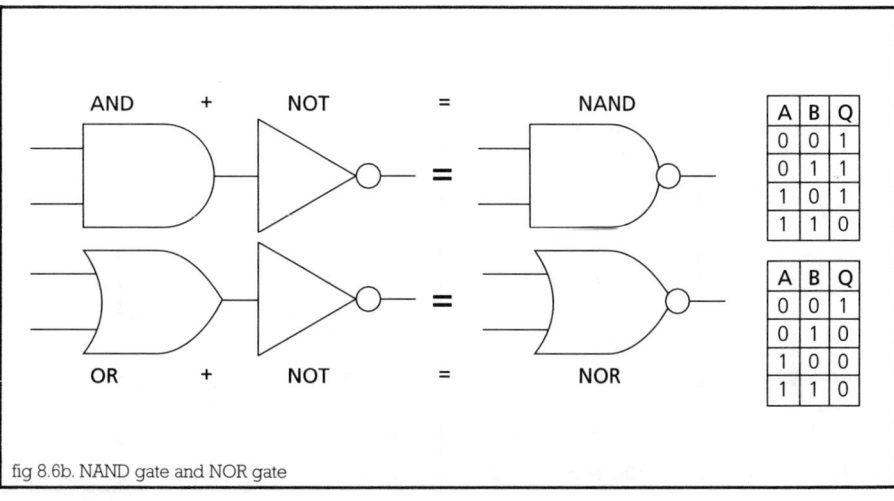

INPUT	OUTPUT
A	Q
1	0
0	1

NOT gate truth table

fig 8.6. NOT gate

A	B	Q
0	0	1
0	1	1
1	0	1
1	1	0

A	B	Q
0	0	1
0	1	0
1	0	0
1	1	0

fig 8.6b. NAND gate and NOR gate

ELECTRONIC LOGIC

On the previous page you looked at logic systems that have inputs and outputs. The inputs were the pressing of switches, the output being either 9V or 0V. If you exchange the mechanical input for an electrical input, you can build transistor switching devices, which can be quite complicated in construction. For example, to build an AND gate with eight inputs as a transistor circuit could well be beyond your capabilities. Help is at hand. Many different types of logic gates have already been built and packaged for use. You need not concern yourselves with what is in each 'chip', but only with how to use them in your circuit building.

LOGIC FAMILIES

There are two distinctly different methods of producing electronic logic devices, giving two families of chips.

One family is TTL (Transistor Transistor Logic) which use bipolar transistors; the other is CMOS (Complementary Metal Oxide Semiconductor) based on field effect transistors (FETs).

TTL integrated circuits are listed as the '7400' series and the CMOS ICs belong to the '4000B' series. The 7400 series is now produced with the advantages of both types and is listed as the '74HCT00' series. The last two numbers referring to the type of gate (e.g. 74HCT08 is a chip with four 2 input AND gates, fig 8.7). Fig 8.8 shows the equivalent CMOS IC, the 4081.

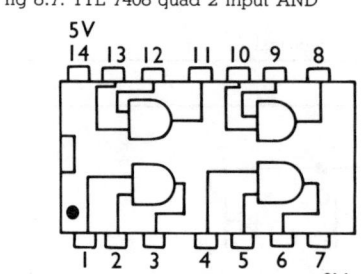

fig 8.7. TTL 7408 quad 2 input AND

fig 8.8 CMOS 4081 B quad 2 input AND

fig 8.9. Properties of TT1 and CMOS ranges

Property	TTL	CMOS	TTLHCT
Power supply	$5V \pm 0.25V$	3V to 15V	$5V \pm 0.5V$
Current required	3mA	$8\mu A$	$20\mu A$
Input impedance	Low	Very high	High
Speed of switching	Fast	Slow	Fast
Fan out	10	50	25
Temperature range	0 to +70	−40 to +85	−40 to +85

POWER SUPPLIES

The power supply for TTL must be stabilized at 5V, while CMOS will work on any voltage between 3V and 15V. Fig 8.10 shows how an IC regulator can be used to provide a 5V, 1A power supply. A 9V battery could power this regulator.

Logic levels

Ideally the 'high' and 'low' voltages should be Vcc (supply voltage, 5V for TTL) and 0V, respectively. In practice this is not possible due to the drop in voltage across the transistors in the chip. These logic chips are arranged to operate on a range of voltages. Logic 'high' (TTL) can be any voltage between 2.4V and 5V, logic 'low' can be any voltage between 0V and 0.8V.

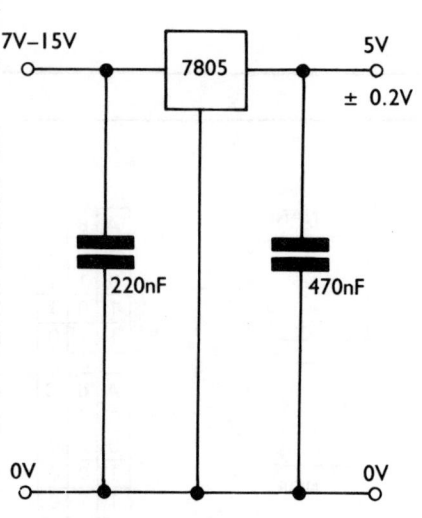

fig 8.10. TTL power supply using regulator 7805

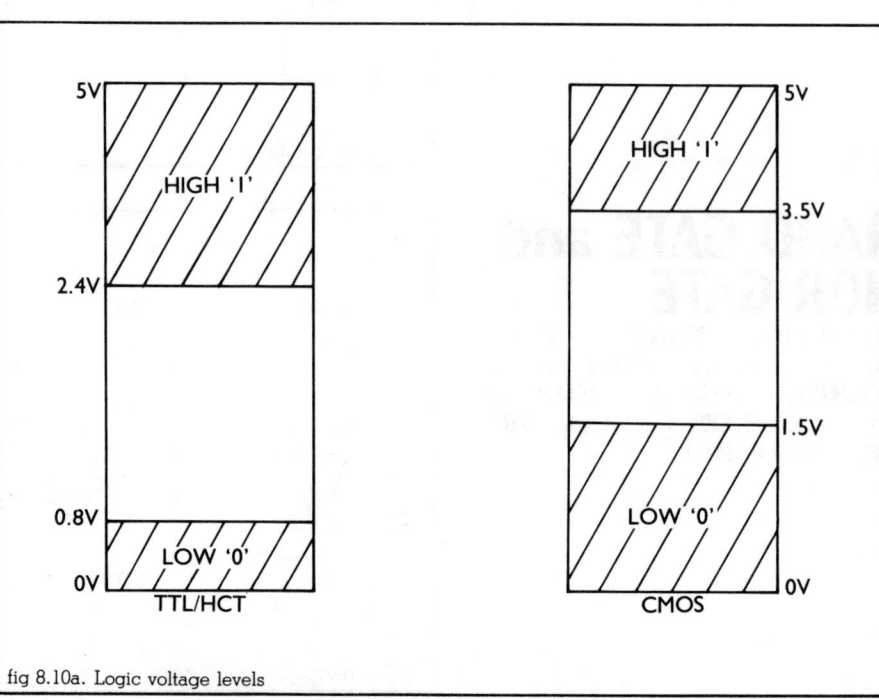

fig 8.10a. Logic voltage levels

NAND GATES

NAND gates were the first type of gate to be developed into integrated circuits and became widely used to build logic circuits. A NAND gate is an AND gate combined with a NOT gate. It is possible to build all the other gates from NAND gates. This means that only one type of integrated circuit need be manufactured.

The fact that NAND gates can be used to build other gates may be of use to us in our project work. ICs or 'chips' that have different types of logic circuits on them are available. Sometimes in our project work we require only a few gates, which may be made from NAND gates combined to form other gates. For example a QUAD, 2 input NAND chip can be used to build one AND gate and two NOT gates. Fig 8.12 shows possible methods of making specific gates from NAND gates.

There is a 13 input NAND gate available. Try to imagine how complicated it would be to build this from lots of 2 input NAND gates! Sometimes it is much easier to use the special chip if it is available.

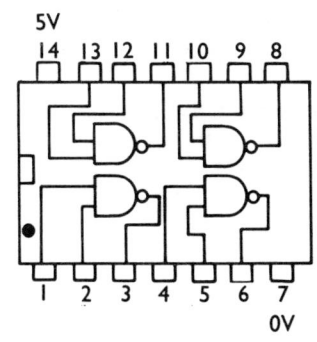

fig 8.11. 7400 quad 2 input NAND

---

fig 8.12. Methods of making specific gates from NAND gates

A	B	Q
0	0	1
0	1	1
1	0	1
1	1	0

2 input NAND gate — NOT

A	Q
1	0
0	1

A	B	Q
0	0	0
0	1	0
1	0	0
1	1	1

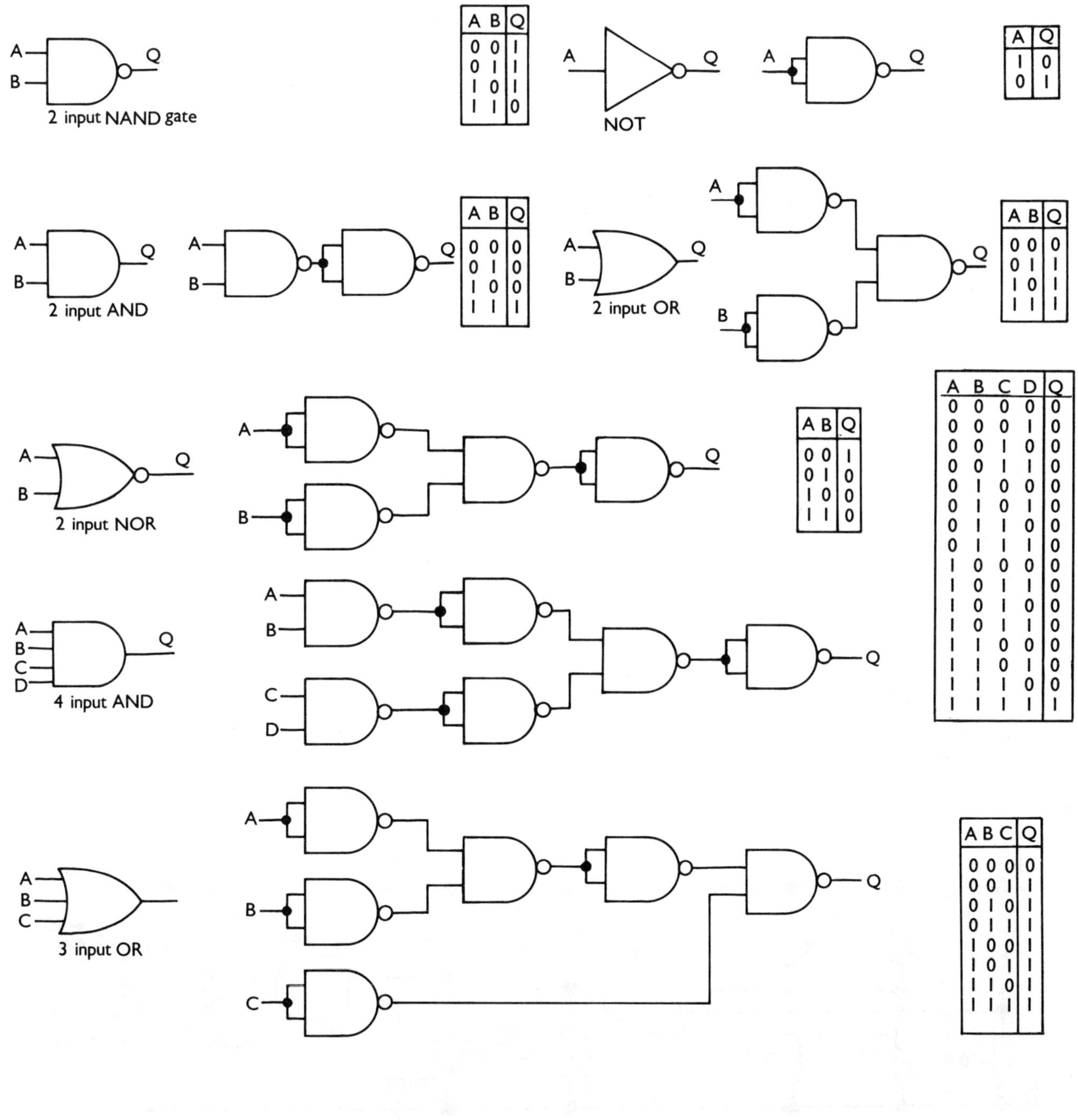

2 input AND

A	B	Q
0	0	0
0	1	1
1	0	1
1	1	1

2 input OR

A	B	Q
0	0	1
0	1	0
1	0	0
1	1	0

2 input NOR

A	B	C	D	Q
0	0	0	0	0
0	0	0	1	0
0	0	1	0	0
0	0	1	1	0
0	1	0	0	0
0	1	0	1	0
0	1	1	0	0
0	1	1	1	0
1	0	0	0	0
1	0	0	1	0
1	0	1	0	0
1	0	1	1	0
1	1	0	0	0
1	1	0	1	0
1	1	1	0	0
1	1	1	1	1

4 input AND

3 input OR

A	B	C	Q
0	0	0	0
0	0	1	1
0	1	0	1
0	1	1	1
1	0	0	1
1	0	1	1
1	1	0	1
1	1	1	1

The American system of symbols has been used as all data sheets for the IC's use this system.

OUTPUT CAPABILITIES

It can be seen from fig 8.13 that to achieve the highest output current we must select a TTL gate. The CMOS or HCT gate will not 'drive' an LED requiring at least 10mA to light it. A TTL gate used to 'drive' an LED is required to act as a **sink** rather than a source.

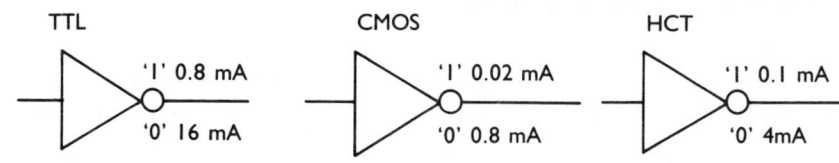

fig 8.13. Maximum possible output currents

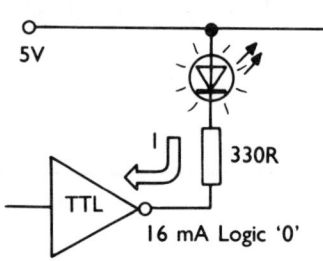

fig 8.14. Output as a current sink LED ON for logic 0

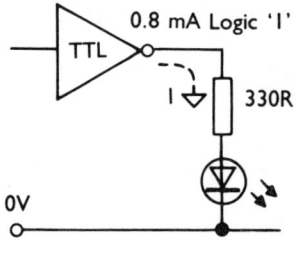

fig 8.15. Current source LED NOT ON for logic 1

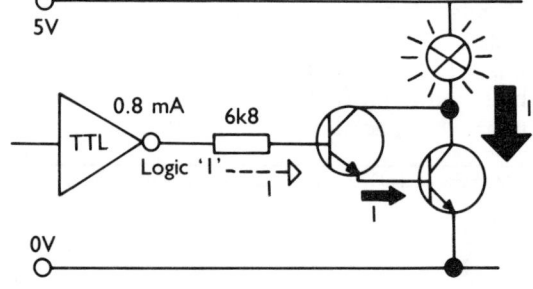

fig 8.16. Lamp ON for logic 1 Darlington amplifier

Circuit design problem

A central heating boiler requires a control circuit to allow it to function safely.

Specifications

1. There must be a room thermostat, to switch on the heating when cold.
2. It must be possible to turn on the boiler if the room is warm.
3. The gas to the boiler must not be turned on if the pilot light is out.

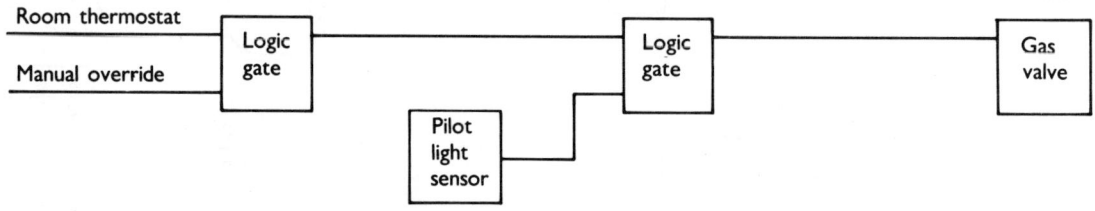

fig 8.17. Block diagram

Logic diagram

The gas valve is turned **on** only if the room is cold **or** is manually switched on **and** the pilot light is lit.
OR
If the room is cold **or** is manually switched on **and** the pilot light is lit then the gas valve is turned **on**.

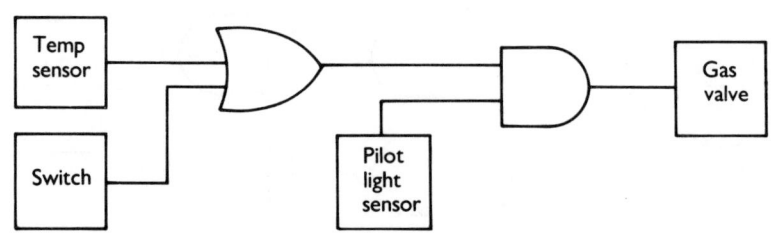

fig 8.18. Logic diagram

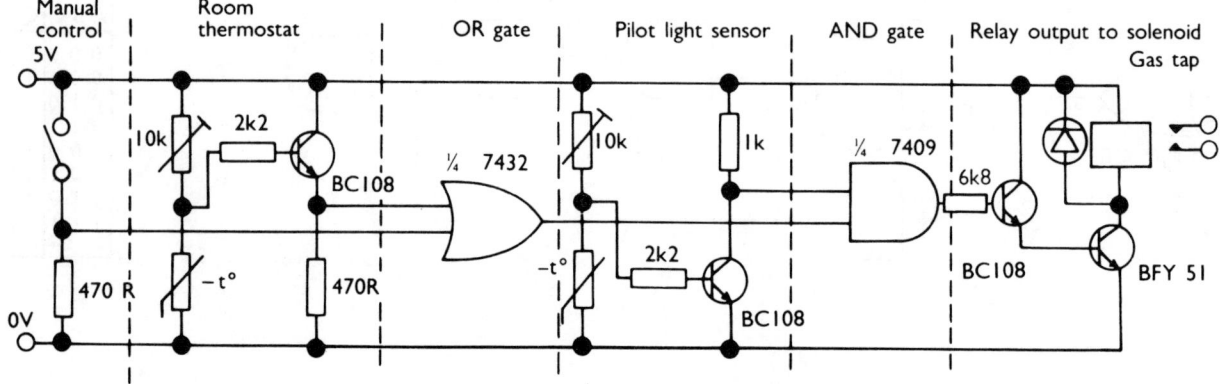

fig 8.19. Circuit diagram

BINARY NUMBERS

Normally a scale of ten is used for counting numbers. There are 10 different digits or symbols 0, 1, 2, 3, 4, 5, 6, 7, 8, 9. When the count exceeds 9, a 1 is carried, or placed in the next column on the left and then counting starts again at 0 in the right column. The values of the columns working from right to left are 1, 10, 100, 1000. They can be written in powers of ten, i.e. 10^0, 10^1, 10^2, 10^3. This type of counting is called **decimal**, counting to base 10.

Binary code uses only two, not ten symbols; 1 and 0. This system of counting is unusual for us, but can represent switches, with two states, 1 and 0, or on and off. As in decimal counting, binary starts in the right hand column. The first symbol is 0 and the second is 1. We now have no more symbols, like reaching 9 in decimal, so we carry into the next column and start again at 0, i.e. 10 (we say one ought, not ten). If you follow fig 8.20 across the chart, the decimal and binary numbers mean the same, but there are many more columns in binary. It is complicated to remember 1111000100 in binary, 1988 in decimal is much easier. However, it is much easier to build electrical circuits with two way switches than to make switches that need ten different voltage levels to count in decimal.

DECIMAL				BINARY				
10^3 1000	10^2 100	10^1 10	10^0 1	2^4 16	2^3 8	2^2 4	2^1 2	2^0 1
			0					0
			1					1
			2				1	0
			3				1	1
			4			1	0	0
			5			1	0	1
			6			1	1	0
			7			1	1	1
			8		1	0	0	0
			9		1	0	0	1
		1	0		1	0	1	0
		1	1		1	0	1	1
		1	2		1	1	0	0
		1	3		1	1	0	1
		1	4		1	1	1	0
		1	5		1	1	1	1
		1	6	1	0	0	0	0

fig 8.20. Decimal and binary numbers

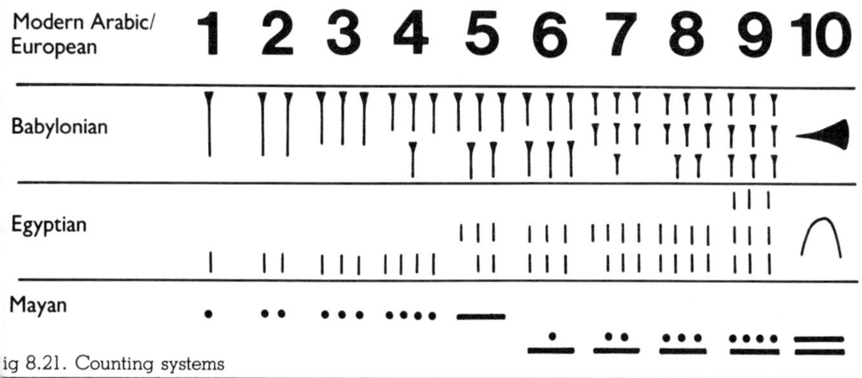

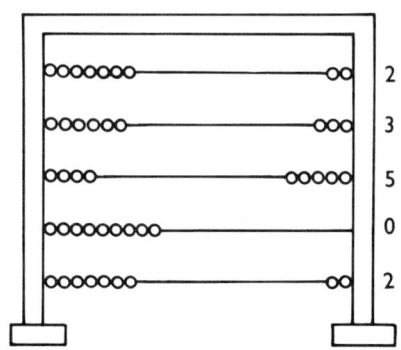

fig 8.21. Counting systems

fig 8.22. Abacus counting system

Decimal to binary

To convert a number from decimal to binary it is necessary only to keep dividing the number by two and recording the remainder.

To convert decimal 13 to binary.
13/2 = 6 rem. 1 Least Significant Bit (LSB)
6/2 = 3 rem. 0
3/2 = 1 rem. 1
1/2 = 0 rem. 1 Most Significant Bit (MSB)
    decimal 13 = 1101 in binary

To convert 23_{10} to binary.
23/2 = 11 rem. 1 (LSB)
11/2 = 5 rem. 1
5/2 = 2 rem. 1
2/2 = 1 rem. 0
1/2 = 0 rem. 1 (MSB)
    decimal 23 = 10111 in binary

Binary to decimal

To change from binary to decimal you must add together the decimal equivalents of the columns that have a 1 in the column. The columns with 0 are disregarded and not added into the sum (fig 8.23).

2^4 16	2^3 8	2^2 4	2^1 2	2^0 1
1	0	1	1	0

$= 16 + 4 + 2 = 22$

2^4 16	2^3 8	2^2 4	2^1 2	2^0 1
1	1	1	0	1

$= 16 + 8 + 4 + 1 = 29$

fig 8.23. Binary to decimal

Binary coded decimal

Another way that is used to represent numbers is for each bit of the decimal number to be coded separately into binary. It requires a 4 bit code for each decimal digit (fig 8.24).

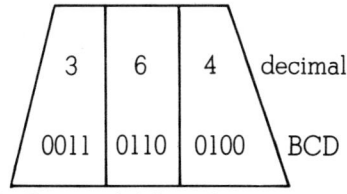

fig 8.24. Binary coded decimal

364_{10} becomes 0011 0110 0100. This is not the same as binary code. The digits are grouped in fours, representing each of the decimal digits. This code is used when converting numbers to be read by humans. Examples include speedometers, counters, and most number readouts on digital devices. If man had developed with 8 or 16 fingers instead of 10 computer counting would have been much easier.

DECODERS

While electronic systems work in binary, humans prefer to count in decimal. To convert from binary to a more common output we can use a decoder.

If you required an output from one digital line that shows not a single on and off, but two outputs, either one or the other on (see fig 8.25), you could use a very simple switch arrangement. A single pole double throw switch will do this (fig 8.26).

To **decode** two lines it is possible to produce four separate lines as output. Two binary digits allow a count of four different states. This is called a **one of four decoder**. It is now not possible to use simple switches in combination. A digital solution is required (figs 8.27 and 8.28). If you follow the lines to the outputs you should see how the decoder works:

Q_0 is 1 when A is 0 AND B is 0
Q_1 is 1 when A is 1 AND B is 0
Q_2 is 1 when A is 0 AND B is 1
Q_3 is 1 when A is 1 AND B is 1

Only one of the lines out is high at any one time. This represents a **binary to decimal** conversion.

From page 94 you know that if we wish to drive LEDs from the decoder, you must sink the current, the output line needs to be low. You can do this by using an invertor or a NOT gate on each line (fig 8.29).

Logic chip number 74139 is produced to do this. In fact there are two, one of four decoders on the chip. You do not need to construct it from individual gates.

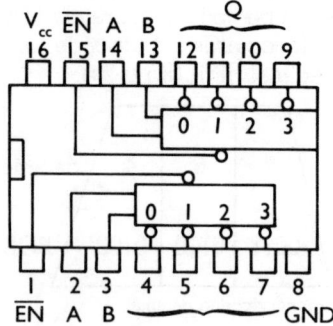

fig 8.29. Dual one of four decoder

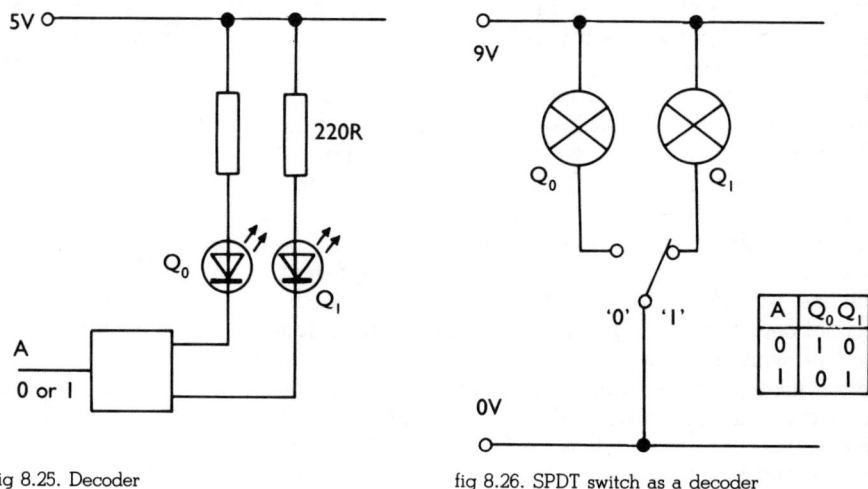

fig 8.25. Decoder

fig 8.26. SPDT switch as a decoder

A	Q_0	Q_1
0	1	0
1	0	1

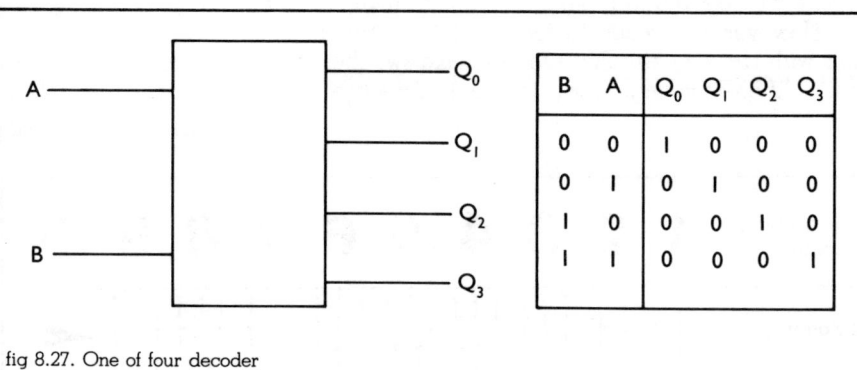

fig 8.27. One of four decoder

B	A	Q_0	Q_1	Q_2	Q_3
0	0	1	0	0	0
0	1	0	1	0	0
1	0	0	0	1	0
1	1	0	0	0	1

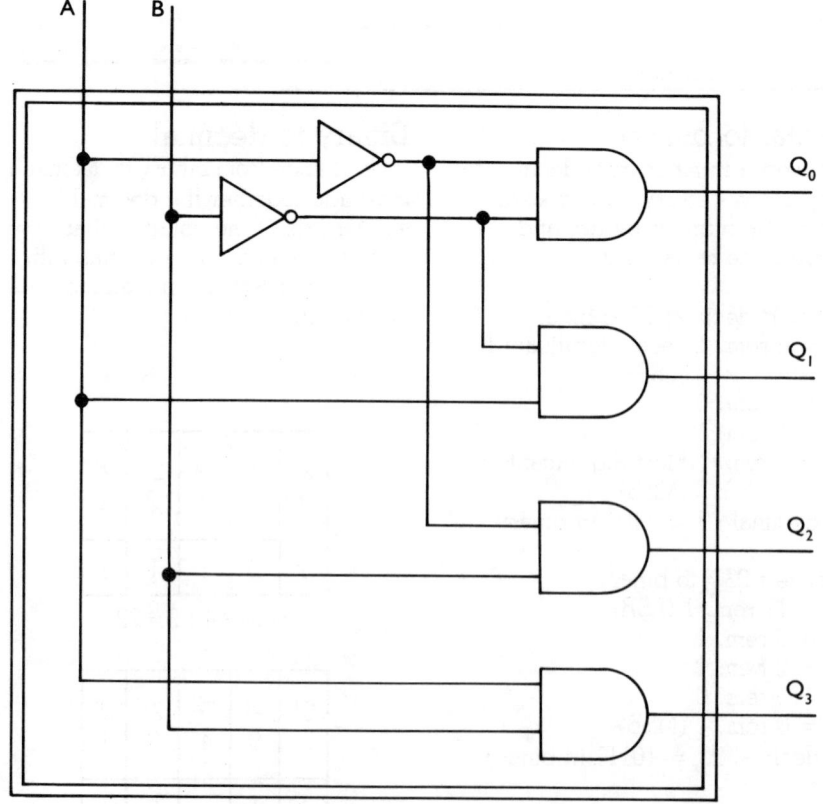

fig 8.28. Digital decoder using gates

CLOCKS

The logic so far in this chapter has been **combinational logic**. This means that the outcome is due to a combination of inputs using AND, OR or NOT gates. This works well if you are testing to see if a door is open or if a person has crossed a security light beam. However, if you wish to count how many times the door has been opened or how many times the light beam has been broken, you need something different. You need **sequential logic**. Here things happen in a sequence, counting the number of door openings or a sequence of events such as the operation of a set of traffic lights. A complex example is a computer; it processes information in a series of simple steps. To make the system 'tick' you need a set of pulses, just as your heart beats so must the logic system. This pulse is called a **clock pulse**.

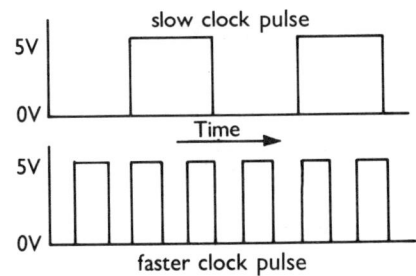

fig 8.30. Clock pulse

The graph of the clock voltage (fig 8.30) shows that the change from 0V to 5V is almost instantaneous and the same back to 0V. This type of voltage is called a **square wave**, from its graphical representation. Clocks for digital systems work at different speeds, depending on their needs. A clock for a traffic light system need only change perhaps every 30 seconds or even each minute. The clock for a BBC computer ticks 2 million times every second!

Fig 8.31 is a circuit that can be used to produce a 1Hz (1 change per second) clock voltage (see page 75 for the operation of the 555 IC chip). The 555 timer can produce clocks as fast as 1MHz to 1 beat every 3 minutes. It is suitable for most operations and will easily drive TTL logic circuits.

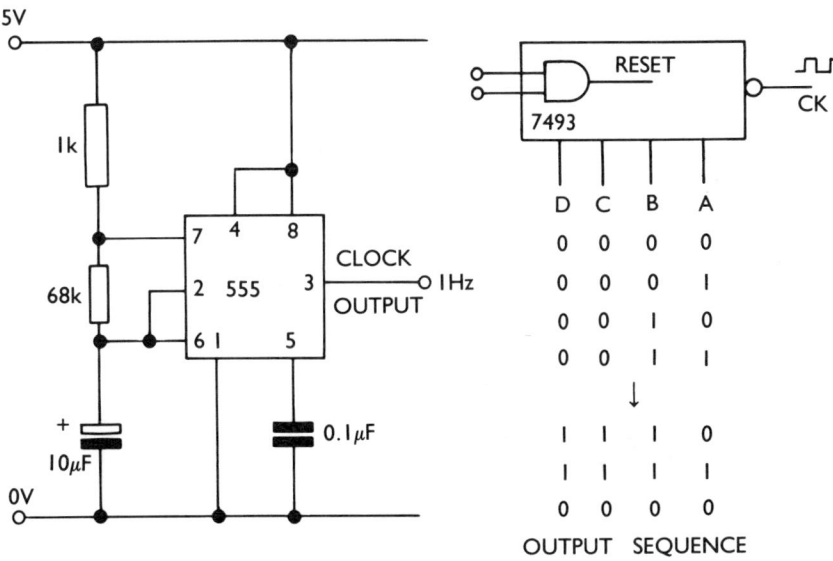

fig 8.31. 1 Hz clock voltage using a 555 timer

fig 8.32. 7493 chip binary 4 bit counter

D	C	B	A
0	0	0	0
0	0	0	1
0	0	1	0
0	0	1	1
↓			
1	1	1	0
1	1	1	1
0	0	0	0

OUTPUT SEQUENCE

COUNTERS

To record the number of times that an event has taken place, you need to be able to record a count. You may use your fingers and even your toes, but a logic counter uses binary. Counters can be made from many transistors or lots of chips called **flip flops**, but complete counter chips are available.

Fig 8.32 shows the block diagram of the 7493 chip which is called a binary counter. The outputs are DCBA, they show the recorded count. If the clock is connected to the CK input, the output counts in binary until 1111 (15), and then starts to count again from 0000. The chip can be reset to 0000 at any time by applying logic 1 to both inputs of the reset AND gate.

The last output can be fed into the clock input of a second counter to make an eight bit binary counter that will count from 0 to 255 (fig 8.33).

Fig 8.34 shows the wiring details of the same 7493 chip. It is constructed as a one bit counter and a three bit counter. The output A must be connected to the second clock to count the full four bits.

A very similar chip, 7490, is a BCD counter. This counter will count to 9 and then start again from 0.

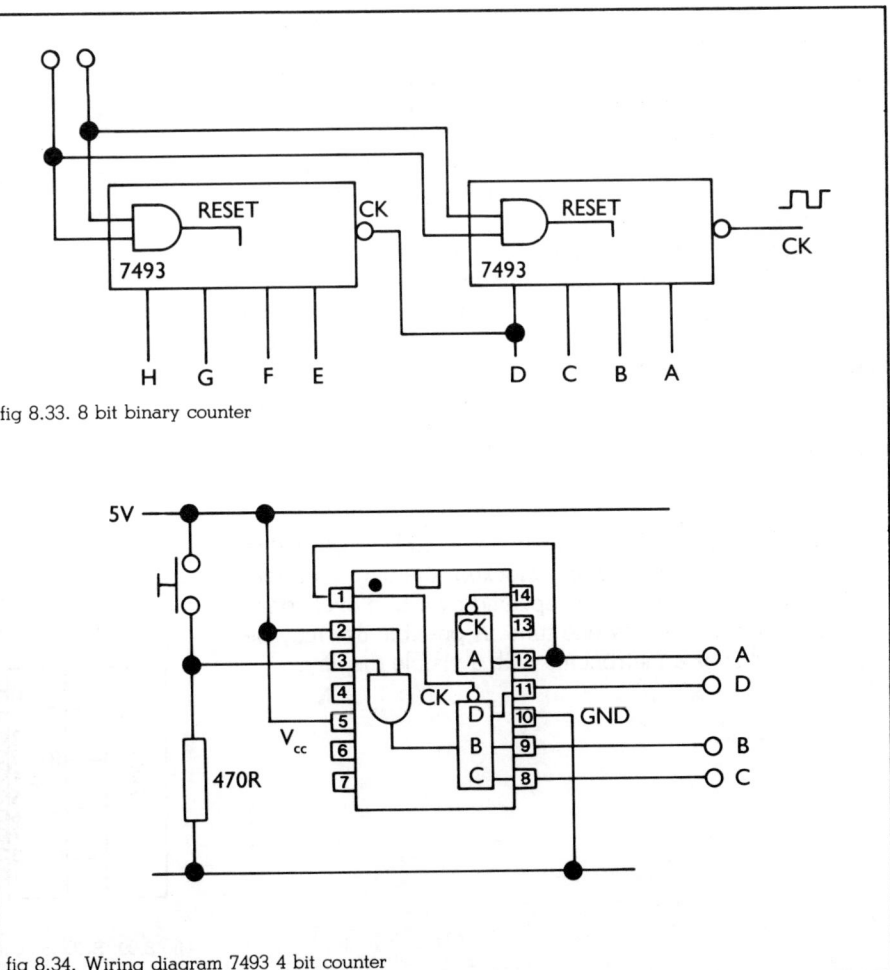

fig 8.33. 8 bit binary counter

fig 8.34. Wiring diagram 7493 4 bit counter

EVENT COUNTING

Counters can be pulsed by a clock, they can also be operated by pulsing a switch. There are problems with a mechanical switch connected to a digital counter. When the internal switch lever moves from one contact to the other, it strikes the contact and will bounce a few times before settling on the contact in the same way that a ball falling to the ground will bounce before it is stationary on the ground. This bounce only takes a millisecond, but as counters will work at speeds far in excess of that, it will register one count for every bounce, and thus will give an incorrect count. If the machine was counting out change it would stop before giving enough change, thinking it had counted out more, not a very good situation to be in!

By connecting two NAND gates in a special way, the switch can be 'de-bounced' (fig 8.35). This arrangement is called a **flip flop** or a **bistable circuit**. The output changes on the lever hitting the contact for the first time. If the lever bounces without going all the way back to the other contact, no more change is made. The output only changes when it goes all the way to the other contact. A single pole double throw spring return push switch could be used to provide a hand operated input to a counter.

A transistor switch input as in fig 8.36 does not need a de-bouncing circuit as it does not have any mechanical components in the switch. In this light operated switch the output is 1 unless the beam is broken when it is 0.

COUNTER DISPLAYS

Inputs to a counter can be in the form of a clock or a switched pulse. The output, if the information is for an operator not a computer, must be in an understandable form. Fig 8.37 shows a binary counter connected to four LEDs. Remember they can only be connected with the output from the chip as a sink (see page 94), i.e. to the 5V rail. This presents problems: the LED is lit when the output is logic 0, which is not as we would wish. The LED display for 0000 would show all lit, very difficult to understand. The solution is to connect a NOT or inverter gate to each output so switching on the LED only when logic 1, as we would expect, is at the counter output. Fig 8.38 shows this circuit. The output is in the form of a binary count which is understandable, but not as easy as if it were decimal. The problem of understanding or not understanding binary is not new. Designers have met and solved the problem in various ways. A special IC called a **BCD to decimal decoder** has been developed (fig 8.39). This chip contains logic gates that decode the BCD count and give an output to one of ten possible lines. Each output has already been inverted in a NOT gate, shown by the bar over the output numbers and the circle on the output line. When zero is at the counter output (0000), LED 0 is lit. When 1000 decimal 8 is on the counter, LED number 8 is lit. When the counter is being clocked, the output on the LEDs is a ripple from 0 to 9 repeated over and over, only one LED being lit at any one time. A possible use may be in disco lights or a metronome with a visible display.

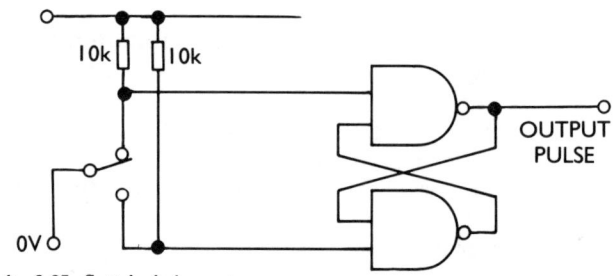

fig 8.35. Switch de-bouncing

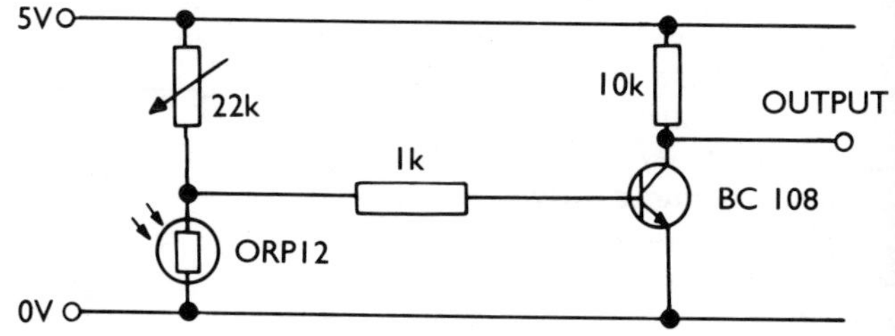

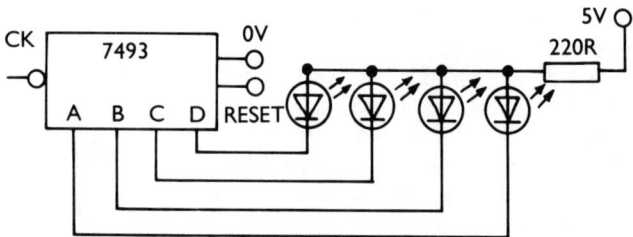

fig 8.36. Light activated input

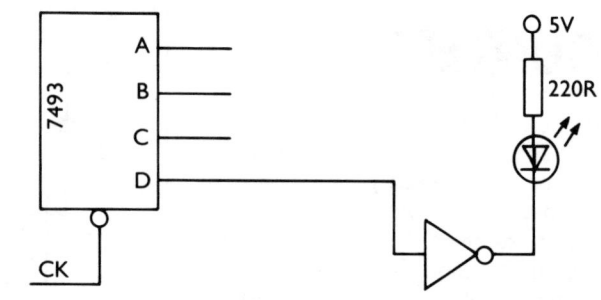

fig 8.37. Binary counter with LED display

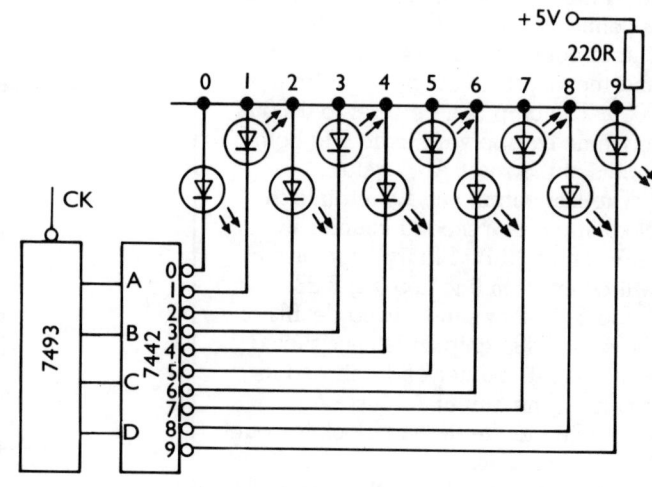

fig 8.38. Display with NOT gate

fig 8.39. BCD to decimal decoder

SEVEN SEGMENT DISPLAY

You are now familiar with seven segment displays (fig 8.40). Digital watches and many electronic displays use them in one form or another. There are seven LEDs in one case, each LED being bar shaped. Each segment is labelled with a lower case letter from a to g and can be switched on as desired, with any number on at one time.

For each binary number from a BCD counter, certain LEDs must be lit to form the numbers (see fig 8.40). Number 1 requires b and c, number 2 requires a, b, d, e and g, but not c and f. It is possible to build logic gates to light the correct bars that are needed for the number required. However, once again it is not a new problem, it has been solved. You do not need to 're-invent the wheel'. An IC, number 7447, called a **BCD to 7 segment decoder** is available, it is produced to do just this job. The output has also been inverted so that it will directly drive an LED, this time shaped like a bar (fig 8.41).

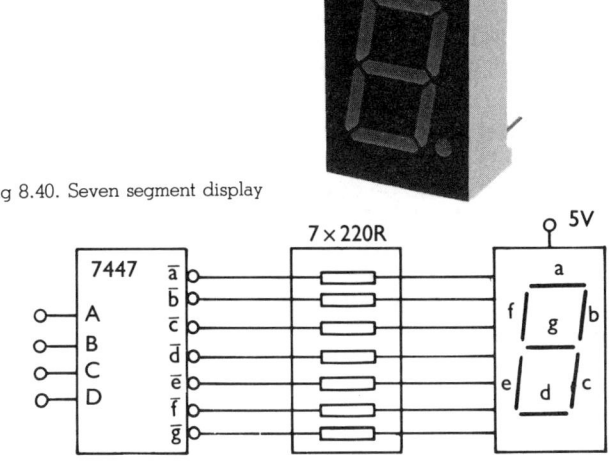

fig 8.40. Seven segment display

fig 8.41. BCD to 7 segment decoder

Project
Problem
A method of counting people entering an art gallery is required. There is a turnstile at the entrance.

Specification
1. It must count people entering the gallery.
2. It must count at or near the turnstile.
3. It must count up to a maximum of 999 people.
4. The system must be electronic so it may be connected at a later stage to the gallery's computer.

fig 8.42. Block diagram

Possible circuits:
Box 1. (a) Light operated switch, people crossing a light beam.
    (b) Pressure pad to walk on.
    (c) Switch connected to the turnstile mechanism.
Box 2. (a) Binary counter, 11 bits required, three 4 bit binary counter chips.
    (b) BCD counter, three required, one per decimal digit.
Box 3. (a) LED display like an abacus showing three columns for the three digits.
    (b) Seven segment displays, three required.
Box 4. (a) Simple press switch.
    (b) Press switched de-bounced.

Selected circuits
1. Switch on turnstile, easy and simple, but will require de-bouncing.
2. Three BCD counters, easier, to decode for human understanding.
3. Three seven segment displays and decoders, easy to read.
4. Simple switch, does not need de-bouncing as resetting does not cause problems if it is reset three or more times with one press.

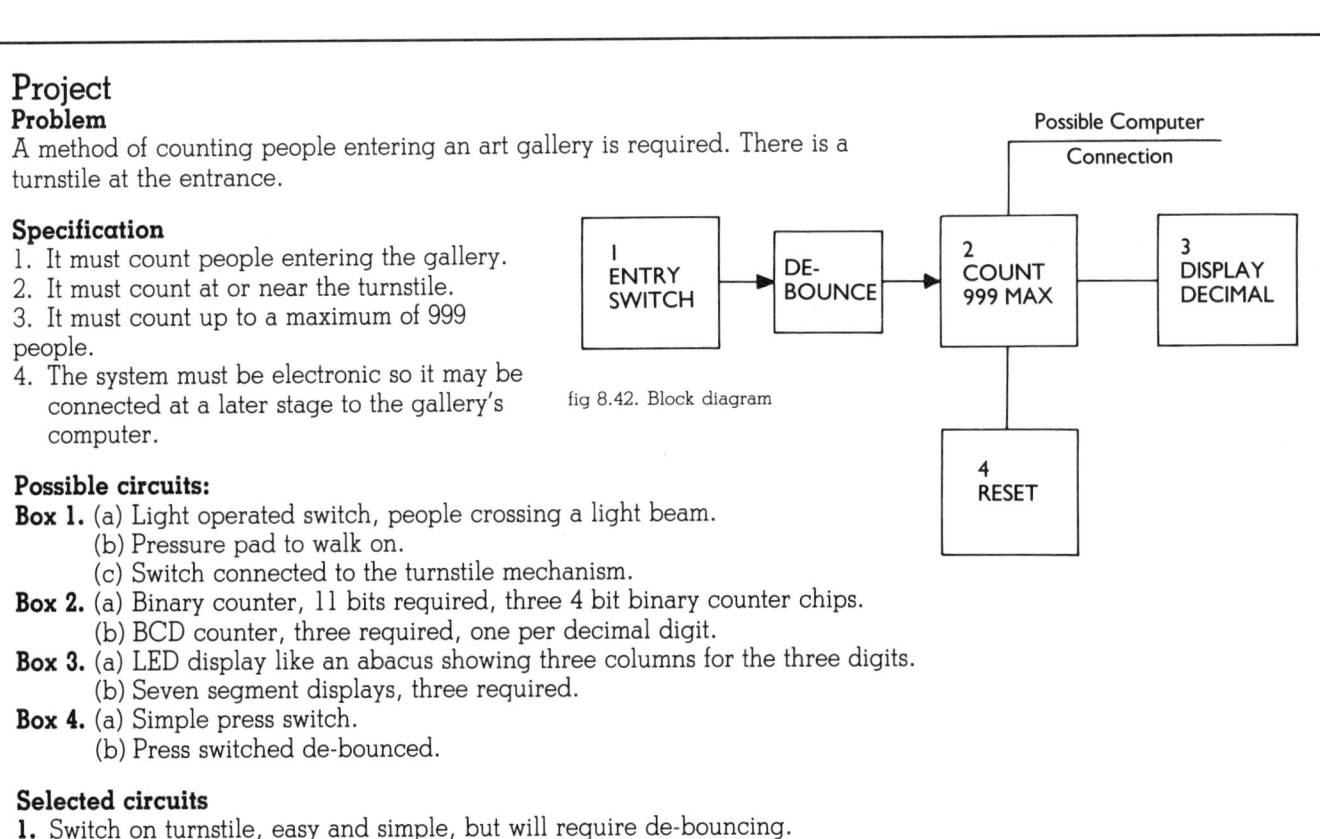

fig 8.43. Circuit diagram-counter 0–999

MEMORY

One very important aspect of logic gates is the fact that they can be assembled in such a way that they can remember logic values even after they have changed. For example, a semi automated vehicle is to be designed that will run on a track between two points. When it reaches one point it will then reverse back to the other point. It will move backwards and forwards until switched off. This is a bistable system (see page 74), with two stable states, i.e. forwards or backwards running of the motor.

On reaching the end of the track, a pulse of information, possibly from a micro switch hitting a stop, will change the state of the bistable and the motor will change its direction of rotation. On reaching the other end, the next pulse of information is received to reverse the motor again. This is called an **SR bistable** (set and reset) and can be built from two NAND gates. It is similar to the de-bouncing circuit on page 98. The output $\overline{Q}$ (NOT Q) is the complement of the output Q, i.e. it is always opposite in value. It is sometimes used, but need not be (fig 8.44).

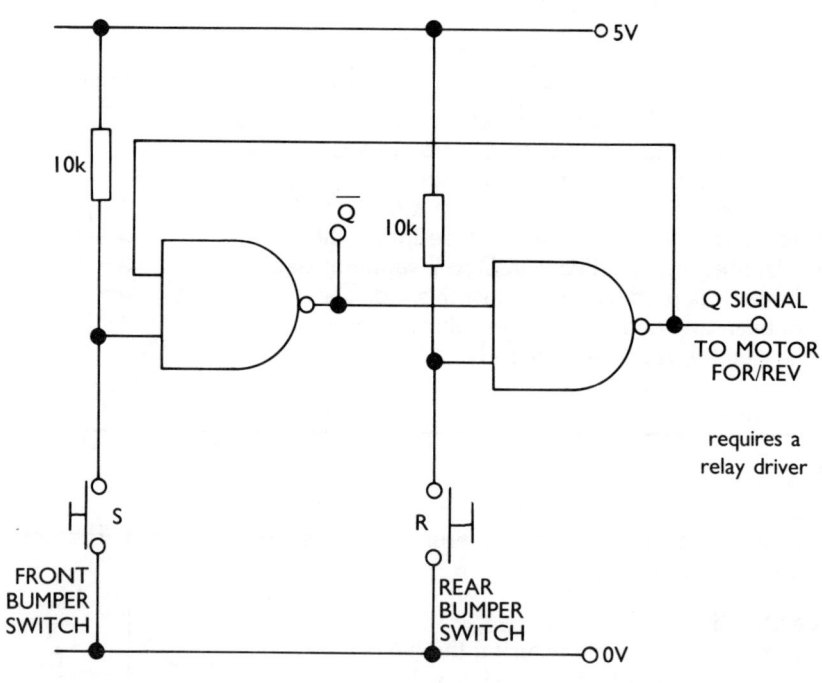

fig 8.44. SR bistable latch

DATA LATCH

When a digital stopwatch is counting in hundredths of a second, the output is a blur, we cannot take in the information quickly enough. If the watch is stopped in the lap mode, the number is displayed, or latched, showing the time when it was enabled (button pressed). The watch is still running in the background counting lapsed time. A second press on the lap time button records the next latched time (fig 8.45). A **data latch** or **D-type bistable** is very similar. It has data inputs and data outputs. The information at the input is only sent to the output when the latch is enabled. The input at the watch can be still changing. The output shows the data at the instance of enabling, and holds this until the next enable signal. Fig 8.46 shows a 7475 chip, a four bit D-type latch. Each line containing information is called a bit. The name comes from the shortened version of **BI**nary digi**T**. The information being in the form of a binary code, each line can be either 5V or 0V. There are 8 outputs on a 7475 chip. Four are the outputs Q and the other four their complements NOT Q. These low active outputs could be used to drive LEDs as a visible latched output.

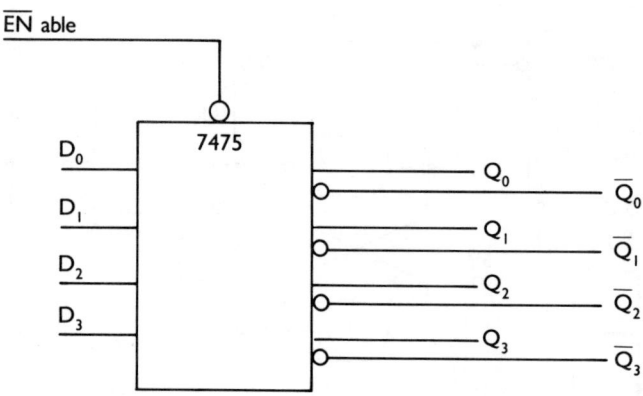

fig 8.46. 7475 4 bit D-type latch

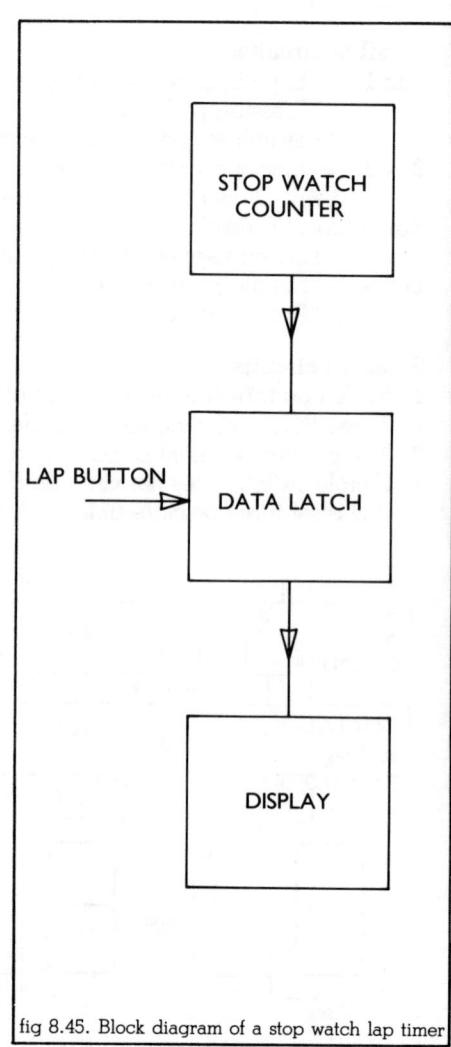

fig 8.45. Block diagram of a stop watch lap timer

DATA REGISTER

A data latch transmits the input to the output at the instance of the enable signal. A data register is a store for information. The input signal is latched into the register, but not out to the output until a signal reaches the output enable. The information, digital signal, can be collected and dispatched to suit the timing of the rest of the circuit. Computers use many thousands of these registers to store information. Fig 8.48 shows chip type 74373, an octal latch with **three state outputs**. This register has outputs that are called three state. The three states are logic 1, logic 0, and a state that appears totally disconnected when not outputting information. These must be used when information is sharing connecting lines, as we shall see in a computer.

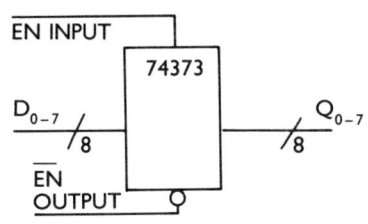

fig 8.48. 74373 octal latch with three state outputs

The temperature collecting system in fig 8.47 is an example of a system using a register. You need to know the temperature at 3am, but do not wish to get up until 8am. A signal from an alarm clock could enable the register to collect the information at 3am. Then, when you get up at 8am you can send a signal to the output enable to extract the information you require. The information could be shown on the display of the alarm clock. The output enable also disables the clock output for the time you require the temperature to be displayed. The clock then continues to count. You have shared an output system with the clock and stored the required information in a register until it was required. If you required the outside and the inside temperatures at 3am, it would be necessary to have two registers. Each output in turn could be enabled at your leisure. The wind speed and direction could be stored on another two registers. This is the basis of a weather logging station.

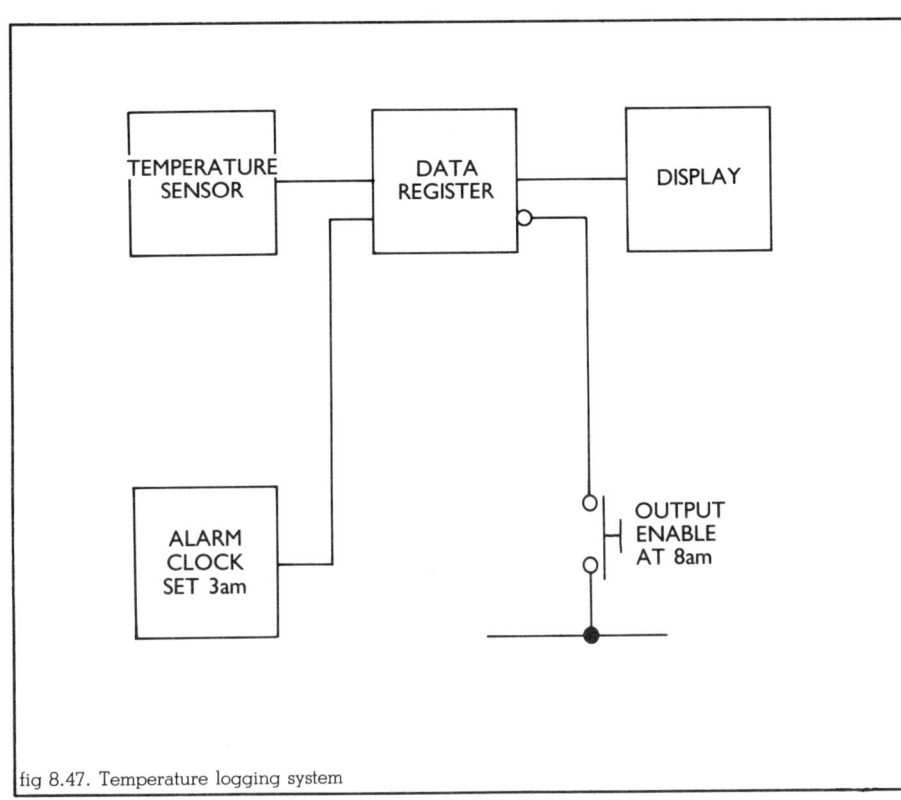

fig 8.47. Temperature logging system

Information transmission

To send logic signals over a distance of less than a metre should cause no problems if the frequency of the signal is 100kHz or less. Some chips and some computers work at speeds in excess of 12MHz, twelve million operations each second. If you could count once a second, it would take you 138 days or 4 and a half months to do 12 million counts! Some computers do this every second. At this speed, the signal is being sent as fast as the electrons and they begin to be confused in only a matter of a few centimetres of wire.

To send digital signals over long distances special chips are needed at each end of the line. **Drivers** are used at the start and **receivers** at the other end. The information is then boosted along the cable and amplified back to normal at the other end. Some voltage is lost along the length of the cable, as all cables are resistive components.

The **open collector gates** have a larger capacity to sink a current, up to 50mA in some cases. They are used as transmitting gates. The **Schmitt trigger** (fig 8.49) takes a weakened signal and boosts it to the proper logic voltage levels. Never attempt to drive long lines directly from bistables or registers without an open collector driver, as the signal becomes very erratic.

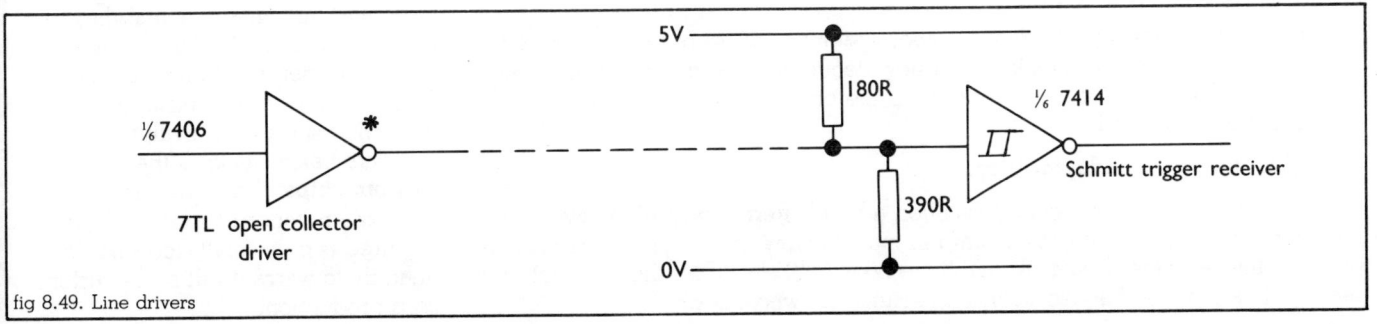

fig 8.49. Line drivers

DIGITAL CONTROL

Washing machines are machines that follow a set sequence of operations. The control system used to be a mechanical electrical system, a very complicated switching device that was very expensive to repair and replace. The washer follows a pattern, turning on taps and motors at set intervals. A simplified set of instructions is set out in fig 8.50. Each event is timed, some taking longer than others. The switch pattern of the taps and motors is set out as a truth table using binary to represent the on and off states required.

Time	Instruction	Empty pump	Water tap	Spin	Wash	Heater
0 min 1 2	Fill with water	0	1	0	0	1
3 4 5 6 7	Heat water	0	0	0	0	1
8 9 10 11 12	Wash	0	0	0	1	0
13 14	Empty and wash	1	0	0	1	0
15	Spin	1	0	1	0	0
16 17 18	Fill with water	0	1	0	1	0
19 20 21	Rinse wash	0	0	0	1	0
22 23	Spin empty	1	0	1	0	0
24 25 26	Fill with water	0	1	0	1	0
27 28 29	Final spin	1	0	1	0	0
30	Stop	0	0	0	0	0

fig 8.50. Washing program

The shortest timed event is 1 minute for the first spin. The cycle is broken into 30 different minutes sized events. The information for each event can be stored in a register as we have just seen. The registers are then enabled in the correct sequence, giving a **digital control system.**

The data is placed into the registers before use. This is called **programming.** If many thousands of washing machines are to be made, a special memory chip with this information set into the registers can be made for a few pence, not pounds as the earlier mechanical switch.

Why digital? .....**Low Cost.**

Digital systems are replacing many other systems because of their low cost and ease of changing the specifications. If I now decide that I require a spin of only 2 minutes at the end, all I do is re-program the chip, all the other control equipment is untouched.

Why digital? .....**Adaptability.**

This type of register or memory is called a ROM (**Read Only Memory**). The program, or data, is built into the chip in manufacture. It cannot be changed to provide different sets of data. A telephone book is like a ROM, the information is read, but cannot be changed without rewriting the whole book.

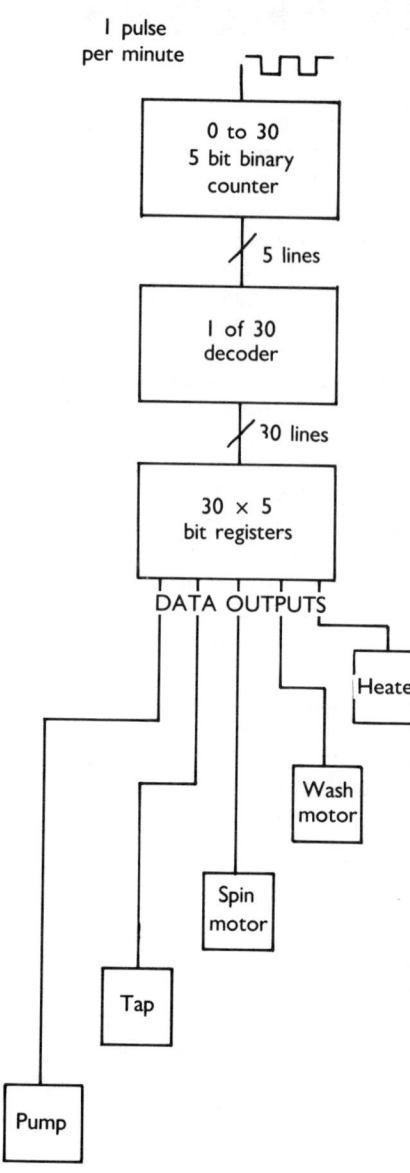

fig 8.50a. Clock set on a 1 minute pulse

Having made the ROM chip for the washing machine it now requires a clock set to a 1 minute pulse. Each register has to be enabled in the correct order. If they are numbered 0 to 30, a 1 of 30 decoder is required. Each register can be enabled as the counter counts and is then decoded as an enable line. Each one of the data lines is connected to the pumps, heaters and motors through an amplifying interface.

Chips called EPROMS (**Erasable Programmable Read Only Memories**) are available. They are more expensive than special ROMS. They can do the same job of holding data, their advantage is that they can be re-programmed. The data is wiped clean and then programmed again. This type of chip is used in the modelling stage, and evaluation of the program before making the special or `custom chips`. They are also used when the number of special chips required is only small, too small a quantity to warrant full scale custom chip production.

MICROPROCESSORS AND COMPUTER CONTROL

A microprocessor is a special electronic device that acts as an intelligent digital circuit. It can remember set sequences of events and repeat them on command. The washing machine cycle on page 102 is an example of a sequence of events required by the machine to clean clothes. The electronic answer is a specific collection of gates wired together to produce the required events. To change the sequence the device must be rebuilt using a different collection of gates. The microprocessor is a chip that is changeable. Events can be predetermined in a 'program' and set to execute the required sequence of events. Changing the 'program' can change the way the sequence is delivered. These chips are therefore very versatile.

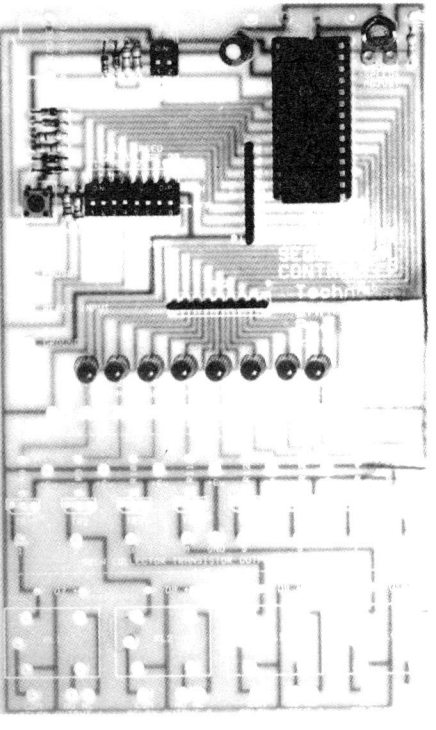

SYSTEM BOARD CONTROLLER

The Technology Enhancement Program has developed an example of a system board controller called a 'bit-by-bit controller' (Fig 8.51). This makes use of a microprocessor chip and simple switches to control its program. The system is capable of remembering 64 steps and can control 8 different output switches. The washing machine cycle in fig 8.50 requires 30 steps and needs to control 5 switches. The 'bit-by-bit controller', therefore, is capable of programming this sequence, as shown below.

fig 8.51. Bit-by-bit contoller

When the control program sequence has been programmed into the chip's memory it will retain the information as long as the battery is connected. To run the program set the run switch to 'RUN'.

The sequence will be repeated forever, or at least until the battery is dead or you switch it off.

The same controller could be reprogrammed to run a sequence of traffic lights without rebuilding electronically. System board controllers can therefore be manufactured identically, but can be programmed to control different sequences. It is therefore possible to make these devices very cheaply. System board controllers will control many different open loop sequences (see page 34).

Control program for a washing machine
1 Set run and program switches to "PROG OFF" and "STOP"
2 Connect battery
3 Set program switch to "PROG ON"
4 Set data switches to first line of sequence 0 1 0 0 1 (switches 6,7,8 will be set at 0 for all the sequences)
5 Press memory button 3 times (for the first 3 steps)
6 Set data switches to 0 0 0 0 1
7 Press memory button 5 times for the next sequence
8 repeat all the lines of the sequence until line 30 has been entered.
9 Set program switch to "PROG OFF"

PROGRAMMABLE LOGIC CONTROLLERS (PLCs)

fig 8.52. PIC programmer

The system board controller uses a PIC (Programmable Interface Controller) as the control chip in a simple circuit. PICs can be used in more complex circuits. The next stage of sophistication in control is to have a PIC that can have control inputs as well as outputs. The device can be programmed by a computer and then left to run on its own. These chips have almost entirely replaced all of the logic gates that we have been using. They can control devices as complex as an automated car spray painting booth, automatically checking the car model and colour of paint to be used. They can be built into touch screen pay points such as in some fast food outlets. All these programmable circuits will however only deliver a small voltage and current at their outputs. These are only signal voltages. To operate a motor, a door latch, a light or a spray gun, further heavy duty switches and solenoids must be used (these are called 'interfaces').

SIMPLE INTERFACES

It is possible to build simple interfaces using only a few basic components. The basis of the circuits is the transistor, used singly or as a darlington pair. The choice of circuit to be used will depend on the current and voltage that needs to be driven. It is for the designer to choose from the data available.

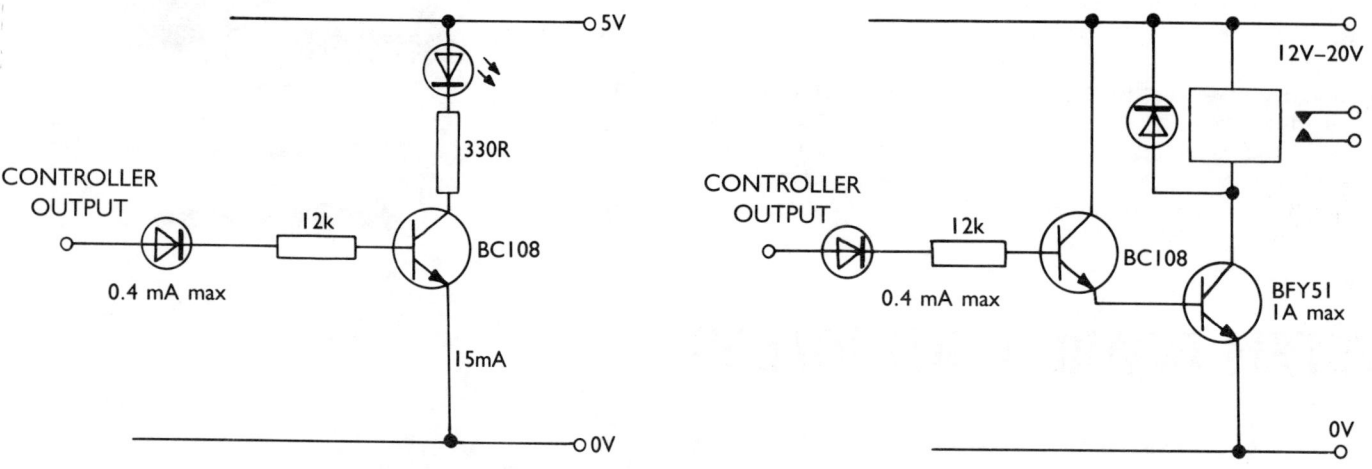

fig 8.53. Driving an LED at 15 mA

fig 8.54. Driving a relay with a darlington pair

A special chip with 8 darlington drivers built into one 18 pin integrated circuit is available (ULN 2803A). It also includes the diodes required for driving inductive loads such as motors and relays. This chip makes interfacing very simple, allowing you to drive motors, lights, relays and many other devices.

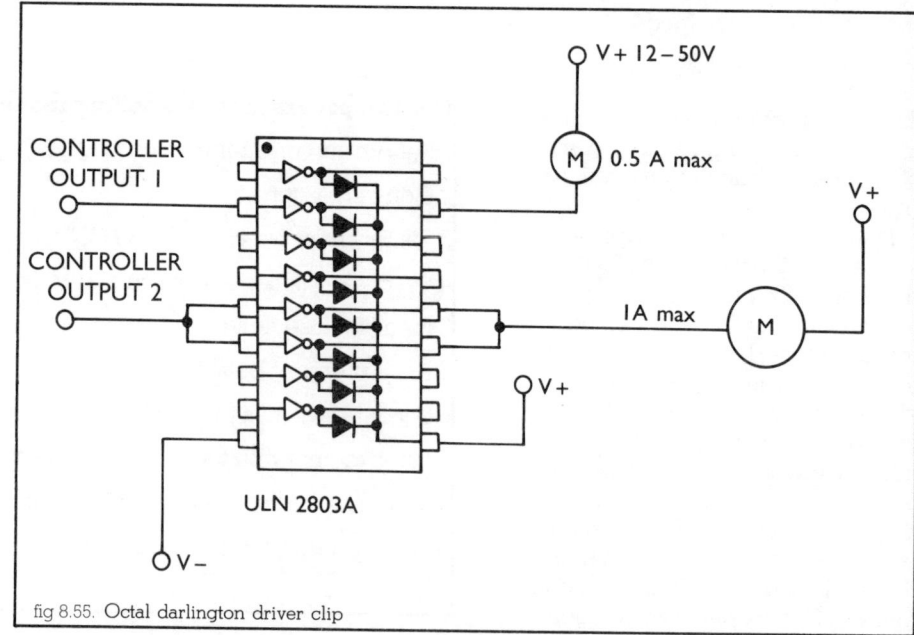

fig 8.55. Octal darlington driver clip

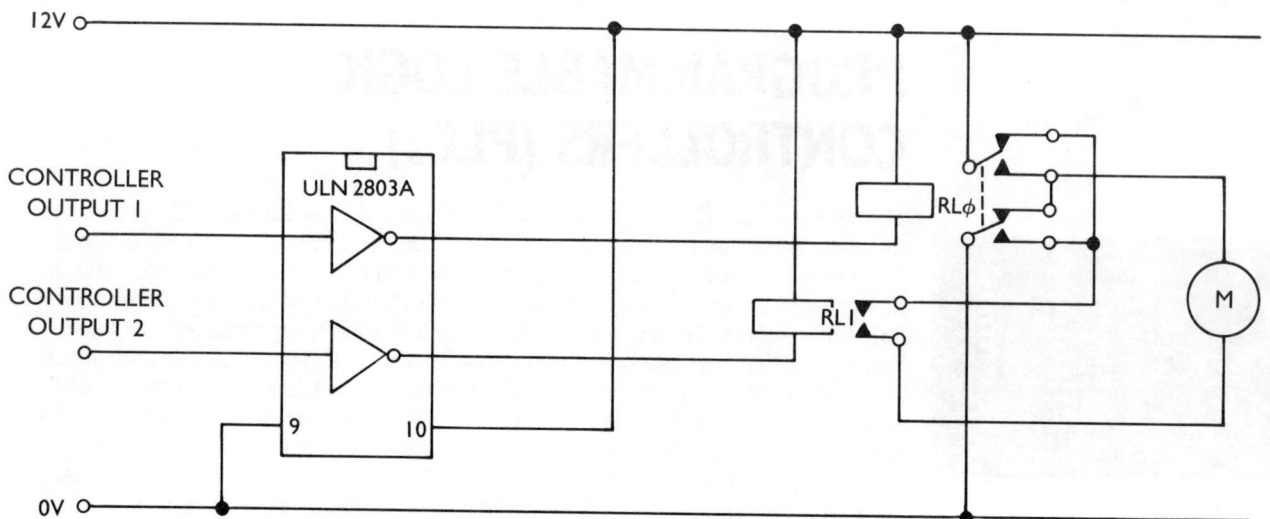

fig 8.56. Motor control RLØ gives directions, RL1 gives ON/OFF

MOTOR CONTROL

A normal dc motor can be switched on and off and even reversed by a computer driving an interface as shown on page 104. To drive a vehicle or 'buggy', you would use two motors (fig 8.58) to drive each wheel. A third wheel or castor is used to balance the buggy. The two motors can be controlled independently to give directional movement to the vehicle. This could be both forward, both backward, or one in each direction giving a spin turn.

fig 8.57. A stepper motor

Unfortunately it is not possible to control exactly how far each motor turns. The motors will vary in power turning one way or the other and vary one with another. To move an exact distance the number of turns of each motor or wheel must be counted and this information fed back into the computer. This information feedback loop will be looked at in more detail on page 113.

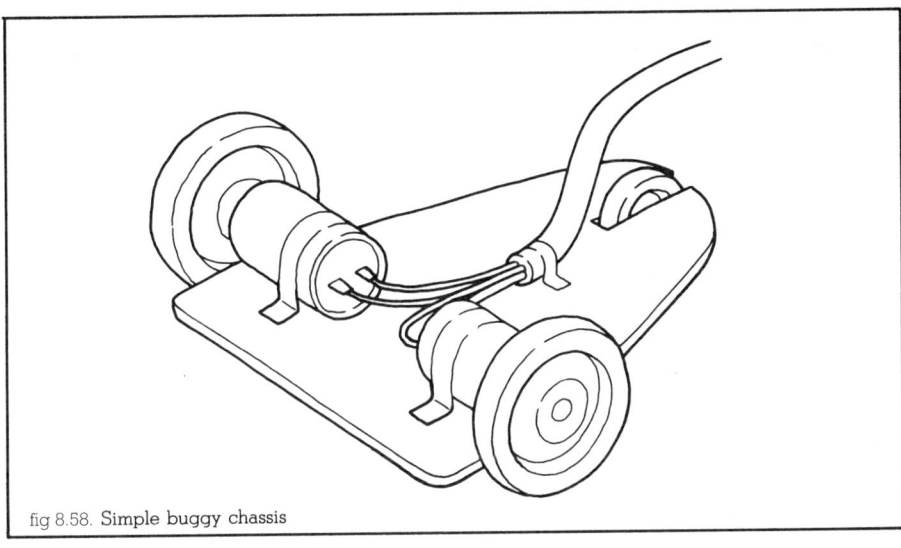

fig 8.58. Simple buggy chassis

STEPPER MOTORS

Stepper motors were designed specifically to ease the problem of accurate control of movement. There is no need for a feedback loop to monitor movement. The motor moves in a series of steps each turning the rotor by 7.5°. They are very useful for movement when accuracy is required in a vehicle or robotic device. However, there is a problem when switching on: they cannot be switched on as a dc motor, but instead must be turned on in a series of pulses to the four coils that make up the motor.

The **rotor** in a stepper motor is constructed from several permanent magnets with north and south poles. The **stator**, the fixed part, is wound into a series of electromagnets, usually four, which can be switched on and off. The two activated electromagnets in fig 8.59 hold the rotor rigidly in the position shown. In fig 8.60 the original windings have been switched off and the other pair activated. This pulls the rotor round by one step, and holds it rigid in this new position. For clarity the diagrams show only two pairs of coils, not the usual four.

fig 8.59. Two activated electromagnets of stepper motor rotor

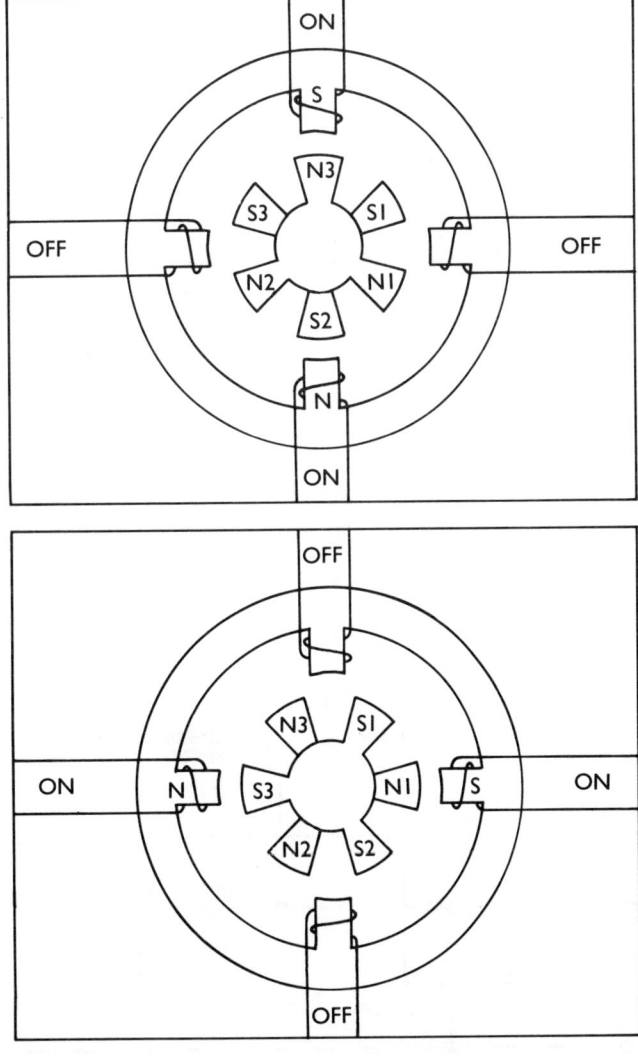

fig 8.60. Other pair of electromagnets activated

Stepper motors provide the power for most low cost buggies and robots and are used in X-Y plotters to give accuracies of up to 0.01mm. However, these very useful motors do have limitations. First, the power output is greatest when stopped, which is not very helpful when lifting a weight. Their power gets less the faster they step. Most small educational robots are limited to lifting a small load, often only 100 g. Second, the speed of stepping is limited to about 100 steps per second, which is about 2 revs per second or 120 rpm. If they are stepped too quickly or under too great a load, they will miss steps and the advantages of accurate control are immediately lost, and with no feedback the computer will not realise the problem. The drive from a stepper motor can be geared down, loosing speed, but gaining torque or power. The BBC buggy uses stepper motors geared at 7.5:1, making a single pulse turn the drive wheels exactly one degree.

fig 8.61. BBC buggy

A special chip that sends the current to the correct coil in the correct order is used to control these motors. Fig 8.62 shows the sequence of pulses required. The four coils can be driven by four parts of an octal darlington driver. The computer would be required to turn on the coils in the required order (see fig 8.64

Using the **motor driver chip**, SAA 1027, the sequence of stepping is inbuilt. All that is required is to send a pulse down a single line to step the motor round. If the motor is required to turn in both directions, another line is needed to give the required direction. Fig 8.65 shows the circuit details.

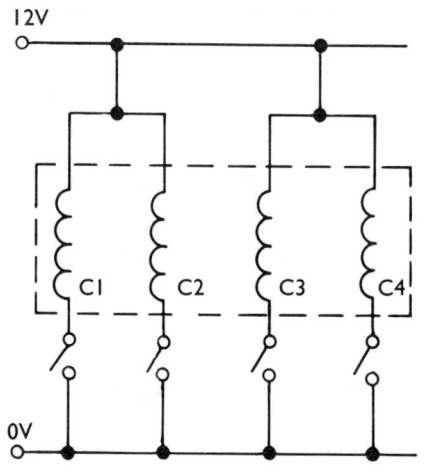

fig 8.63. Stepper motor controlled by individual switches

	C1	C2	C3	C4
Step 1	1	0	1	0
Step 2	0	1	1	0
Step 3	0	1	0	1
Step 4	1	0	0	1

fig 8.62. Truth table for stepper motor sequence

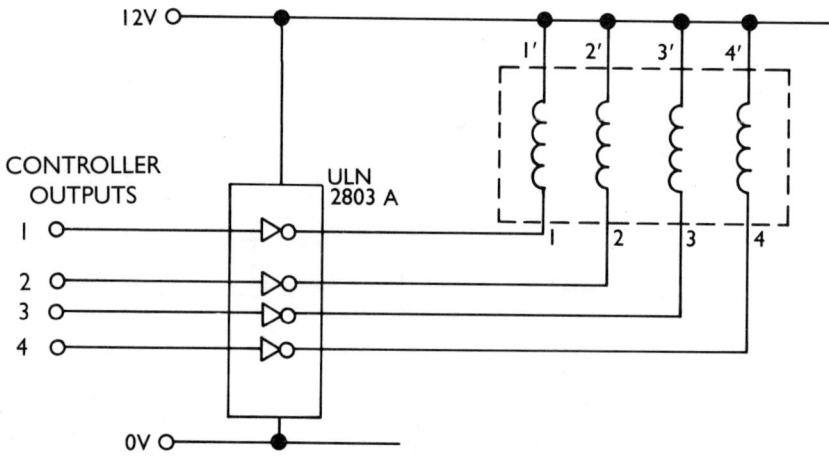

fig 8.64. Direct control of stepper motor

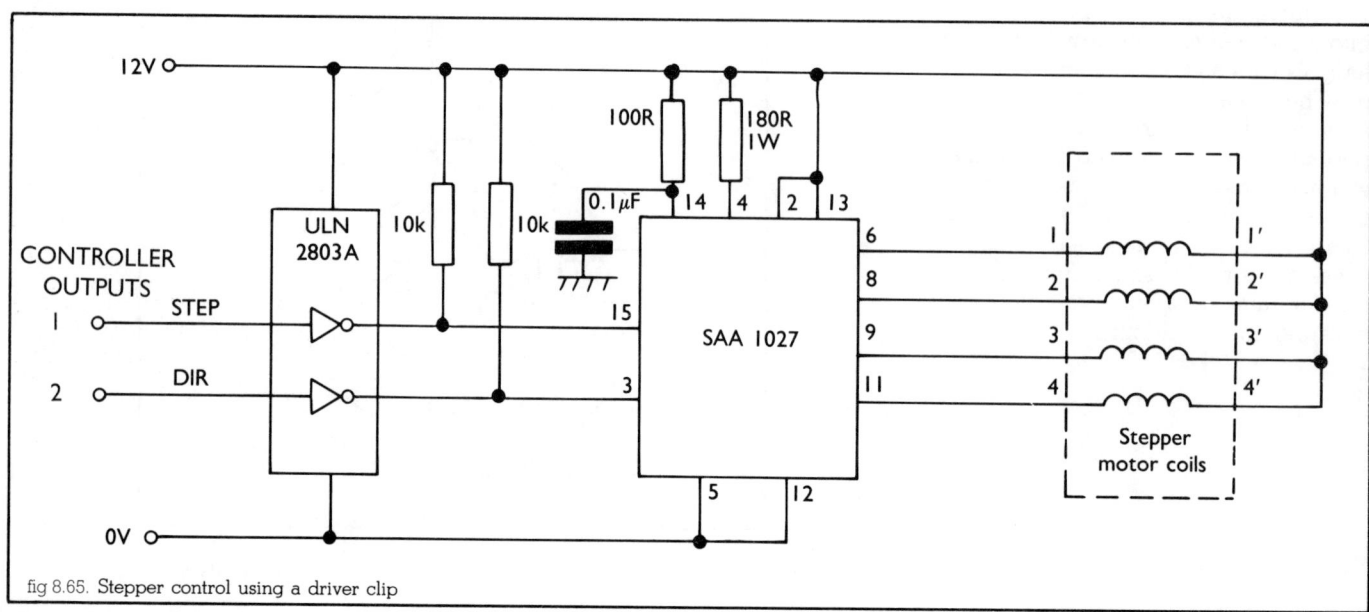

fig 8.65. Stepper control using a driver clip

106

DIGITAL COMPUTERS

A digital control system such as the washing machine program is a very good way to organise a set series of happenings or program. The information is set into the memory when manufactured and then used in the machine time and time again. The computer takes this system a few steps further. The data can be programmed at will into its memory and operated upon as a washing machine program. It can also run a program to operate a sewing machine, a welding machine or a word processing machine. The computer is a device with memory connected to inputs and outputs. The great advantage of the computer is that it is versatile. It is a **non specific digital controller**. It can be programmed to control many different devices. All computers are built with the same basic parts. 'Larger' computers have more memory, they can run larger programs or hold more data.

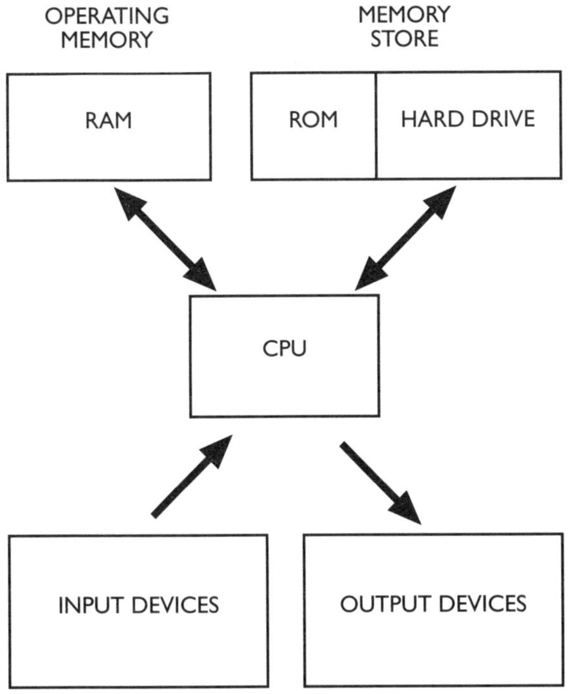

```
OPERATING          MEMORY
MEMORY              STORE

  RAM          ROM    HARD DRIVE

              CPU

 INPUT DEVICES    OUTPUT DEVICES
```

fig 8.66. Block diagram of a computer system

PICs

PICs (Programmable Interface Controllers) are used to control many different industrial processes and pieces of electrical equipment. They are complete computers on a single small board. They contain a processing unit, an operating instruction set, and a store for the instructions. They are programmed using a larger computer and then they are set to run completely independently. They can be programmed to follow, for example, a set of instructions like the washing machine program on page 102, or any other similar instruction set. They are also relatively cheap.

Central processing unit (CPU)
The CPU is the brain of the computer (fig 8.66). It organises the sequence of events and adds and subtracts numbers. It has a selection of registers for temporary storage of data. The CPU is run by the clock, running at 200MHz in a Pentium processor. It is connected to all the other parts of the computer by wires called a bus. The information is sent along this **communications bus** to all other parts and returns with more information along the same track.

Memory
The computer memory consists of various types. Operating memory is only used whilst the computer is switched on, while stored memory can hold information when the computer is switched off.

Operating memory (RAM [Random Access Memory]) is the store that can be both read from and written to. It can be changed in a fraction of a second from representing one bit of information to representing another, different bit of information. The size of this memory store is the restricting factor in manipulating huge amounts of information very quickly. The problem with this type of store is that when the computer is switched off the data is lost forever from this area. The data must be transferred to a permanent data store before switching off. When a program is loaded into a computer it is placed in the RAM store while it is in use. The text for this book was written on a word processor, and was stored in RAM as digital data while it was being written and manipulated.

ROM, as we have seen, can only have its data set during manufacture and never changed. This is the memory holding the computer's operating systems.

The hard drive stores all the other information. This includes the data and the programs that are required, such as letters, databases, word processors or control sequences.

fig 8.67. A computer system

COMPUTER INPUTS

KEYBOARD

The keyboard is the **general purpose interface** between the operator and the computer. The keys are connected to logic circuitry that changes the finger press to a special code. This code is given the name **ASCII** (American Standard Code for Information Interchange). There are 256 different codes representing numbers, letters, upper and lower case, punctuation marks and various other codes. The code is fed into the computer where it can be operated upon by the computer language and the program.

fig 8.68. **Key pad**

MAGNETIC DISC

If the data required to run a program or the information upon which a program operates is held in the RAM, it must be stored in a permanent form, otherwise it would be lost every time the computer is switched off. To rewrite a list of data every time you wish to use it in the computer is pointless. To store data it is sent to one of a number of devices. The most common device is the hard disc, which can store large amounts of data (programs and all the associated files that you have produced). Information can be sent at high speed between the computer processor and the hard disc whenever it is required. A hard disc is both an **input device** and an **output device**. To move information to another computer a smaller version is used, called a floppy disc. All computers need a floppy drive, or similar, to allow data and programs stored on floppy discs to be loaded.

fig 8.69. **Keyboard**

fig 8.70. Disc drive and floppy disc

MAGNETIC TAPE

Cassette tapes can store information from a computer. They are a cost-effective way of storing information that is required infrequently. As with an audio tape you cannot jump quickly to any place on the tape, it must run through to the correct spot. Tapes are used to store backup data. Most computer systems in the workplace store invaluable information on tape. Each day a tape is run to store all the information that has been collected that day. If a problem arises later the tape can be replayed to find the stored data. Speed is not important.

CDs

Compact discs are familiar to us as a storage system for music. They can also store vast quantities of data. A fifteen-volume encyclopaedia could be stored on a single disc. Pictures, text, video, games, programs, music and speech can all be stored. CDs are an extremely effective way of storing large data files in a small package.

fig 8.71. A DAT tape

fig 8.72. CDs can store images, text, sound and programs

JOYSTICK

A device looking a little like an aeroplane joystick is used to input information to the computer. It sends positional information, used in games. Joysticks may have a knob grip design as shown here (fig 8.73) or a grip handle (see page 41).

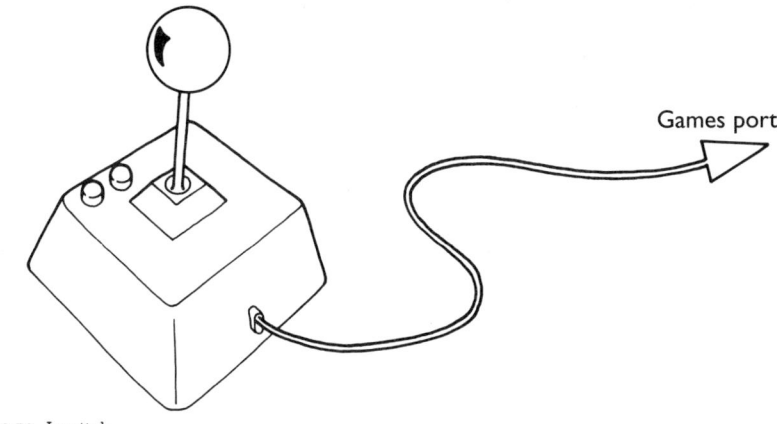

Games port

fig 8.73. Joystick

MOUSE

The ball underneath the mouse box is rolled on a table top. These devices are used to move the cursor around the screen. Drawing programs use them. It is possible to write on the screen as though you were writing or sketching a picture on a piece of paper. There are three buttons that are also to be used to input information from the mouse.

TRACKER BALL

A tracker ball is used to input information rather like a mouse, but the ball is on the top. The fingers are used to rotate the ball. Positional control can be closely monitored and adjusted with this device.

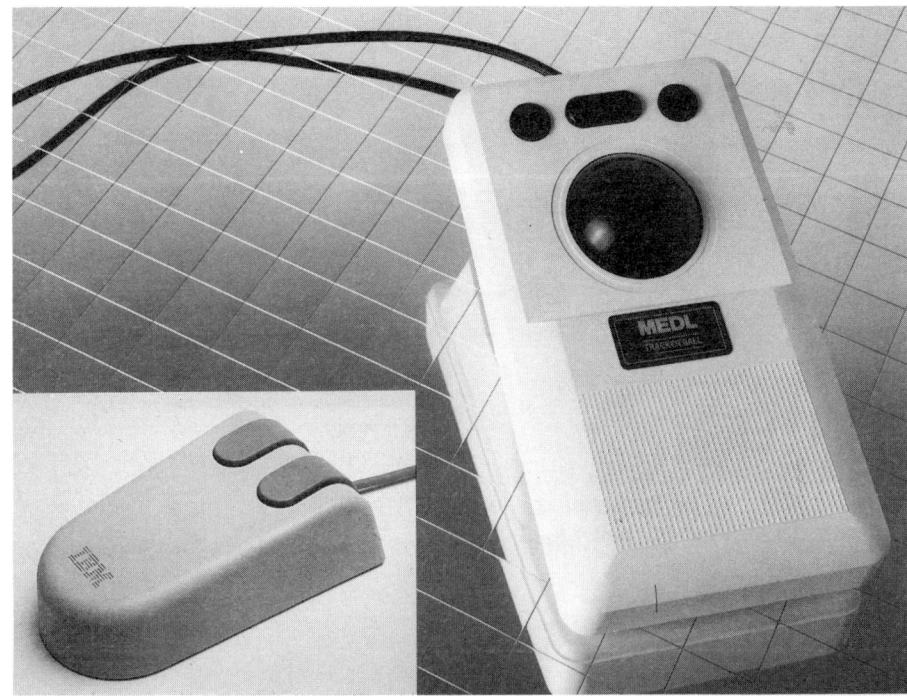

fig 8.74. **Mouse**   fig 8.75. **Tracker ball**

GRAPHICS TABLET

A pencil like probe, or a puck with a crossed hair targetting device is used on a special drawing area to send positional control signals to the computer. The advantage of this input can be seen when a drawing is to be traced into the computer. It is easier to trace a map of the British Isles than to draw it freehand.

SPECIAL INPUT DEVICES

The computer can be connected to any manner of digital and analogue devices that can send information to be processed. Sensors of temperature, pressure, light, dark, wet, dry or position can be connected. These will be investigated later in this chapter.

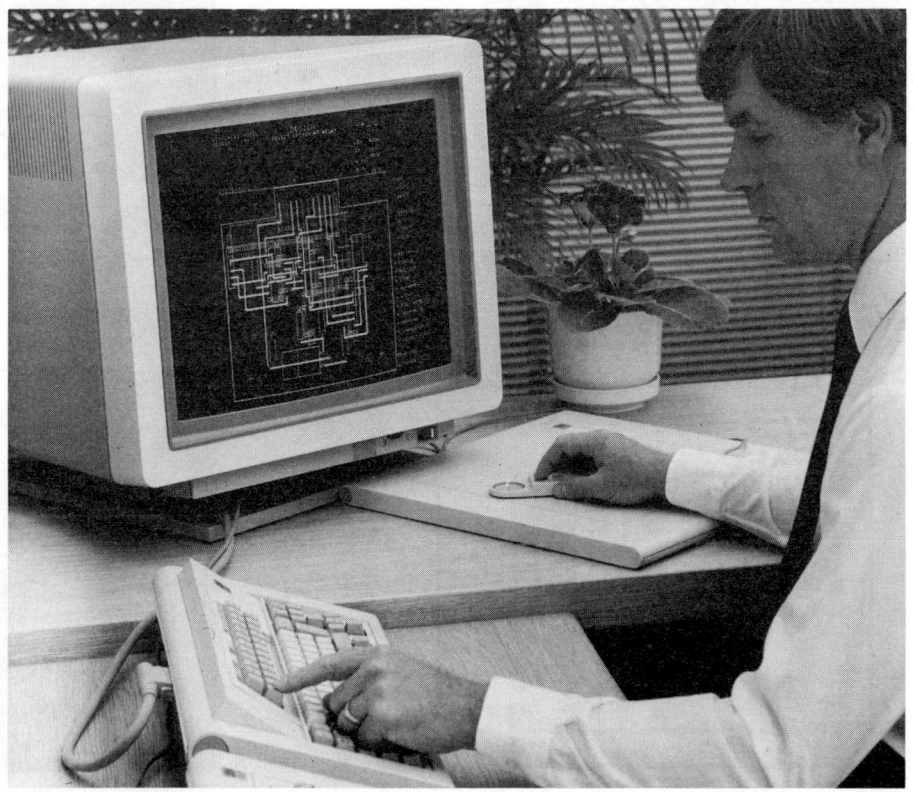

fig 8.76. **Graphics tablet**

COMPUTER OUTPUTS

VISUAL DISPLAY UNIT

The computer can send information to many different devices. The most common device is called a visual display unit or VDU. This looks like a television screen. The information sent to the screen is in the form of a code turning on a coloured spot at a predetermined place on the screen. The screen is divided up into small areas, or spots, called pixels. A modern small VDU screen is divided into 480,000 pixels (800 pixels across, 600 high) each of which can be any one of 16.8 million different colours. Letter shapes, drawings and photographs are all displayed using these coloured pixels.

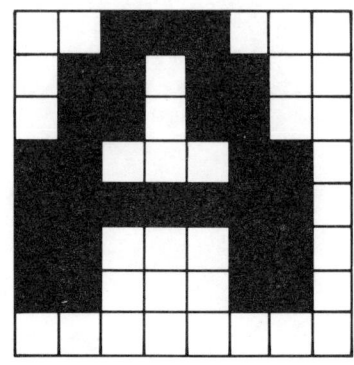

fig 8.77. Letter A showing 8 × 8 grid

DISC AND TAPE

These can be used to receive information from the computer. It can then be stored until required. These give a permanent method of storing data.

fig 8.78. **VDU**

PRINTER

The data displayed on screen can be made into 'hard copy' by sending it to a printer. There are two main types of printer in use.

Laser printers produce text and graphics of the highest quality. They are usually black printers but coloured laser printers are available at a much higher cost. Laser printers are similar in operation to photocopiers and can print quickly 20 pages a minute are easily possible.

Ink jet printers, as the name suggests, squirt a jet of ink at the paper. They use coloured inks to produce colour pictures and text. Their operating speeds are considerably slower than laser printers, in the region of 2 pages a minute, but their initial cost is lower.

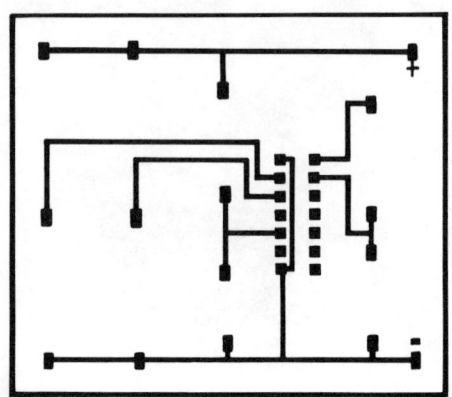

fig 8.79. **PCB design**

fig 8.80. A laser printer

PLOTTERS

Plotters are used to produce drawings and charts of very high quality. Different colours and types of pens can be used and can plot on paper, card, acetate sheet, drafting film or directly onto copper clad board to produce PCB masks.

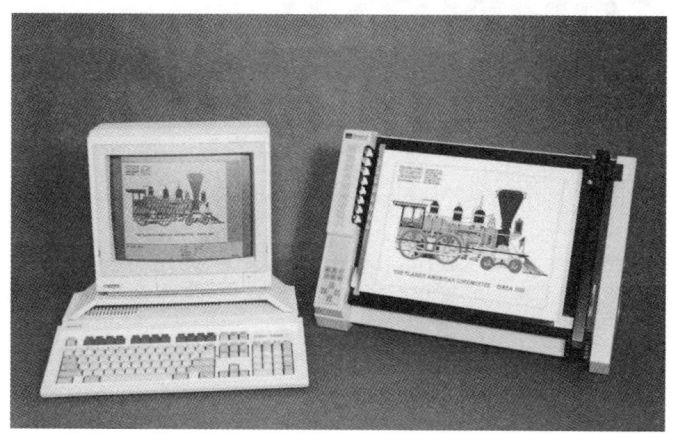

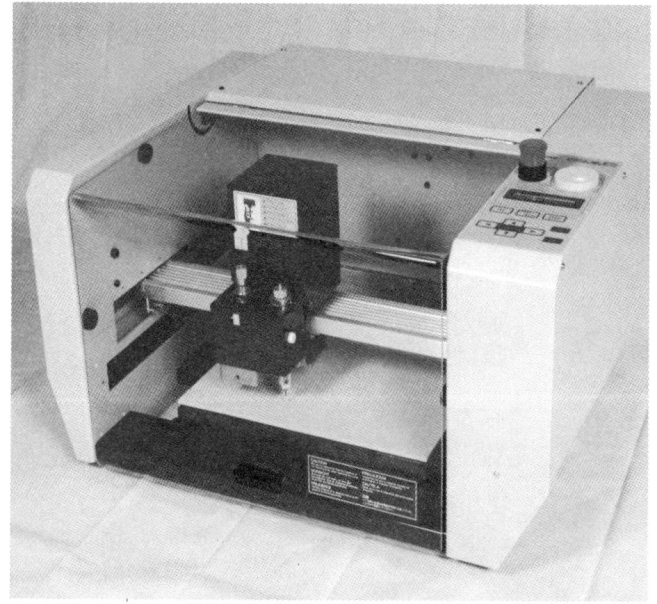

fig 8.82. Computer-controlled cutting and engraving machine

fig 8.81. Flat bed plotter (top) , Roller bed cutter and plotter (bottom)

MACHINES

Robots are machines that can be controlled by a computer. The early idea of robots looking like humans is now seen only in science fiction films. Robots today are very powerful manipulating machines that can be used in many industrial situations. Automatic welding machines and handling devices are commonplace in many manufacturing industries.

Many machines are now controlled by computer, from washing machines to machine tools such as a lathe or milling machine. **Computer Numeric Controlled** (CNC) lathes are now to be found in school workshops. Fig 8.83 shows an example of a CNC lathe.

fig 8.83. **CNC lathe**

COMPUTER CONTROL

The computer, as you have seen, can send information to various devices. We are now going to look in more detail at some of the ways and means of controlling machines using a computer. The computer has various ways of connecting to devices such as a mouse, printer, VDU. They are all connected to various 'ports' at the back of the computer. The one we use for control is called the serial port, similar to the one used to connect the mouse. This port can send and receive information. It is also used to connect to a modem, allowing the machine to talk to other computers on a telephone line, to send Email and to connect to the World Wide Web as part of the Internet.

Computers can only send and receive signals that are quite small. Such signals would not in themselves be able to directly turn on a light or a motor. An **interface** is required. An interface is an electronic device that can turn a small signal into a more powerful electrical current that can directly turn on lights, relays and motors.

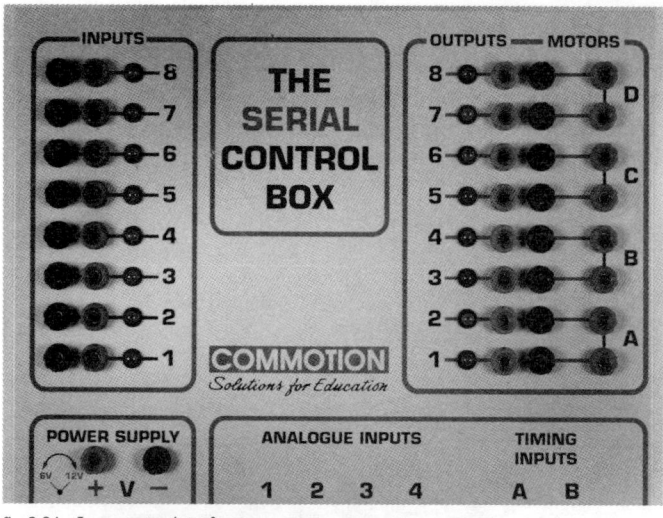

fig 8.84. A computer interface

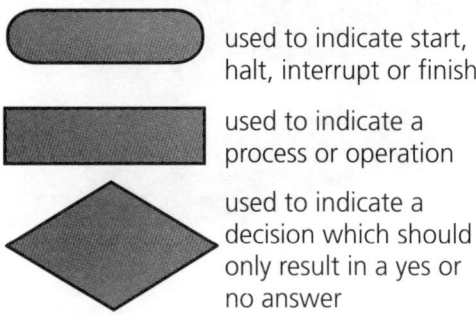

used to indicate start, halt, interrupt or finish

used to indicate a process or operation

used to indicate a decision which should only result in a yes or no answer

fig 8.85. Flowchart symbols

CONTROL PROGRAMS

The computer talks to an interface through the user selecting a suitable program. There are many programs available, for example word processing programs for writing letters, and database programs for collecting information. Before writing the program details it is first necessary to organise the control requirements into a simple format. Flowcharts are used to plan the sequence of events and commands. They use symbols similar to those in the block diagrams used when designing electronic circuits. The most common symbols are shown in fig 8.85.

A flow chart for a simple pelican crossing sequence is shown in fig 8.86.

When the program has been written, the commands will be carried out in the sequence that has been set. In industrial situations this computer-controlled interface is used to model the system. Changes can easily be made to the program on screen and tested until the instructions to the machine system are correct. At this point it is usual to send a set of coded information instructions to a PIC which will take over the process in the real system. Using a PIC costing a few pounds is much cheaper than using a computer, monitor system to repeat a set of control instructions.

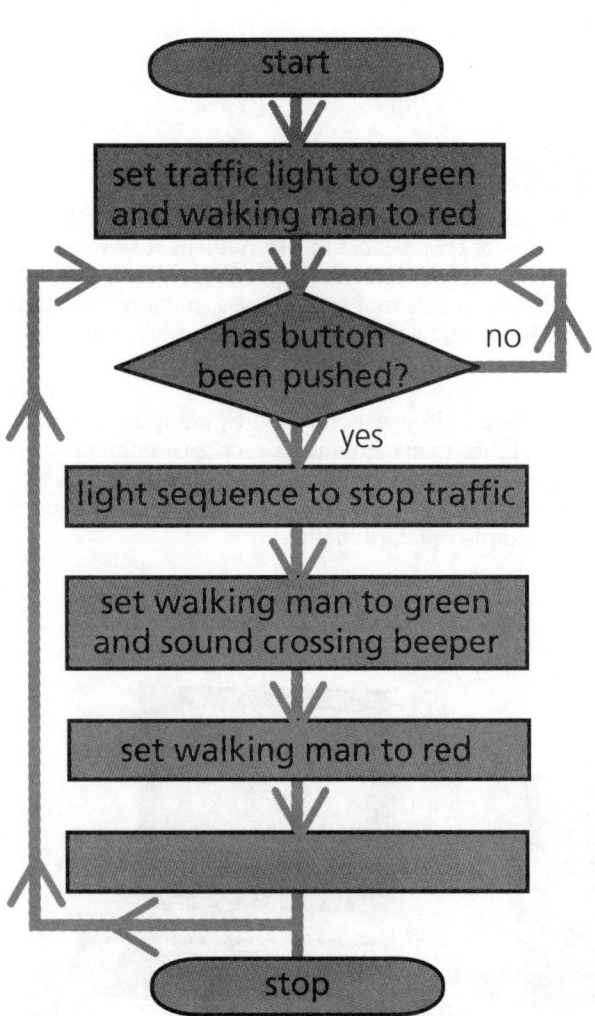

fig 8.86. A flowchart for a pelican crossing

CONTROL INPUT

Computers can not only send signals to devices, but also can receive and process incoming information. An interface can be used for these incoming signals. Information from digital devices can only be on or off. The interface recognises this information as 0 volts and a higher voltage, often 5 volts (this could also be 12 volts). The simple sensors in the diagrams below would send signals of 0V or 5V to the interface.

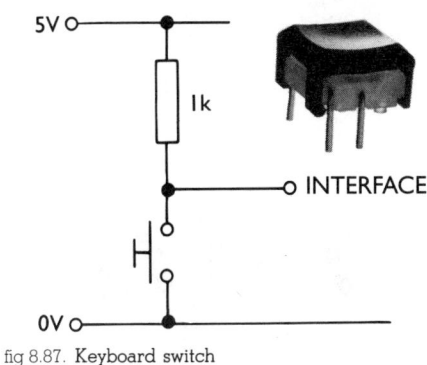

fig 8.87. Keyboard switch

SIMPLE SENSORS

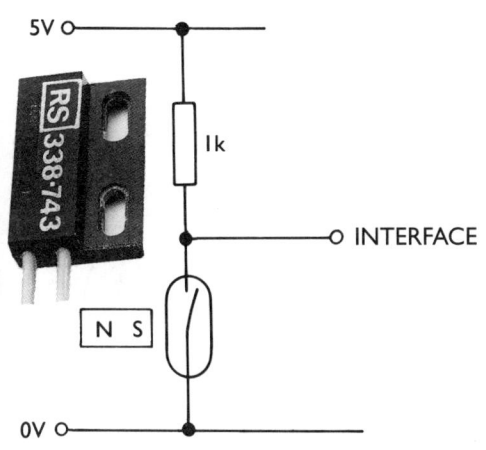

fig 8.89. Reed switch

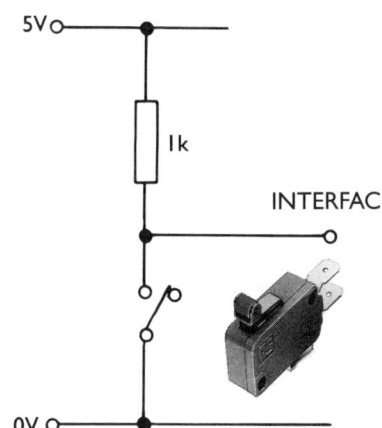

fig 8.90. Micro switch

fig 8.88. Float switch

fig 8.91. Mercury positional sensor

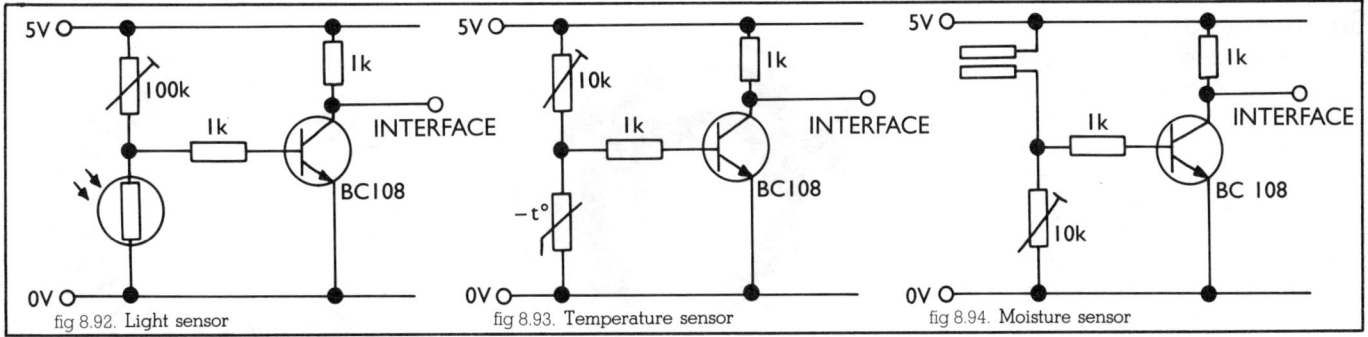

fig 8.92. Light sensor

fig 8.93. Temperature sensor

fig 8.94. Moisture sensor

SCHMITT TRIGGER GATE

A Schmitt inverter can be used to clean up a signal. This type of gate will change a slightly varying voltage output from a transistor switch and make it change abruptly from one logic state to another. All transistor switched sensor circuits would benefit from the use of this gate as a logic interface.

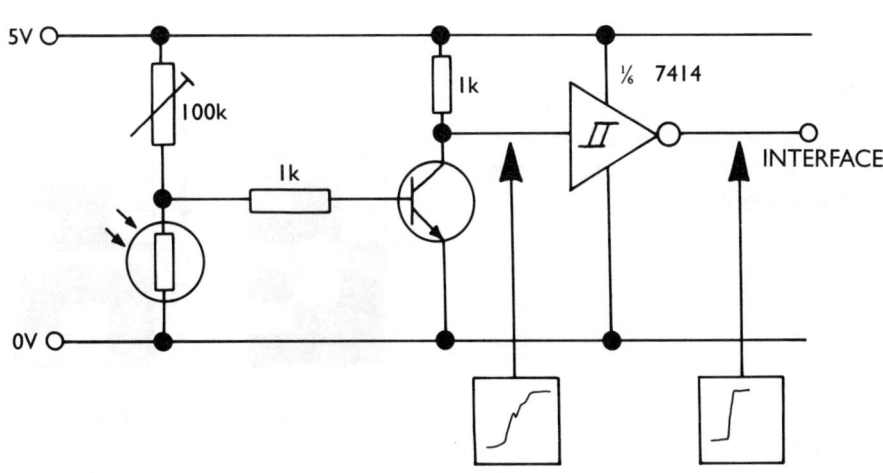

fig 8.95. Trigger circuit

113

POSITIONAL CONTROL

Machines that move and are controlled by computer, will benefit from having their physical position reported back to the computer. Imagine that you are a vehicle with no eyes. Move forward until you almost hit the wall in front of you. Impossible, you say, but if you could detect the presence of the wall or know how far away it is and how far you have travelled, then it is possible. We have many sensors that we take for granted, a simple machine may only have a few.

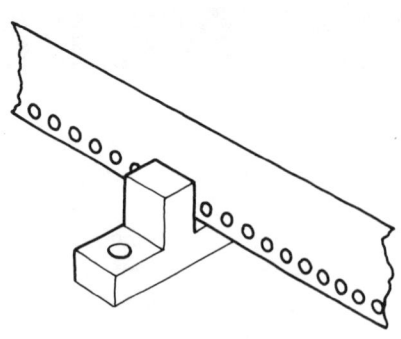

fig 8.96. Counting holes in a strip

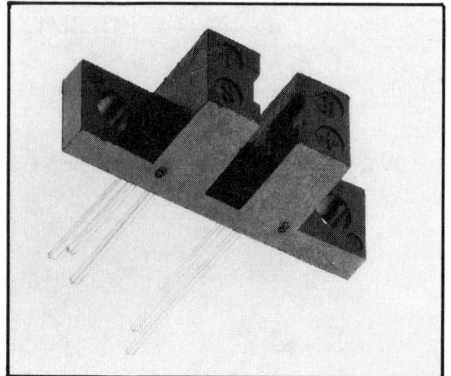

fig 8.97. Slotted opto switch with logic

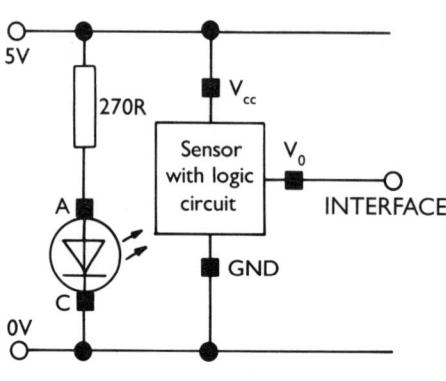

fig 8.98. Slotted opto switch circuit

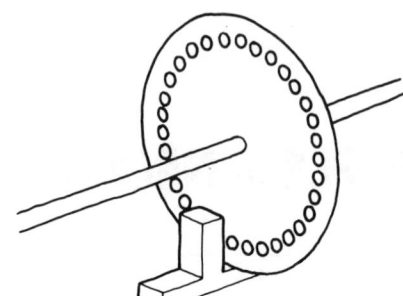

fig 8.99. Rotary counting

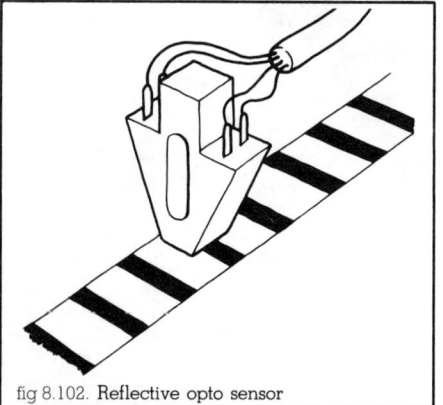

fig 8.102. Reflective opto sensor

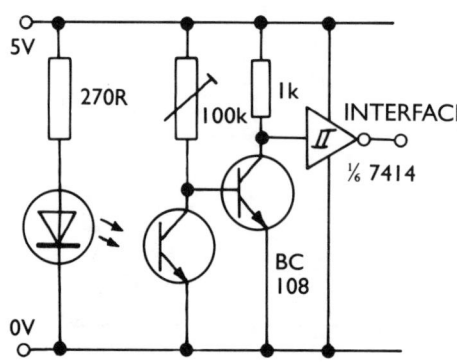

fig 8.101. Reflective opto sensor circuit

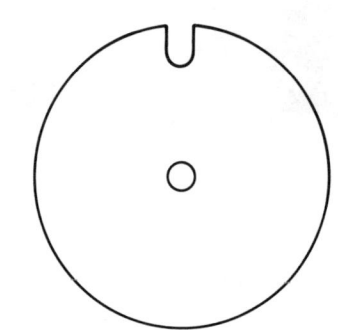

fig 8.100. Slotted disc

Binary encoders

The obvious choice of coding system is to use a binary code, as this is the code used by all computer systems. Reflective strips or bars are used on strips or discs. The problem is that when used in this manner, the code is subject to errors. If the code is sensed on the change of digit, an error can give a position a long way away from the truth because several bits can change at this point.

Gray code

Here only one digit changes at each change. Therefore, the maximum amount of error can only be either side of the change line. This type of code is used in preference to a binary code in many situations.

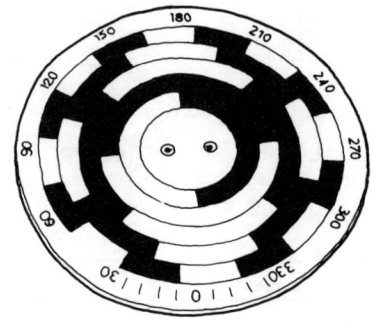

fig 8.103. Binary coded disc

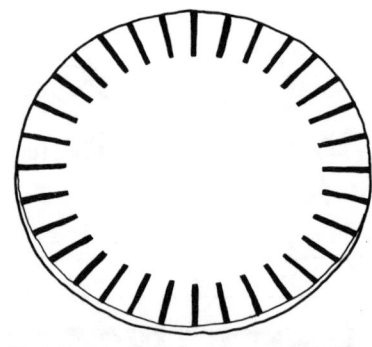

fig 8.104. Strip coded disc

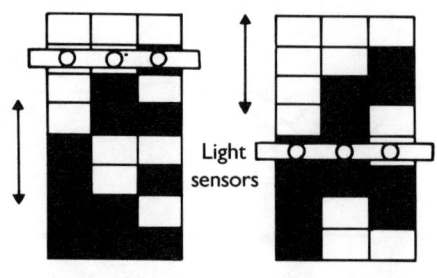

Binary coded strip   Gray–coded strip

fig 8.105. Binary and Gray-coded strips

fig 8.106. Bar coding

ANALOGUE CONTROL

So far we have only considered switches and sensors that register two states, on and off, hot and cold, wet and dry or light and dark. The real world is composed of signals that are not always on or off, but may be a little on or almost dark or slightly wet or slightly wetter. They have values that can be any value between two extremes.

Taking light as an example. How light is light? Is it above or below a certain level when it can be regarded as light or dark? Do you need more information about light levels varying from light to dark? Do you need more stages; dark, almost dark, almost light and light? Can you, for simplicity, call them light levels 1, 2, 3 or 4? Camera light meters are used to convert a light level into a number value that will give us a setting for the shutter speed and aperture size at the lens.

Sound is another signal that can vary in strength or loudness. Your mother's voice is much easier to hear when she is calling mealtimes than when she requires a dishwasher! Sound levels mean different things to different people. What may be gentle pop music to you may be too loud for others. These signals that can have many different levels are called **analogue** signals. A computer cannot understand them directly, it can only deal with numbers and codes. The levels must be changed to numbers as we did with the light levels.

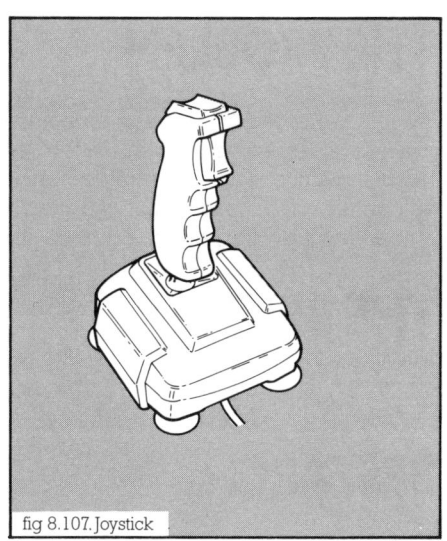
fig 8.107. Joystick

ANALOGUE TO DIGITAL CONVERTER

An analogue to digital converter (ADC) is a piece of electronic circuitry that changes these varying signals into varying number levels. It does this by taking samples at small time steps and giving a number for the value of the signal voltage. The speed with which the sample can be taken and into how many different levels the signal can be split will depend on how complicated and accurate the circuit is.

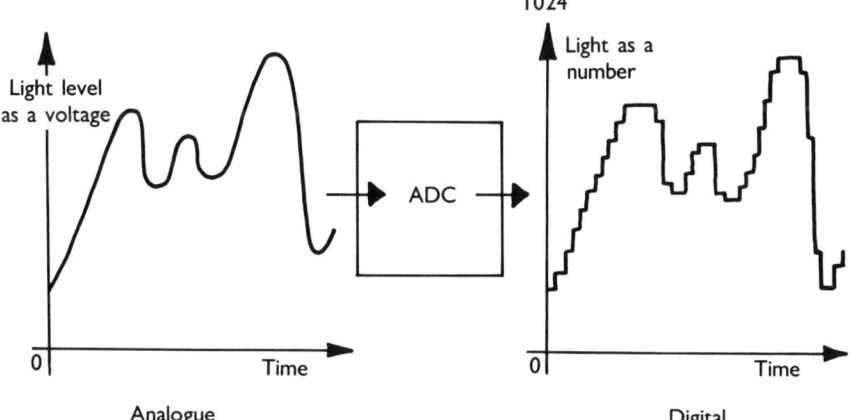
fig 8.108. Analogue to digital convertor

LIGHT SENSOR

Using a light dependent resistor you can obtain a continually varying signal from a varying light source. Previously you have used an LDR to switch a transistor at a preset level. Here we use an analogue voltage signal to give more information about the actual level of light present at the sensor

The circuit can be used to measure the light reflected from different coloured markers or paper or paints, giving a numeric value for each colour. A colour sorting device could develop from this.

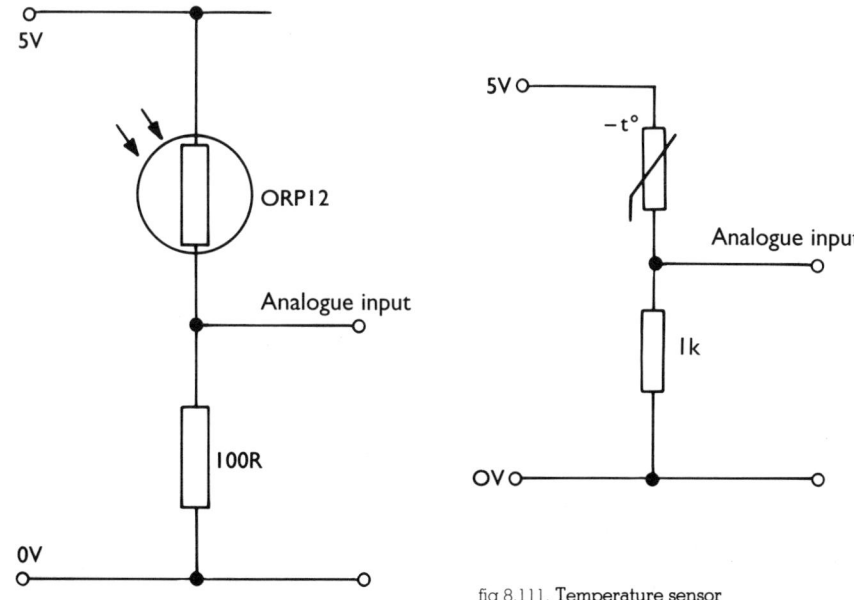
fig 8.110. Light sensor circuit

fig 8.111. Temperature sensor

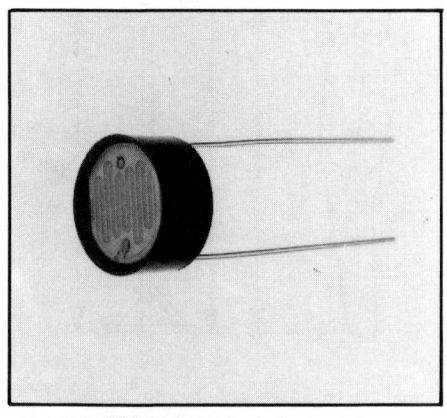
fig 8.109. ORP 12 light dependent resistor

TEMPERATURE SENSOR
The resistance of R-T matched thermistors can be established from data charts and used in a manner similar to the LDR, this time measuring temperature. The normal temperature range for a thermometer in a greenhouse would be from -20 to $+40°C$.

DESIGNING WITH ELECTRONICS

THE PROBLEM

The children in my local primary school learn about traffic and road safety.
They have a road 'traffic' system painted out on the playground and the school
driveways. They use this course for learning to ride a bicycle and how to cross
roads in safety. There is a traffic light system with coloured card stuck onto a
box on a post. This does not change colour and could be much better. There is
a need for a properly working traffic light system.

THE BRIEF

Design and make a set of traffic lights to use as a teaching aid both in the
classroom and outside in the 'traffic area'.

RESEARCH

Light sequence

I took a stop watch to a local traffic
light junction and timed each light
change. The results are shown in fig
8.113. The shortest timed section or
event was the amber time, it was found
to be about three times shorter than
either the green or red.

To make the system easier to
simulate, I have broken the sequence
into equally timed sections. This makes
the amber one count and the red and
green three counts each.

Traffic light systems must have the
correct sequence of lights. At a
standard crossroad junction there are
two styles of lights in operation. The
first set (fig 8.114). has both sets of
lights changing to amber at the same
time. The second (fig 8.115). has only
one set changing to amber at a time,
the other remains at red a little longer.
This second system takes longer to
complete its cycle, but is safer at very
busy crossings in a town or city.

The time for an event is about 20
seconds, giving a cycle time of 2
minutes 40 seconds or 3 minutes 20
seconds. This would be rather long for
a teaching aid. Perhaps one second
for indoors and 10 seconds in the
'traffic' per event would be better.

Power source

The lights could be mains operated,
but need not be. The distance to a
socket is about 5 metres. The lights
could be run from a battery; 6V, 0.3A
bulbs should be bright enough
outside. A 6V battery is available in a
large size called a lantern battery, and
should last quite a long time with the
light system.

fig 8.112. Traffic lights

Time secs	Light
63	Red
21	Red/Amber
60	Green
19	Amber

fig 8.113. Timed sequence

Stage No	Main Road	Side Road
0	Red	Green
1	Red	Green
2	Red	Green
3	Red/Amb	Amber
4	Green	Red
5	Green	Red
6	Green	Red
7	Amber	Red/Amb

fig 8.114. Simple sequence

Stage No	Main Road	Side Road
0	Red	Green
1	Red	Green
2	Red	Green
3	Red	Amber
4	Red/Amb	Red
5	Green	Red
6	Green	Red
7	Green	Red
8	Amber	Red
9	Red	Red/Amb

fig 8.115. Complex sequence

Size

The children using the light will be
both boys and girls in the 8–11 year
age range. From the British Standards
Anthropometric data I found the sizes
required for a light on a stand (see fig
8.116).

From BSI data sheet PP7310

Age 8 to 11

Eye height A	Boys	1050 to 1385mm
	Girls	1095 to 1395mm

Reach B	Boys	1460 to 1830mm
	Girls	1440 to 1865mm

Maximum height of switch	1440mm
Minimum lower light height	1395mm
Approx height of stand	1400mm

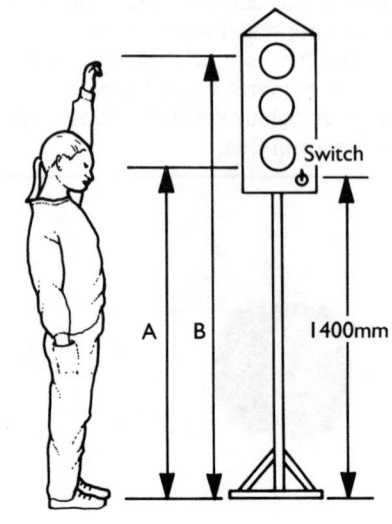

fig 8.116. Sizes of lights from BSI data

SPECIFICATION

1. The sequence of lights will be as fig 8.119, the short version.
2. The time for class use is to be set at 1 sec per event (8 sec cycle).
3. The time for 'traffic' use is to be 10 sec per event (80 sec per cycle).
4. The packaging should be suitable for safe and simple use.
5. Lights should be 6V, 0.3A bulbs.
6. Power will be from a 6V lantern battery.
7. The stand must be stable in use.
8. The light should not be a hazard if used in the rain.
9. The general size to be suitable for children aged 8 to 11 years.

IDEAS

1. Use a computer with suitable interfacing and software to drive the lights (fig 8.117). This should be possible, as your school should have a computer and interface boxes that could be used for control projects. It could be made more versatile with varying speeds as required. The computer would need to be involved all the time the lights are in use. This may create problems if the computer is already in very great demand.

2. Simulate industrial control and develop a program on the computer and then download to a specific system board controller using a PIC as the control computer (fig 8.118). These micro boards do not need a screen and are becoming more cost effective as a solution. It may be possible to make use of an older computer, such as an Acorn or BBC micro computer. Use them as a dedicated control system, like a large version of a PIC, as they have control connections that can be used directly. You can switch off and remove the monitor once the program is running.

3. Build a dedicated logic controller to perform the required task. This could be more complicated electronically, but should be lower in cost. The device could be small and fit totally inside the light box, a big bonus (fig 8.119).

DEVELOPMENT

I have chosen to develop the third idea, a dedicated logic controller. I think this will be lower in cost and not involve the already overused micro. The sequence has been broken down into eight equally timed sections (fig 8.114). The lights will be used in the classroom as a teaching aid before they are used in the 'traffic' area outside. I will need some sort of timing device and a digital controller to interface with the more powerful lights. The block diagram (fig 8.120) is the basis of the circuit. I will now need to look at the requirements for each block in detail.

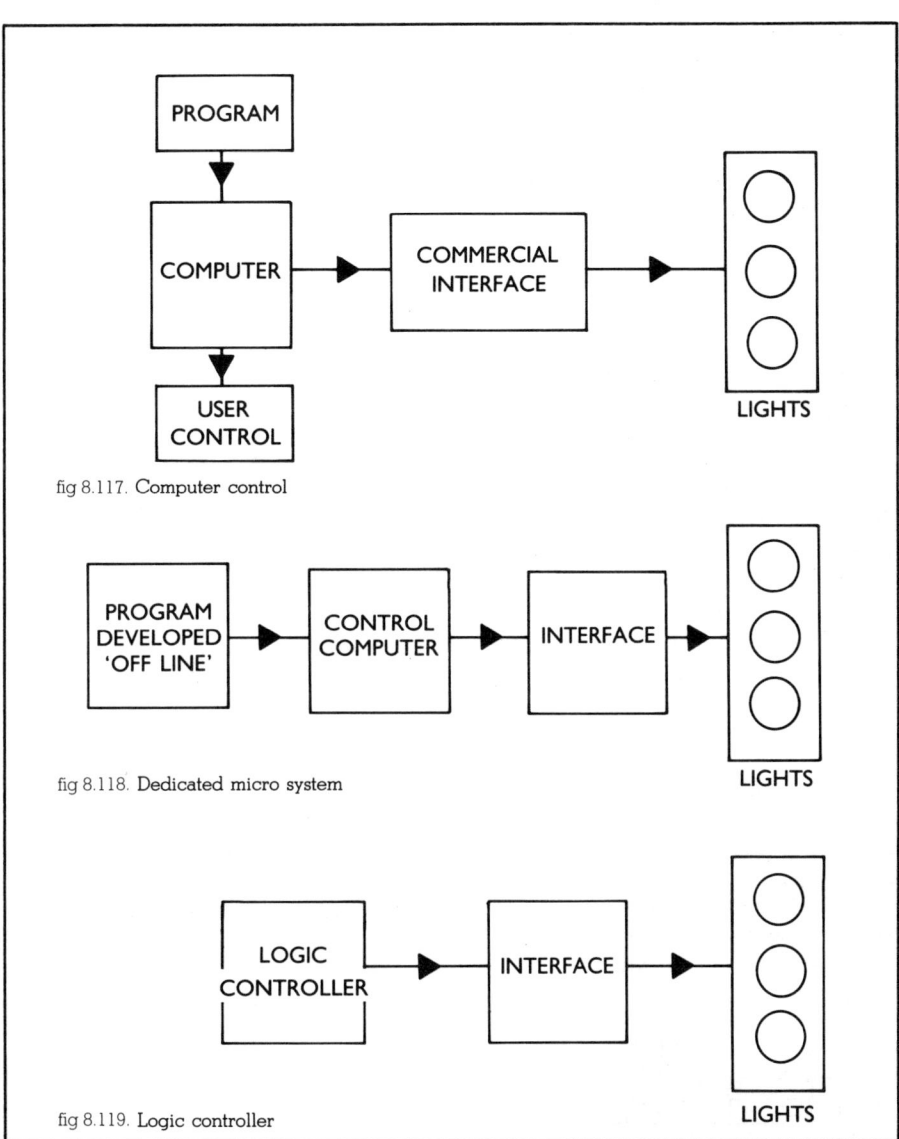

fig 8.117. Computer control

fig 8.118. Dedicated micro system

fig 8.119. Logic controller

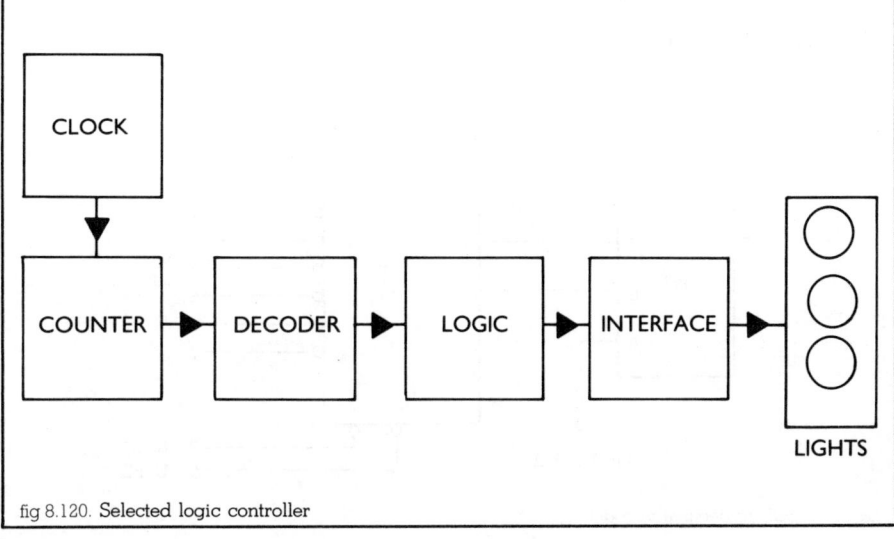

fig 8.120. Selected logic controller

Clock

The device is required to run with a clock speed of 1 second and at 10 seconds per beat or tick. A 555 timer should be suitable if I use two switchable resistors to select the time bases (8.121).

Counter

There are eight stages in the complete sequence. A counter that counts in binary is needed. Three binary bits are required for a count of eight. Chip 7493 is a binary counter that can be used as a 3 or 4 bit counter (0 to 7 or 0 to 15). As only three lines or bits are required, bit A is ignored and B becomes the least significant bit of the three bits, B, C, and D (fig 8.122).

This is a TTL chip so it will need 5V to operate properly. A 6V battery will damage it.

The reset is not required as we are counting to the maximum number possible, therefore, the counter will automatically change to 000 (0) after 111 (7). The reset pins (2,3) are connected to 0V.

Power source

If I now require a 5V logic supply to run the TTL chips I must regulate this carefully. Two possible ideas spring to mind. First, I could use a zener diode, which is cheap, but set at 5.6V, hardly near enough to 5V (fig 8.123). Alternatively, I could use a 5V regulator chip, which is not too expensive and very simple from a battery supply. This would also allow me to use a 9V, or even 12V battery for power. This extra bonus makes the regulator a better choice (fig 8.124). A battery supply should not need the usual capacitors here, saving components and cost.

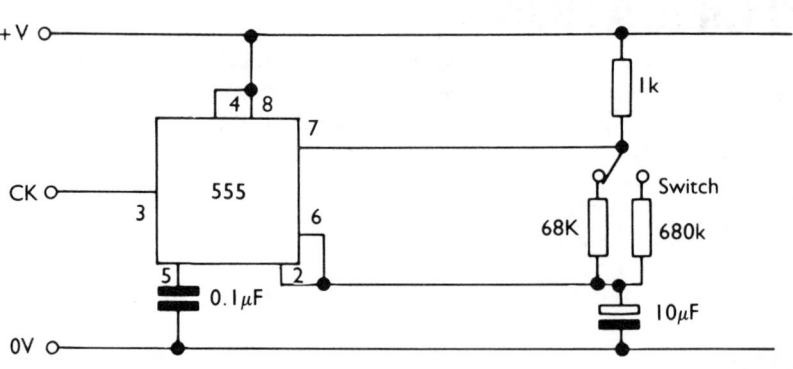

fig 8.121. Timer circuit

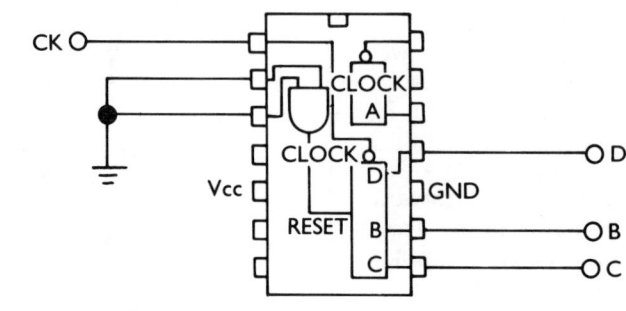

fig 8.122. 7493 counter

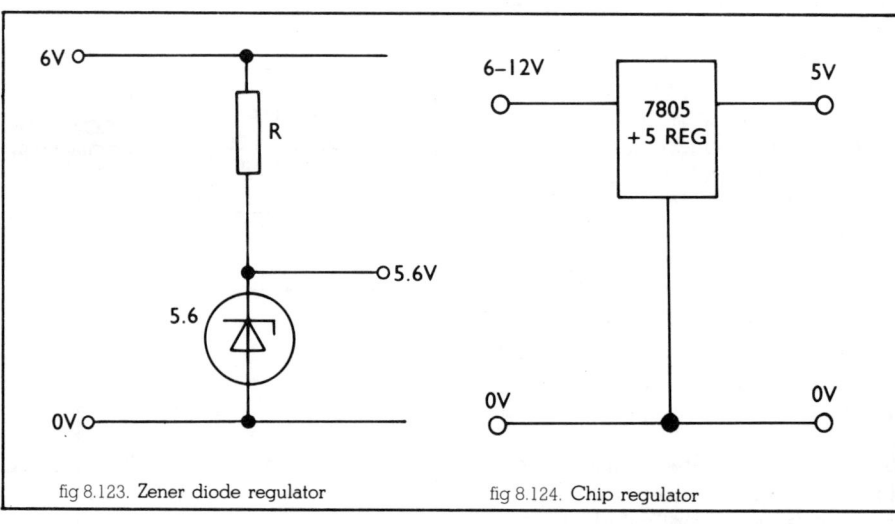

fig 8.123. Zener diode regulator   fig 8.124. Chip regulator

Decoder

At the present stage in the circuit there are three lines called B, C and D with either 5V or 0V present, representing a coded count from 000 to 111 (0 to 7). These three lines must now be 'decoded' to suit the conditions that we require.

To make the decoding easier I am going to change these three lines B, C and D and their complicated code into a simple version, operating eight separate lines, only one of which is on at any time. A BCD to decimal decoder 74141 is required. This BCD (binary coded decimal) decoder will decode all the 4 bits from the counter 7493. I will only need 3 binary bits as I have only three lines of information to decode.

I must use the three least significant bits on the decoder chip, A, B and C and not use line D. This means that line B from the counter is connected to line A on the decoder (see fig 8.125).

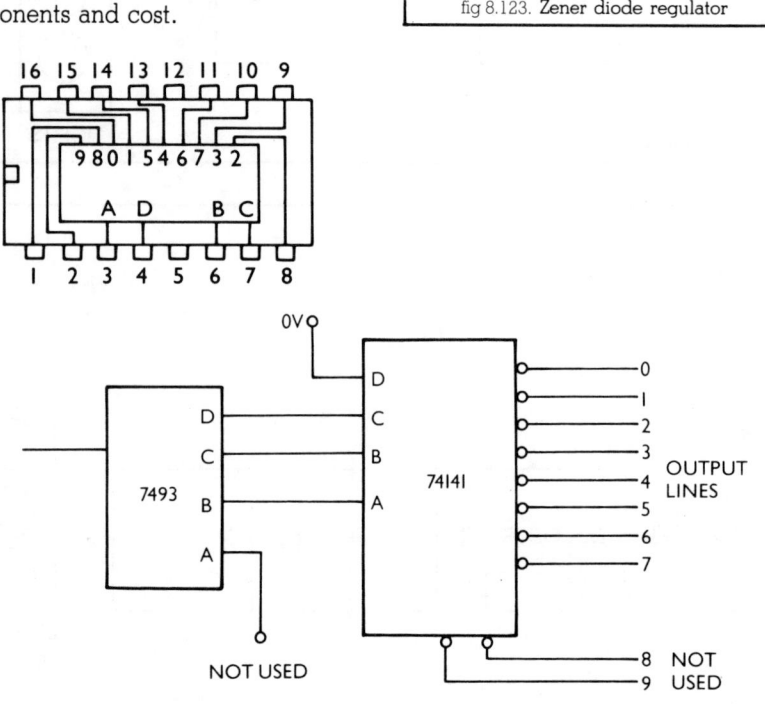

fig 8.125. BCD to decimal decoder

Output logic

I now have eight lines that will be 0V in turn, one after the other. Fig 8.114 shows the lights required for each of the eight parts of the sequence. The lines from the decoder need to go to more than one bulb. If they were wired directly the wrong lights would be lit at various times. This cannot be allowed to happen. The main red for example must be on when line 0 **OR** line 1 **OR** line 2 **OR** line 3 is turned on. Fig 8.126 shows the logic requirements for the lights separately, and fig 8.127 shows them linked together sharing OR gates where possible.

The circuit uses all **NAND** gates, they have been chosen to keep the logic as simple as possible.

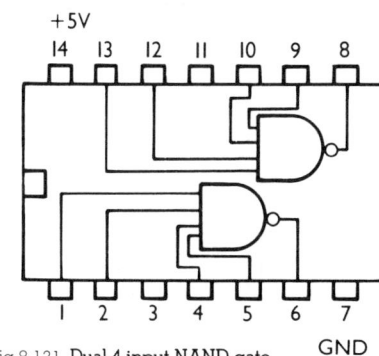

fig 8.131. Dual 4 input NAND gate

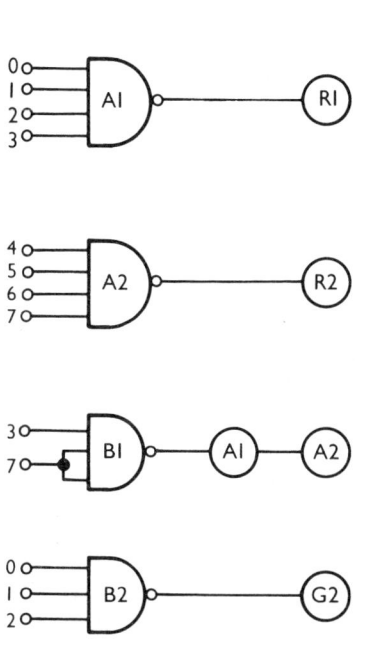

fig 8.126. Logic functions

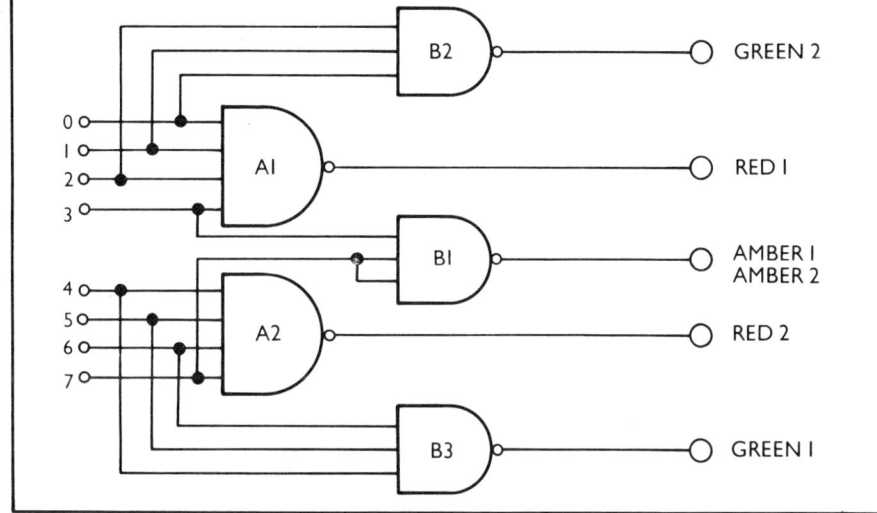

fig 8.127. Logic circuit

Power amplification

The bulbs being used are 6V 0.3A MES type. The output from a NAND gate can be 8mA as a source or 16mA as a sink (see page 94). A boost to power is required.

Idea 1: use a transistor to switch the bulb. Will it work? 8mA into the base of a transistor requires a gain in power of about 40 to give sufficient current. A BFY51 has a gain of 40 and will sustain a current of 1A, therefore, it should work (fig 8.128).

Idea 2: use a darlington driver ULN2803A. This has a very high gain and will deliver 0.5A. This will take less current from the NAND gate and is a simple one chip solution. This will be my choice (fig 8.129).

The block diagram is now filled with possible components (fig 8.130). The circuit (fig 8.127) should now be built on a breadboard and tested for problems.

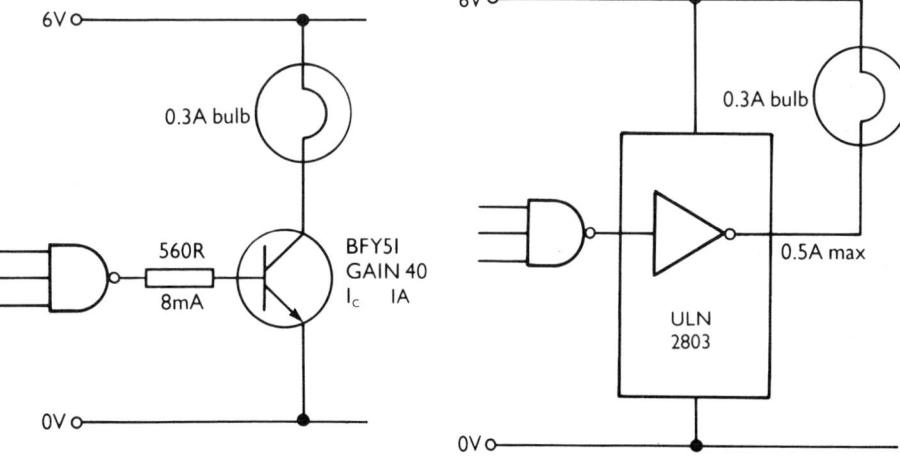

fig 8.128. Transistor drive

fig 8.129. Darlington driver

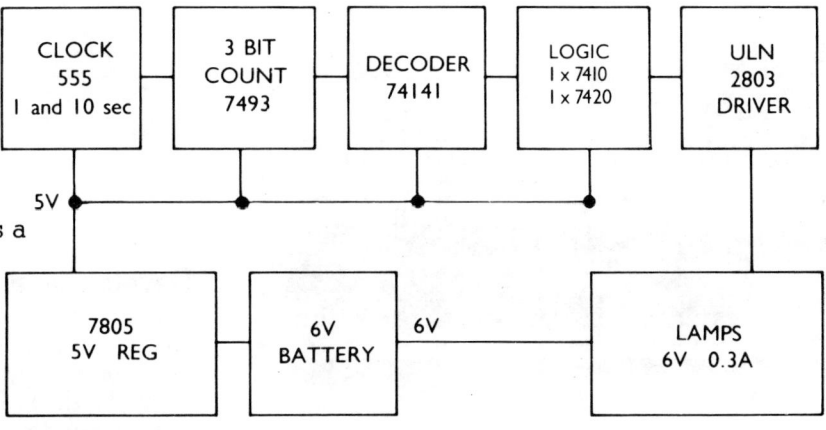

fig 8.130. Circuit block diagram

119

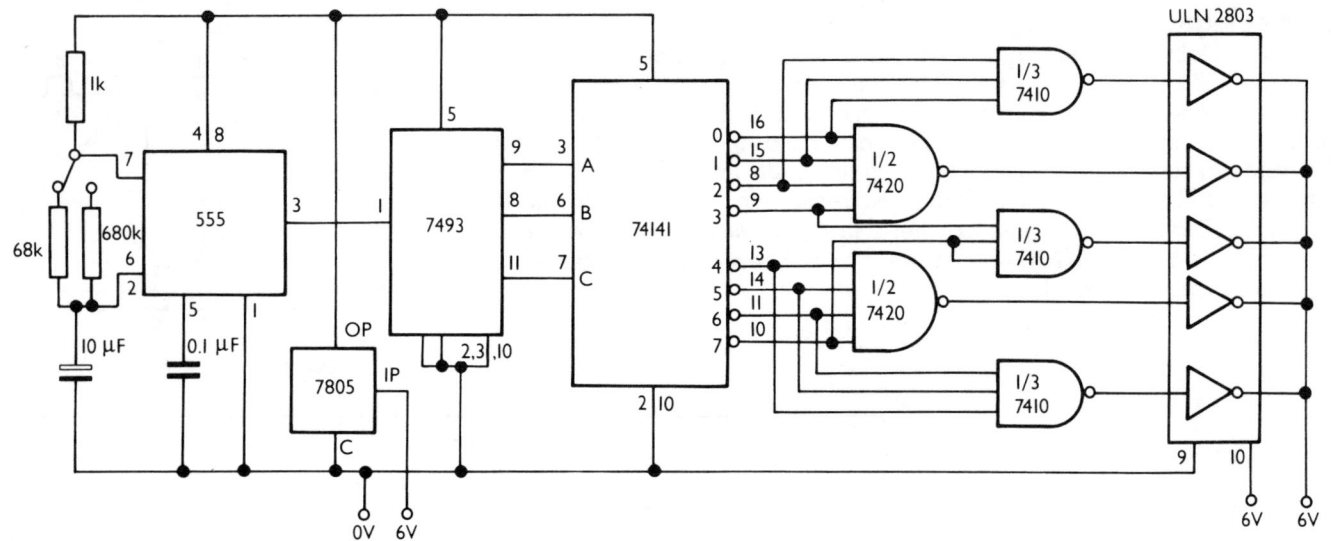

fig 8.132. Traffic light controller

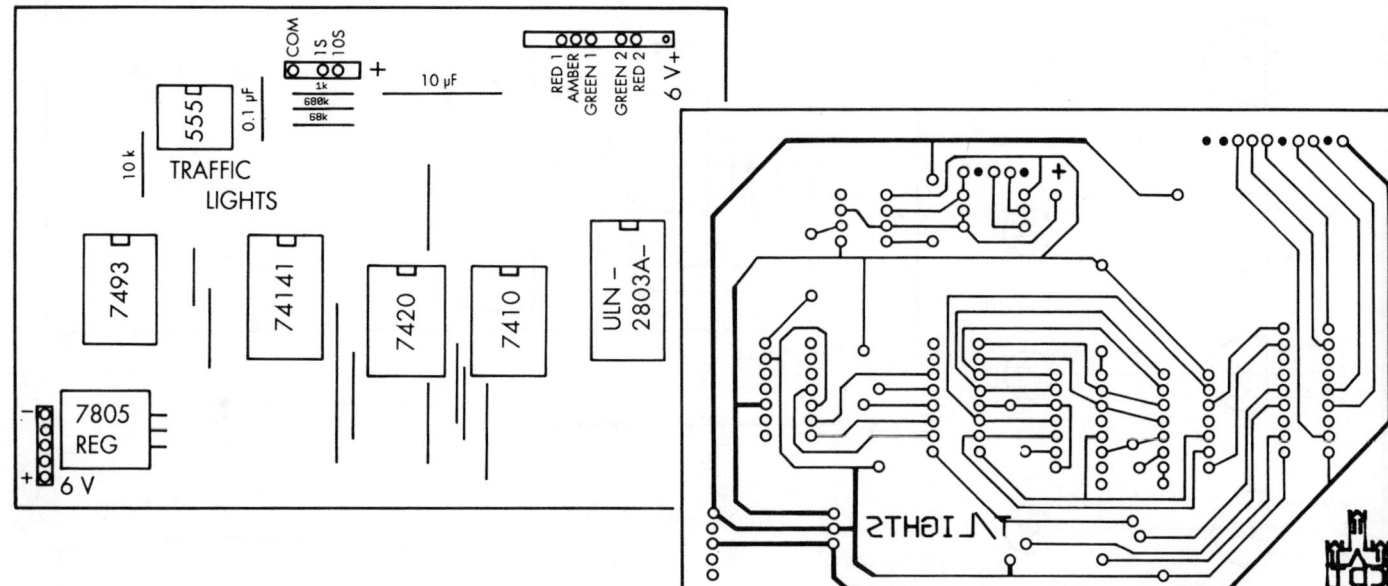

fig 8.133. PCB mask and component overlay

Circuit production stages

1. Design circuit.
2. Model circuit on breadboard.
3. Test and modify if required.
4. Design PCB mask.
5. Etch PCB.
6. Drill holes.
7. Fit components and connectors.

fig 8.135.

Development of light box and stand

It must be remembered that a product, not just a circuit board, is being designed. Possible box materials include wood, either solid or manufactured board, aluminium sheet or acrylic sheet. It may also be possible to use manufactured boxes or make use of extrusions such as rainwater pipe.

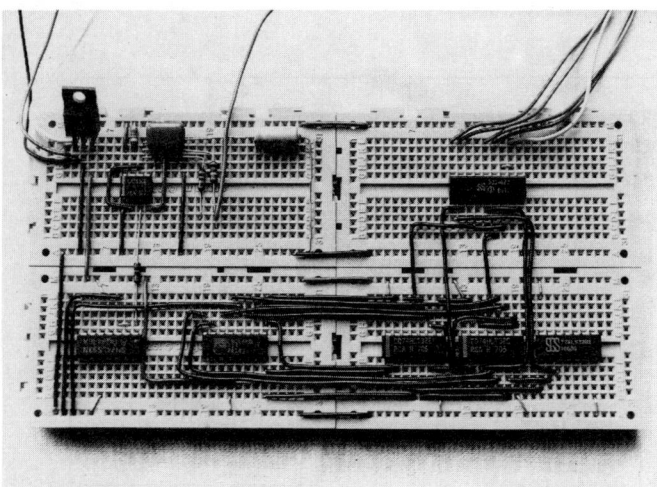

fig 8.134. Circuit on test breadboarding

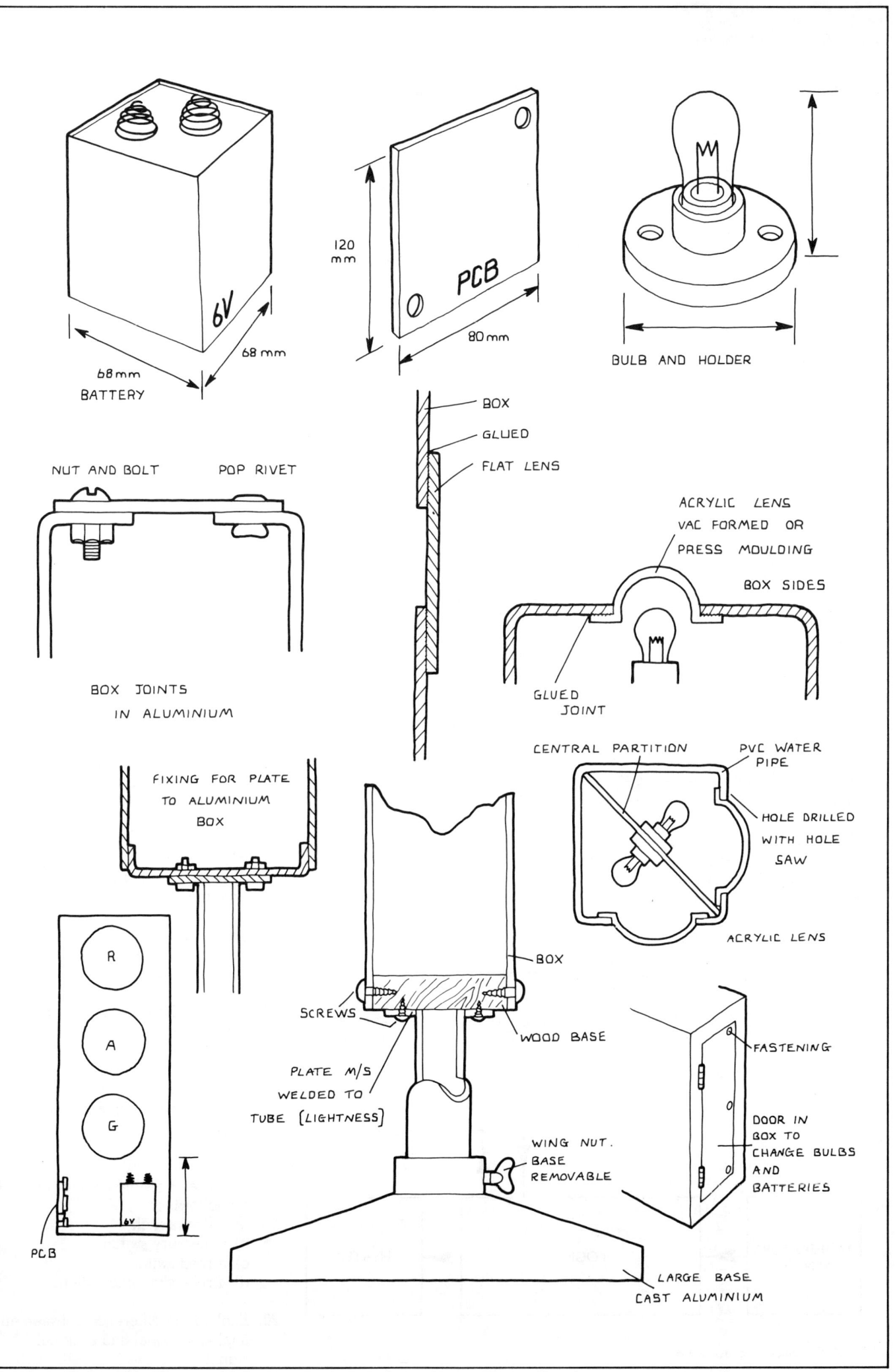

BATTERY
6V
68mm
68 mm

PCB
120 mm
80 mm

BULB AND HOLDER

NUT AND BOLT POP RIVET

BOX JOINTS
IN ALUMINIUM

BOX
GLUED
FLAT LENS

ACRYLIC LENS
VAC FORMED OR
PRESS MOULDING

BOX SIDES

GLUED
JOINT

FIXING FOR PLATE
TO ALUMINIUM
BOX

CENTRAL PARTITION
PVC WATER
PIPE

HOLE DRILLED
WITH HOLE
SAW

ACRYLIC LENS

R
A
G

PCB

BOX

SCREWS

WOOD BASE

PLATE M/S
WELDED TO
TUBE (LIGHTNESS)

WING NUT.
BASE
REMOVABLE

FASTENING

DOOR IN
BOX TO
CHANGE BULBS
AND
BATTERIES

LARGE BASE
CAST ALUMINIUM

EXERCISES

1. Digital electronics is based on manipulation of **bits** of information. What is a **bit**?

2. List ten items used in the home that are both electronic and digital.

3. A logic circuit can be built to act as a safety device on a microwave cooker. It is required that the cooker works only if the door is closed and the timer is set. Explain how this could be accomplished, giving a suitable circuit diagram.

4. Draw a logic diagram for your answer to question 3.

5. Draw the logic symbols for OR, AND, NOR and NOT gates.

6. Draw a truth table for:
   a) a three input AND gate.
   b) a three input NOR gate.

7. Explain why a regulated power supply must be used with TTL logic gates.

8. Using NAND gates draw the logic diagram for a three input AND gate.

9. Explain why a TTL gate cannot directly drive a 5V bulb or a 4.5V Lego motor.

10. An automated greenhouse system controls the temperature of the greenhouse. The block diagram in fig 8.136 shows a possible method. There are two sensors, temperature and window position (open or closed). The logic system controls the window opener and the heater.
    a) Write a set of logic sentences that explain how this system works.
    b) Draw a logic diagram to do the same.

11. Change the following decimal numbers into binary: 4, 9, 21, 56 and 38.

12. Change the following binary numbers into decimal: 1101, 1001, 10, 100, and 10010111.

13. The output from a three bit counter is required to reset at 6, giving a count from 0 to 5. This is required to operate an electronic die (fig 8.137). What logic circuit will recognise this event, i.e. number 110 in binary? Draw the logic diagram.

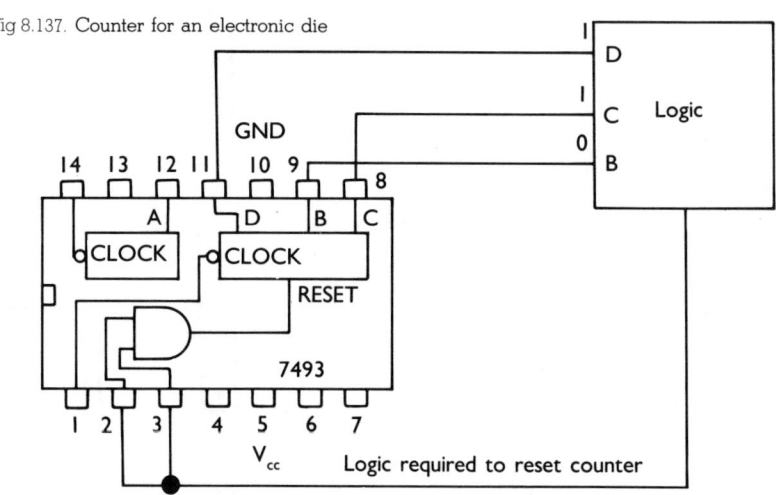

fig 8.137. Counter for an electronic die

14. Draw the possible numbers that can be shown on a seven segment display.

15. Draw the possible shapes of letters, a to f, on a seven segment display. They must be different from the numbers in question 14.

16. List as many different methods of sending information by code as you can (e.g. morse code).

17. List your three favourite computer games and explain what skills they demand of you.

18. Design a device that can register how far a wheel on a simple buggy, as on page 110, has travelled.

19. List three possible uses for each of the following:
    a) a float switch
    b) a mercury switch
    c) a reed switch
    d) a reflective opto switch.

20. Explain the difference between an analogue signal and a digital signal.

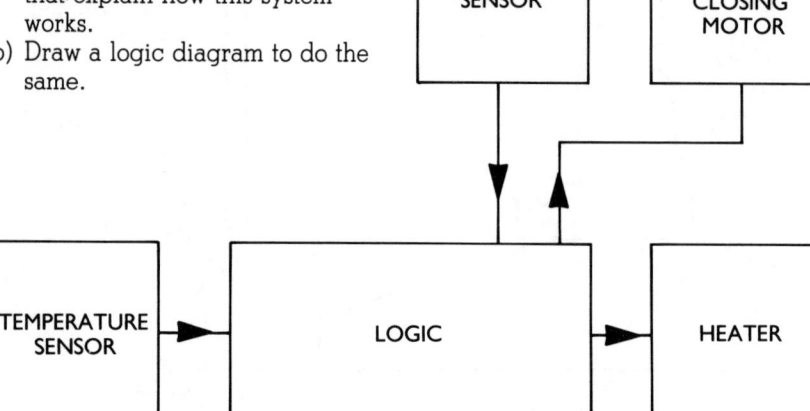

fig 8.136. Greenhouse control system

MECHANISMS

Most of the mechanisms you use are so familiar that you never think about them. Simple things like light switches, door handles and tin openers are just a few of the many mechanisms you use every day. Each one has been designed to do a particular job and most of the time they do it perfectly. It is probably only when they go wrong that you think about them.

Some mechanisms are easy to see and understand, but many are hidden away behind glossy panels and covers. In the past mechanisms were much easier to see, as with an old steam engine for example, but as people became more concerned about safety it was necessary to fit guards over the moving parts. Today, styling of these covers makes it even harder to see what is going on inside, but whether you can see them or not, mechanisms are a vital part of everyday life. They allow you to do the simple things like switch on lights, turn taps and open doors. They also make it possible for you to travel in cars, take fairground rides and use escalators.

Mechanisms play a vital role in industry. While many industrial processes now have electronic control systems, it is still mechanisms that provide the muscle to do the work! They provide the forces to press steel sheets into the shape of car body panels, to lift large components from place to place, to force plastic through dies to make curtain tracks and window frames. The list of jobs is endless. It is only by using mechanisms that industry can make the products you use every day. Perhaps next time you use a door handle, a tin opener or a coffee machine you might stop to think about how their mechanisms work or the mechanisms that were used to produce them.

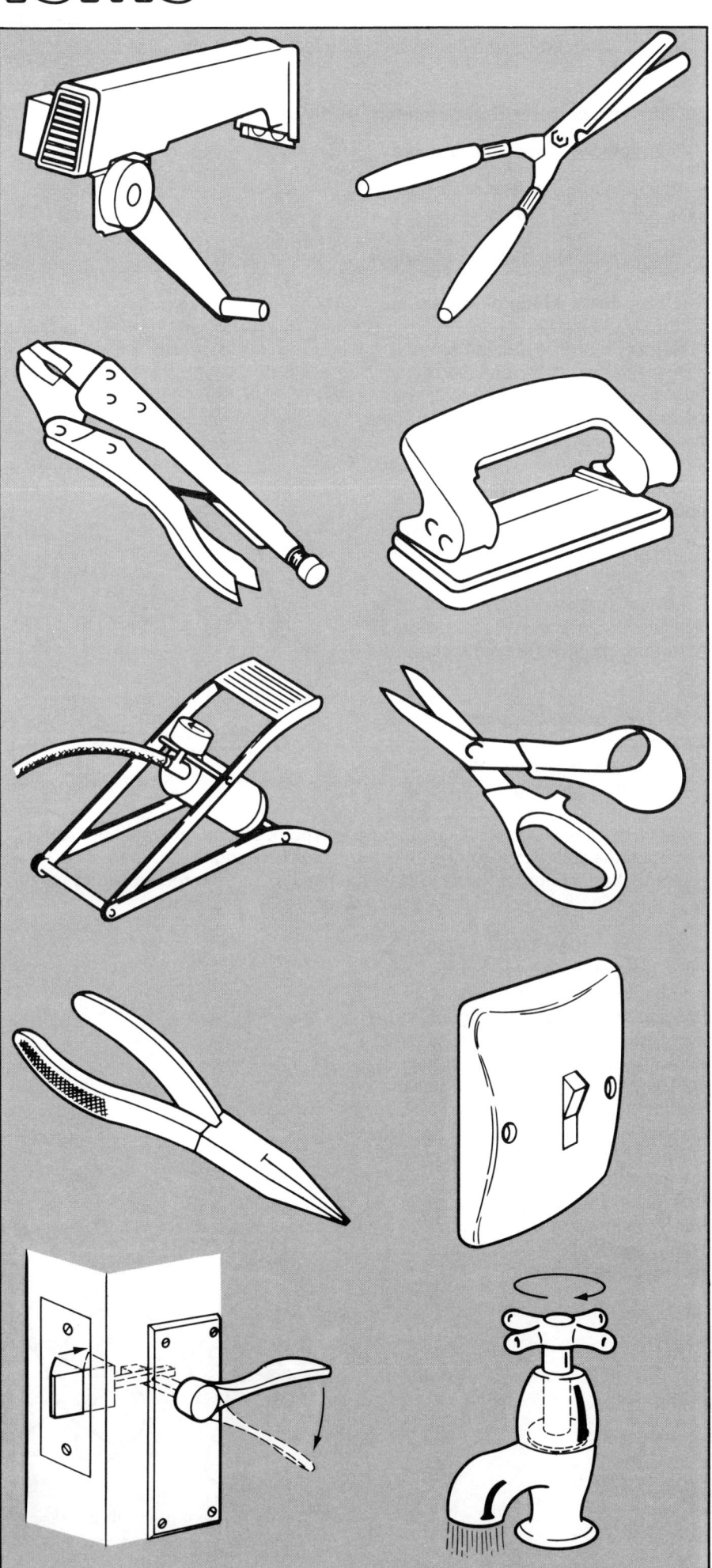

MOTION AND FORCE

Although designed to do different jobs, all mechanisms have some things in common:

— they involve some kind of **motion**

— they involve some kind of **force**

— they make a job **easier** to do

— they need some kind of **input** to make them work

— they produce some kind of **output.**

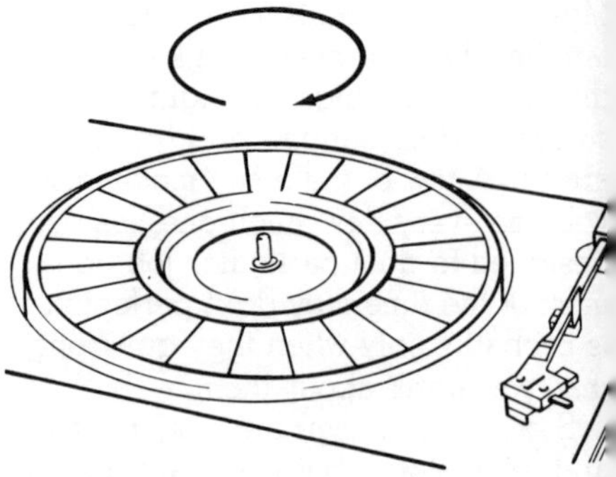

The **four basic kinds of motion** are:

Rotary, going round and round. This is the most common kind of movement. Think about wheels, cogs, record decks and tape decks. All these involve rotary movement (fig 9.1).

Oscillating, swinging backwards and forwards. A swing oscillates, so does the pendulum of a grandfather clock (fig 9.2).

Linear, in a straight line. The linear movement of a paper trimmer is used to cut a straight edge on the paper (fig 9.3).

Reciprocating, backwards and forwards in a straight line. A sewing machine needle reciprocates, so does a piston in a car engine (fig 9.4).

Many mechanisms involve changing one type of motion into another type. For example, the rotary motion of an electric motor is changed into the reciprocating motion of a sewing machine needle, and the reciprocating motion of a piston is changed into the rotary motion of a crankshaft.

These four basic kinds of motion are often used on their own, but they can also be combined to produce more unusual types of motion. Fig 9.5 shows a pull along toy which, because of its eccentric wheels, rolls from side to side as it moves forward. The linear motion of the toy being pulled along combined with the rotary motion of the wheels causes the whole body of the toy to oscillate. It is not only in toys that this mixing of types of motion goes on. Many designers use similar techniques to get machine parts to move just as they want.

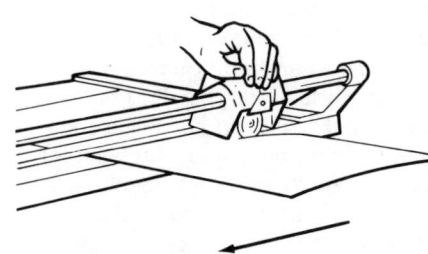

fig 9.1. Rotary motion of record deck

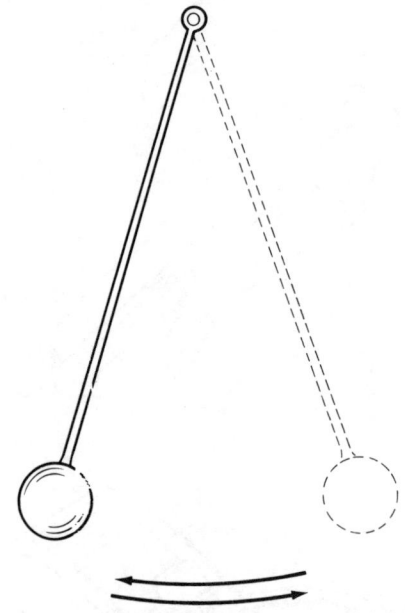

fig 9.2. Oscillating motion: pendulum

fig 9.3. Linear motion: paper trimmer

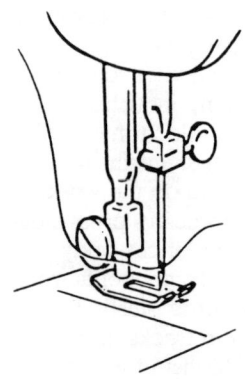

fig 9.4. Reciprocating motion: sewing machine

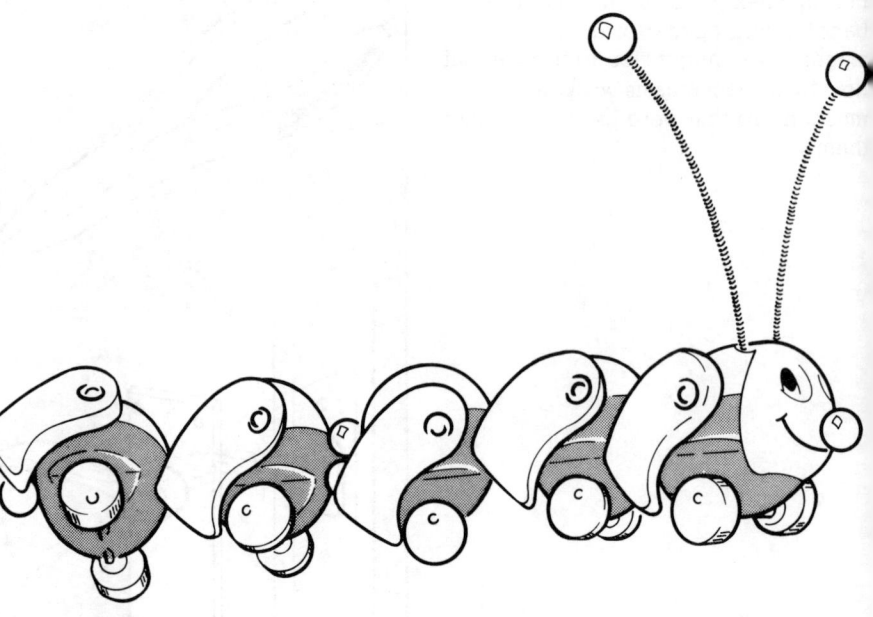

fig 9.5. Combined motion in child's pull-along toy

FORCES

Most mechanisms are designed to make jobs easier, which means only a small force is needed to apply a larger force to the load. For example, the vice shown in fig 9.6 makes the job of holding work firmly very easy. By turning the handle, using quite a small force, you can apply quite large forces to hold your work in place. It is the length of the handle that makes this possible. The longer the handle, the smaller the force needed to turn it. A simple calculation will explain why.

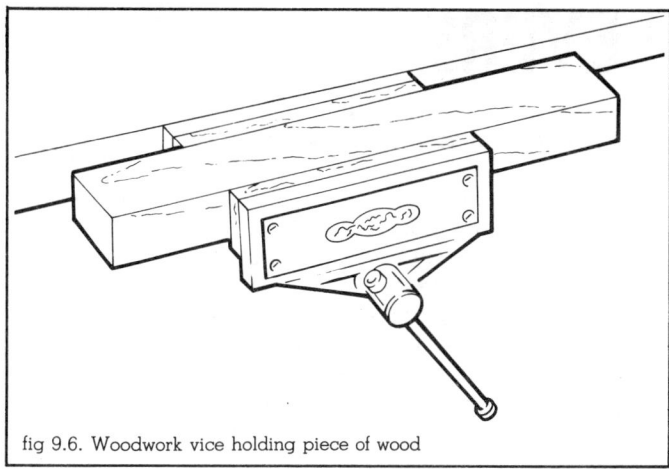

fig 9.6. Woodwork vice holding piece of wood

MOMENTS

'Moment' is the name given to the product of Force × Distance. For example, the moment for the vice shown in fig 9.7 is calculated as follows:

Moment = **Force** (Newtons) × **Distance** (metres)

$$= 15 \times 0.2$$
$$= 3 \text{ Newton metres (Nm)}$$

If the handle is longer, shown dotted in fig 9.7, a smaller input force can be used to apply the same force to the work:

If **Moment** = **Force** × **Distance**

$$\textbf{Force} = \frac{\textbf{Moment}}{\textbf{Distance}}$$

Input force needed $= \dfrac{3}{0.3} = 10N$

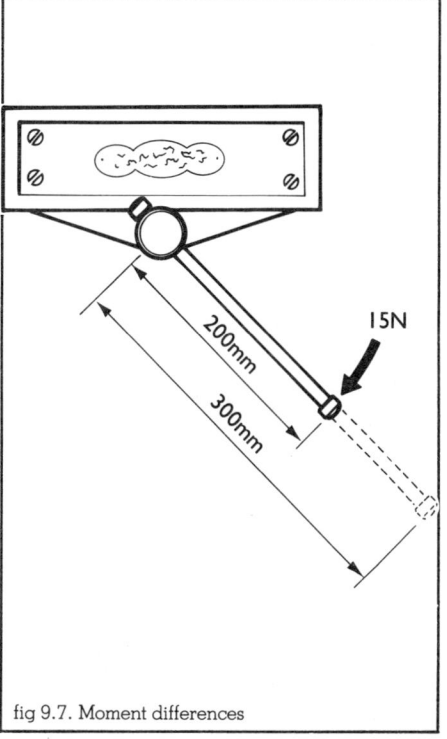

fig 9.7. Moment differences

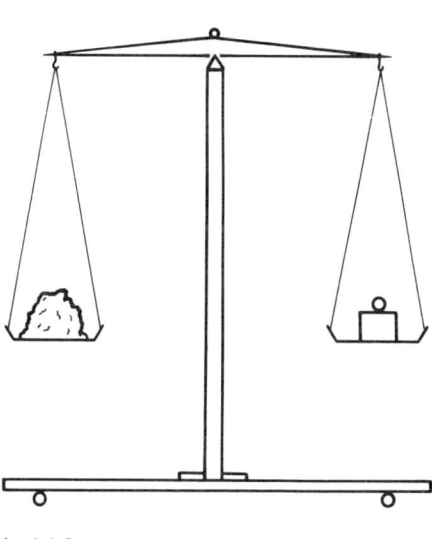

fig 9.8.Simple balance type weighing scales

EQUILIBRIUM

Moments are very important in the design of some mechanisms. From simple examples like weighing scales (fig 9.8), to more complex examples like the dockside crane shown in fig 9.9. Both rely on the moments on either side of the vertical support being balanced. This state of balance is called **equilibrium**. If the weighing scales are in equilibrium then the anti-clockwise and clockwise moments acting on them must be equal. This can be shown by calculating both moments:

Anti-clockwise moment **Balances** **Clockwise** moment
6N × 0.3m = 1.8Nm = 12N × 0.15m = 1.8Nm

Similar calculations could be done for the dockside crane. The complication when designing the crane, is that the equilibrium is disturbed as the load being lifted moves in or out on the jib. To keep the whole crane in equilibrium, and so stop it falling over, the counterbalance has to be designed to move in or out as the load moves in or out.

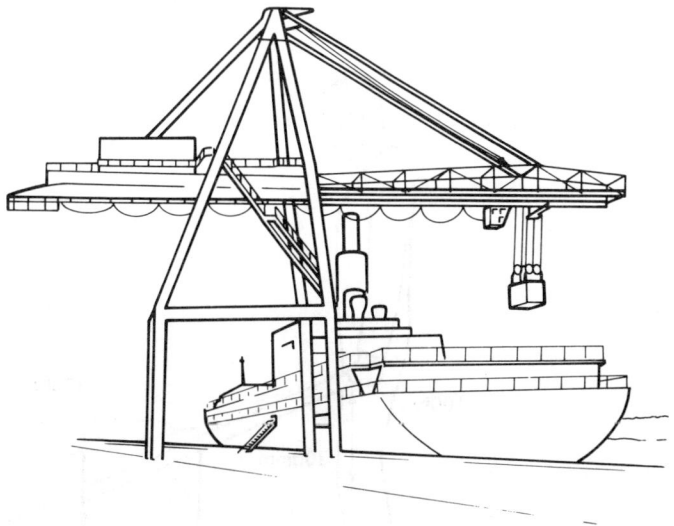

fig 9.9. Dockside crane

LEVERS

Levers were probably the first kind of mechanisms to be used, to help move large rocks or prise open shells. They were used in much the same way that we might use a crowbar to open a crate (fig 9.10), or a tyre lever to take off a tyre. These are very obvious levers, but there are many other less obvious levers you use every day, things like knives and forks, switches, door handles and bike brakes. All levers are one of the three basic kinds, often called **Classes**, shown in fig 9.11. They can be used individually (e.g. a spanner or hammer), in pairs (e.g. pliers or scissors), or connected together to form a linkage (e.g. lazy tongs).

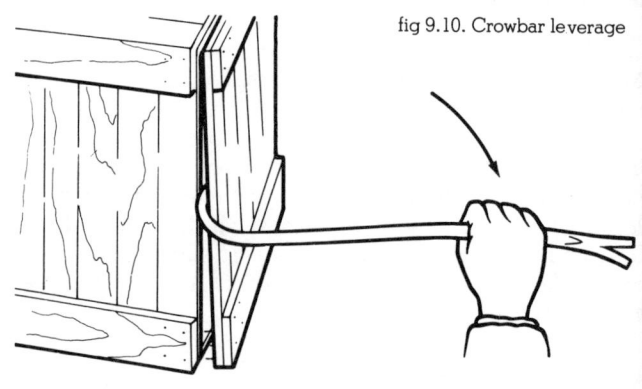
fig 9.10. Crowbar leverage

MECHANICAL ADVANTAGE

Most levers that you come across will be examples of either Class 1 or Class 2. They are the most common because they give you a **mechanical advantage (MA)**, which means that you can move a large load using a small effort. The mechanical advantage of the Class 2 lever shown in fig 9.12 is found by comparing the weight of the load with the effort needed to move it:

$$MA = \frac{\textbf{Load}}{\textbf{Effort}} = \frac{300N}{100N} = \frac{3}{1} \text{ or 3:1 or 3}$$

The mechanical advantage of any mechanism can be calculated in the same way. The larger the number, the greater the mechanical advantage.

Class 3 levers are used less often because their mechanical advantage is less than one. This means that the force needed to use them is greater than the force they can move. When

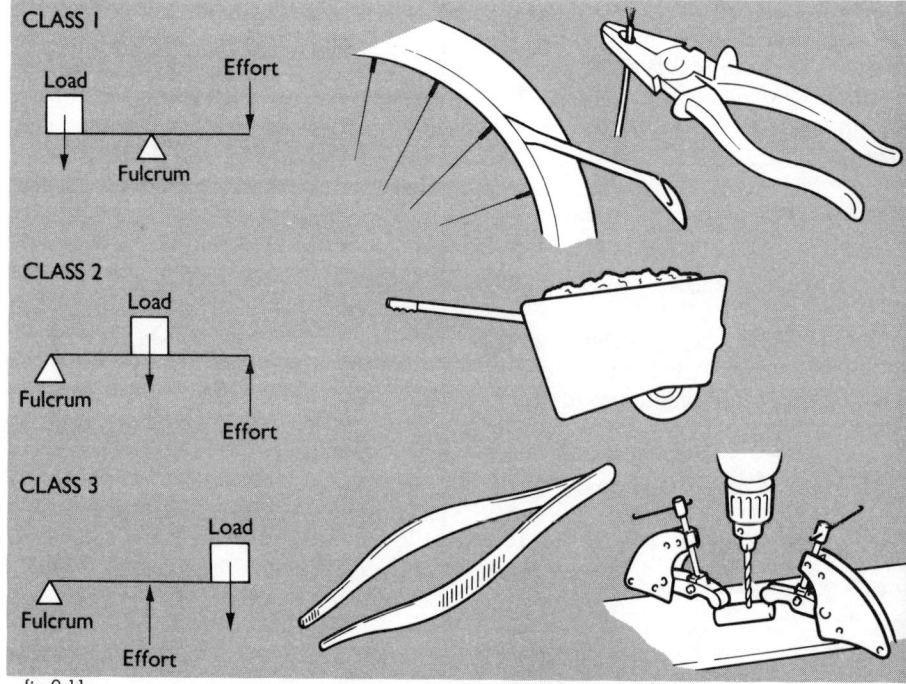

CLASS 1
Load Effort Fulcrum

CLASS 2
Load Fulcrum Effort

CLASS 3
Load Fulcrum Effort

fig 9.11.

they are used it is for things like tweezers which only need very small forces or for situations where Class 3 levers offer some other advantage, such as saving space.

fig 9.12. Loaded wheelbarrow being lifted

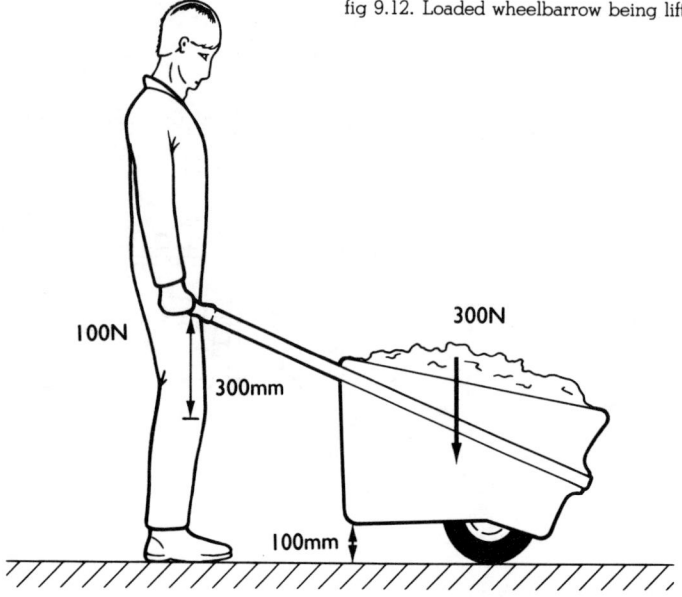

100N

300mm

300N

100mm

VELOCITY RATIO

When calculating mechanical advantage it seems as if you are getting something for nothing, you are moving a large load using a small effort! If you look at how far your effort is having to move to raise the load, you will see it has to move much further than the load is moved. By comparing the two distances you get the **velocity ratio (VR)**. Using the same Class 2 lever as in fig 9.12:

$$VR = \frac{\substack{\text{Distance} \\ \text{moved} \\ \text{by \textbf{effort}}}}{\substack{\text{Distance} \\ \text{moved} \\ \text{by \textbf{load}}}} = \frac{30cm}{10cm} = \frac{3}{1} \text{ or 3:1}$$

This shows that in order to move a load 3 times greater than your effort, you have to move your effort 3 times as far as the load moves. So you are not getting something for nothing.

EFFICIENCY

All this assumes that a mechanism is 100% efficient. It never is! In practice, parts bend, twist and rub against each other, making them less efficient.

The **efficiency** of a mechanism can be calculated using the formula:

$$\text{Efficiency} = \frac{\textbf{MA}}{\textbf{VR}} \times 100\%$$

For example, for a mechanism which has a mechanical advantage of 4 and a velocity ratio of 5 (fig 9.13):

$$\text{Efficiency} = \frac{4}{5} \times 100\% = 80\%$$

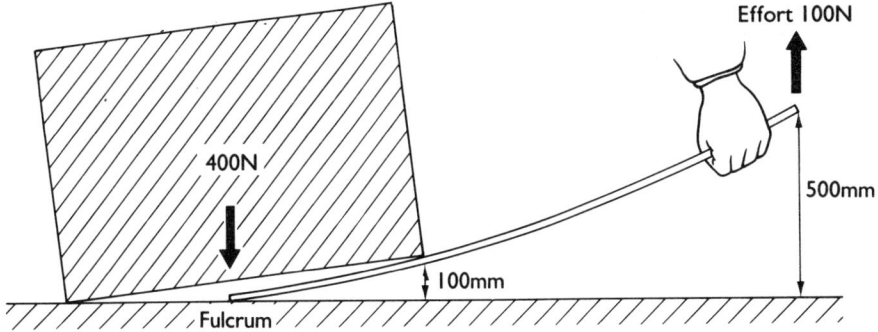

fig 9.13. Lifting a crate by lever

SIMPLE LEVERS IN ACTION

Car tyre foot pump

Pneumatic tyres need to be kept at the correct pressure. A foot pump is one way of inflating them to the correct pressure. A look at the pump (fig 9.14) will show you that it works by using a Class 2 lever. Your foot provides the input motion and force at one end of the lever and the pivot is at the other end. The load, in this case the pneumatic pump, is connected to the lever quite close to the fulcrum.

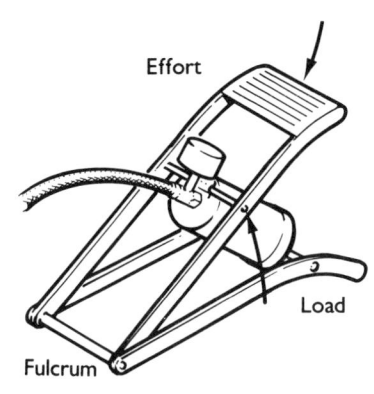

fig 9.14. Car foot pump

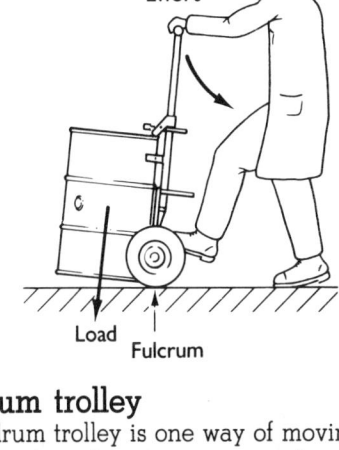

fig 9.15. Drum trolley

Engine hoist

The hydraulic engine hoist shown in fig 9.17 has to be able to lift a car engine out of the car body. Most car engines are made of cast iron and can weigh up to half a ton. In order to keep clear of the car bodywork, the lifting arm has to stick out a long way from the vertical frame. The pivot at the top of that vertical frame is the fulcrum for the lifting arm. The engine is lifted by the force produced by the hydraulic piston pushing upwards. So, with the effort between the fulcrum and the load, this is a Class 3 lever. A Class 3 lever has been used, even though Class 3 levers are inefficient, because the shape of a car body makes it difficult to use other types. Another good reason for using this type of lever system is that the whole thing can be folded up neatly into a small space when not in use (fig 9.18).

fig 9.16. Drum trolley in travelling position

Drum trolley

A drum trolley is one way of moving heavy liquid containers around a factory or warehouse. When the man first tips it back, so lifting the drum off the floor, the trolley is acting as a Class 1 lever (fig 9.15). The wheels are the fulcrum, positioned between the load and the effort. Once it has been lifted, most of the load is positioned between the fulcrum and the effort, making it a Class 2 lever (fig 9.16). Because the load is carried so close to the fulcrum very little effort is needed to keep it in position. Most of the man's effort can be used to move the load forward.

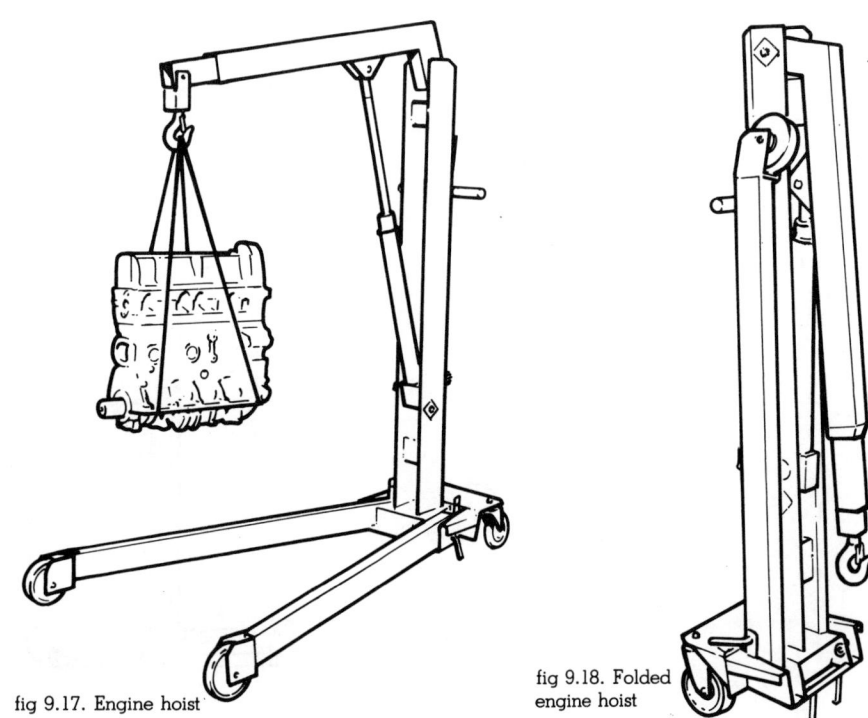

fig 9.17. Engine hoist

fig 9.18. Folded engine hoist

LINKAGES

Linkages are very important in mechanical control systems because they allow forces and motion to be transmitted where they are needed. They can change the direction of a movement, the size of a force, or make things move in a particular way. They usually do several of these things at once.

Bell cranks and **reverse motion linkages** can be used to change the direction of motion (fig 9.19a). This is useful for taking motion round a corner or changing a push into a pull. Fig 9.19b shows the caliper brakes on a bicycle. The mechanism is made up of two bell cranks, both pivoting on the same bolt.

By changing the position of the fulcrum or lengthening one side of the lever, the amount of movement or the size of the forces produced by the linkage can also be changed. In fact you cannot change one without changing the other. The brakes used on mountain bikes (fig 9.19c) have much longer levers on the effort side of the fulcrum than those in fig 9.19b. This allows larger forces to be applied to the wheel when braking.

PARALLEL LINKAGES

Linkages based on a parallelogram can be used to make two or more parts move together or stay parallel to each other as the linkage moves. The important feature of this linkage is that **opposite sides stay parallel** to each other as they move (fig 9.20). Linkages based on this idea can be seen in use on tool boxes, keeping the trays level, and lift safety gates, allowing the gate to fold back into a small space (fig 9.21). The scissor lift table (fig 9.22) shows the same principle being used to keep the table surface parallel with the ground as the table is raised or lowered. Fig 9.23 shows a few of the many other industrial uses of this type of linkage.

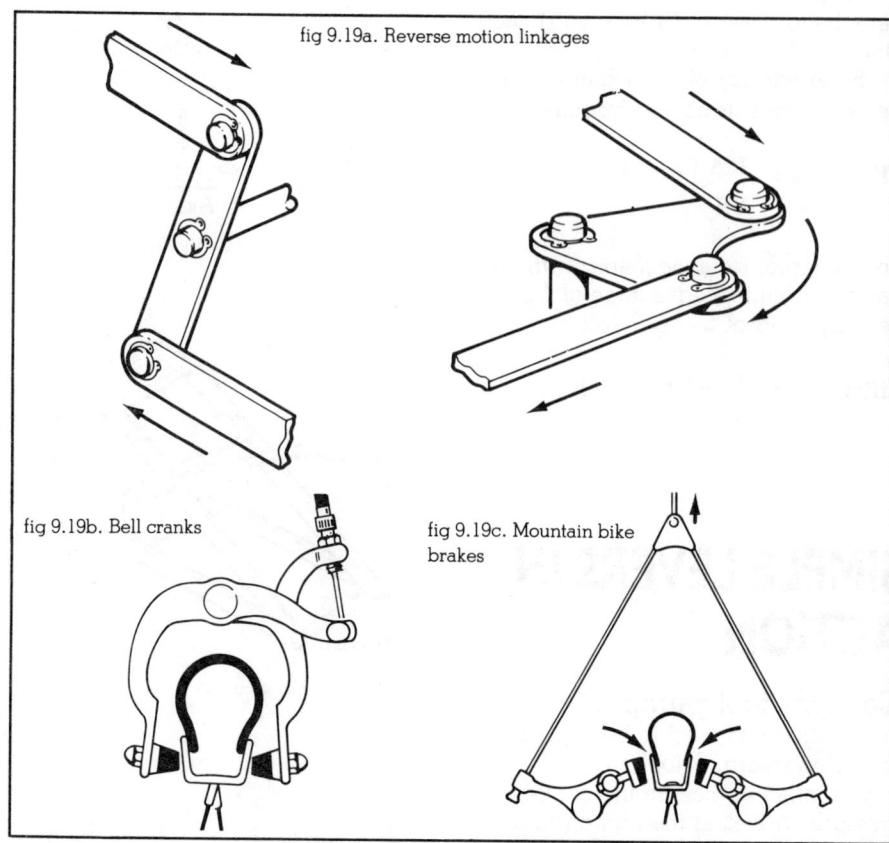

fig 9.19a. Reverse motion linkages

fig 9.19b. Bell cranks

fig 9.19c. Mountain bike brakes

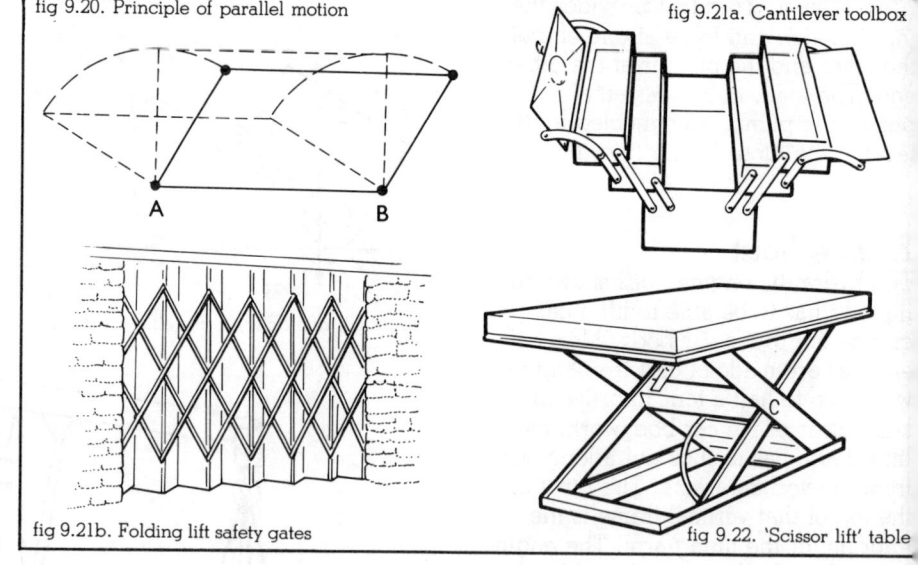

fig 9.20. Principle of parallel motion

fig 9.21a. Cantilever toolbox

fig 9.21b. Folding lift safety gates

fig 9.22. 'Scissor lift' table

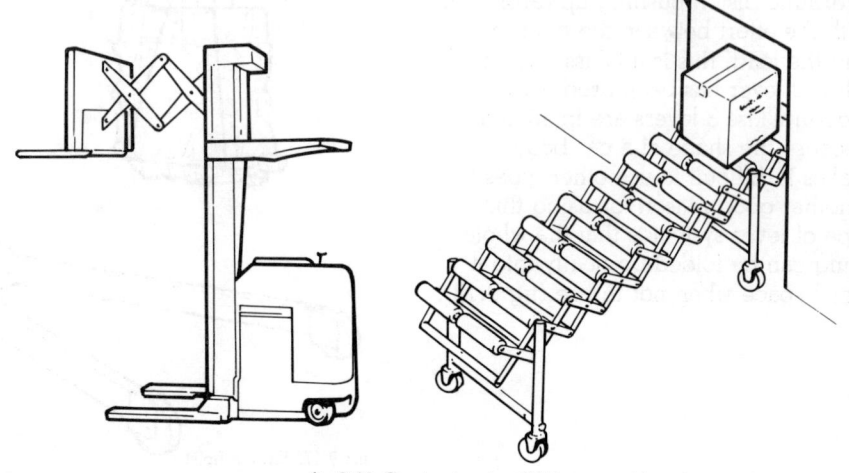

fig 9.23. Stacker truck with 'deep reach', and expanding conveyor

TREADLE LINKAGE

This linkage involves changes between rotary and oscillating motion (fig 9.24a). It can be used to power a rotary machine by working the treadle by foot. You may have seen a sewing machine that works like this. It is also used the other way round to change rotary into oscillating motion. Some car windscreen wipers use this kind of linkage, combined with a parallel motion linkage (fig 9.24b).

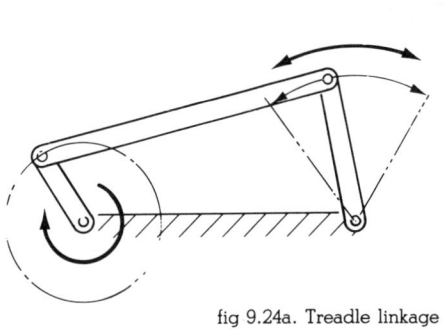

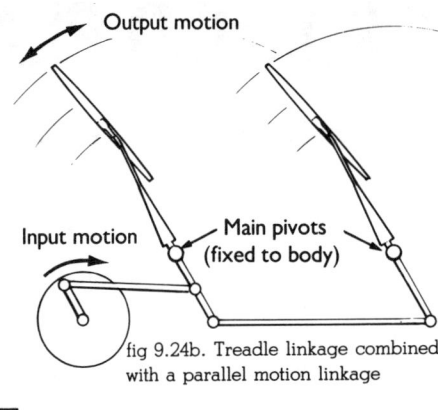

fig 9.24a. Treadle linkage

fig 9.24b. Treadle linkage combined with a parallel motion linkage

TOGGLE CLAMP

One very useful linkage is the toggle clamp which is used to lock things into position. It **holds very firmly** and is **very quick to use**. Mole grips (fig 9.25) and louvre window catches are examples you may have come across.

Toggle clamps are used a great deal in industry to hold work in position while it is worked on. They work rather like the knee joint. If you have to stand on one leg for a while, you will lock your leg by pushing the knee back. Fig 9.26 shows how the toggle clamp works by pushing the middle of the three joints slightly 'over centre' against a stop. Once in that position it is locked. Any force created by the load trying to open it only pushes it into the locked position even more. Only by applying a force from the side can the clamp be unlocked. Fig 9.27 shows some of the many types of toggle clamp used in industry to hold work in place. Although they may look different when you first look at them, they all work on the same principle. If you look carefully you should be able to identify the three pivot points on each one.

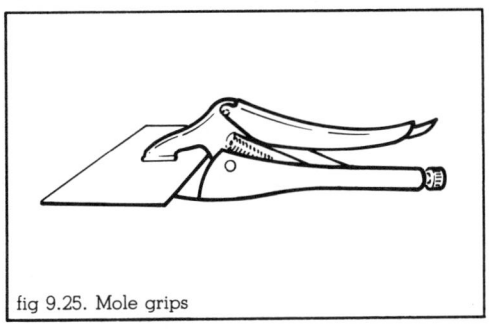

fig 9.25. Mole grips

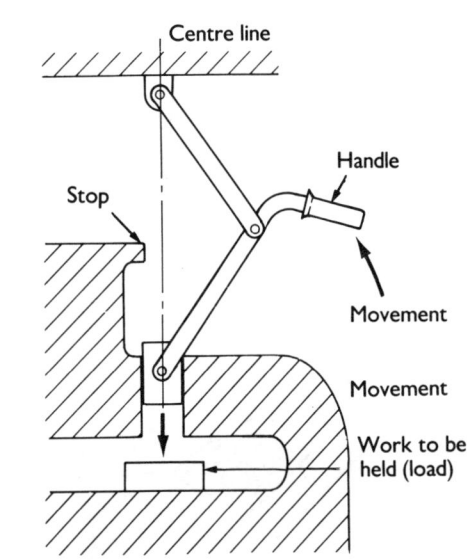

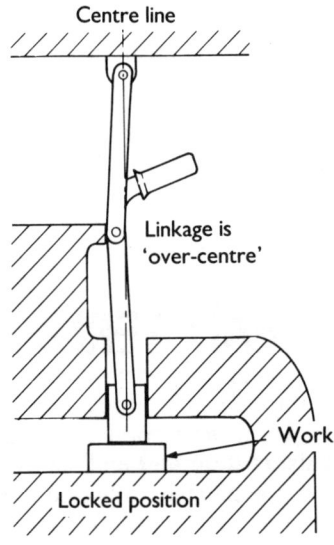

fig 9.26. Toggle clamp

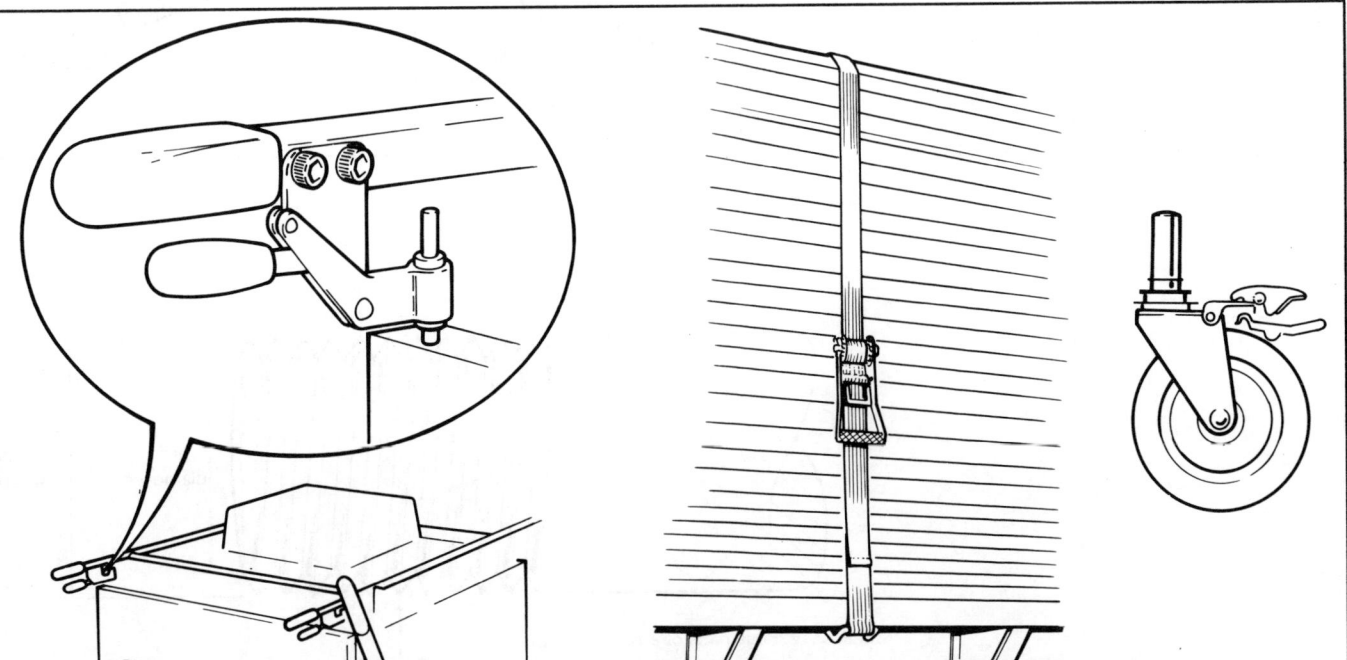

fig 9.27. Toggle clamps used in industry

INCLINED PLANE

An inclined plane is simply a **sloping surface**. It can be used in many ways to make a job easier. In the simple example shown in fig 9.28a, the load can be raised to the top either by pulling it up the slope or by lifting it vertically. Which would you find easiest? Some of the other uses of an inclined plane are a wedge, car ramps and the tips of many cutting tools.

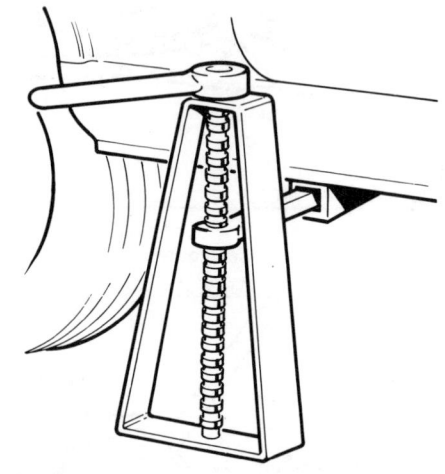

fig 9.28a. Inclined plane

fig 9.28b. Screw thread

SCREW THREADS

Screw threads make use of the inclined plane principle. Fig 9.28b shows how by wrapping an inclined plane around a cylinder you get the same **helix** form as on a screw thread. Screw threads are used in several different ways:

1. To provide powerful movements (e.g. car jacks, fig 9.29).

2. To hold things in place (e.g. screws and bolts, fig 9.30).

3. To position things accurately (e.g. binoculars, fig 9.31).

fig 9.29. Car jack that uses screw thread

THREAD TERMS

The names of the various parts of a screw thread can be seen in fig 9.32. In use, the most important parts are the **outside diameter** and the **pitch**. The outside diameter tells you the size of the thread (e.g. 10mm). The pitch of a thread is the distance from the top of one crown to the next. One turn of the screw thread will move it into the nut by that distance. Some threads have a very small pitch (e.g. those for adjusting positions accurately), while others have quite a large pitch (e.g. those used where quicker movements are needed).

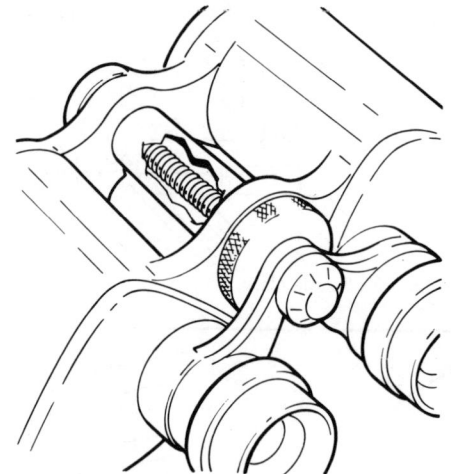

fig 9.31. Fine adjustment of binocular lens using a screw thread

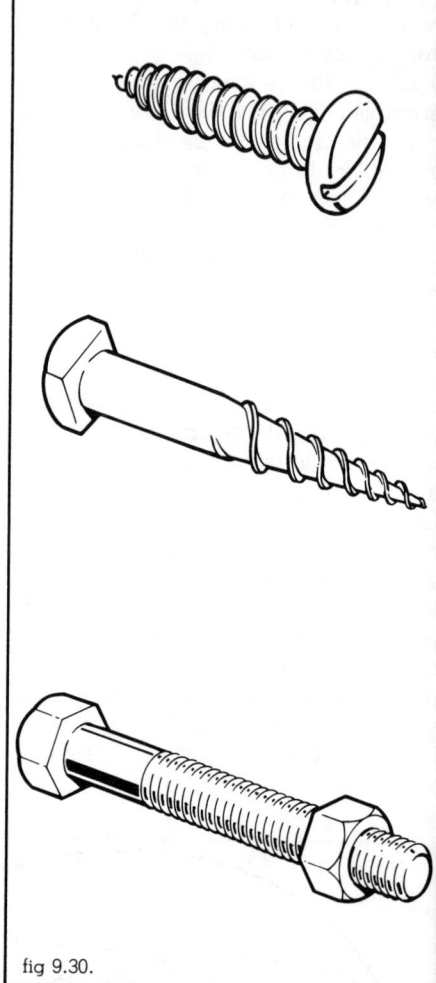

fig 9.30.

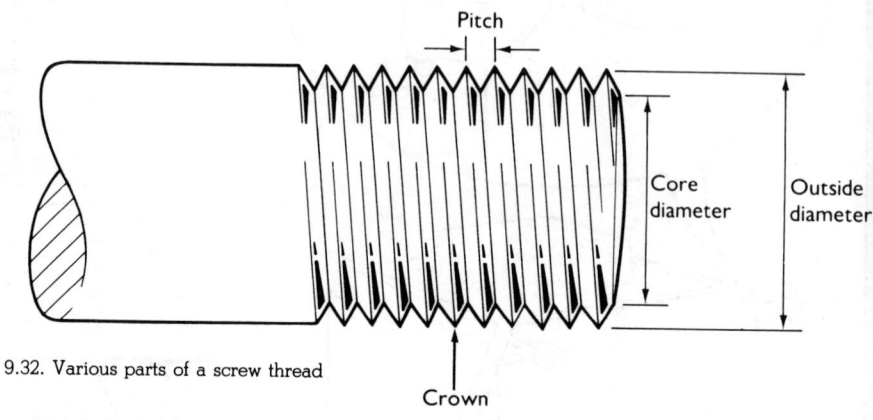

fig 9.32. Various parts of a screw thread

Pitch

Core diameter

Outside diameter

Crown

SCREW THREAD TYPES

There are several different types of screw thread, each designed for a particular job. The most common one is the **V-thread** (fig 9.33), as used for nuts, bolts and screws. Its 'V' shape causes a lot of friction, making it ideal for holding things in place. There are many variations of the V-thread (e.g. BSW, BSF, BA, BSP, UNC). Although still in use these are all gradually being replaced by the **metric** thread. As its name suggests, it is based totally on metric measurements.

When specifying a V-thread you need at least two pieces of information. They are the type of thread and its outside diameter. For example, M8 means it is a metric thread of 8mm outside diameter, while $\frac{1}{4}$" BSW means it is a British Standard Whitworth thread of $\frac{1}{4}$ inch outside diameter.

The **square thread** is used for the moving parts of things such as cramps, vices and jacks (fig 9.34). It does not cause as much friction as a V-thread, but allows you to apply large forces to your work.

Vices with quick release mechanisms use a **buttress thread**. This allows you to apply forces in one direction to hold your work in place, then to open the vice quickly by releasing the half nut (fig 9.35).

The **acme thread** is used on machines where the nut which engages with the thread has to do so while the thread is turning. The leadscrew of a lathe is an example of this (fig 9.36). The leadscrew is turned by the same electric motor that turns the lathe chuck. By engaging the nut the cutting tool is made to travel along the work automatically. It has to be disengaged when the tool reaches the end of the cut. The acme thread is ideal for this job because of its sloping sides which allow the nut to engage and disengage easily with the rotating thread.

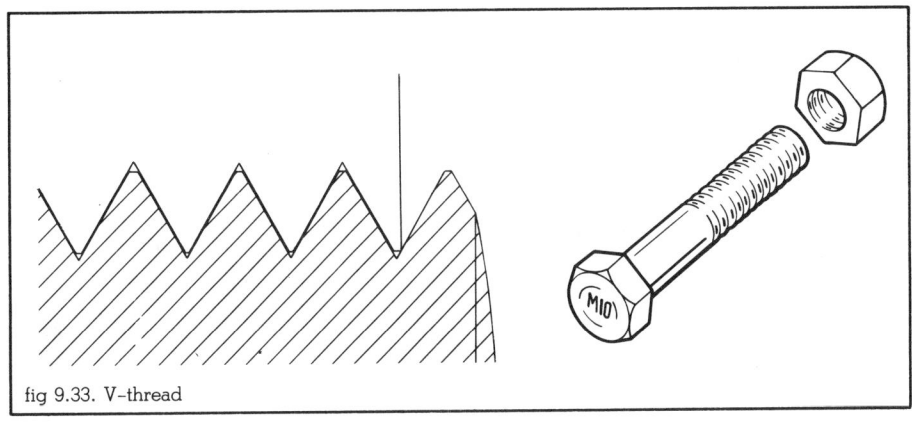

fig 9.33. V-thread

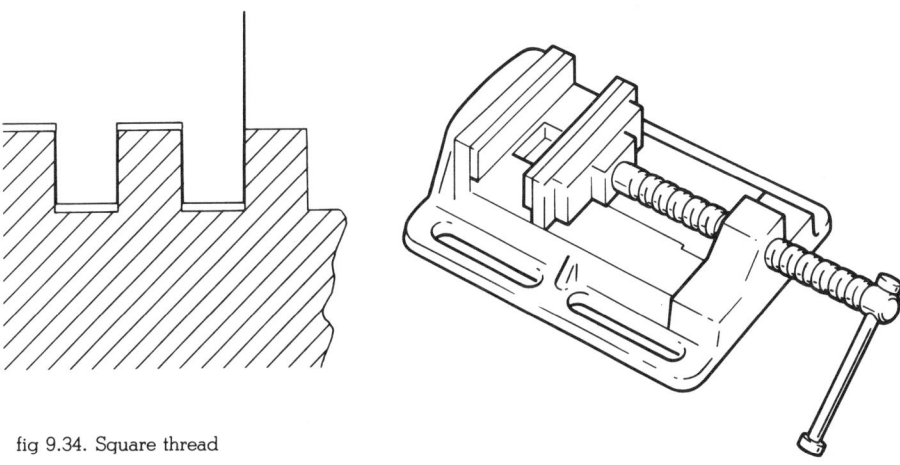

fig 9.34. Square thread

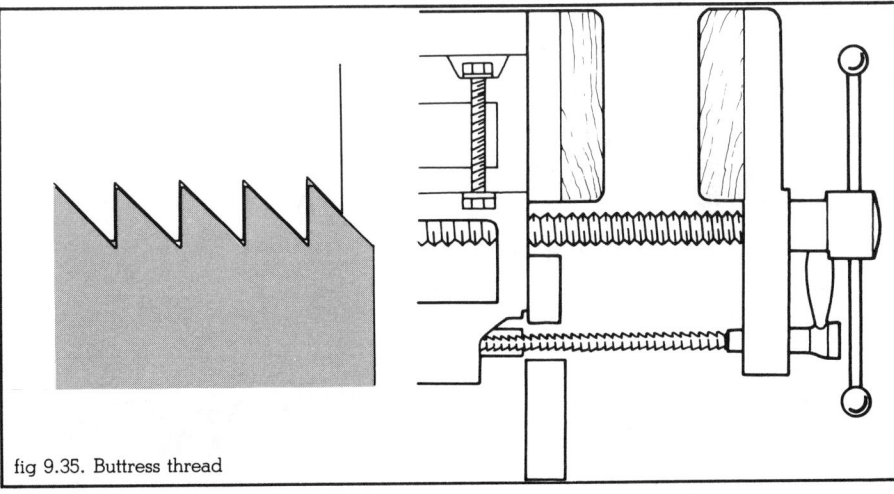

fig 9.35. Buttress thread

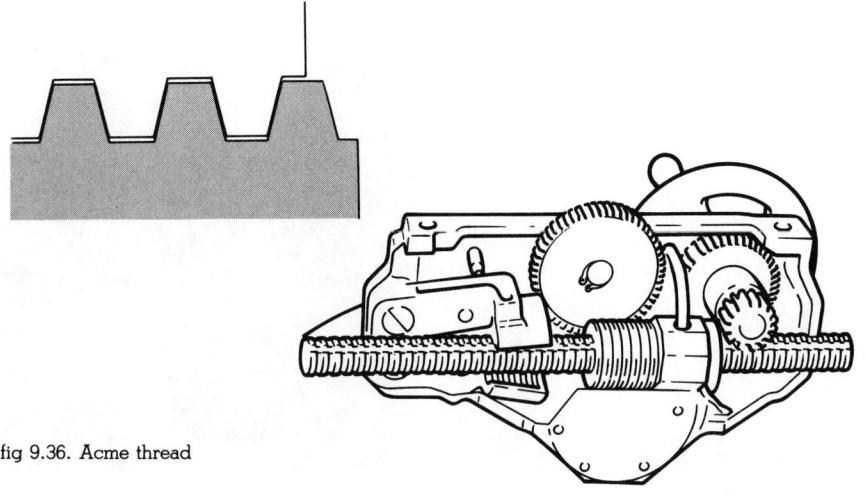

fig 9.36. Acme thread

LIFTING SYSTEMS

Pulleys have been used for thousands of years to make the job of lifting large weights easier. The simplest systems use just one pulley. This allows a load to be lifted up by pulling down. It is much easier and safer to be on the ground pulling down rather than up on a roof pulling a load upwards. Fig 9.37 shows such a pulley in use on a building site today. In theory, to raise the load, your effort force has to be equal to or just greater than the load. There is no mechanical advantage. In practice, with a heavy load, your effort has to be quite a bit more than the load in order to overcome friction.

By using more than one pulley you can create a mechanical advantage. The more pulleys you use, the greater that advantage is. Fig 9.38a shows a pulley system with a mechanical advantage of 2:1. It allows a load to be raised using only half the effort needed using a single pulley, but twice as much rope must be pulled in. Fig 9.38b shows pulley systems with mechanical advantages of 4:1 and 6:1. Note that you can work out the mechanical advantage simply by counting the number of ropes passing between the top and bottom sets of pulleys. As more pulleys are used, less effort is needed to raise the load, but much more rope has to be pulled in. Yachts and other sailing vessels make great use of these kind of pulley systems to cope with the large forces involved in controlling large sails (fig 9.38c).

Fig 9.39 shows a modern 'Midget Hoist' pulley system, with a mechanical advantage of 8:1. This compact, durable system is capable of lifting up to 300kg. Notice that the pulleys are side by side, rather than as shown in the earlier diagrams. Note also the safety device which makes sure that the load does not drop when the rope is released.

fig 9.37. Single pulley in use

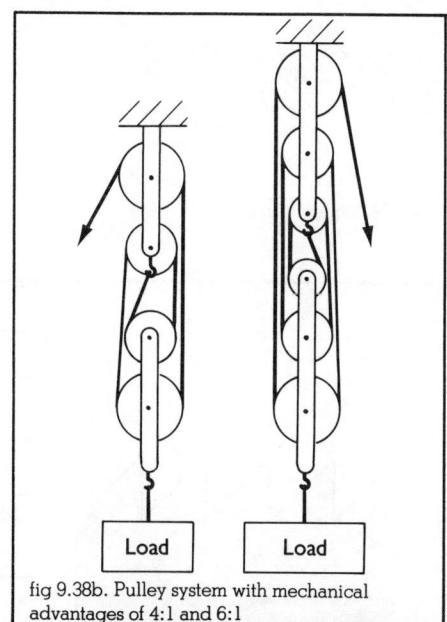

fig 9.38b. Pulley system with mechanical advantages of 4:1 and 6:1

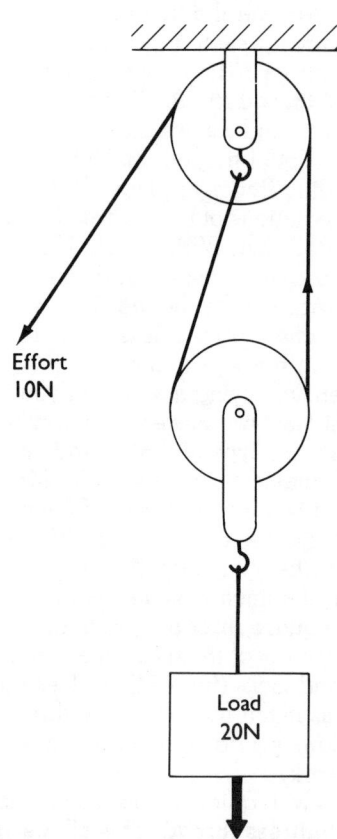

fig 9.38a. Pulley system with a mechanical advantage of 2:1

fig 9.38c. Pulley system used on a sailing dinghy to control mainsail

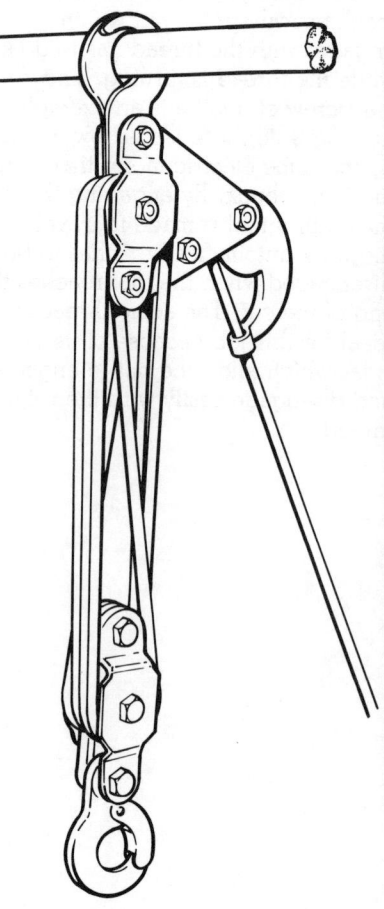

fig 9.39. Midget hoist

The **Weston differential pulley** (fig 9.40) is a special kind of pulley system which has a continuous chain going round two upper pulleys and a lower one which moves up and down with the load. The two upper, different sized, pulleys are firmly fixed to each other. To prevent the load slipping, toothed pulleys and a continuous chain are used (fig 9.41). The mechanical advantage of this system is not quite so obvious. You can calculate it using the sizes of the two top pulleys:

$$\text{Large pulley radius (R)} = 100\text{mm}$$
$$\text{Small pulley radius (r)} = 60\text{mm}$$

$$MA = \frac{2 \times R}{(R-r)} = \frac{200}{(100-60)} = \frac{200}{40} = 5:1$$

The velocity ratio, assuming 100% efficiency, will always be equal to the mechanical advantage.

This kind of system is used a great deal in industry, both in its manual and powered form. Its advantages are that it is compact, it uses a continuous loop (no loose ends) and it always holds the load safely. Fig 9.42a shows a typical electrically powered hoist. Others use compressed air or hydraulics as a power source.

An additional safety feature on this, and many other lifting systems, is a **ratchet** and **pawl** (fig 9.43a). It allows rotation in one direction, but stops any rotation in the other direction. Ratchet and pawl mechanisms are also found in some screwdrivers, socket sets and winches (fig 9.43b).

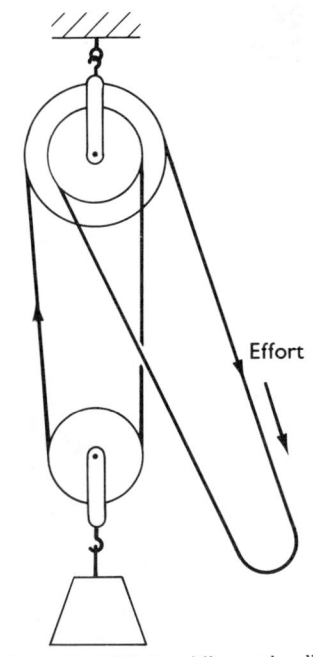

fig 9.40. Principle of Weston differential pulley

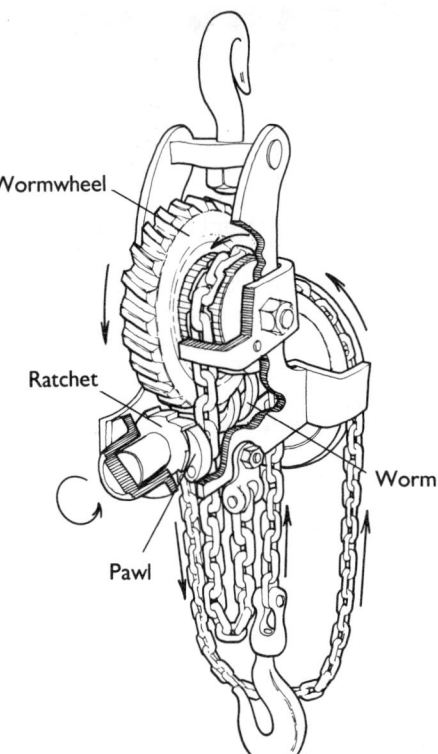

fig 9.41. Chain and toothed pulley

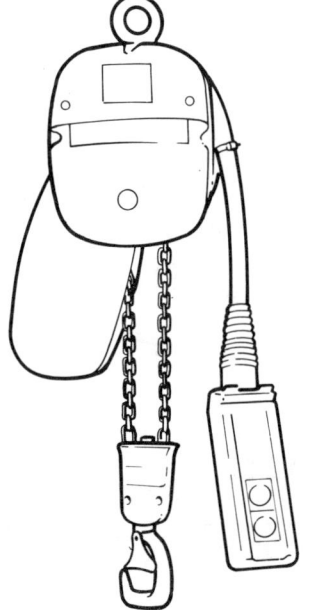

fig 9.42. A modern powered hoist

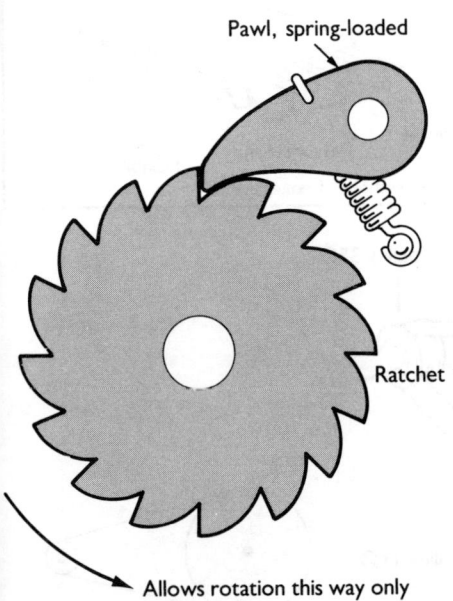

fig 9.43a. Principle of ratchet and hoist

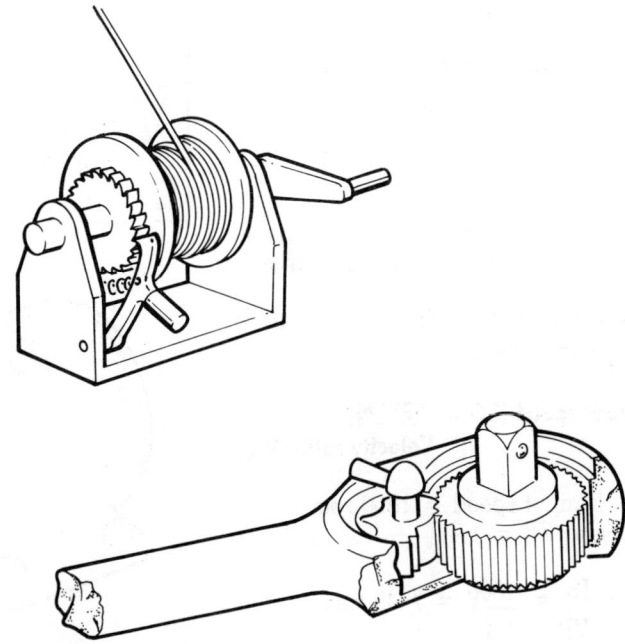

fig 9.43b. Applications of ratchet and pawl

133

ROTARY SYSTEMS

The majority of machines use some kind of rotary movement. Some, like the bicycle, are totally based on rotating parts. Others use a rotary input motion which they change into a different output motion, a graphic plotter is a good example of this (fig 9.44). The rotary motion of its stepper motors is turned into linear movement of the pen. A car engine does things the other way round, it changes the reciprocating motion of the pistons into a rotary motion of the wheels (fig 9.45).

Internal combustion engines or electric motors are often used to provide a rotary input movement and force to a machine. The input speed is rarely the one needed for the output. A means has to be found of connecting the input and output while also changing the speed. It may also be necessary to reverse the direction of rotation at the same time. These things can be done using either **pulley** systems, **chain** and **sprocket** systems or **gear** systems, or a combination of the three.

fig 9.44. Plotter showing motor mechanism

PULLEY SYSTEMS

Pulley systems use a belt to transmit motion and force from the driver shaft to the driven shaft. The continuous **V-belt** is the one most often used. It fits tightly into the groove on the pulley wheels to keep slipping to a minimum (fig 9.46). V-belts come in a variety of widths and thicknesses. Fig 9.47 shows some of the more common shapes and shows two alternative methods of construction. Construction 2 allows the belt to bend around the pulleys more easily which means that it will last longer.

Speed changes are made by using different size pulleys on the driver and driven shafts. By comparing the size of the two pulleys you can calculate the **velocity ratio** of the system.

For example in fig 9.48:
- Driver pulley = 140 diameter
  Driven pulley = 35 diameter

$$\text{Velocity ratio} = \frac{\textbf{Driven} \text{ pulley diameter}}{\textbf{Driver} \text{ pulley diameter}}$$

$$= \frac{35}{140} \quad \frac{1}{4} \text{ or } 1:4$$

In other words, one turn of the driver shaft will give four turns of the driven shaft.

The speed of the driven shaft can be calculated using:

$$\textbf{Output speed } (OS) = \frac{\textbf{Input speed } (IS)}{\textbf{Velocity ratio } (VR)}$$

e.g. If input speed = 1860 rpm
  and VR = 1:4

$$OS = \frac{IS}{VR} = \frac{1860}{1:4} = 7440 \text{ rpm}$$

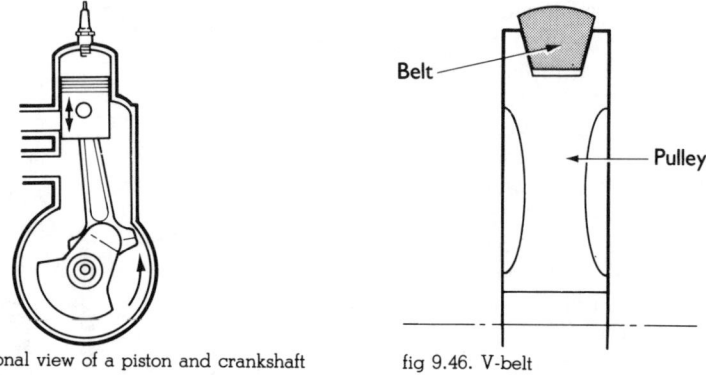

fig 9.45. Sectional view of a piston and crankshaft

fig 9.46. V-belt

Belt

Pulley

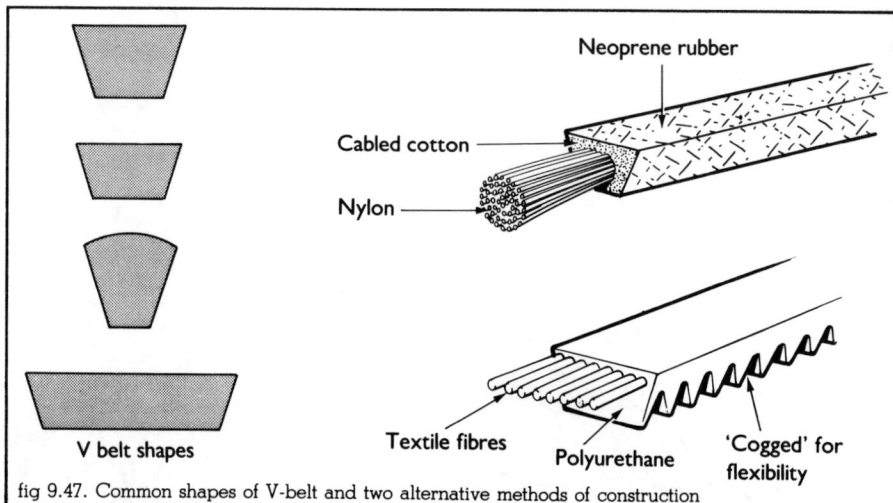

Neoprene rubber

Cabled cotton

Nylon

V belt shapes

Textile fibres

Polyurethane

'Cogged' for flexibility

fig 9.47. Common shapes of V-belt and two alternative methods of construction

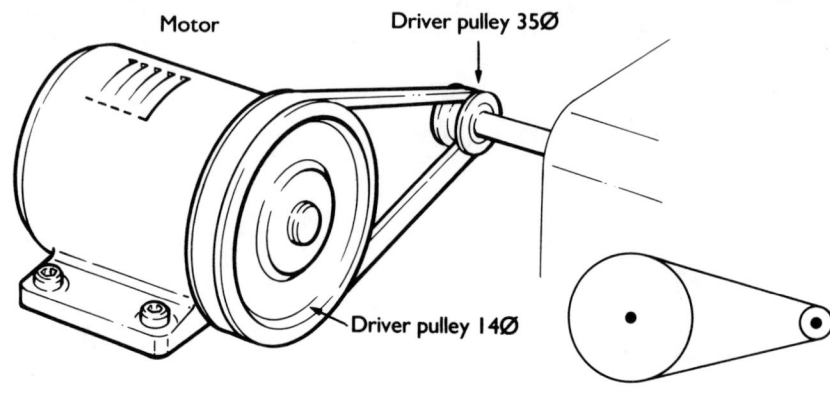

Motor

Driver pulley 35Ø

Driver pulley 14Ø

fig 9.48. Different sized pulleys

TORQUE

In rotary systems the **turning force**, known as torque, produced by the driver shaft is transferred to the driven shaft. Any speed changes also cause changes in the turning forces produced by the system. If the system speeds things up, the torque at the output shaft is reduced.

For example, fig 9.49 shows an electric motor, used to power a drilling machine, which produces a torque of 100Nm. The torque produced at the driven shaft is given by:

Output torque = **Input torque** × **VR**
= $100 \times 1/2$
= 50Nm

If the system slows things down, the torque at the output shaft is increased. The slower a shaft is going the more torque it is likely to have. Cranes only lift things slowly, but their slow running lifting winches produce very large forces.

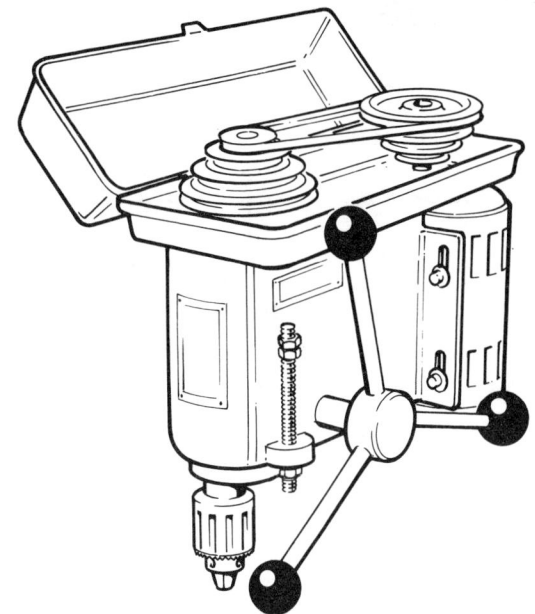

fig 9.49. Electric motor used to power a drilling machine

LINKED BELTS

In some situations it is impossible or very difficult to fit a continuous belt. **V-link** belts have been developed to overcome this problem (fig 9.50). These can be made to any desired length very simply, as shown in fig 9.51.

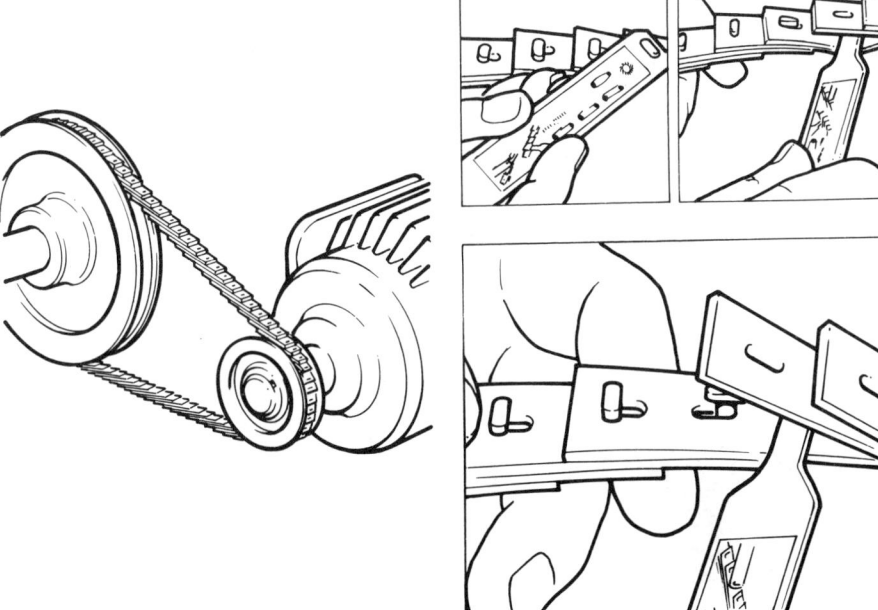

ROUND BELTS

Round belts are used where small forces are involved or where the belt has to **twist**. Fig 9.52 shows a round belt used to drive the brushes of a vacuum cleaner. Note that the driver and driven shafts are at 90° to each other. A V-belt would not stay on if used like this, but the round belt does the job very well. Round belts are also used on sewing machines, where only small forces are involved.

fig 9.50. Linked belt

fig 9.51. Changing belt length

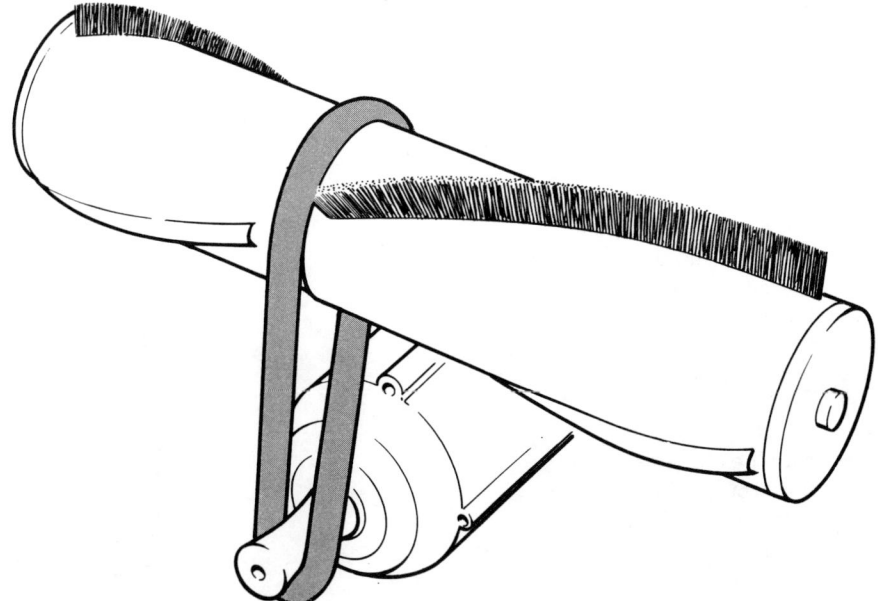

fig 9.52. Vacuum cleaner drive belt

MULTIPLE BELTS

The forces acting on a belt are often at their highest when a machine starts up. Where **large forces** are involved, more than one belt has to be used (fig 9.53). By using several V-belts alongside each other the forces are shared. It is important that the belts are exactly the same length. If they are not, some belts will be taking much more of the load than others, and so are more likely to break.

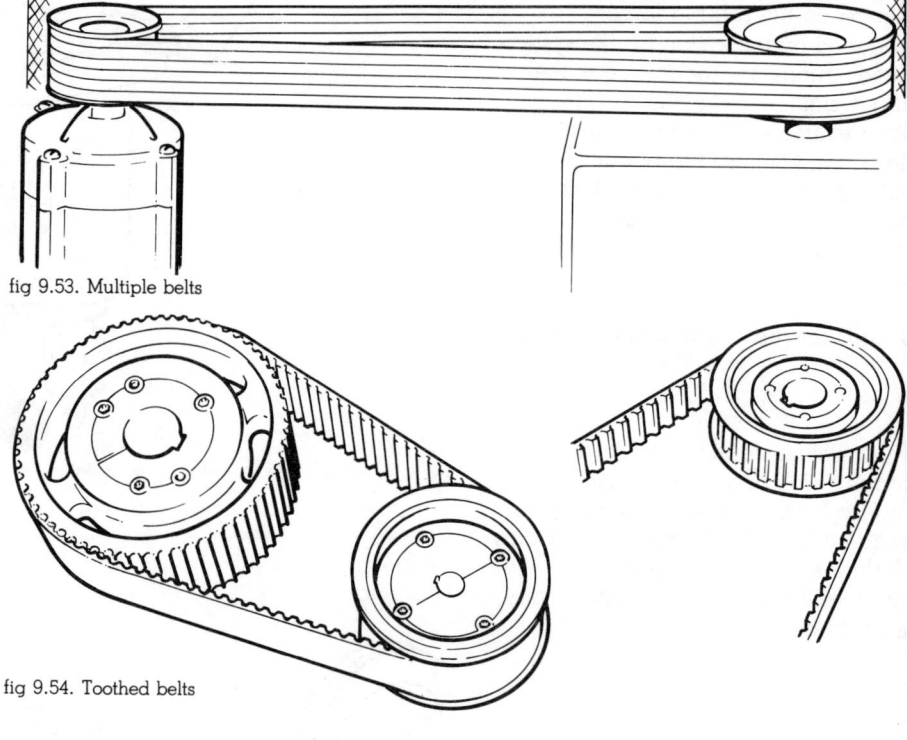

fig 9.53. Multiple belts

TOOTHED BELTS

One of the problems with belt drives is that the belts can slip, causing the driven shaft to rotate slower than expected. Where it is vital that the rotation of the driven shaft is kept in sequence with the driver shaft, a toothed belt can be used (fig 9.54). They are used for such things as timing when the valves open and close in a car engine or for moving the pen of a graphic plotter accurately.

fig 9.54. Toothed belts

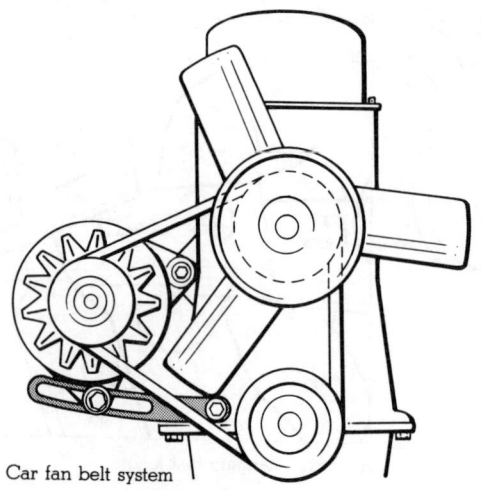

fig 9.55. Car fan belt system

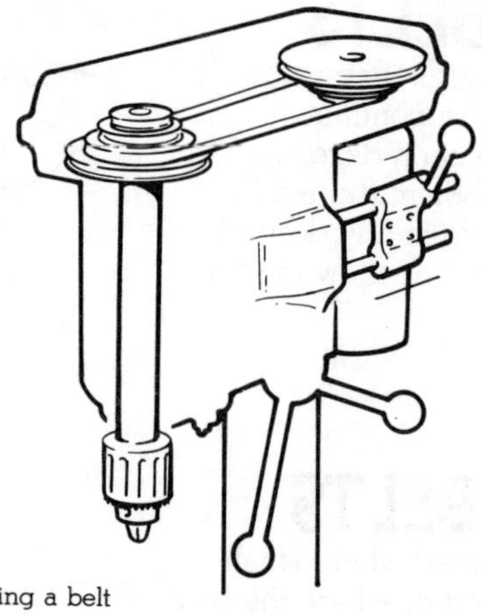

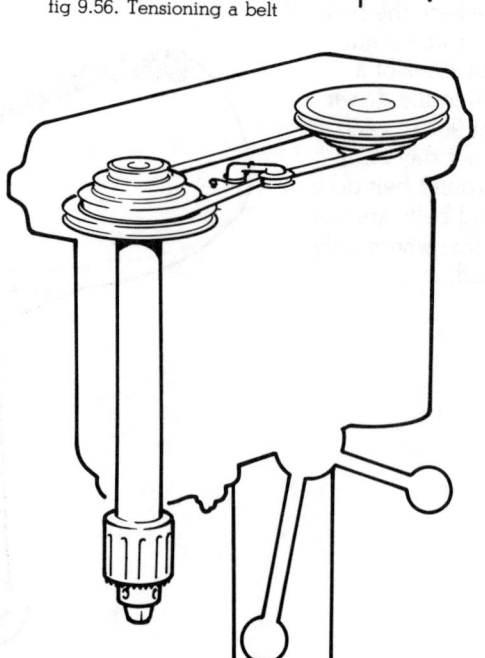

fig 9.56. Tensioning a belt

BELT TENSION

Whatever kind of belt is used, it must be tensioned correctly, not too tight and not too slack. If it is too tight, it will apply bending forces to the pulley shafts. If it is too slack it will slip or come off the pulleys. Most systems are tensioned by having the position of one of the pulleys adjustable. There are various ways of doing this, but the most common is the one shown in fig 9.55. Once the tension is correct, the adjustable pulley is locked in position. Figs 9.56 and 9.57 show two other ways of tensioning a belt.

Most machines in your school workshop use pulley systems. As well as the system on a drilling machine, there are similar systems on lathes and milling machines.

fig 9.57. Tensioning a belt using a spring loaded jockey wheel

CHAIN AND SPROCKET SYSTEMS

Chain and sprocket systems use a chain to transmit rotary motion from the driver shaft to the driven shaft. **Sprockets** are the toothed wheels on which the chain runs. Unlike some pulley systems the chain and sprocket cannot slip. Bicycles and motorbikes use a chain and sprocket system (fig 9.58), because of its strength and because it will not slip. Like a pulley and belt system, a chain needs to be correctly tensioned. On a bicycle, this is done by moving the position of the back wheel or, if the bike has derailier gears, by the spring loaded **jockey wheels** (fig 9.59). One of the disadvantages of a chain and sprocket system is that it needs to be well oiled, particularly on a bicycle, if it is not to go rusty.

When it comes to working out speed changes, sprocket and chain systems are very similar to pulley systems. The only difference is that instead of using the pulley diameters you use the number of teeth on the sprockets.

For example looking at the system shown in fig 9.59:

Driver sprocket has 60 teeth
Driven sprocket has 15 teeth

$$VR = \frac{\text{No of teeth on } \textbf{driven} \text{ sprocket}}{\text{No of teeth on } \textbf{driver} \text{ sprocket}}$$

$$= \frac{15}{60} = \frac{1}{4} \text{ or } 1:4$$

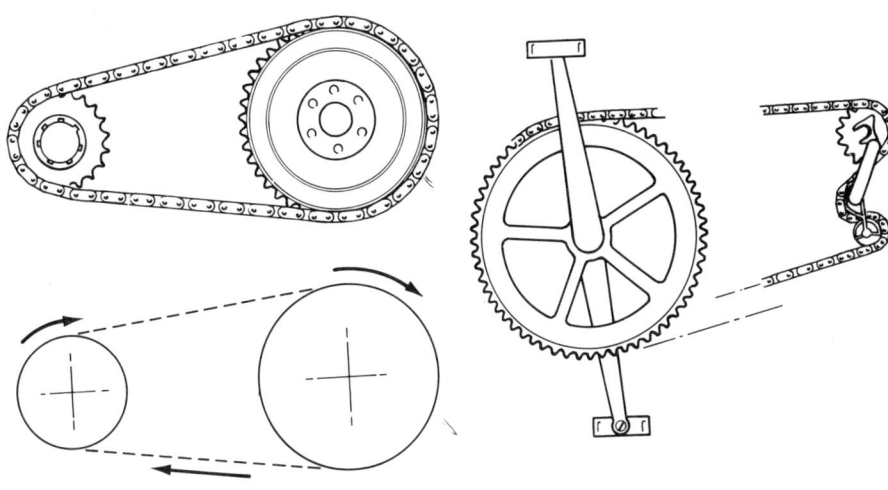

fig 9.58. Chain and sprocket system on a push bike and motor bike

fig 9.59. Derailleur gears on a bike

fig 9.60. Conveyor belt taking parts to an industrial robot

CONVEYORS

Chains and belts can also be used to move things from place to place. You have probably seen a conveyor belt at a supermarket checkout. Conveyor belts and chains are used widely in industry to move parts and finished goods around a factory. Fig 9.60 shows a conveyor belt being used to transport manufacture products for packing. Fig 9.61 shows a conveyor chain being used to move car bodies along the production line. The stacker truck shown in fig 9.62 uses a chain and sprocket system to raise and lower the load.

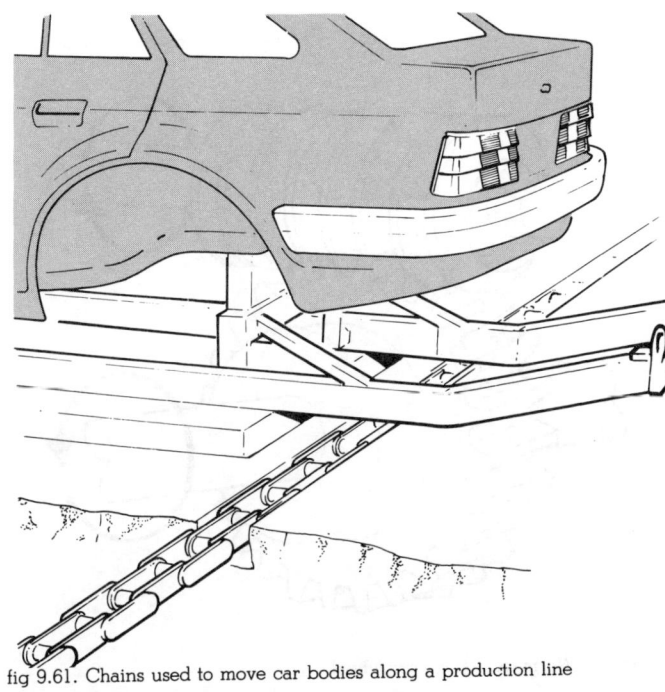

fig 9.61. Chains used to move car bodies along a production line

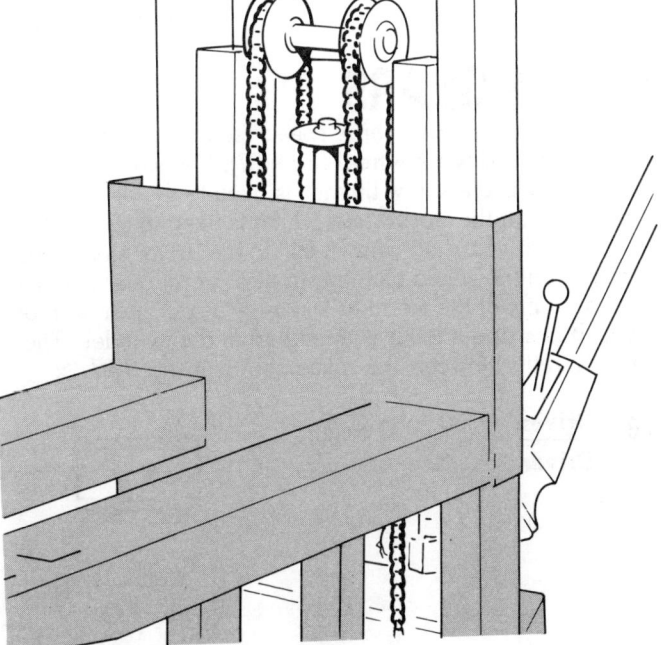

fig 9.62. Chains used on a small stacker truck

GEAR SYSTEMS

Gears are toothed wheels, fixed to the driver and driven shafts, which mesh together. A number of gears connected together is called a **gear train**. Fig 9.63 shows a pair of **spur gears**, fixed to parallel shafts, forming a **simple gear train**. The shafts will turn in opposite directions and, because the gears are different sizes, at different speeds. The difference in their speeds (velocity ratio) can be calculated from the number of teeth on each gear:

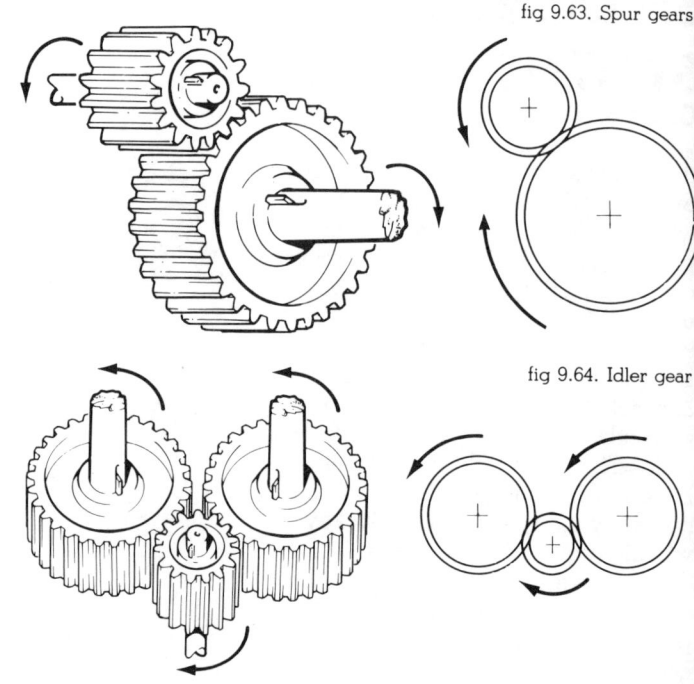

fig 9.63. Spur gears

$$VR = \frac{\text{No of teeth on } \textbf{driven} \text{ gear}}{\text{No of teeth on } \textbf{driver} \text{ gear}} = \frac{30}{15} = \frac{2}{1} \text{ or 2:1 or 2}$$

In other words, two turns of the driver shaft are needed to give one turn of the driven shaft.

To get them to turn in the same direction, a third gearwheel has to be fitted between them, as in fig 9.64. This **idler gear** has no effect on the speeds of the other two shafts, whatever its size. It simply makes the driver and driven shafts rotate in the same direction.

fig 9.64. Idler gear

Compound gear trains (fig 9.65) involve several pairs of meshing gears. They are used where it is necessary to make large speed changes or to get different outputs moving at different speeds. Fig 9.66 shows a speed reducer connected directly to an electric motor. It uses a compound gear train to give an output speed 72 times slower than the motor input speed. There are three pairs of gears involved. To work out the total velocity ratio you have to know the velocity ratio of each pair. In this reducer the three are:

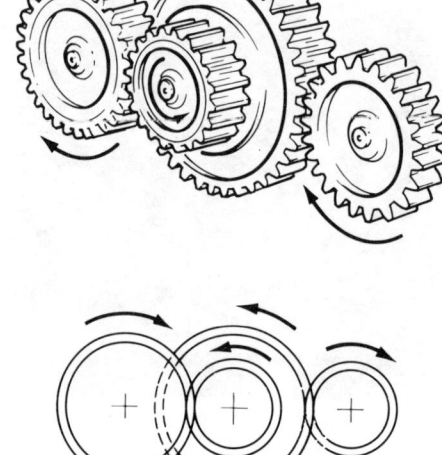

VR1 = 3:1, **VR2** = 4:1, **VR3** = 6:1

$$\textbf{Total VR} = \textbf{VR1} \times \textbf{VR2} \times \textbf{VR3}$$
$$= 3:1 \times 4:1 \times 6:1$$
$$= 72:1$$

fig 9.65. Compound gear train

fig 9.66. Speed reducer connected to motor

WORM GEARS

Another way of making large speed reductions is to use a **worm gear** and **wormwheel** (fig 9.67). The worm, which looks rather like a screw thread, is fixed to the driver shaft. It meshes with the wormwheel which is fixed to the driven shaft. The driven shaft runs **at 90°** to the driver shaft. When considering the speed changes in most worm gear systems, you can think of the worm as if it were a spur gear with one tooth. It is a single tooth wrapped around a cylinder. The velocity ratio between the gears shown in fig 9.67 is:

$$VR = \frac{\textbf{Driven}}{\textbf{Driver}} = \frac{30}{1} = 30 \text{ or 30:1}$$

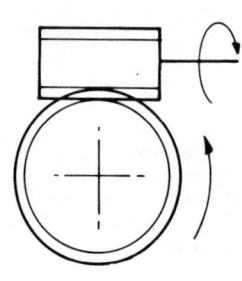

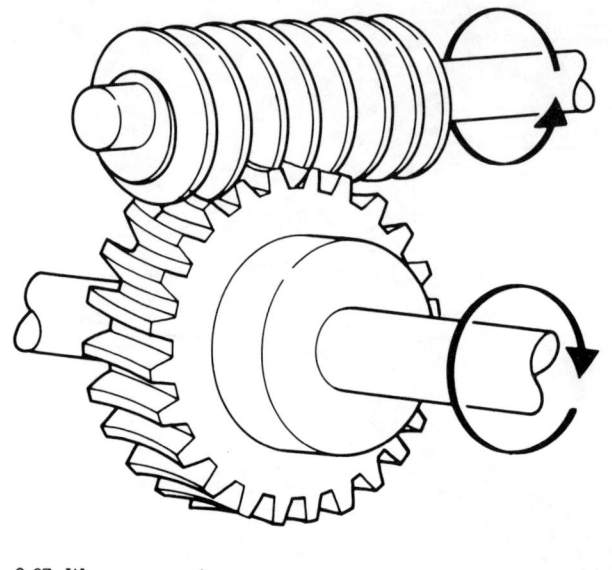

fig 9.67. Worm gear and wormwheel

The industrial worm gear unit shown in fig 9.68 gives a speed reduction of 60:1. Because they are so compact and give such large speed changes, these systems are widely used in industry. Their only disadvantage is that they change the direction of drive through 90°, which is not always wanted. One way of getting the input and output shafts running parallel again is to use two worm gear systems connected together, as in the unit shown in fig 9.69. This double arrangement can also be used to give very large speed reductions. The one shown would give a speed reduction of 640:1.

fig 9.68. Industrial worm gear unit

fig 9.69. Double worm gear units

Total VR = VR1 × VR2
$$= 16:1 \times 40:1$$
$$= 640:1$$

BEVEL GEARS

Bevel gears, like worm gears, use **shafts at 90° to each other** (fig 9.70). The hand drill shown in fig 9.71 uses them not only to change the rotary motion through 90°, but also, by using different sized gears, to increase the speed of rotation. The one shown gives a speed increase of 1:5.

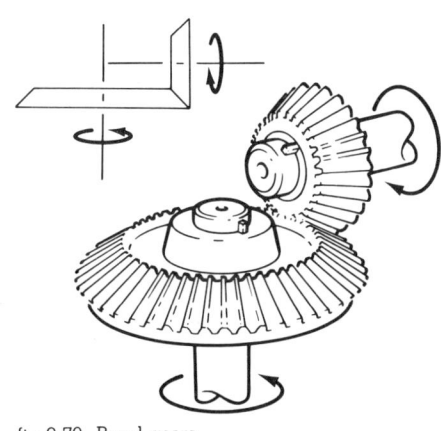

fig 9.70. Bevel gears

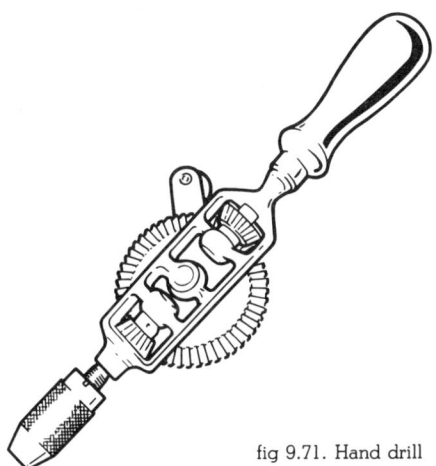

fig 9.71. Hand drill

HELICAL GEARS

Helical gears have their teeth **at an angle** across the gearwheel (fig 9.72). Each tooth is very slightly curved. Its shape is part of a helix, a type of spiral. Helical gears are quieter and more efficient than normal spur gears. They are used in things like gearboxes where smooth, quiet, efficient transfer of power is important. By angling the teeth even more, they can also be used to change the direction of drive through a 90° angle (fig 9.73).

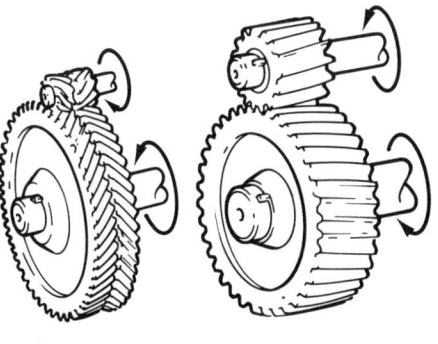

fig 9.72. Helical gears

fig 9.73. Crossed helical gears

RACK AND PINION SYSTEMS

Rack and pinion systems (fig 9.74) involve changes between rotary and linear motion. They can be used either way round. You use this type of system on a drilling machine to bring the drill down into the work (fig 9.75).

Gearwheels are normally made of metal or plastic. Plastic gears have the advantage that they are much quieter running and need less lubrication. Many gear trains include one plastic gear to reduce noise.

fig 9.74. Rack and pinion

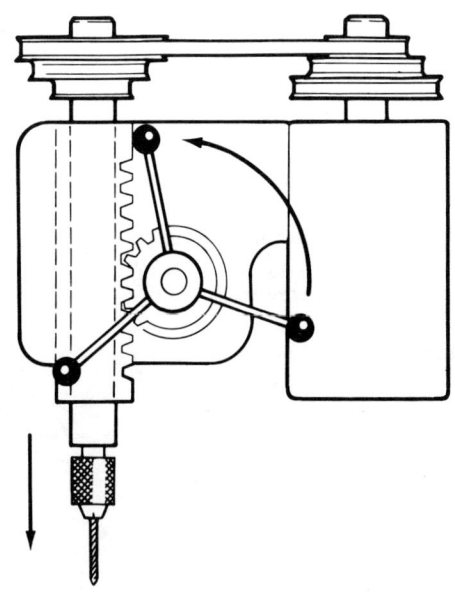

fig 9.75. Drill feel system

SHAFTS AND COUPLINGS

FIXING TO SHAFTS

Many parts, including pulleys and gears, have to be fitted to shafts. It is important that they are firmly fixed together so that they rotate together. The forces acting on them as they rotate are trying to separate them. Some of the more common ways of fixing them together are given below.

Grub screws can be used where relatively small forces are involved. These headless screws go down a threaded hole and press onto the side of the shaft. Their holding effect can be improved by having either a 'flat' or shallow hole on the shaft (fig 9.76). One way of preventing a grub screw from coming loose is to put a second grub screw in to press down the first one (fig 9.77).

Keys and **keyways** give a much more secure fixing method (fig 9.78). Keyways are simply grooves cut into the shaft and the part to be fixed to it. With the two grooves lined up, the key, which is slightly tapered, is tapped firmly into place.

Cotter pins also give a secure fixing method. They are used on bicycles to connect the pedal cranks to the crankshaft (fig 9.79). The tapered pin is tapped firmly into place, its flat side fitting into the flat on the shaft. It is pulled in and held in place by the nut on the other end.

An **interference fit** is another way of fixing parts onto a shaft. The hole is made slightly too small for the shaft (fig 9.80). When the two parts are pressed together the friction between them holds them firmly together. Many toys use this method because it is cheaper and easier than the other methods.

COUPLINGS

A common problem in rotary systems is that of how to connect two shafts together. Some of the more common ways of connecting them are given here.

Aligned shafts

Where the shafts are perfectly aligned (in line with each other) either a **flanged coupling** or a **muff coupling** can be used (fig 9.81). Notice that they both use a key and keyway to fix them firmly to the shafts. Fig 9.82 shows a flanged coupling used to connect a gearbox to a metal conveyor drive.

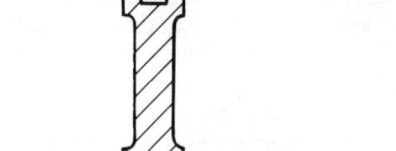
fig 9.77. Second grub screw pressing on first one

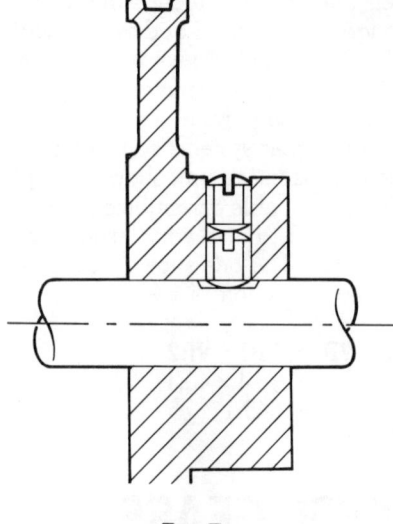

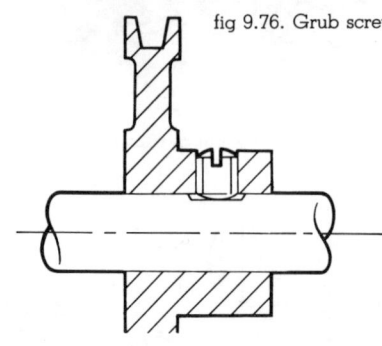
fig 9.76. Grub screw

fig 9.78. Keys and keyways

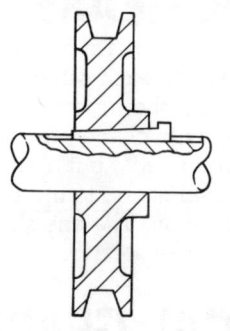
Nylon pulley wheel

Shaft 4mm ∅

Motor

pressed on

Hole 3.9mm ∅

fig 9.80. Interference fit

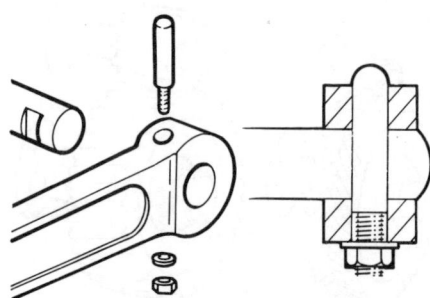
fig 9.79. Cotter pin

fig 9.81. Flanged coupling and muff coupling

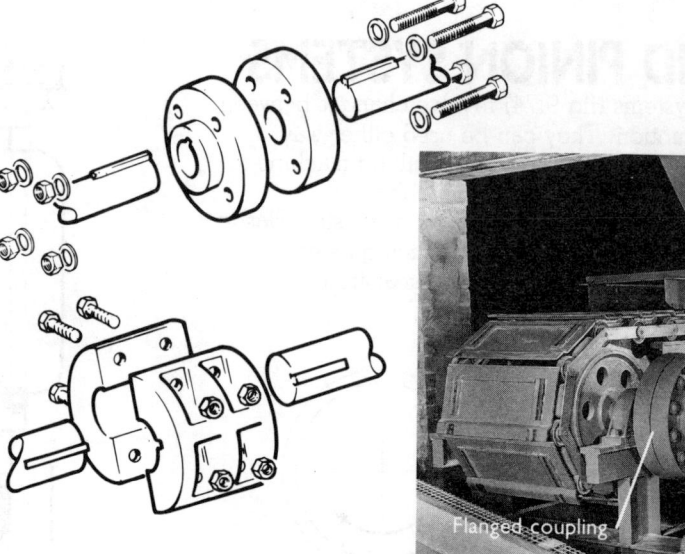
Flanged coupling

fig 9.82. Flanged coupling used to connect a gearbox to a metal conveyor drive

There are situations where shafts are aligned, but the length of one of them needs to be able to change. For example, as the rear wheels of a rear wheel drive car move up and down, due to bumps in the road, the prop shaft needs to be able to alter slightly in length. This is made possible by using a **splined coupling** (fig 9.83). The interlocking splines give a positive drive, while allowing the splined shaft to slide in and out of the coupling.

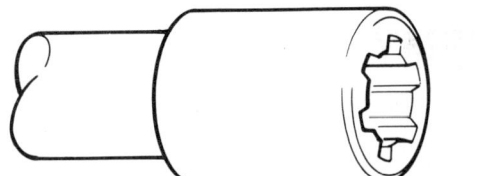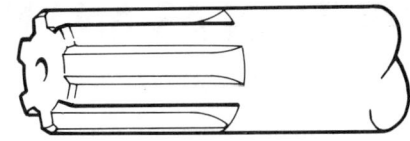

fig 9.83. Splined coupling

Non-aligned shafts

Many shafts are not perfectly aligned, they meet at a slight angle. Even while they are running, the angle between them may change, due to the effects of vibration and changing loads. The couplings used to join them have to be flexible. The simplest type of **flexible coupling** (fig 9.84) is a rubber disc to which each of the shafts is bolted. The flexibility of the disc allows for the changes in angle between the two shafts. Joints like this are quite cheap and easy to make, but can only be used where shafts are slightly out of line.

Where shafts are more than a few degrees out of alignment, a **universal joint** is often used. It can transmit drive through angles of up to 20°. A simple universal joint, the Hooke's type, is shown in fig 9.85. The two yokes, connected to the shafts, pivot on the central 'spider'. Fig 9.86 shows a better quality universal joint which uses lubricated roller bearings at the pivots. This type is often used on transmission systems because of the high speeds and large forces involved. One problem with universal joints is that because the two shafts are at an angle to each other, the output shaft does not rotate at a constant speed. The output shaft goes alternately faster then slower than the input shaft. This problem is overcome by always using universal joints in pairs (fig 9.87). The changes in speed taking place at one joint are cancelled out by those at the other. If this is to work correctly, the driver and driven shafts must be at the same angle to the intermediate shaft and the yokes of the two joints must be in line with each other (fig 9.88).

If the angles are greater than 20° or there is not room to use two universal joints, a **constant velocity joint** is used (fig 9.89). With this type of joint the driven shaft goes at a constant speed whatever the angle of the driver shaft. Front wheel drive cars use them to connect the drive shafts from the engine to the wheels.

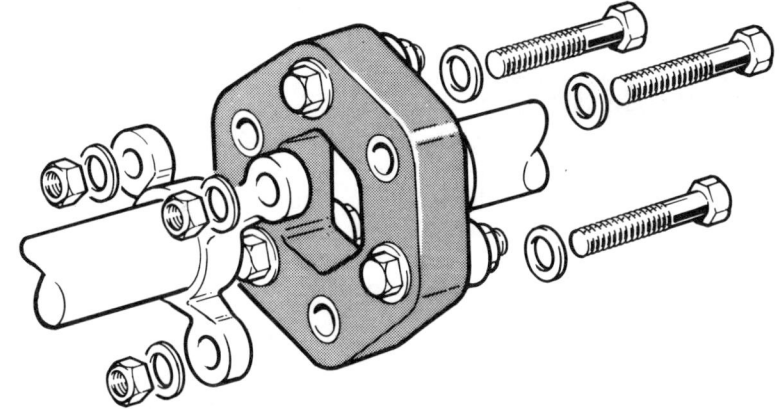

fig 9.84. Flexible coupling

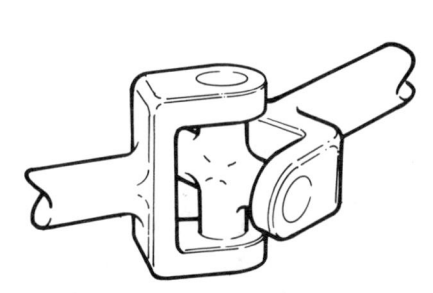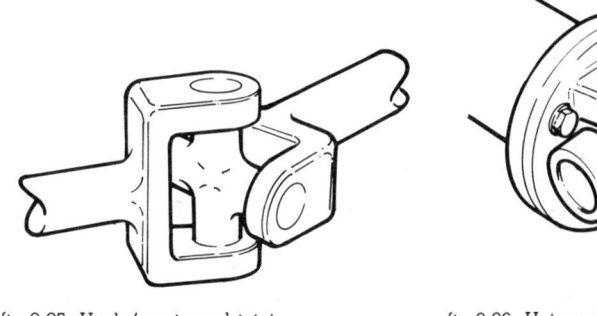

fig 9.85. Hooke's universal joint

fig 9.86. Universal joint with lubricated roller bearing at the pivots

Universal joint   Universal joint

Driver shaft   Intermediate shaft   Driven shaft

fig 9.87. Pair of universal joints

fig 9.88. Pair of universal joints with driver and driven shaft at the same angle to the intermediate shaft

fig 9.89. Constant velocity joint

BEARINGS AND LUBRICATION

FRICTION AS AN ENEMY

Any rotating shaft has to be supported in some way. Bearings are used to give that support. Bearings may have to cope with one or two different types of load, **radial** and **axial** (fig 9.90). Radial loads are those caused simply by the shaft rotating. **Radial bearings** have to allow the shaft to rotate as freely as possible while still being a good fit so as to stop it moving from side to side. Axial loads are those acting along the shaft, pushing the shaft into the bearing. **Axial**, or **thrust bearings** have to allow the shaft to rotate as freely as possible while resisting the axial forces.

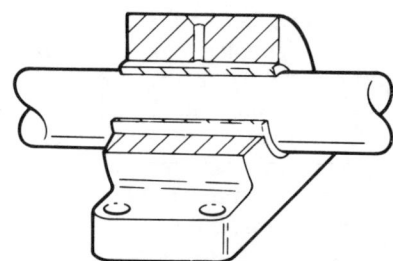

fig 9.90. Radial and axial loads

PLAIN BEARINGS

The simplest form of bearing is a **plain bearing** or **bush** (fig 9.91). This is fixed firmly into the component. The shaft rotates in the bearing, rubbing against it. The problem of friction between the rubbing surfaces is overcome either by using materials which slide against each other, or by using lubricating oil. Bearings may be made of some sort of plastic, bronze or white metal. The type of material used for any particular job will depend on what the shaft is made of and what loads it has to take.

Plastic, particularly nylon, is used where light loads are involved. Plastic bearings are cheap, quiet running, easy to make, do not corrode and do not need lubricating. They are often used on such things as printers, tape decks and food mixers. The one shown in fig 9.92 is designed to take both radial and axial loads.

Bronze bearings can take greater loads than plastic ones. Phosphor-bronze and sintered-bronze bearings, like plastic ones, need little or no lubrication. They are used on small electric motors (e.g. electric shavers, drills and sanders).

Split bearings are used in combustion engines to support the crankshaft (fig 9.93). Because of the shape of the crankshaft, plain bearings cannot be used. The steel bearing is coated with a thin layer of white metal. It is much softer than the steel crank it is supporting and so must be well lubricated. Fig 9.93 shows the grooves cut into the surface of the bearing, into which oil is pumped to spread it all over the bearing surface. The advantage of using white metal for this job is that if the lubrication does fail, it will be the cheap white metal that is damaged and not the expensive crankshaft.

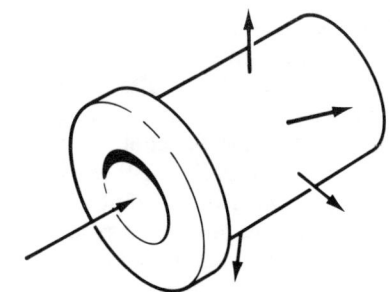

fig 9.91. Plain bearing

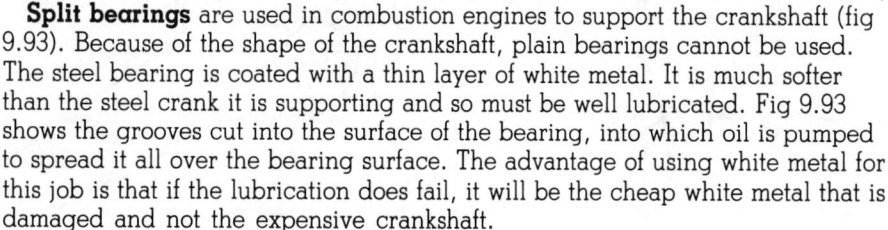

fig 9.92. Plastic bearing

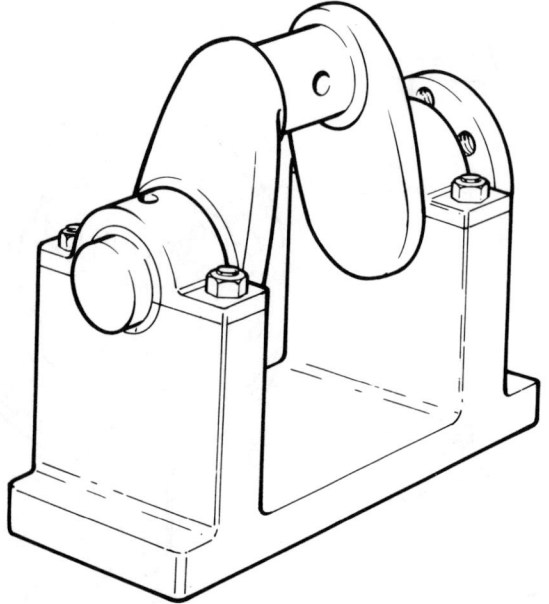
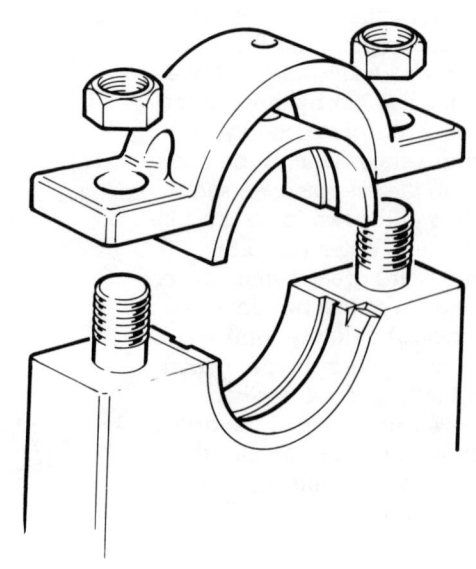

fig 9.93. Split bearing

BALL BEARINGS

Ball bearings get over the problem of having a shaft rubbing against a bearing. A row of hardened steel spheres, fitted between the shaft and the component, roll round as the shaft spins. Fig 9.94 shows the basic parts of a 'ball bearing'. The outer race is firmly fixed to the component and the inner race is firmly fixed to the shaft. They are usually a tight 'press' fit. The spheres, running in grooves inside the two races, are kept from rubbing against each other by the cage. The cage can be made of steel, brass or plastic. Ball bearings are usually lubricated using grease. Some have shields fitted to keep the grease in and dust out. Ball bearings can also be used to take axial loads or both radial and axial loads at the same time (fig 9.95). They are widely used on all types of machinery, including lawn mowers, washing machines and lathes.

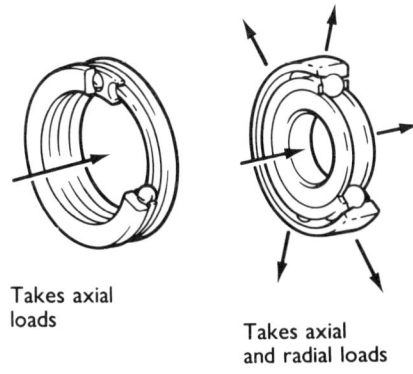

fig 9.94. Basic parts of a 'ball bearing'

ROLLER BEARINGS

Roller bearings are similar to ball bearings in many ways. The basic difference is that they use steel rollers instead of balls. They are used where greater loads have to be carried or where space is limited. Where space is really limited needle roller bearings are used (fig 9.96). Taper roller bearings, which will cope with radial and axial loads, are used for the wheel bearings of most vehicles (fig 9.97).

LUBRICATION

Any two parts of a machine which rub together, get hot due to friction. The aim of all lubrication is to separate moving parts, so stopping them from rubbing together. The lubricant, in the form of oil or grease, gets between the moving parts, so reducing friction, heat and wear. Different oils and greases are made to cope with different conditions. The thin oils, those with a low **viscosity** (e.g. SAE 20), flow easily into small gaps, but are also easily squeezed out. They are used on such things as sewing machines and door hinges where light loads are involved. The thicker oils, those with a high viscosity (e.g. SAE 75), do not penetrate small gaps so easily, but do resist being squeezed out. Therefore, they can cope with higher loads, such as in a car gearbox.

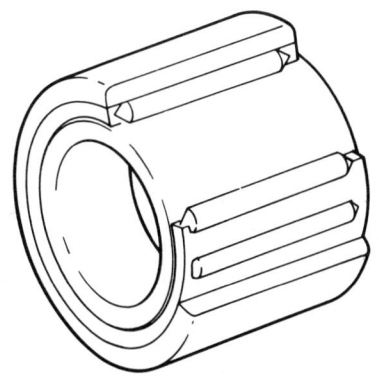

Takes axial loads

Takes axial and radial loads

fig 9.95. Ball bearing taking loads

EFFECTS OF HEAT

All oils are affected by heat. Most get thinner as they get hotter, their viscosity decreases. This has to be taken into account when deciding what type of oil to use for a particular job.

Multigrade oils get over this problem and are used where high temperatures and loads are involved. Most car engines use SAE 20/50 multigrade oil because of its special properties. When cold, it has a viscosity of SAE 20, which means that it flows easily. This is very important in a cold engine. As the engine warms up, instead of getting thinner, the oil gets thicker, having a viscosity of SAE 50. This makes it more able to cope with the high temperatures and large forces. It changes state again when the oil cools.

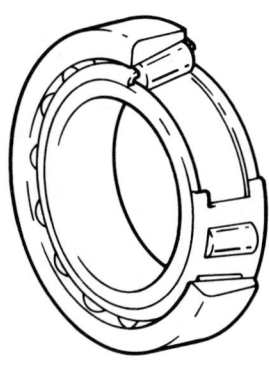

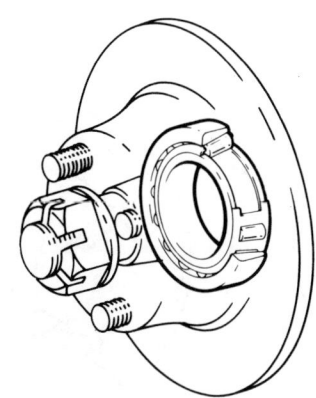

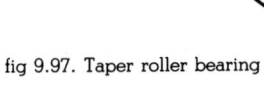

fig 9.97. Taper roller bearing

fig 9.96. Needle roller bearing

CLUTCHES AND BRAKES — FRICTION AS A FRIEND

Clutches are a kind of coupling which allow shafts to be connected or disconnected easily. The two basic types of clutch are the **positive clutch** and the **friction clutch**. The **dog clutch** (fig 9.98) is the most common of the positive clutches. It uses interlocking metal parts fitted to the ends of the two shafts. Positive clutches are only used where both shafts can be stopped before the clutch is engaged.

Friction clutches can be disconnected and reconnected while either or both shafts are rotating. To transmit the drive from one shaft to the other, they rely on the friction between the 'plates' on the end of each shaft as they are pressed together (fig 9.99). Drivers use a clutch to connect or disconnect the engine driveshaft from the wheels. By disconnecting it, they can keep the engine running while the car is not moving. Then, when needed, it can be quickly reconnected. If brought together carefully, any difference in the speeds of the two shafts is soon equalled out, both rotating at the speed of the driver shaft. Most vehicles use a single plate friction clutch. Fig 9.100 shows the two most common types. The diaphragm type is the better one because it is compact, simpler and easier to use. When connected, the clutch plates are pressed together by strong springs. To disconnect the clutch, the driver presses a pedal which, using either a cable or hydraulic system, separates the two plates, working against the springs.

Where larger forces need to be transmitted, such as on big lorries, or where there would only be room for a small single plate clutch, as on a motorbike, **multiple plate clutches** are used. They are simply a number of friction plate clutches working together (fig 9.101).

A different type of friction clutch is used on some motor mowers and mopeds. They are called automatic or **centrifugal clutches** (fig 9.102). The shoes, linked to the driver shaft, are thrown outwards as the speed of the driver shaft increases. As they move out they press against the drum which is fixed to the driven shaft, so connecting the two shafts, without switching off the engine.

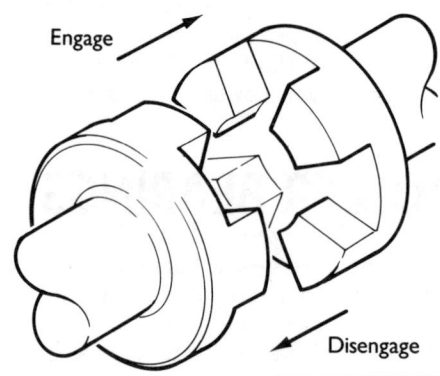

fig 9.98. Dog clutch

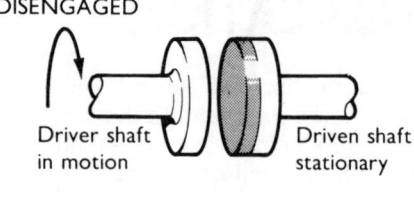

DISENGAGED

Driver shaft in motion — Driven shaft stationary

ENGAGED

fig 9.99. Friction clutch

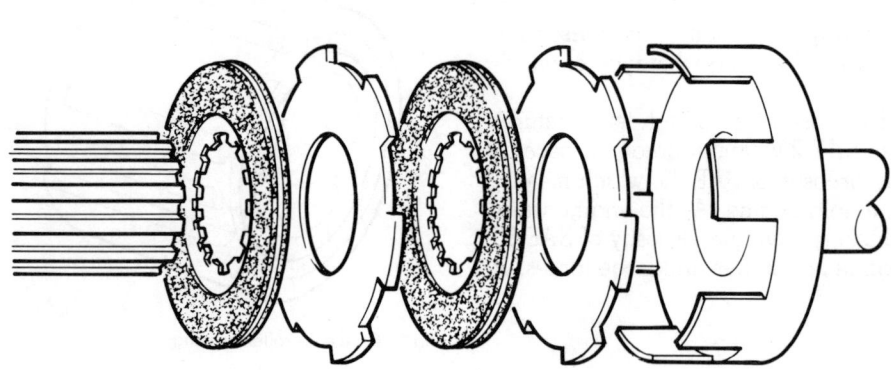

fig 9.101. Multiple plate clutch

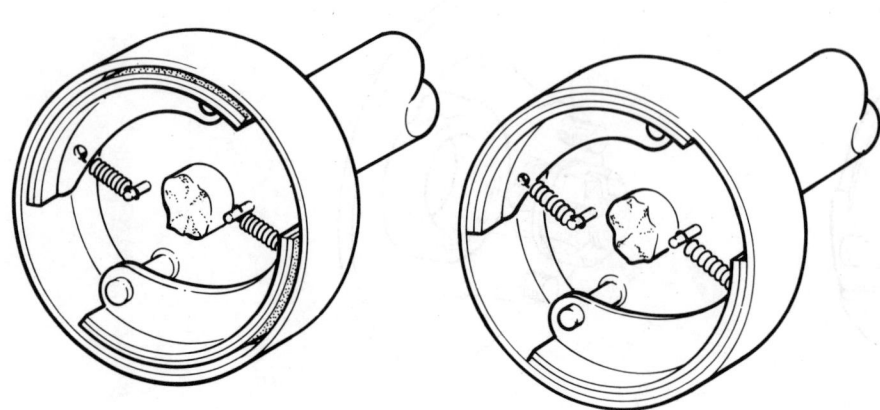

fig 9.102. Centrifugal clutch

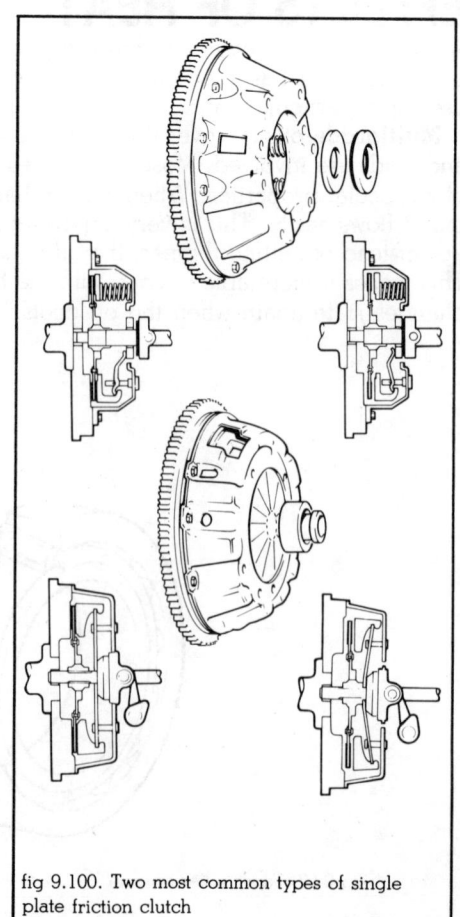

fig 9.100. Two most common types of single plate friction clutch

BRAKES

Brakes use friction to slow down or stop a rotating shaft. A very simple type of brake is the **band** brake (fig 9.103). You may have seen one in use on a fairground ride, the operator using it to get 'cars' to stop at the right place for you to get in or out. However, most brakes are fitted to a moving vehicle of some sort. Bicycle brakes (fig 9.104) show the basic idea. By pulling the brake lever the brake blocks are pressed against the wheel. The harder they are pressed against the wheel, the greater the friction and the sooner you stop. What you have to avoid is causing so much friction that the wheel 'locks' suddenly, causing you to skid. Hard braking generates a great deal of heat. The better a breaking system is at getting rid of this heat, the more efficient it is. The brakes on cars and other vehicles work on exactly the same principle, but have to make use of hydraulics to be able to cope with the much larger forces involved. For an explanation of a hydraulic braking system, see page 177.

The two basic types of vehicle brakes are **disc** and **drum** brakes (fig 9.105). Both types press a friction pad or shoe against a rotating metal part which is fixed to the wheel or axle. Most cars have disc brakes on the front wheels and drum brakes on the back wheels. There are good reasons for this. When a car brakes most of its weight is thrown forward onto the front wheels, so the front brakes need to be very good. Disc brakes are more efficient than drum brakes because of their larger area and the way the air flow around the disc gets rid of heat quickly. Drum brakes are not so good at getting rid of heat, but are suitable for the back wheels because they take less of the braking load. Their added advantage over disc brakes is that it is very easy to link a mechanical handbrake to them.

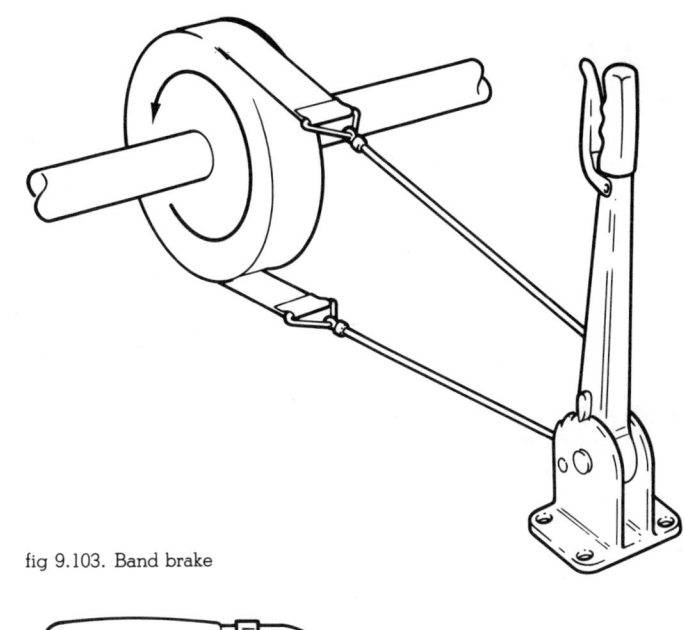

fig 9.103. Band brake

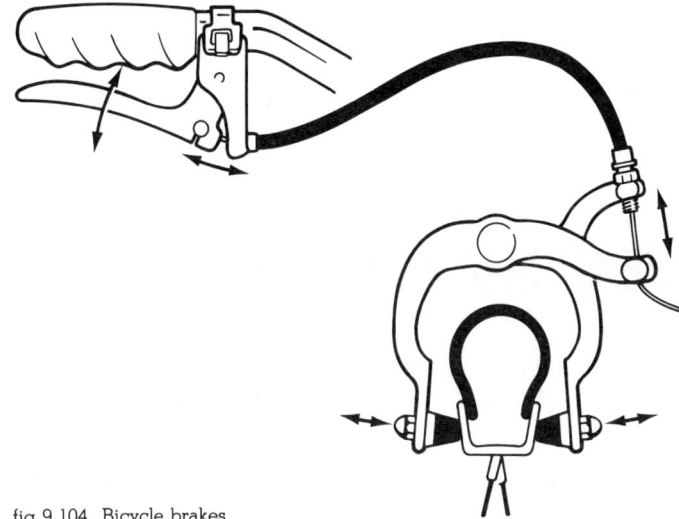

fig 9.104. Bicycle brakes

fig 9.105. Disc and drum brakes

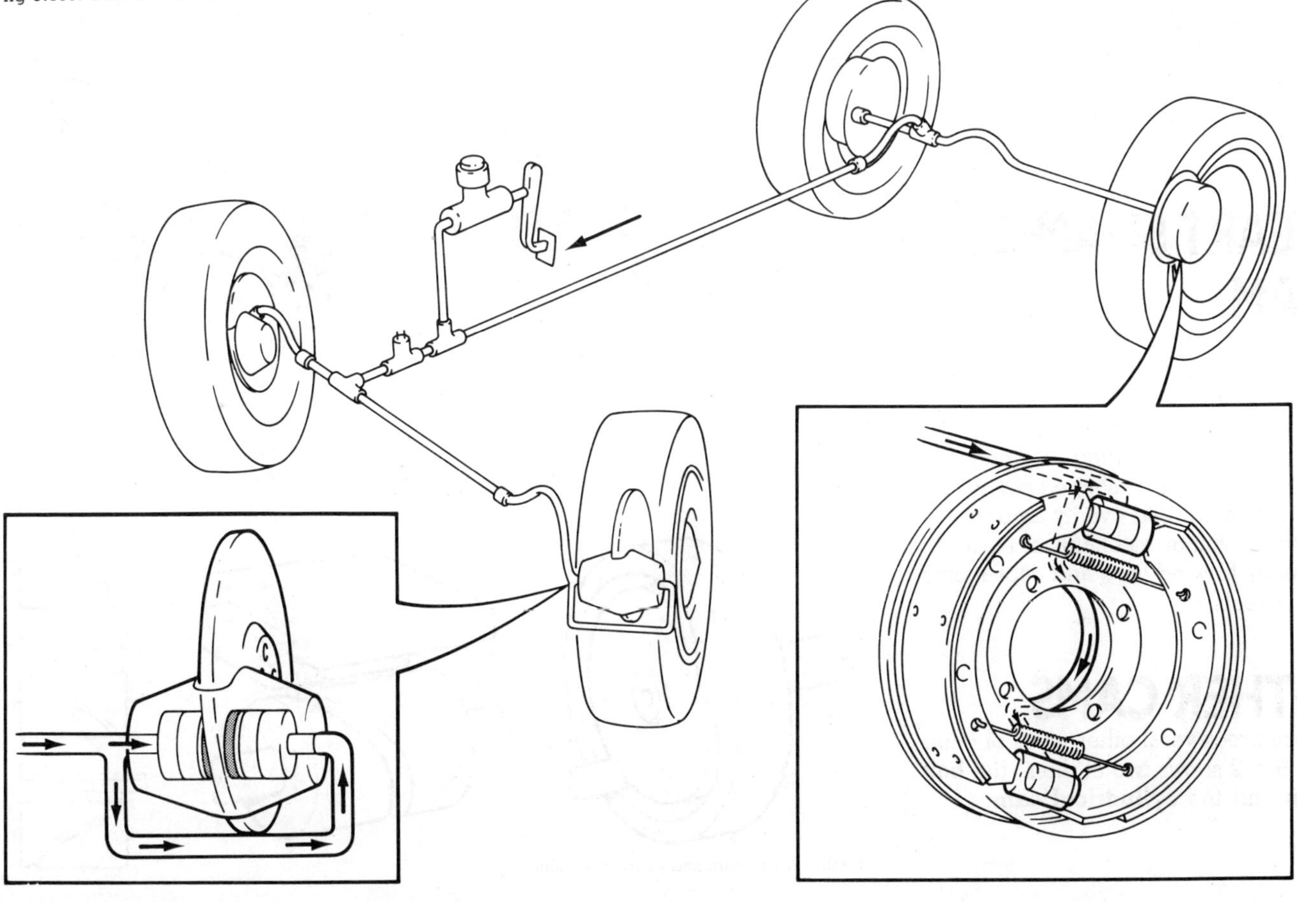

CAMS

Although there are both **linear** and **rotary** cams (fig 9.106), rotary cams are far more common. They are used to change rotary motion into either reciprocating or oscillating motion. Cams are shaped pieces of metal or plastic fixed to, or part of, a rotating shaft. A 'follower' is held against the cam, either by its own weight or by a spring. As the cam rotates, the follower moves (fig 9.107). The way in which it moves and the distance it moves depends on the shape of the cam. Fig 9.108 gives the terms used when describing cam movement.

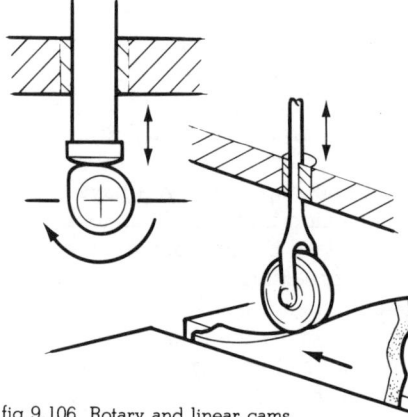

fig 9.106. Rotary and linear cams

CIRCULAR CAM

This is the simplest form of rotary cam, also known as an 'eccentric' cam because the circle is fitted 'off-centre' on the driving shaft. This type of cam gives the follower a smooth continuous movement known as **simple harmonic motion**. The mechanical fuel pump shown in fig 9.109 uses a circular cam. It produces an oscillating motion in the follower. This moves the diaphragm up and down, so pumping fuel to the engine.

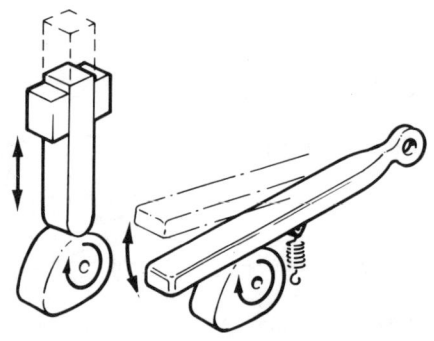

fig 9.107. Follower moves as cam rotates

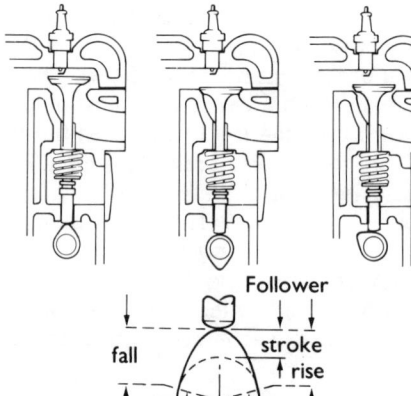

fig 9.108. Terms used to describe cam movement

PEAR SHAPED CAM

Fig 9.110 shows an overhead camshaft which, as it rotates, opens and closes the inlet and exhaust valves in an engine. Each of the pear shaped cams controls the movement of one valve, opening and closing it at the correct time in the firing sequence. With pear shaped cams there is quite a long dwell period, more than half the cycle, during which the follower does not move. When the follower is moving, the rise and fall times are equal because of the symmetrical shape of the cam. The distance the follower moves depends on the stroke of the cam.

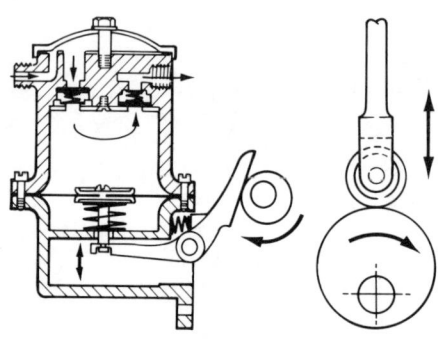

fig 9.109. Mechanical fuel pump using a circular cam

HEART SHAPED CAM

The heart shaped cam gives the follower a continuous uniform motion. It moves smoothly, at a constant speed. The bobbin winding mechanism on a sewing machine (fig 9.111) uses a heart shaped cam so as to wind the thread evenly onto the bobbin. Similar mechanisms exist in industry to wind wire and cables onto large reels.

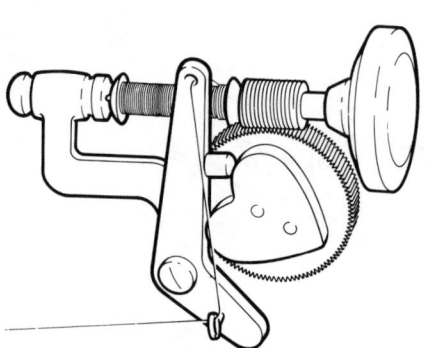

fig 9.111. Bobbin winding mechanism using heart shaped cam

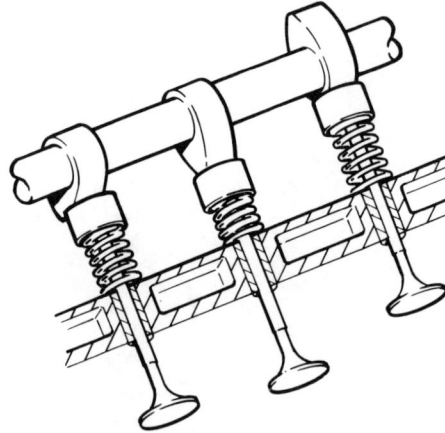

fig 9.110. Overhead camshaft with pear shaped cams

OTHER CAMS

There are several other types of cam. Fig 9.112 shows two of them, the **box cam** and the **cylindrical cam**.

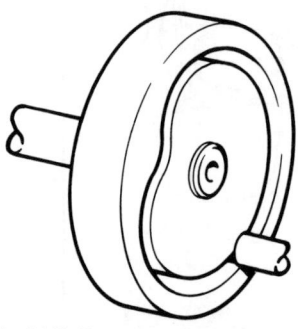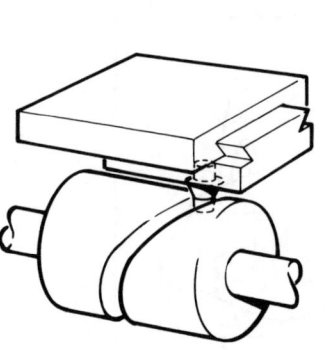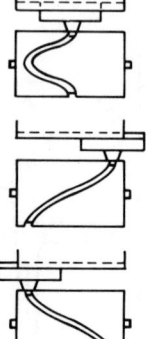

fig 9.112. Box cam and cylindrical cam

CRANK-SLIDERS

Crank slider mechanisms involve changes between rotary and reciprocating motion. Fig 9.113 shows the basic principles. The crank rotates while the slider reciprocates. The longer the crank the further the slider will move. Crank sliders can be used in several ways, but the two main ways are:

1. To change **reciprocating motion** into **rotary motion**, as in a car engine. The reciprocating pistons are connected to the crankshaft by connecting rods. As the pistons move up and down the connecting rods push the crankshaft round (fig 9.114). Each piston moves down in turn, so keeping the crankshaft turning.

2. To change **rotary motion** into **reciprocating motion**, as in a power hacksaw (fig 9.115). An electric motor powers a crank which is connected to the saw frame. The saw frame is free to slide on the 'arm'. As the crank rotates it causes the frame to slide backwards and forwards on the arm. The longer the crank the further the saw frame will move.

A compressor (fig 9.116) also uses this idea to provide compressed air for pneumatic systems. The rotary motion of an electric motor is used to make a piston reciprocate. As it reciprocates, it draws in air and then forces it, through a one-way valve, into the receiver tank.

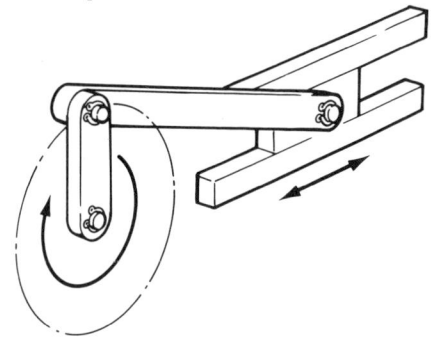

fig 9.113. Crank slider mechanism

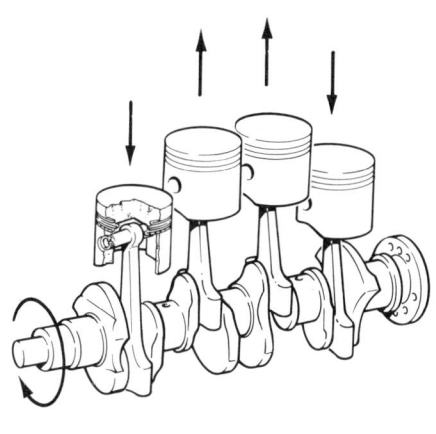

fig 9.114. Pistons pushing crankshaft round

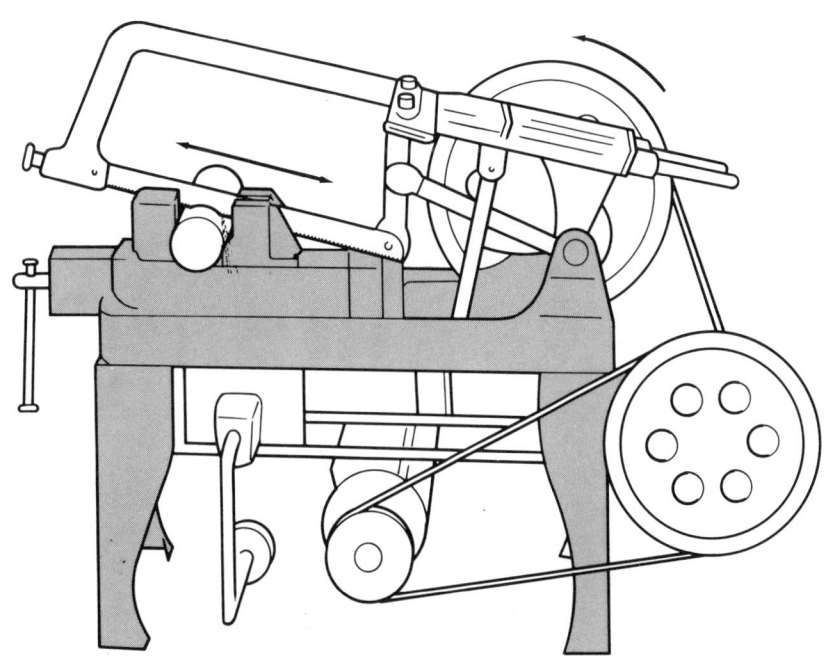

fig 9.115. Power hacksaw showing crank and slider mechanism

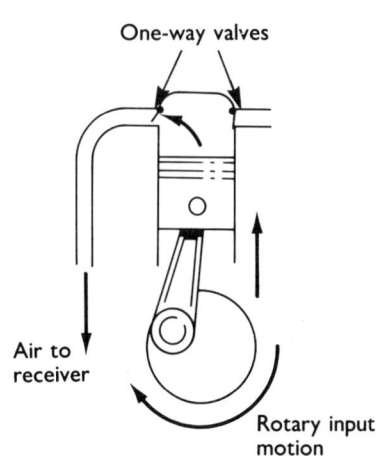

One-way valves

Air to receiver

Rotary input motion

fig 9.116. Compressor

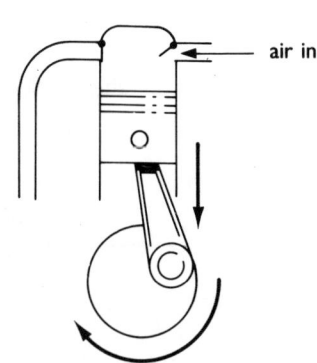

air in

MECHANISMS IN ACTION

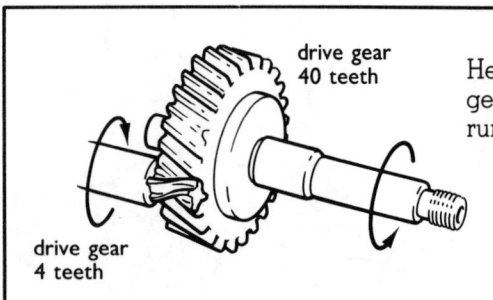

drive gear 40 teeth

drive gear 4 teeth

Helical teeth are used on these gears to give smooth, quiet running

$$VR = \frac{40}{4} = 10:1$$

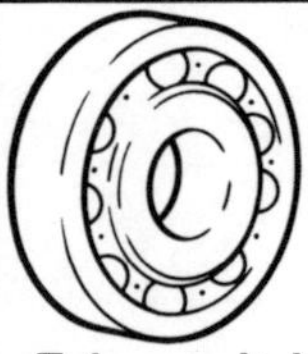

ball bearing

Takes radial loads caused by 30,000 rpm motor speed.

Sintered bronze bearing takes both radial and axial loads.

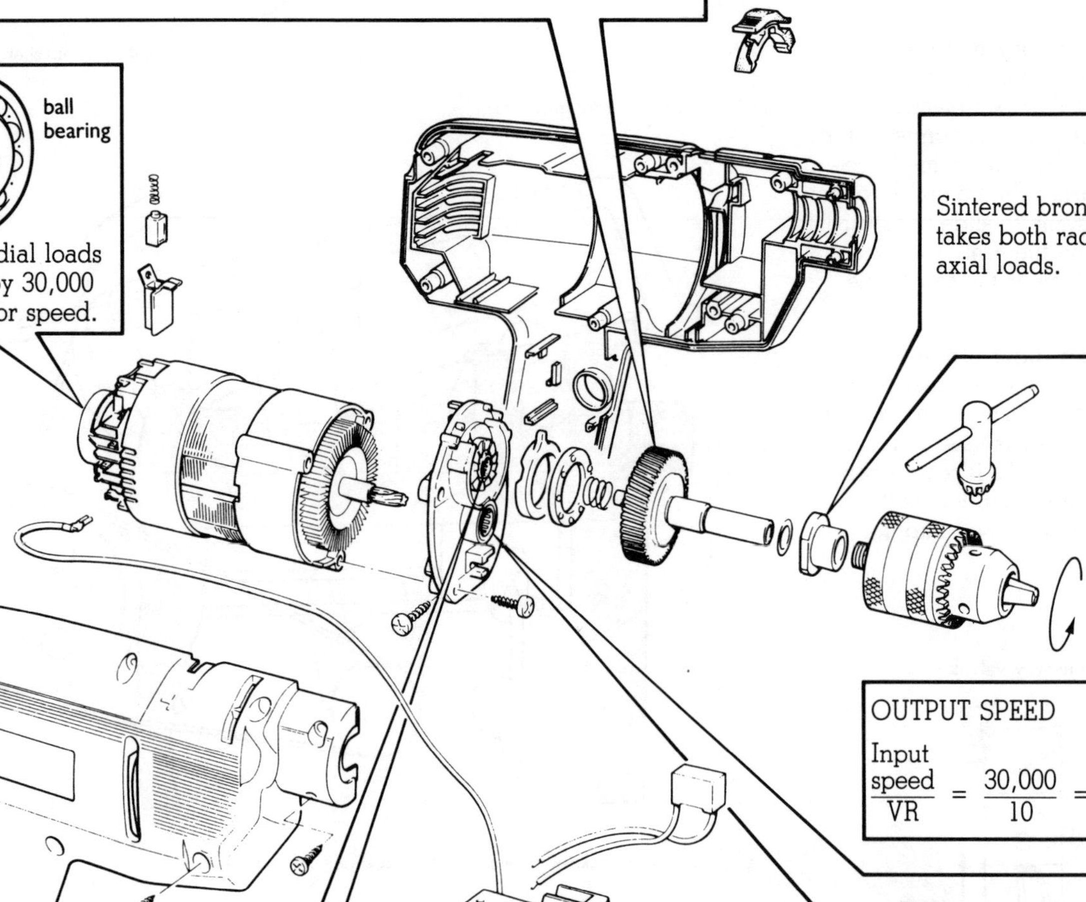

OUTPUT SPEED

$$\frac{\text{Input speed}}{VR} = \frac{30,000}{10} = 3,000 \text{ rpm}$$

Needle roller bearing

Supports front end of motor shaft. You can also see this type used to support the output shaft inside the fixed ratchet.

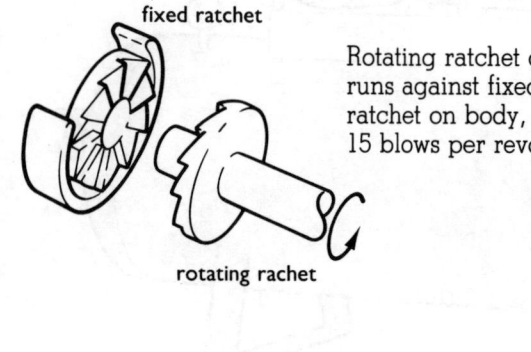

fixed ratchet

Rotating ratchet on shaft runs against fixed ratchet on body, giving 15 blows per revolution.

rotating rachet

FOOD MIXER

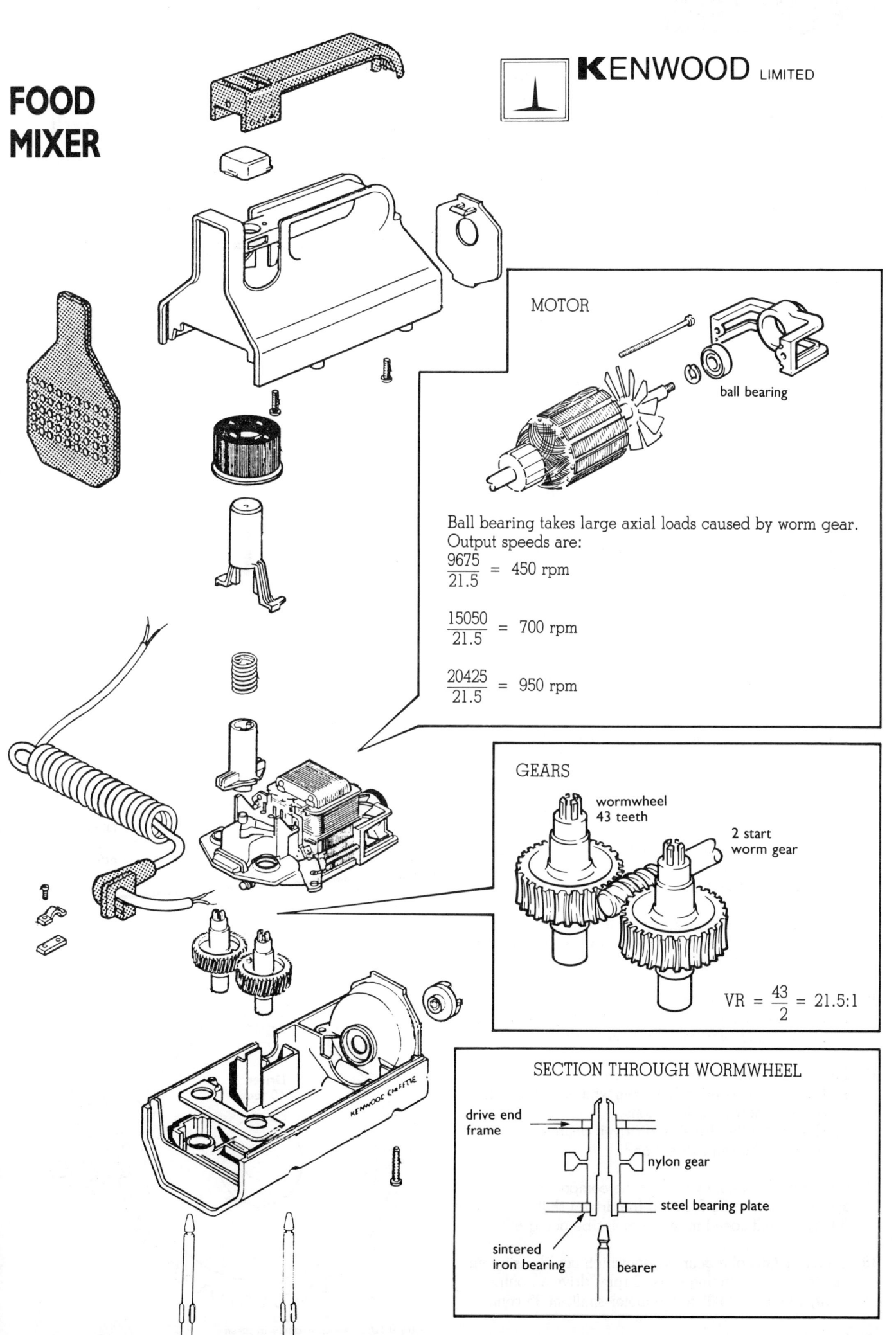

KENWOOD LIMITED

MOTOR

ball bearing

Ball bearing takes large axial loads caused by worm gear.
Output speeds are:

$$\frac{9675}{21.5} = 450 \text{ rpm}$$

$$\frac{15050}{21.5} = 700 \text{ rpm}$$

$$\frac{20425}{21.5} = 950 \text{ rpm}$$

GEARS

wormwheel
43 teeth

2 start
worm gear

$$VR = \frac{43}{2} = 21.5:1$$

SECTION THROUGH WORMWHEEL

drive end
frame

nylon gear

steel bearing plate

sintered
iron bearing   bearer

EXERCISES

1. Give two examples of mechanisms which change one type of input motion into another kind of output motion.

2. Use a simple sketch to help explain the principle of moments.

3. If a man weighing 560N is sitting 3 metres from the pivot of a see-saw, how far from the pivot must a man weight 700N sit to give a state of equilibrium (fig 9.117)?

4. Give examples of the use of each of the three classes of lever.

5. Calculate the mechanical advantage of the lever shown in fig 9.118. What class of lever is it?

6. What is the velocity ratio of the lever shown in fig 9.119? What class of lever is it?

7. If a mechanism has a mechanical advantage of 6:1 and a velocity ratio of 8:1, what is its efficiency?

8. Explain why mechanisms are always less than 100% efficient.

9. Show how, by using a lever operated system, you could apply a force of 50N to squeeze the juice out of a lemon using an input force of only 10N.

10. Use notes and sketches to explain how a reverse motion linkage works.

11. Give two examples of the use of parallel motion linkages.

12. Explain how toggle clamps work and why they are so useful in industry.

13. Name two types of screw thread and give an example of the use of each.

14. a) Sketch a pulley lifting system that has a mechanical advantage of 4:1.
    b) Why will the system be less than 100% efficient?

15. Calculate the mechanical advantage of a Weston Differential pulley system which uses pulley diameters of 150mm and 90mm.

16. a) Sketch a cross-section of a V-belt and pulley.
    b) Give two examples of the use of pulley systems, one in the school workshop and one at home.

17. Fig 9.120 shows two gears in mesh.
    a) If the input speed is 1250 rpm what is the output speed? Show all calculations.
    b) How could the driver and driven gears be made to rotate in the same direction?

18. Calculate the velocity ratio of the compound gear system shown in fig 9.121. If the driven shaft is going at 800 rpm what speed must the motor be going at?

19. Sketch details of a gear system which could, using an electric motor running at 1400 rpm, drive an output shaft, which is at 90° to the motor shaft, at 35 rpm.

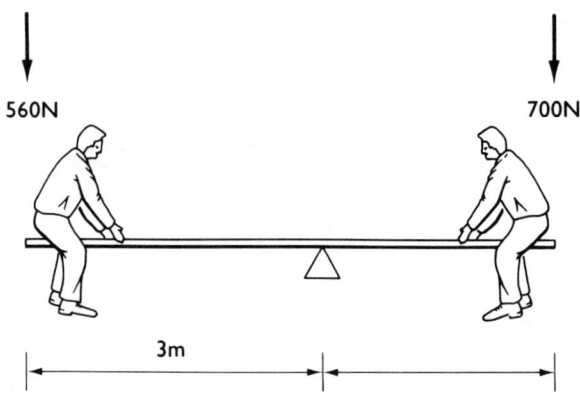

fig 9.117. See-saw in equilibrium

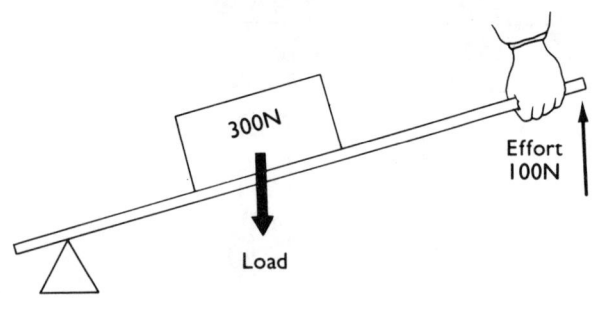

fig 9.118. Lever

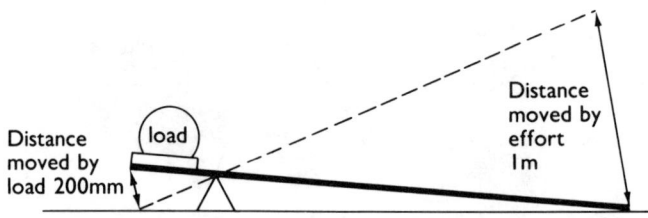

fig 9.119. Lever

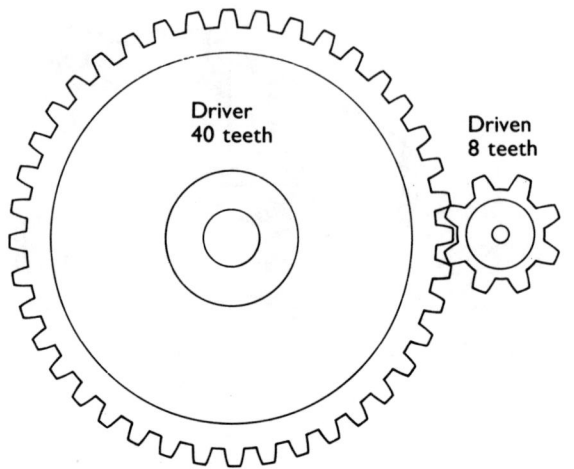

fig 9.120. Two spur gears in mesh

20. Name two types of gears that can be used to change the drive through 90° and sketch one of them.

21. What are the advantages and disadvantages of chain and sprocket systems compared to pulley systems?

22. Show how the turning of a door handle could be used to cause the linear motion of the sliding bolt. Your solution should return the bolt and handle to their original position when the handle is released.

23. If a motor producing a torque of 60 Nm is connected to the gear system shown in fig 9.121, what torque will the output shaft produce?

24. Sketch three different ways of connecting a pulley or gearwheel to a shaft.

25. Explain how a ratchet and pawl works and give an example of its use.

26. Name two different couplings that can be used to connect aligned shafts and draw one of them.

27. Sketch and name a coupling that could be used to join two shafts which are at a slight angle to each other. Label the parts clearly.

28. Explain why 'constant velocity' is a problem with universal joints and how it is overcome.

29. Use notes and sketches to explain the difference between positive clutches and friction clutches.

30. Give two reasons why disc brakes are used on the front wheels of most cars.

31. Explain the difference between the radial and axial loads that bearings may have to take.

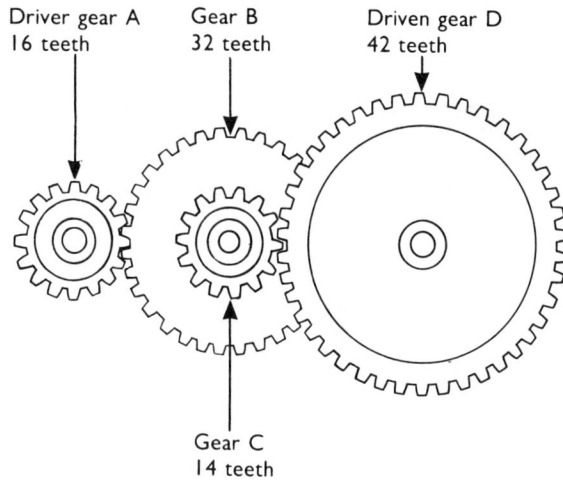

fig 9.121. Gear system

32. Use notes and sketches to explain what is meant by:
a) a ball bearing
b) a thrust bearing.

33. What is the purpose of lubrication?

34. Fig 9.122 shows a motor driven pulley system in a lathe.
a) Give one disadvantage of pulley systems.
b) Name a type of bearing that could be used at X and Y.
c) Coupling C is a flexible coupling. Sketch a cross-sectional view through it, labelling parts clearly. Include a means of fixing the coupling to the shafts.
d) If the motor speed is 3600 rpm, what is the speed of the output shaft for the belt position shown?
e) If the motor torque is 300Nm, what torque will be produced in the driven shaft?
f) With the aid of sketches, describe a system that could be used to slacken the belt to allow speed changes to be made.

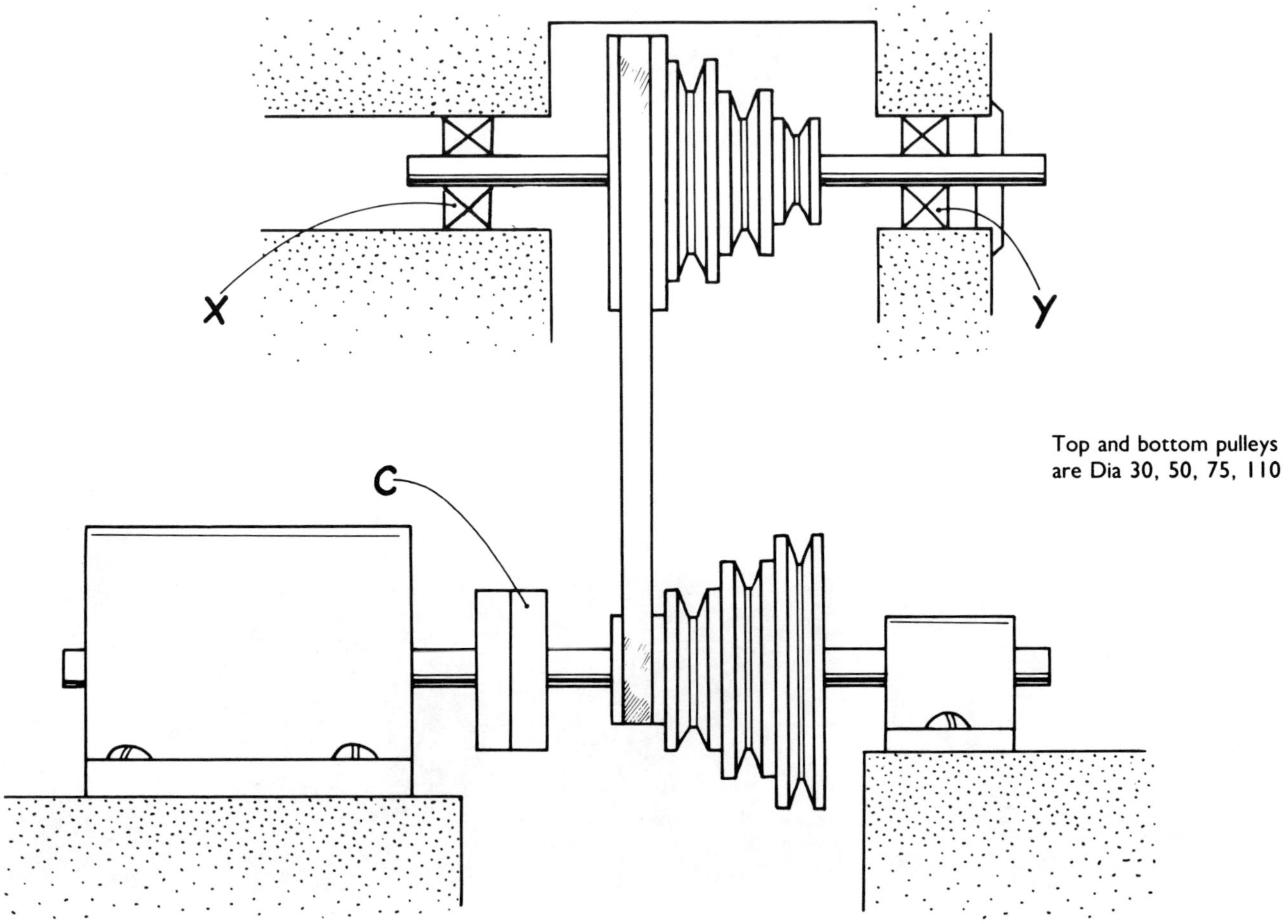

Top and bottom pulleys are Dia 30, 50, 75, 110

fig 9.122. Motor driven pulley system

Pneumatics in industry — lifting a component

PNEUMATICS

Pneumatics is all about using compressed air to do useful work. Compressed air is simply the normal air you breathe forced into a small space. You use compressed air when you inflate a bicycle tyre or blow up a balloon. To compress the air into the tyre or balloon takes energy, which can be released by undoing the tyre valve or by letting go of the inflated balloon. Pneumatic systems use the energy stored in compressed air to do useful work. Compressed air has many common uses, you come across them every day, but its main use is in industry where it is used to power machinery or automate production lines.

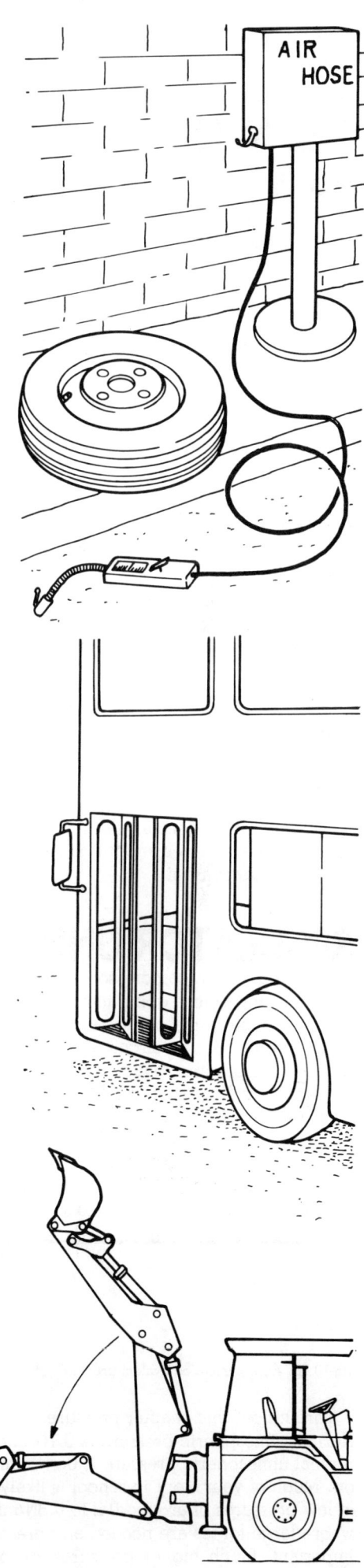

PNEUMATICS IN EVERYDAY LIFE

Tyres
Car tyres are inflated by compressed air. Their correct name is 'pneumatic' tyres.

Road drills
'Pneumatic drill' is the name you probably use to describe these tools used to dig up roads.

Automatic doors
The doors on tube trains and many buses are 'pneumatic'. If you listen carefully you will hear air escaping as the door moves.

Lorry brakes
The braking systems of many large vehicles are 'air assisted'. You often hear high pressure air escaping when they apply their brakes.

Wheel and tyre changing
'Pneumatic' tools are used in tyre fitting bays to undo and tighten wheel nuts as well as to remove old tyres from wheels.

Dentists drills
Next time you are at the dentist you might like to think about the fact that the drill being used is going round at about 500,000 rpm. The best way to acheive this is by using a 'pneumatic' turbine.

PNEUMATICS IN INDUSTRY

Paint spraying
Pneumatic spraying is used not only for cars and other vehicles, but also for cookers, washers, fridges and many other industrial products.

Moving components
Pneumatic cylinders are ideal for pushing, pulling, clamping and positioning parts ready for further work to be done.

Punching and pressing
Many manufacturing processes use sheet materials (e.g. metal, card, plastic and aluminium foil). These need to be cut out, pressed into shape or folded. Pneumatic cylinders, because they give reliable linear movements, are ideal for these jobs.

Dangerous environments
Some working environments are dangerous because the air in them contains large amounts of combustible particles (e.g. paint spraying areas and mines.) It is dangerous to use electrical equipment in these areas because small sparks from it could start fires. Pneumatic equipment is ideal in these areas because it does not cause sparks.

Any pneumatic system uses compressed air, which is normally supplied by a compressor. Compressors come in many shapes and sizes, but they all work on the same basic principle. A pump, driven by an electric motor or combustion engine, sucks in the air through filters and forces it into a strong metal tank called a receiver (fig 10.1).

THE NEED FOR A SAFETY VALVE

When the air inside the tank has reached a certain pressure (e.g. 1 newton/mm^2) the motor is stopped automatically. If for any reason the motor did not switch off, pressure would continue to build up inside the receiver tank, eventually reaching dangerous levels. Compressors are fitted with safety valves to prevent this from happening. Above a certain pressure (e.g. 1.5 newtons/mm^2), the safety valve opens, allowing excess air to excape into the open air. It must be emphasised that this does not normally happen, because the motor usually cuts out once the correct pressure is reached in the receiver tank.

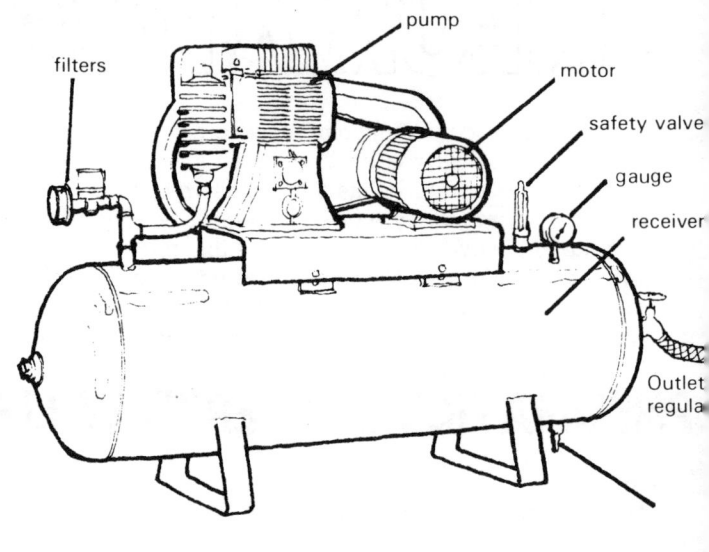

fig 10.1 A typical compressor

AFTER USE

When you have finished using a compressor and switched off the motor, any air left in the receiver is let out through the drain tap. It is not considered safe to leave the receiver 'under pressure' for long periods, doing so may shorten its useful life. The drain tap is positioned under the receiver so that any water in the receiver will drain out as the air comes out.

Question:
Why should there be any water in the receiver?
Answer:
All air contains moisture. When air is compressed into the receiver moisture goes in with it. It tends to collect in the bottom of the receiver and if it is left there causes it to go rusty, so weakening it.

REGULATOR UNIT

Air coming from a compressor unit normally goes to a regulator unit before going into the pneumatic circuit (fig 10.2a). The basic job of the regulator unit is to supply compressed air to the circuit at a constant pressure that you choose. The pressure gauge on the regulator unit may be marked off in N/mm^2 or in BAR. To convert from one to the other use the chart shown in fig 10.2b.

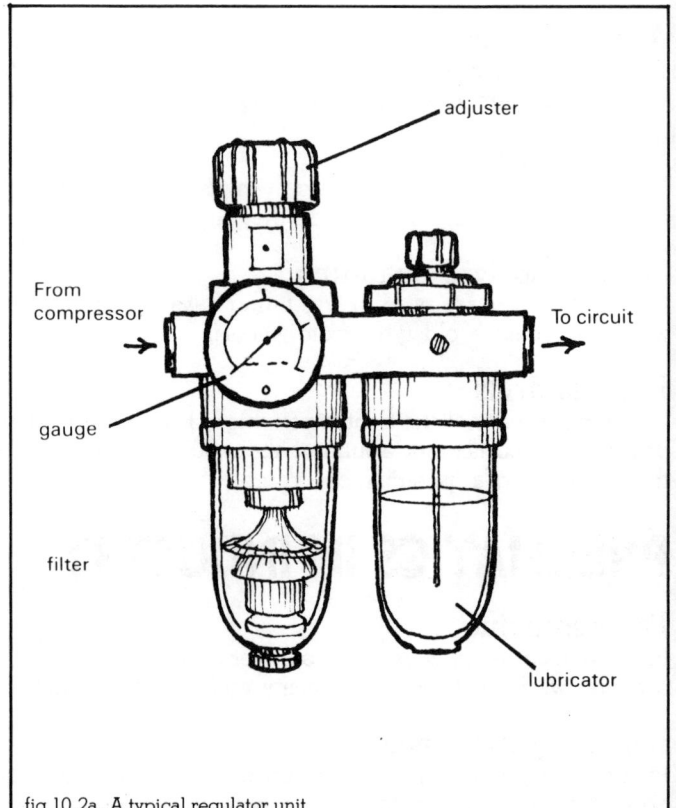

fig 10.2a. A typical regulator unit

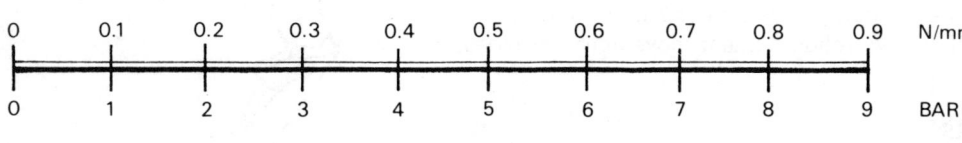

fig 10.2b. Approximate equivalent pressure values

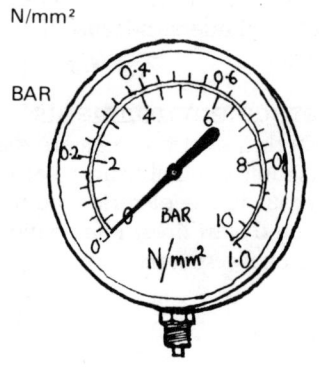

Throughout this chapter pressures are quoted in N/mm^2. Normal atmospheric pressure is 0.1N/mm^2 (1 BAR). Twice normal atmospheric pressure is 0.2N/mm^2 (2 BAR), and so on. Much of your work in school is likely to be carried out at low pressures of around 0.2 to 0.3N/mm^2. In industry, where large forces are needed and greater efficiency is important, much higher pressures are often used.

The regulator unit may also include two other parts. First, a **moisture filter** which removes moisture from the compressed air by passing it through a filter (fig 10.3). If moisture is allowed to get into pneumatic components some parts will go rusty, making them inefficient and shortening their lives. It may also contain a **lubricator**, which puts very small droplets of oil into the compressed air, so lubricating moving parts of pneumatic valves and cylinders. In schools this stage is often not included as it would make things oily to touch. In industry, however, it is vital if components are to work well and last a long time.

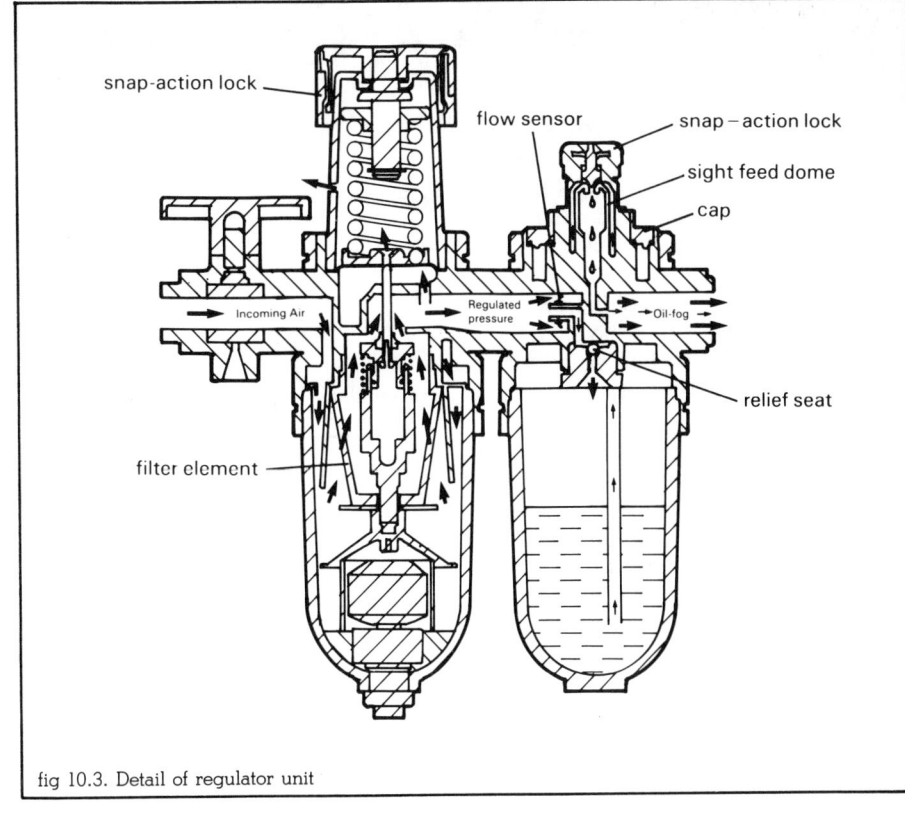

fig 10.3. Detail of regulator unit

MAKING CONNECTIONS

Pneumatic components are connected together using plastic piping. The pipes are fitted to the components in one of two ways, either by a screw on fitting or a push in fitting. Fig 10.4a shows both types. The push in type grips the pipe in a metal ring. If fitted correctly, it is just as secure as the screw in type, but much quicker to use.

To connect a number of pipes together 'T', 'Y', or 'X' connectors are used (fig 10.4b). 'T' and 'Y' connectors are both 3-way connectors, but 'Y' shaped connectors allow smoother air flow. The 'X' connector is a 4-way connector.

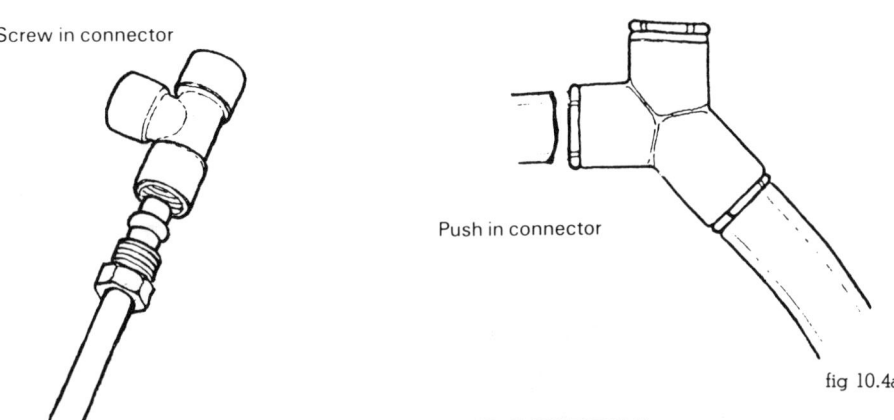

Screw in connector

Push in connector

fig 10.4a.

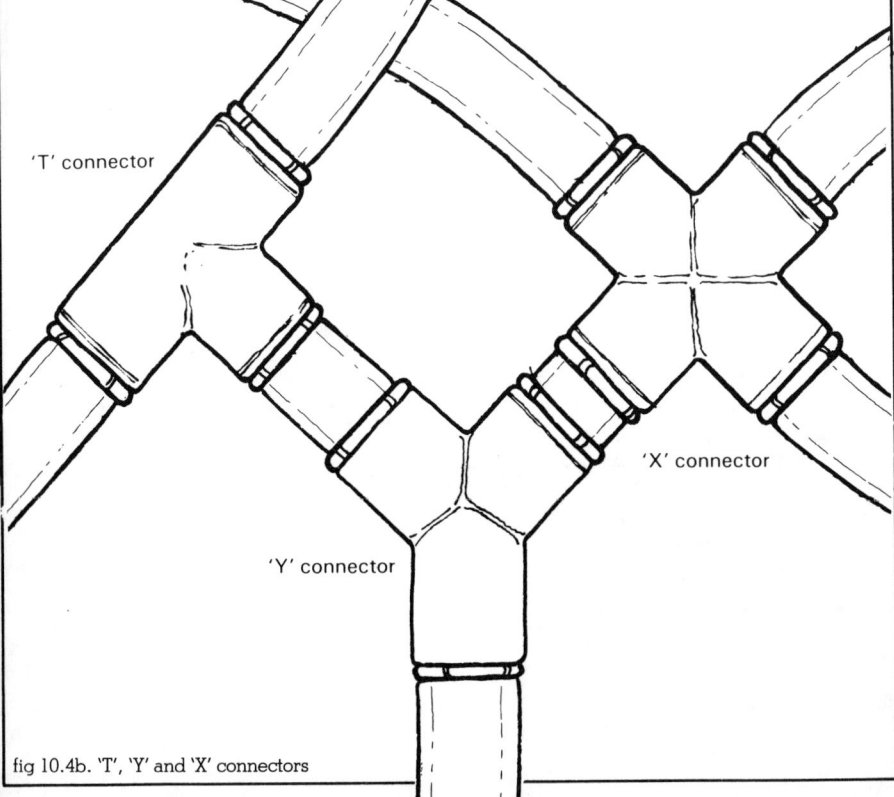

'T' connector

'Y' connector

'X' connector

fig 10.4b. 'T', 'Y' and 'X' connectors

SAFETY

Because pneumatic equipment uses high pressure air and has powered moving parts there are some dangers in using it. However, a few simple precautions should prevent accidents.

1. Do not blow compressed air at anyone, even yourself.

2. Connect all the components in your circuit before switching on the air supply. An air line left unconnected will often writhe about in a dangerous way.

3. If you need to alter a circuit, turn off the air supply before making any changes.

4. Keep your hands out of the way of moving parts to avoid trapping your fingers. This will mainly be the pistons and anything attached to them.

5. Avoid having air lines going across the floor or between tables where people may fall over them.

COMPONENTS AND SYMBOLS

The most basic components you will use in pneumatics are a single acting cylinder and a three port valve (3PV). The two can be connected together to produce the basic circuit shown in fig 10.5.

When the push button of the 3PV is pressed, the piston rod of the single acting cylinder will be pushed out (it goes **positive**) by compressed air which comes in through the 3PV. When the push button is released the piston rod will go back in (it goes **negative**) as the air that was keeping the piston pushed out is allowed to escape. It goes back down the airline to the 3PV and out of the **exhaust port** into the open air. The word **port** means an airway into or out of a valve.

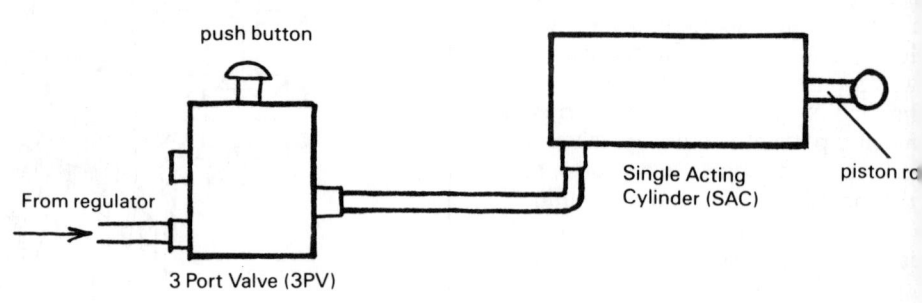

fig 10.5. Basic circuit

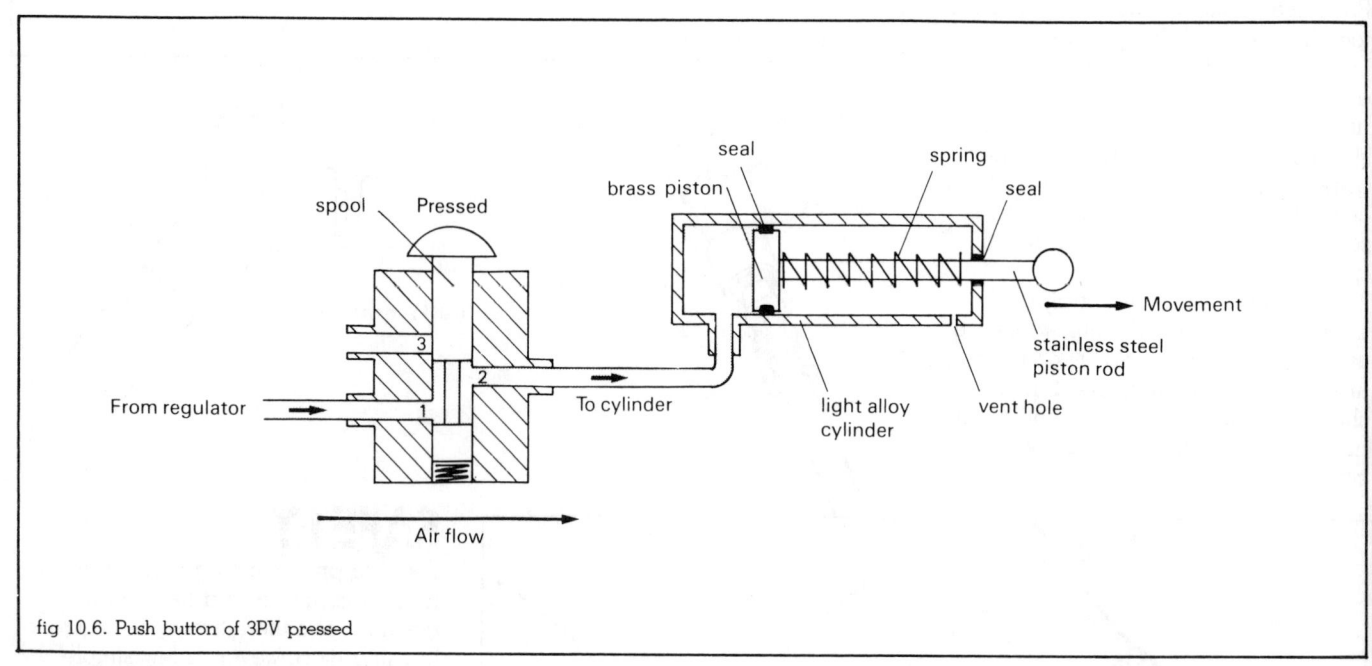

fig 10.6. Push button of 3PV pressed

By looking inside the two components we can see in detail what is happening. Fig 10.6 shows the push button of the 3PV pressed. This pushes the spool down, so connecting port 1 to port 2. Note that port 3 is blocked by the spool. This allows compressed air to flow through the valve to the single acting cylinder (SAC). It pushes the piston along, compressing the spring as it does so. Any air trapped behind the piston (on the spring side) goes out through the vent hole.

Fig 10.7 shows the push button on the 3PV released. This allows the spring underneath the spool to push it up, so connecting port 2 to port 3. The air that was pushing the piston out can now escape through port 3, the exhaust port, of the valve into the open air. The spring in the cylinder pushes the piston back in.

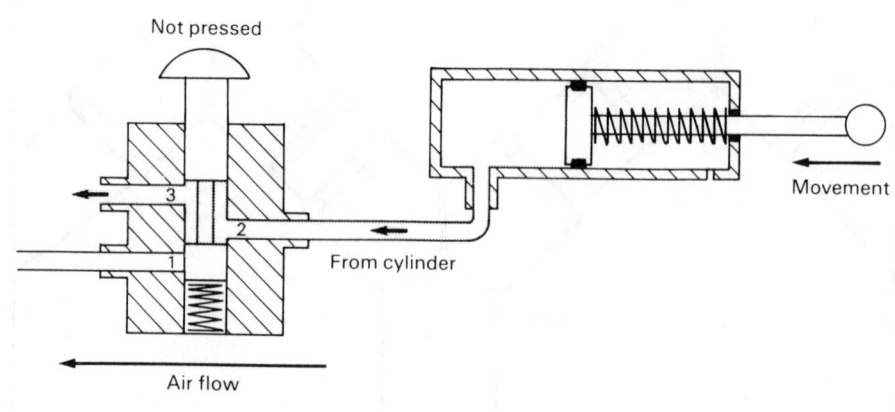

fig 10.7. Push button of 3PV released

Note that used air is not returned to the receiver tank, it is simply released into the open air through 'exhaust' ports. Therefore the term 'pneumatic circuit' does not have exactly the same meaning as the term 'electric circuit'. For an electrical circuit to work, the components must be connected to the power source, the battery, in a loop (fig 10.8). A pneumatic circuit is not connected in a loop, there is no return connection to the power source, the compressor (fig 10.9).

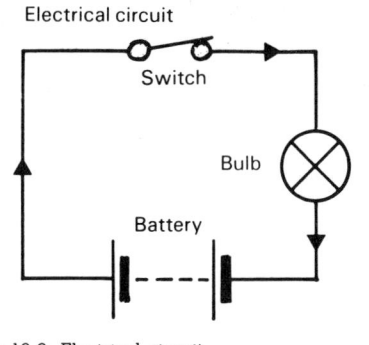

fig 10.8. Electrical circuit

fig 10.9. Pneumatic 'circuit'

SYMBOLS

So far we have used drawings and diagrams to represent the pneumatic components. In order to simplify the drawing of circuits the International Standards Organisation (ISO) has agreed a set of symbols for pneumatic components. Using these symbols, the circuit shown in fig 10.5 looks like fig 10.10a.

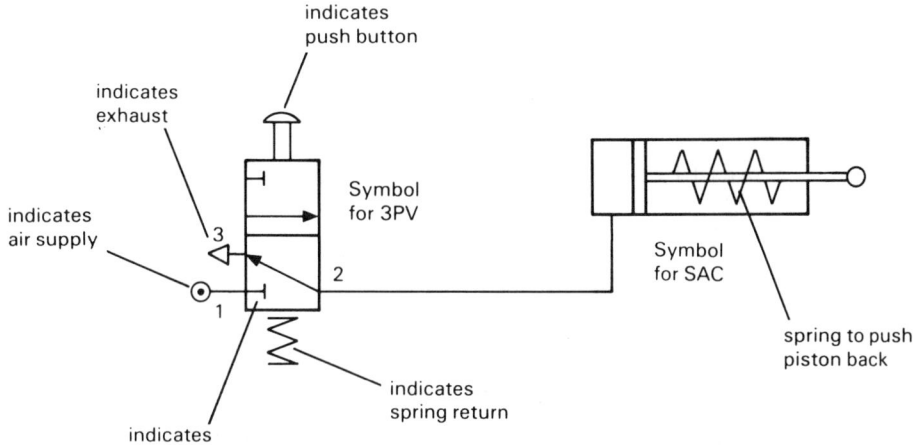

fig 10.10a. Basic circuit using symbols

The symbol for the SAC is easy to understand, but the 3PV symbol needs some explanation. It is made up of two halves. The top half shows the connections in the valve when the button is pressed and the bottom half shows the connections when the button is not pressed. Both halves of the symbol are always shown, the air lines normally being shown connected to the bottom half (fig 10.10b).

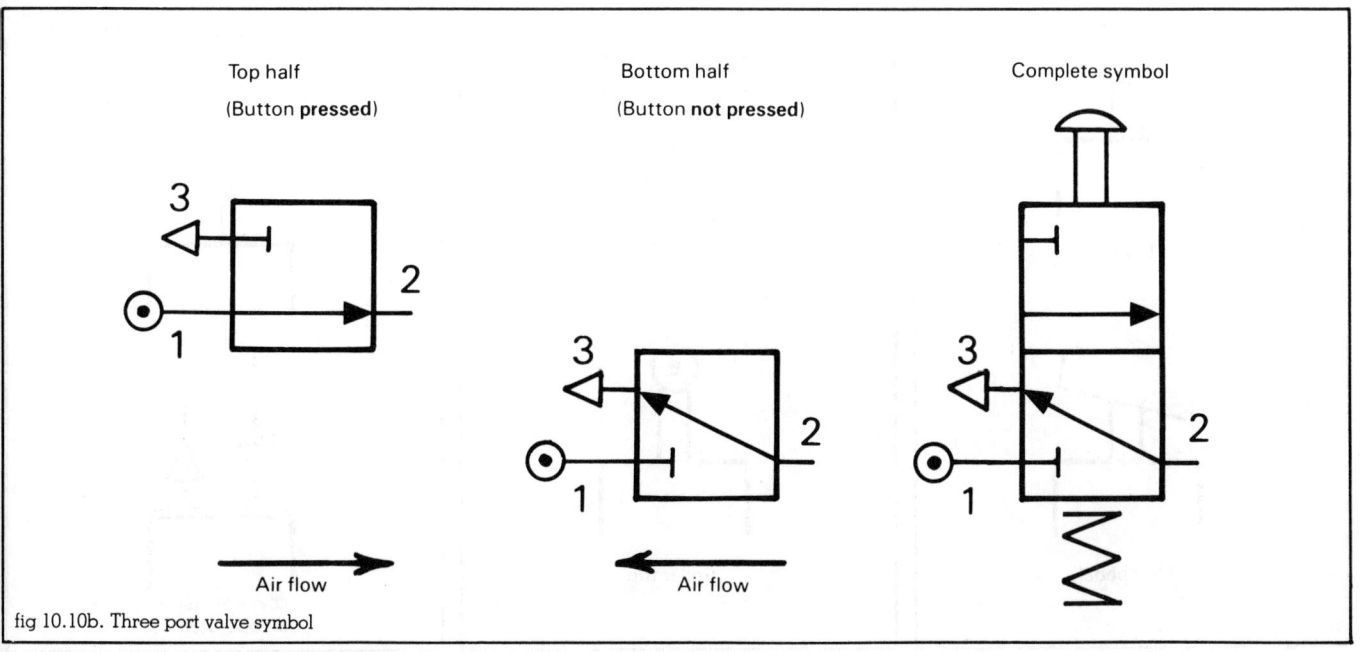

fig 10.10b. Three port valve symbol

Other components you are likely to use are shown in figs 10.11 to 10.15. Those you have not come across before will be explained later in the chapter.

CYLINDERS

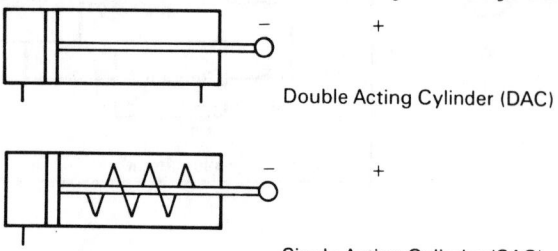

fig 10.11. Single and double acting cylinder symbols

Double Acting Cylinder (DAC)

Single Acting Cylinder (SAC)

VALVES

The valves shown in fig 10.12 are typical examples of three and five port valves. The symbols on the ends show the means by which the spool inside the valve is switched from one position to the other. They could be replaced by any of the end symbols shown in fig 10.13.

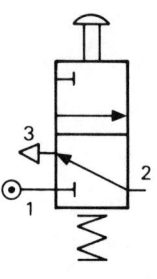
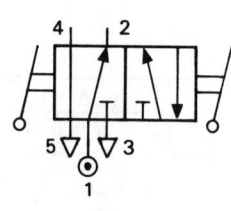

fig 10.12. 3PV and 5PV symbols

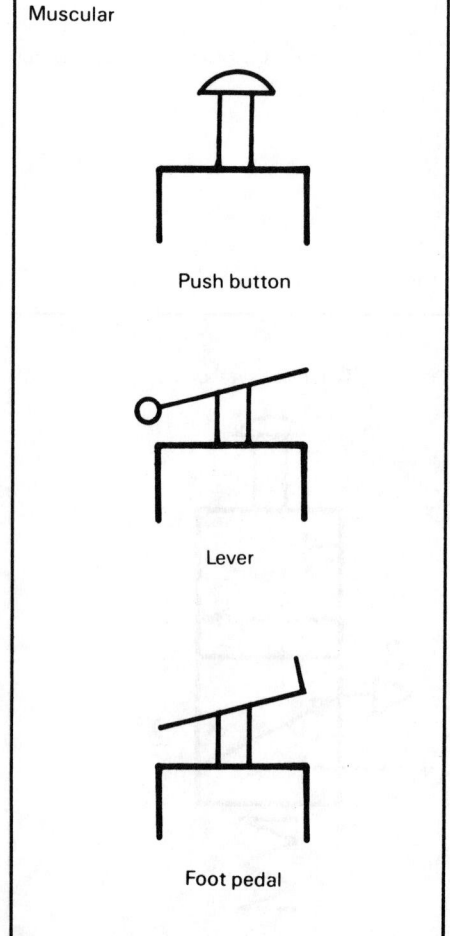

Muscular

Push button

Lever

Foot pedal

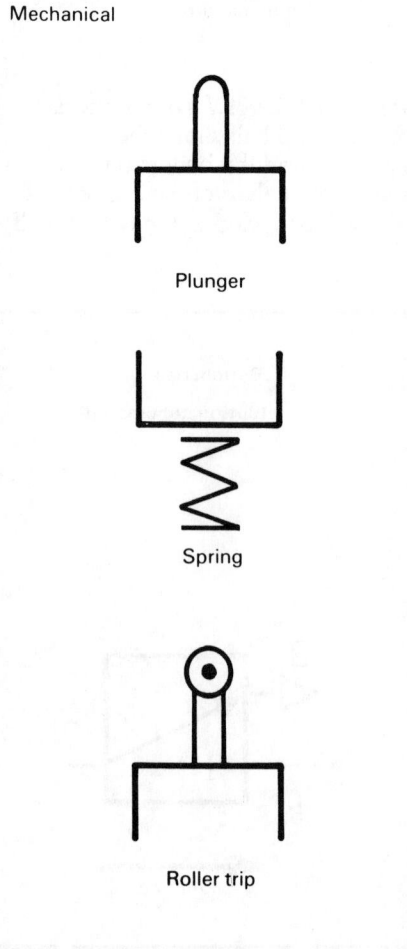

Mechanical

Plunger

Spring

Roller trip

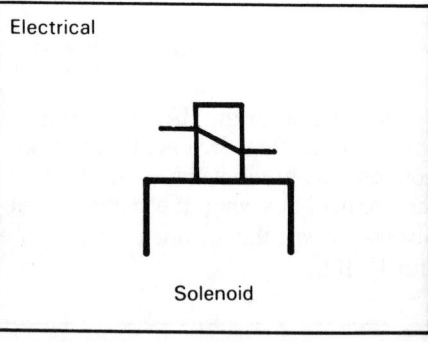

Electrical

Solenoid

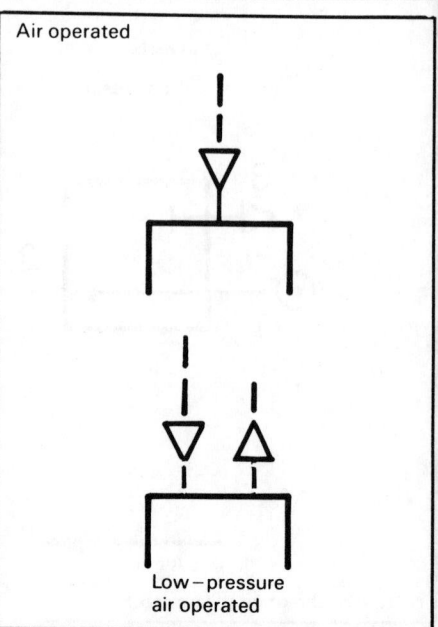

Air operated

Low−pressure air operated

fig 10.13. Valve operation symbols

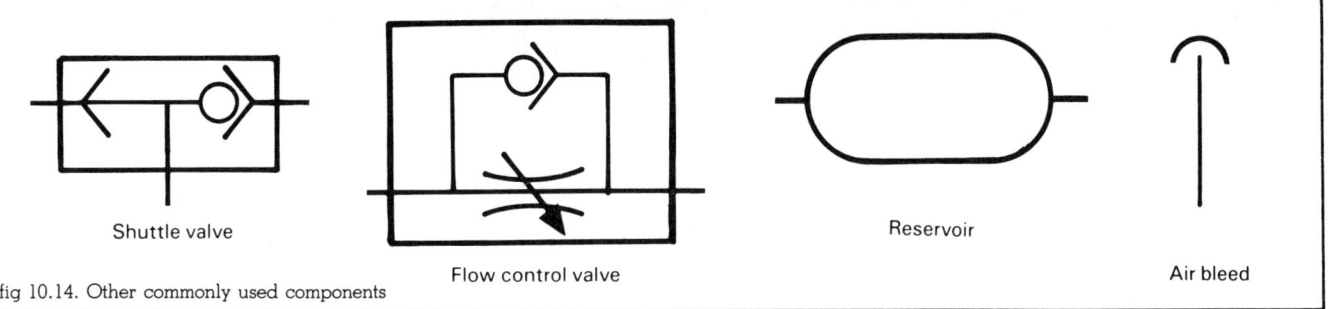

Shuttle valve

Flow control valve

Reservoir

Air bleed

fig 10.14. Other commonly used components

Symbols for compressor and regulator unit (fig 10.15) are rarely shown on circuit diagrams, but may appear on full system diagrams.

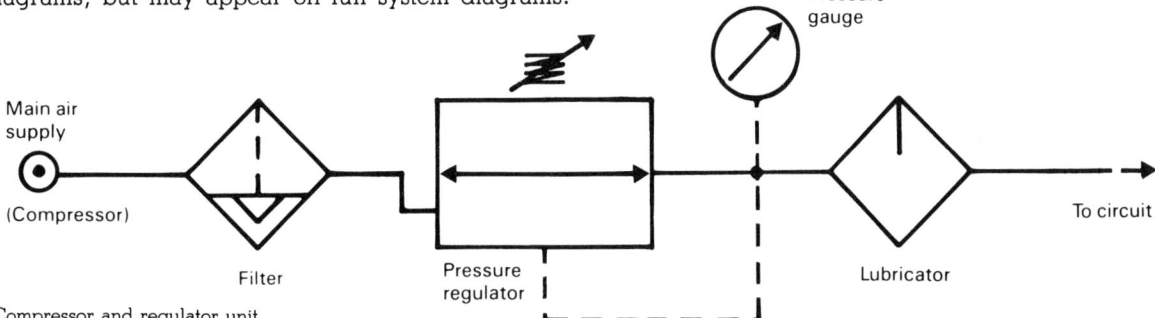

Main air supply

(Compressor)

Filter

Pressure regulator

Pressure gauge

Lubricator

To circuit

fig 10.15. Compressor and regulator unit

Five port valve (5PV)

The 5PV is commonly used to control a double acting cylinder (DAC), as shown in fig 10.16. This valve is supplied with air to port 1. From there it can be switched to either port 4 to send the piston positive, or to port 2 to send the piston negative. The internal connections of the 5PV, shown in fig 10.17a, shows air going from port 1 to port 4 and on to the cylinder, so holding the piston positive. Port 5 is acting as the exhaust port. When the valve is switched (fig 10.17b), air is supplied to the cylinder from port 2, sending the piston negative. Port 5 now becomes the exhaust port. As with the 3PV symbol there are two halves, connections normally being shown to put the piston in the negative position, as in fig 10.16. The lever symbol on both ends of the valve is rather confusing. It indicates that it has to be switched both ways. In reality it works just like a light switch. That is, there is just one lever which can be switched to one of two positions, as in fig 10.17.

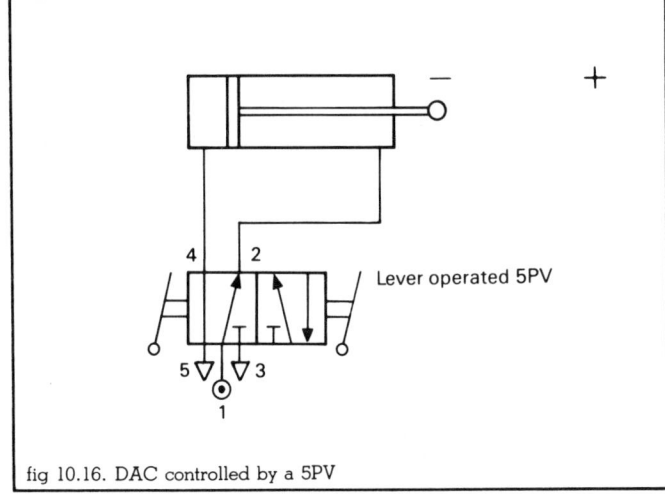

Lever operated 5PV

fig 10.16. DAC controlled by a 5PV

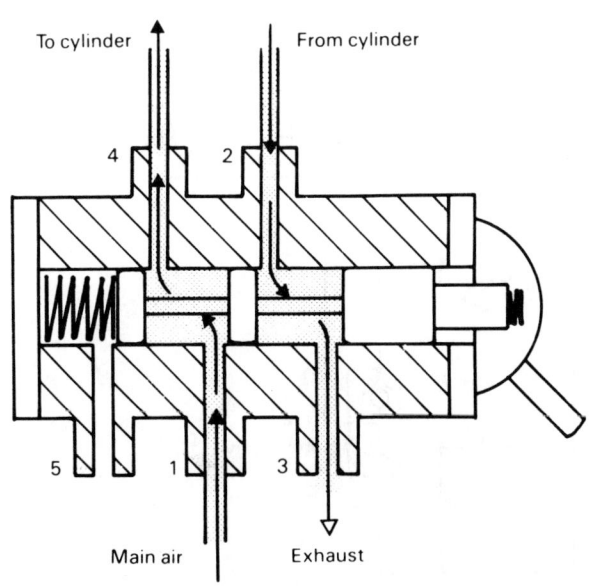

To cylinder

From cylinder

Main air

Exhaust

fig 10.17a. Supplying air to port 4

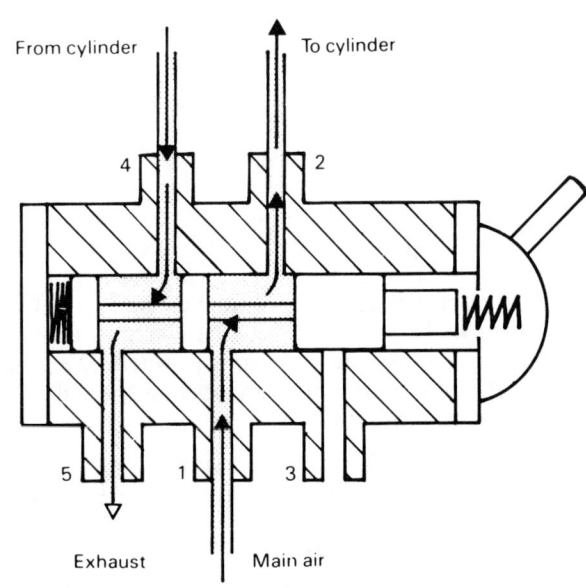

From cylinder

To cylinder

Exhaust

Main air

fig 10.17b. Supplying air to port 2

BASIC CONTROL CIRCUITS

AND Circuit

Connecting two valves in series, as shown in fig 10.17c, means that both valve A **and** valve B have to be pressed before the piston will go positive. This type of circuit is often used to prevent a machine being switched on accidentally or to ensure that guards are in place before the machine will operate. The press shown in fig 10.17c has valves placed so that they are not pressed until the safety guard has been pulled down. Only then can the operator switch on the machine.

The behaviour of this circuit can be shown as a **truth table** (fig 10.18).

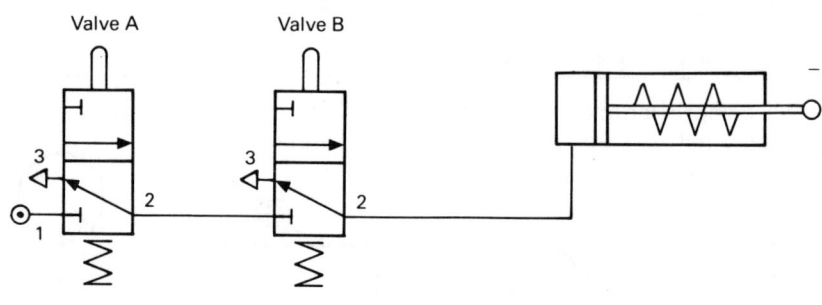

fig 10.17c. AND circuits

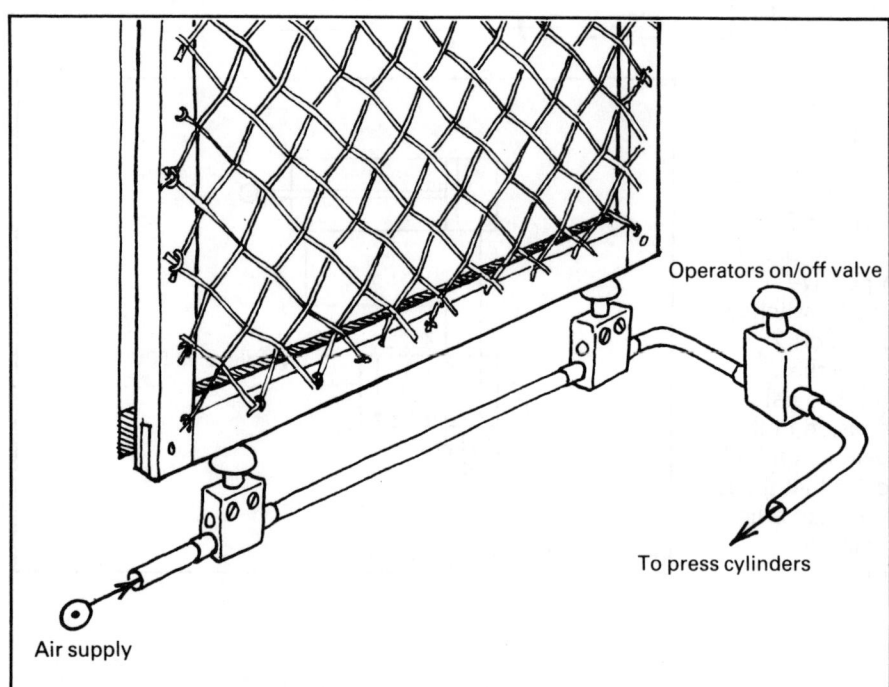

Air supply

Operators on/off valve

To press cylinders

fig 10.19. OR circuit

VALVE A	VALVE B	PISTON
OFF	OFF	– VE
OFF	ON	– VE
ON	OFF	– VE
ON	ON	+ VE

fig 10.18. AND truth table

OR Circuit

Connecting two valves in parallel, as shown in fig 10.19, means that pressing either valve X **or** valve Y will make the piston go positive. This type of circuit is often used to allow a machine to be switched off from a number of different positions. This might be to make life easier for the operator or it may be a safety feature allowing other people to switch off the machine in an emergency. You have come across an electrical OR circuit in your school workshops. The STOP buttons, on the walls around the room, are connected as an OR circuit. Pressing any one of them will switch off all the machines in the workshop. The truth table for the OR circuit is shown in fig 10.20.

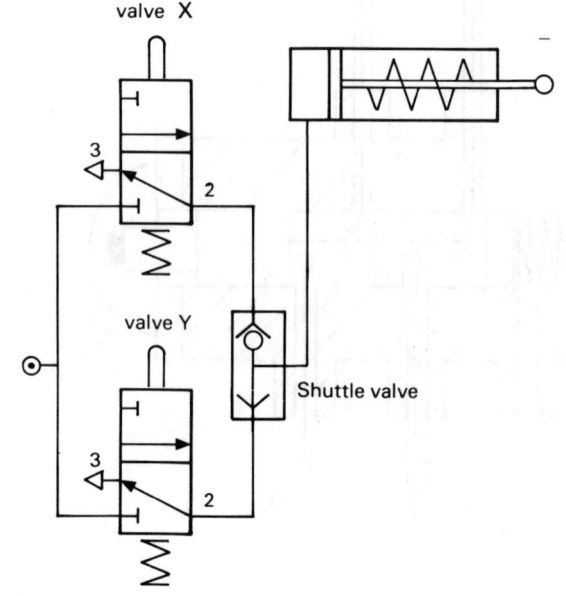

VALVE X	VALVE Y	PISTON
OFF	OFF	– VE
OFF	ON	+ VE
ON	OFF	+ VE
ON	ON	+ VE

fig 10.20. OR truth table

The **shuttle valve** shown in the OR circuit is a very simple device, but it is essential to the circuit. If there were just a 'T' connector at that point the circuit would not work. Instead of air going from valve X to the cylinder, it would take the easy route, going out through the exhaust port of valve B. Exactly the same sort of thing would happen if valve Y were pressed. A closer look at the shuttle valve will show how it works.

As fig 10.21 shows, the shuttle valve has three connections, one from each 3PV and one to the cylinder. The only moving part is a ball, made of plastic. The ball is free to move between two positions, 'a' and 'b'. If air comes from valve X the ball is pushed to position 'b', so directing the air to the cylinder. Air coming from valve Y would push the ball to position 'a'. If both valves are pressed together the piston will go positive, as normal, because air entering the shuttle valve has only one way out, to the cylinder.

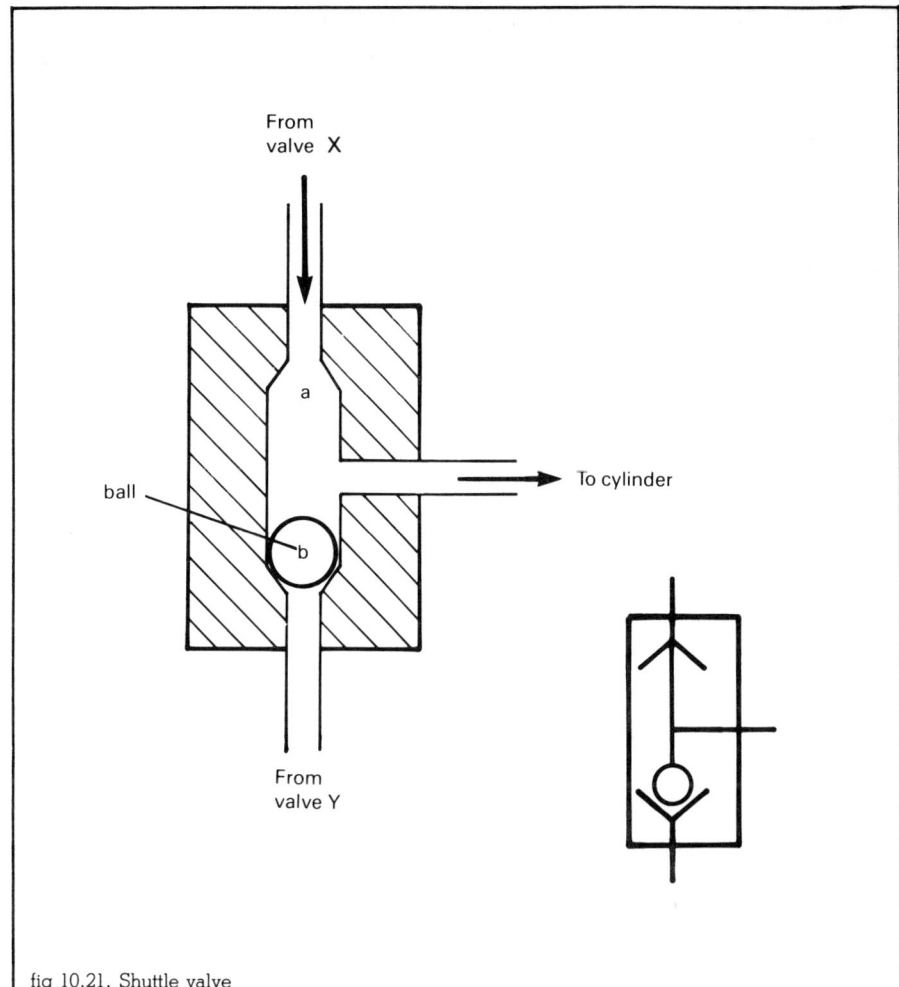

fig 10.21. Shuttle valve

NOT Circuit

Fig 10.22 shows the air supply connected to port 3, the exhaust port, instead of to port 1 as normal. This means that the piston will normally be positive because with the button 'not pressed', air is being supplied to the cylinder. It is only when the button is pressed that the air supply to the cylinder is cut off. In other words, the circuit behaves in the opposite way to normal. This might be used where you need to have something held most of the time, only being released when the valve is pressed. The truth table for the NOT circuit is shown in fig 10.23.

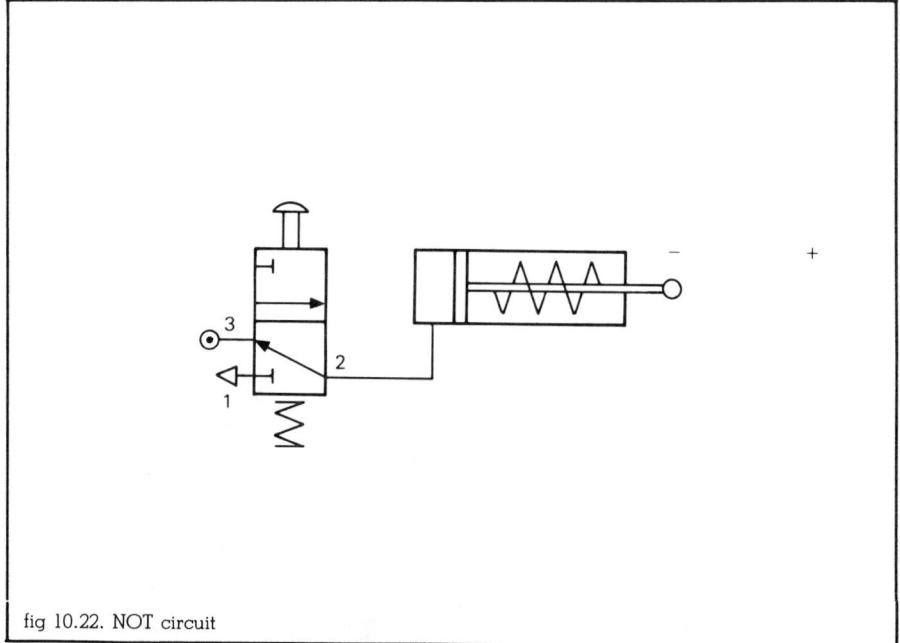

fig 10.22. NOT circuit

VALVE	PISTON
OFF	+ VE
ON	– VE

fig 10.23. NOT truth table

FORCES AT WORK

POSITIVE

The force produced by a single acting cylinder as it goes positive depends on two basic things, the **air pressure** being supplied to the cylinder and the **surface area** of the piston.

For example, if air is coming into the cylinder at a pressure of $0.3N/mm^2$, there is a force of 0.3N pressing on each square millimetre of the piston (see fig 10.24). If the piston is 40mm in diameter, then its area in square millimetres can be worked out using the formula πR^2, where $\pi = 3.14$ and R is the radius of the circle. In this case R = 20mm.

$$\text{Area of piston} = \pi R^2$$
$$= 3.14 \times 20 \times 20$$
$$= 1256mm^2$$

Therefore, the **force** produced by the piston as it goes positive will be given by:

$$\text{Force} = \text{Pressure} \times \text{Area}$$
$$= 0.3 \times 1256$$
$$= 376.8N$$

The formula can also be used to calculate the air pressure needed to produce a certain force. For example, a force of 540N is needed to move a packing case, using a cylinder of 60mm diameter. Fig 10.25 shows how the air pressure required is calculated.

NEGATIVE

The force produced as the piston of a single acting cylinder goes negative is not affected by pressure or area. The force comes from the spring which pushes the piston back to its negative positon.

$$\text{Force} = \text{Pressure} \times \text{Area}$$
$$\text{Therefore, Pressure} = \frac{\text{Force}}{\text{Area}}$$

$$\text{Area} = \pi R^2$$
$$= 3.14 \times 30 \times 30$$
$$= 2826 \ mm^2$$

$$\text{Therefore, Pressure} = \frac{540}{2826}$$
$$= 0.19N/mm^2$$

fig 10.25. Calculation of air pressure

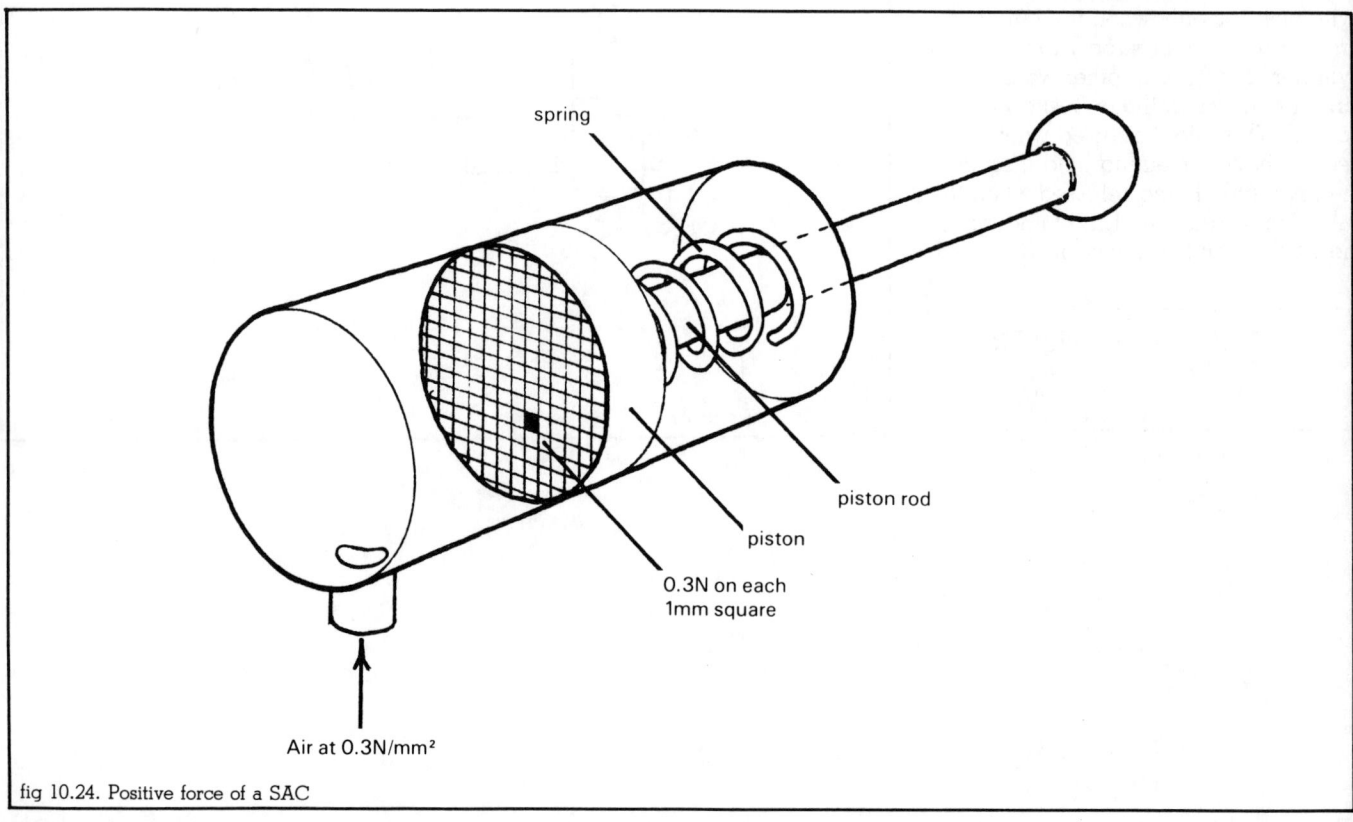

spring

piston rod

piston

0.3N on each 1mm square

Air at $0.3N/mm^2$

fig 10.24. Positive force of a SAC

FORCES IN A DOUBLE ACTING CYLINDER

The **double acting cylinder (DAC)** differs from the SAC in that it has two air inlets (ports) and does not have a spring. This means that the piston movement has to be powered in both directions. Air coming in through port 'A' will push the piston positive. Air coming in through port 'B' will push the piston negative (see fig 10.26).

POSITIVE AND NEGATIVE FORCES

The forces produced by a DAC as it goes positive and negative are different. **Why should this be so?**

A closer look at the two faces of the piston will show why this happens: fig 10.27 shows the face of the piston being pushed against to send the piston positive. It is a full circle, its area being calculated by using area = πR^2 as in the example shown in fig 10.24.

Fig 10.28 shows the face of the piston being pushed against to send the piston negative. It is not a full circle because the centre is taken up by the piston rod where it attaches to the piston. In calculating the area you have to allow for the missing part. The way to do this is to work out the total area, work out the area of the missing part, then take one away from the other (fig 10.29)

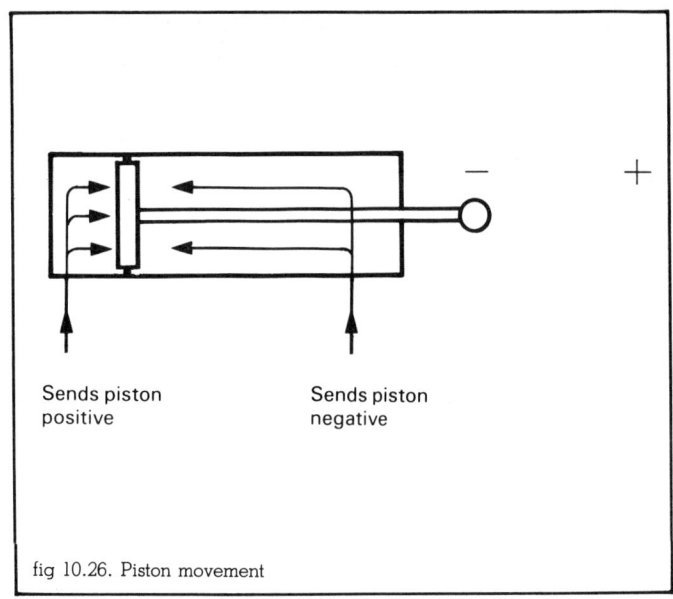

Sends piston positive

Sends piston negative

fig 10.26. Piston movement

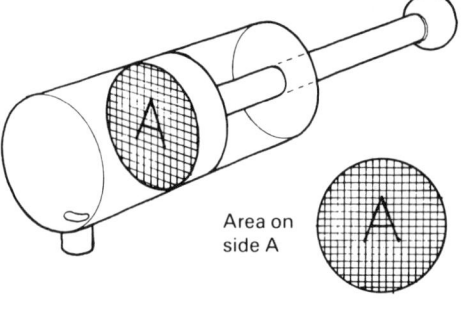

Area on side A

fig 10.27. Area producing POSITIVE force

AREA FOR POSITIVE FORCE	AREA FOR NEGATIVE FORCE
Area = πR^2	Total area = πR^2
$\quad$ = $3.14 \times 25 \times 25$	$\quad$ = $3.14 \times 25 \times 25$
$\quad$ = $1962.5 mm^2$	$\quad$ = $1962.5 mm^2$
	Piston rod area = πR^2
	$\quad$ = $3.14 \times 5 \times 5$
	$\quad$ = $78.5 mm^2$
	Effective area = $1962.5 - 78.5$
	$\quad$ = $1884 mm$
Positive force	Negative force
$\quad$ = Pressure × Area	$\quad$ = Pressure × Effective Area
$\quad$ = 0.4×1962.5	$\quad$ = 0.4×1884
$\quad$ = $785 N$	$\quad$ = $753.6 N$

fig 10.28. Area producing NEGATIVE force

To sum up, a double acting cylinder produces larger forces when going positive than when going negative. This is because the area of the piston to be pushed against by the air is less on the negative going side. As Force = Pressure × Area, if the area on one side is less, then the force on that side must be less.

EFFICIENCY

A pneumatic cylinder is quite effective when working at high pressures, in that the actual push or pull force will be about 95% of the calculated value. At low pressures, however, the efficiency is much reduced, especially below 0.2N/mm². This is due to friction, leaks and back pressure in the system. Back pressure is caused by escaping air trapped behind a piston which incoming air has to push against.

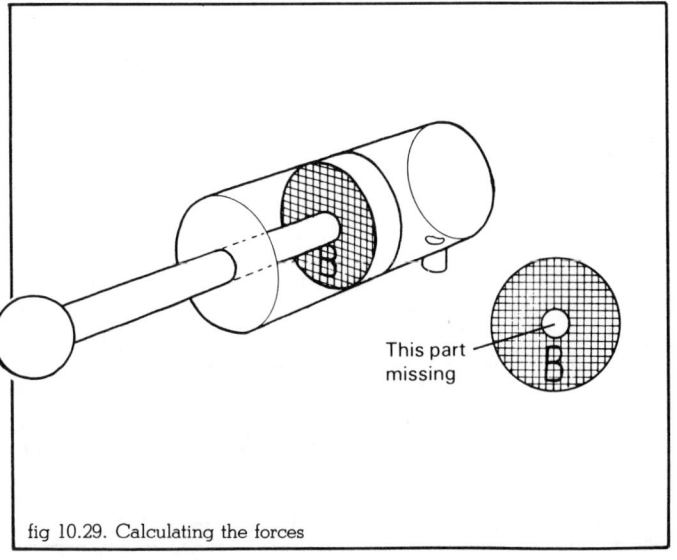

This part missing

fig 10.29. Calculating the forces

PISTON SPEED CONTROL

Controlling the speed at which a piston moves can be very important. For example, if it were being used to close a door, it could be dangerous to close it too quickly. Moving boxes of glassware also needs to be done carefully. Sudden piston movements against the box could be disastrous. This problem can be avoided by using a **flow control valve** (FCV) (fig 10.30). It allows you to control the rate of air flow into or out of a cylinder.

The flow control valve allows free flow of air in one direction and controlled flow in the other. The symbol, shown in fig 10.30, gives a good idea of how it works.

FREE FLOW
Air entering at port 2 blows open the ball valve and passes freely out of port 1, ignoring the restriction.

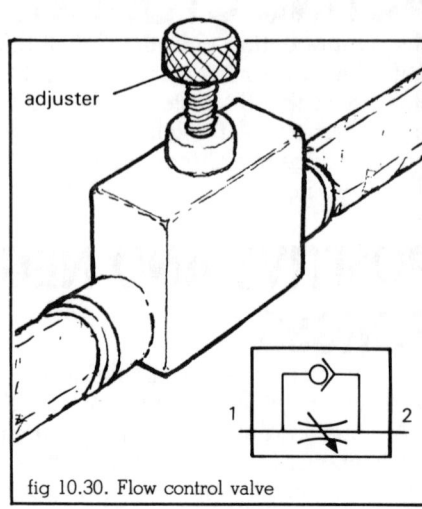

fig 10.30. Flow control valve

RESTRICTED FLOW
Air entering at port 1 closes the ball valve and so has to pass through the restriction before going out of port 2. By adjusting this restrictor, the volume of air passing through can be controlled. This is done by changing the size of the hole that the air has to pass through.

SPEED CONTROL FOR A SAC
The simple circuit shown in fig 10.31, will make the piston go positive slowly. Valve A would have to be held down while it went positive. The speed at which it went positive could be controlled by adjusting the restrictor. The more restricted the flow, the slower the piston would move. Depending on the pressure and the size of cylinder being used, it should be possible to get movements lasting over several seconds.

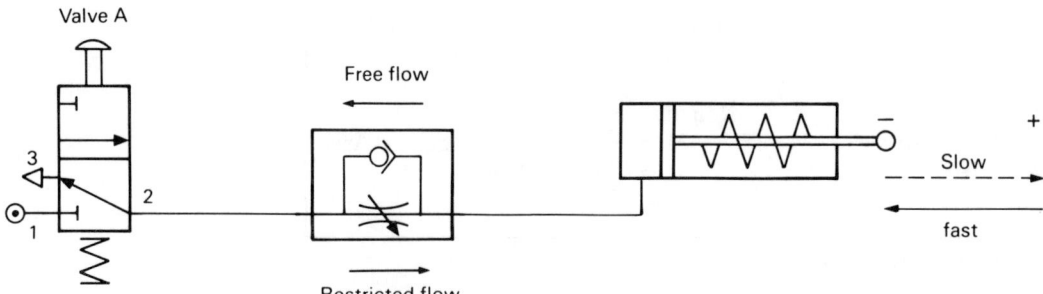

fig 10.31. Controlling positive piston speed, SAC

By adding a second flow control valve fitted the opposite way round, the speed of both the positive and negative movements of the piston could be controlled (fig 10.32).

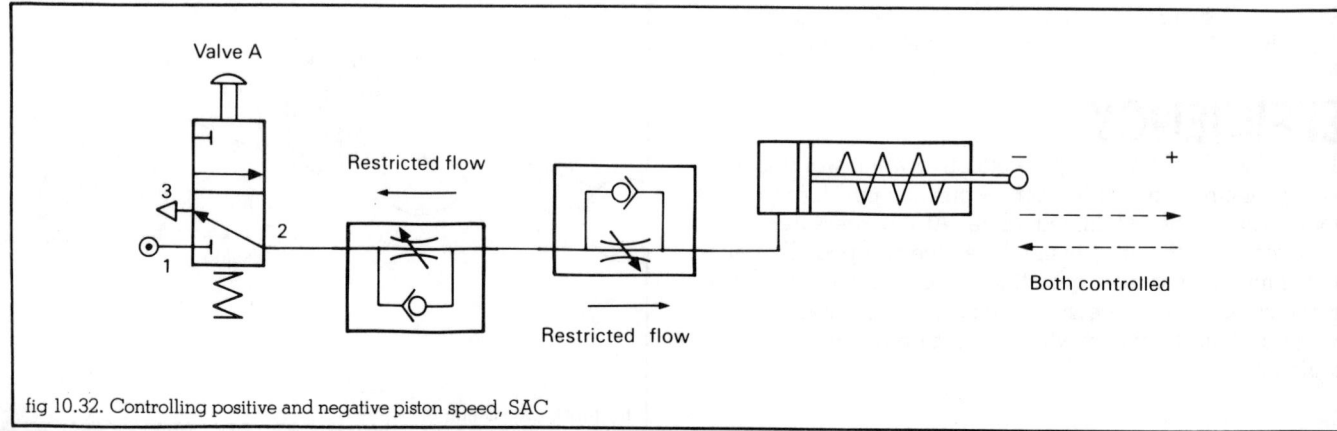

fig 10.32. Controlling positive and negative piston speed, SAC

SPEED CONTROL FOR A DAC

Fig 10.33 shows a DAC being controlled by a five port lever operated valve. A flow control valve has been added to allow the positive movement of the piston to be controlled. To understand how it does this, it is important to realise that, as shown in fig 10.33, there is air in the right hand part of the cylinder. This air is coming from port 2 of the 5PV, flowing freely through the FCV and so into the cylinder, holding the piston negative.

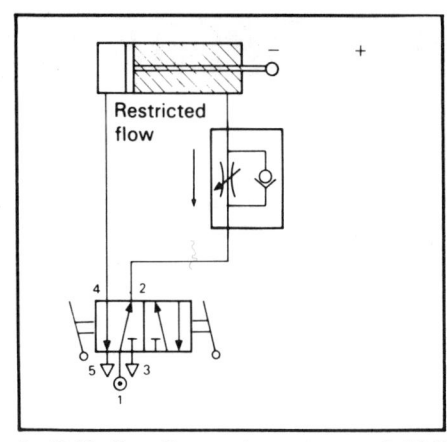

fig 10.33. Controlling positive piston speed, DAC

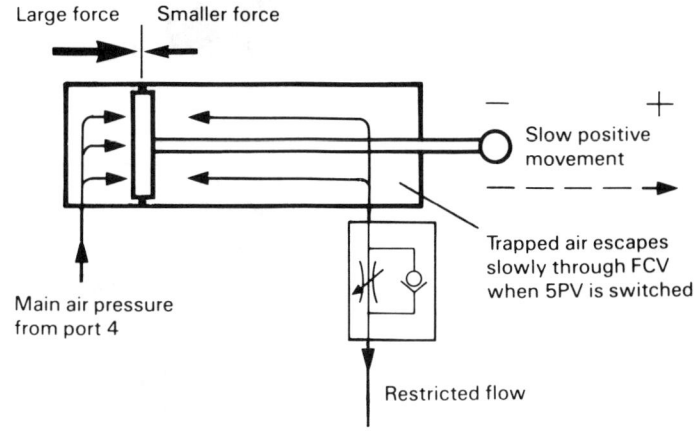

fig 10.34. Opposing forces give slow piston movement

When the 5PV is switched, air is supplied from port 4 to send the piston positive. The positive movement of the piston is slowed by the air trapped behind it. The trapped air escapes slowly through the restrictor in the FCV, allowing the piston to go positive slowly. This method, restricting the exhausting air, is the best way of controlling piston speed. Restricting air going into a cylinder often causes a jerky motion of the piston. Restricting exhausting air gives smooth movement and a constant speed. Fig 10.35 shows both positive and negative piston speeds controlled. This circuit could be used to control the pneumatic vice so that it closes slowly and opens quite quickly. In each case it is the 'exhausting' air that is restricted.

Not all flow control valves look the same. Fig 10.36 shows four different ones. Two of them fit into an air line, the other two fit directly into other components.

fig 10.35 Controlling piston speed

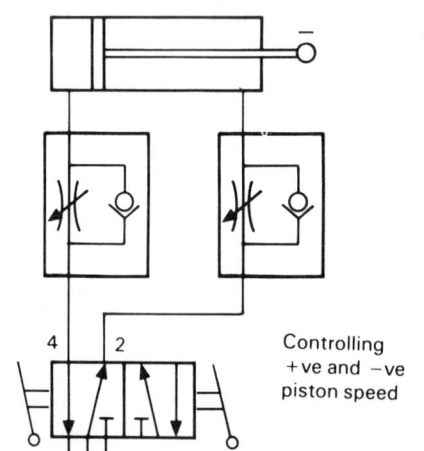

fig 10.36. Flow control valves

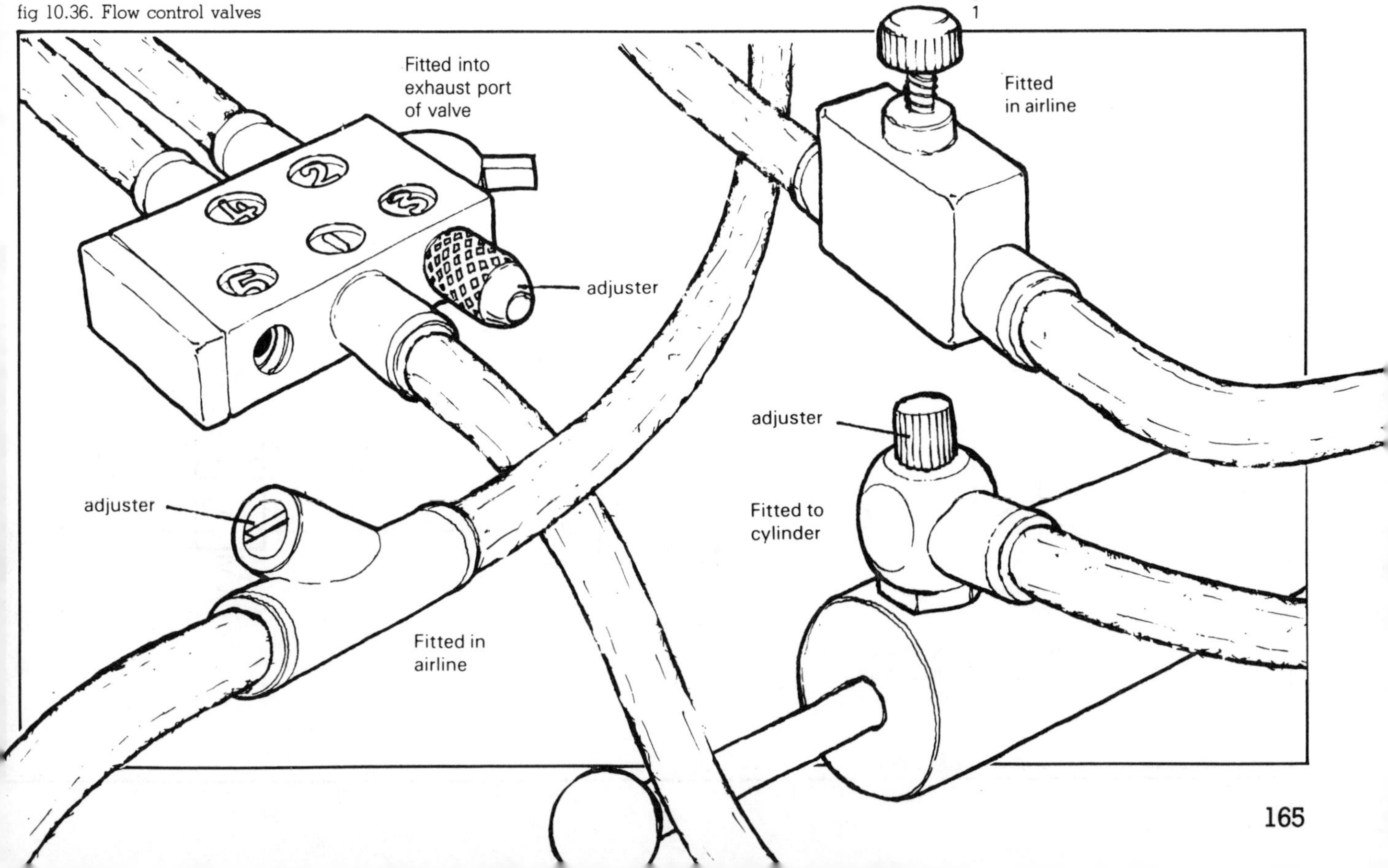

AIR OPERATED VALVES

Air operated valves which, instead of being switched by a push button, lever or other manual means, are switched by air pressure. The three you are most likely to come across are **air operated three port valves** (fig 10.37); **double air operated five port valves** (fig 10.38); **low pressure air operated valves**, sometimes called a 'pressure sensitive valve', which may be either three or five port (fig 10.39).

In these valves, the spool which switches the connections, is moved by air pressure. The normal three and five port valves need a pressure of at least $0.2N/mm^2$ to move the spool. The low pressure valve operates at much lower pressures, around $0.05N/mm^2$. The most commonly used of these valves is probably the double air operated 5PV.

fig 10.37. Air operated 3PV

fig 10.39. Low pressure air operated 3PV

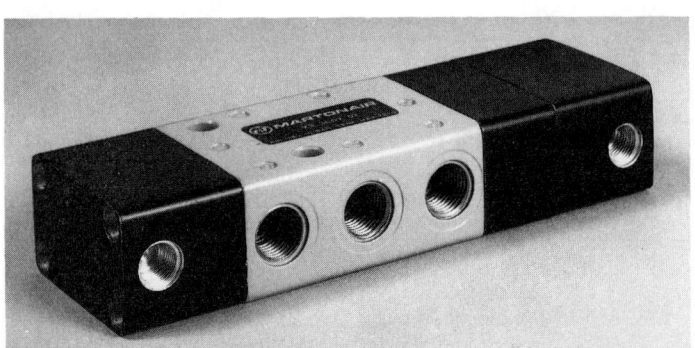

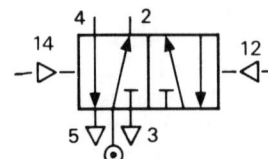

fig 10.38. Double air operated 5PV

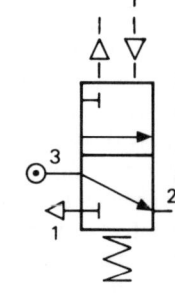

DOUBLE AIR OPERATED 5PV

Fig 10.40 shows a basic circuit for this type of valve. The 3PVs which signal the DAO5PV are known as 'pilot valves', the DAO5PV itself being known as the 'control valve' because it controls the air going to the DAC. The most important thing to understand is that **the air which actually goes to the cylinder is quite separate from the air signals which move the spool**. This has the advantage that the two can operate at different pressures, and normally means using relatively low pressure signals to control high pressure air going to the cylinder.

The equivalent in electronics would be using a low voltage circuit to control a high voltage circuit.

If the DAO5PV receives air signals at both ends at the same time, the spool will not move, provided the signals are of equal pressure. Therefore, the connections to the cylinder will not change.

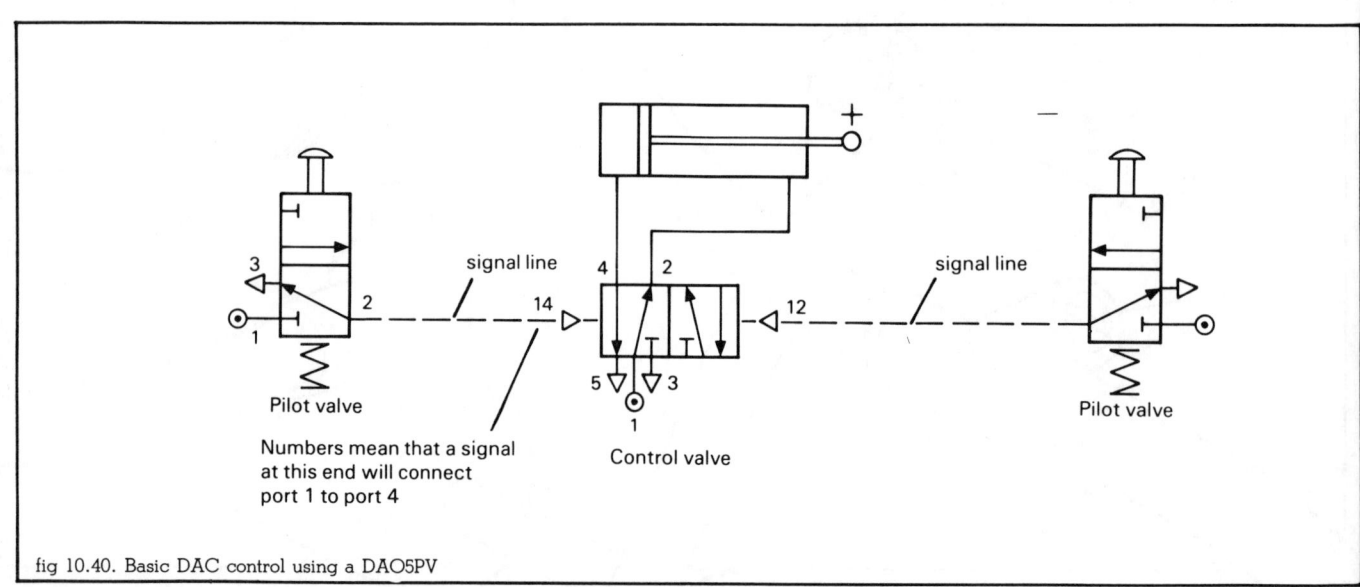

fig 10.40. Basic DAC control using a DAO5PV

LOW PRESSURE AIR OPERATED VALVES

Fig 10.41 shows the arrangement inside the valve. The key to understanding this valve is to remember that Force = Pressure × Area. Therefore, a larger force can be produced by increasing either the pressure or the area. With this valve it is the area that is increased. The spool is connected to a large rubber diaphragm in the top of the valve and low pressure air entering above the diaphragm presses on its large area pushing it down. Because the diaphragm has such a large area, even a low pressure signal is enough to produce the force needed to move the spool.

One use of this valve is shown in fig 10.42a, an **air bleed** circuit. Low pressure air is being supplied through the flow control valve. At the 'T' connector it takes the easy route, leaking out through the air bleed into the open air. If the air bleed is blocked, the air has to go to the valve, so switching it to send the piston positive. As soon as the blockage is removed the valve will switch back, so allowing the piston to go negative. Fig 10.42b shows an air bleed circuit in use, detecting packages and then moving them. Another use for this valve is shown in the section on automatic circuits.

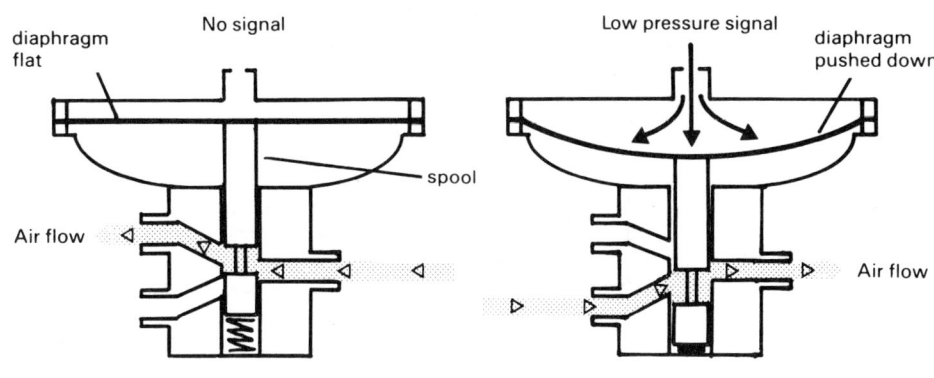

fig 10.41. Low pressure air operated 3PV

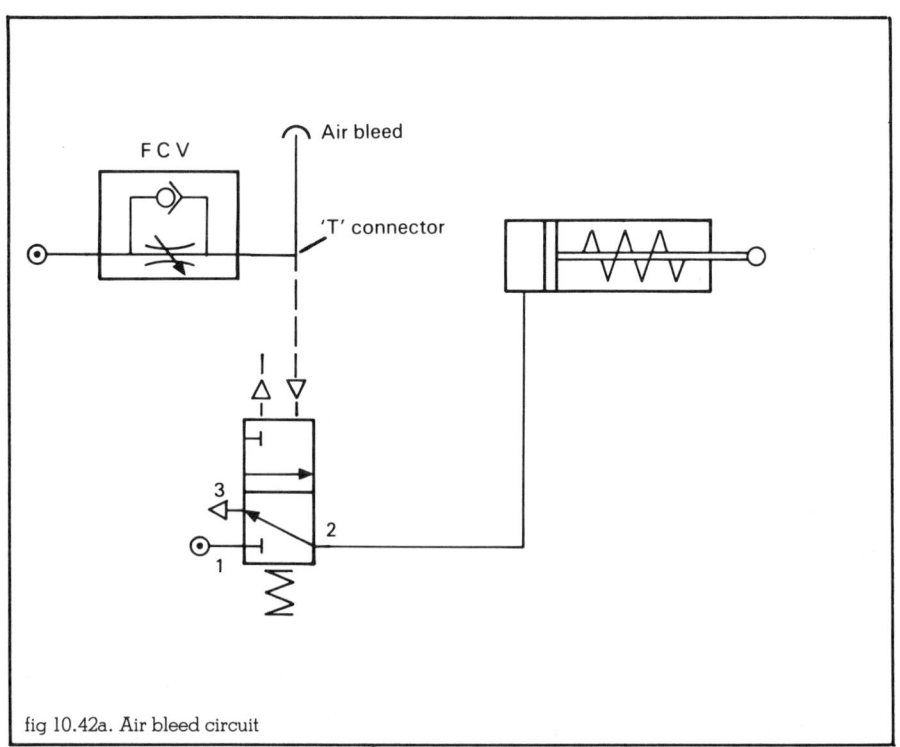

fig 10.42a. Air bleed circuit

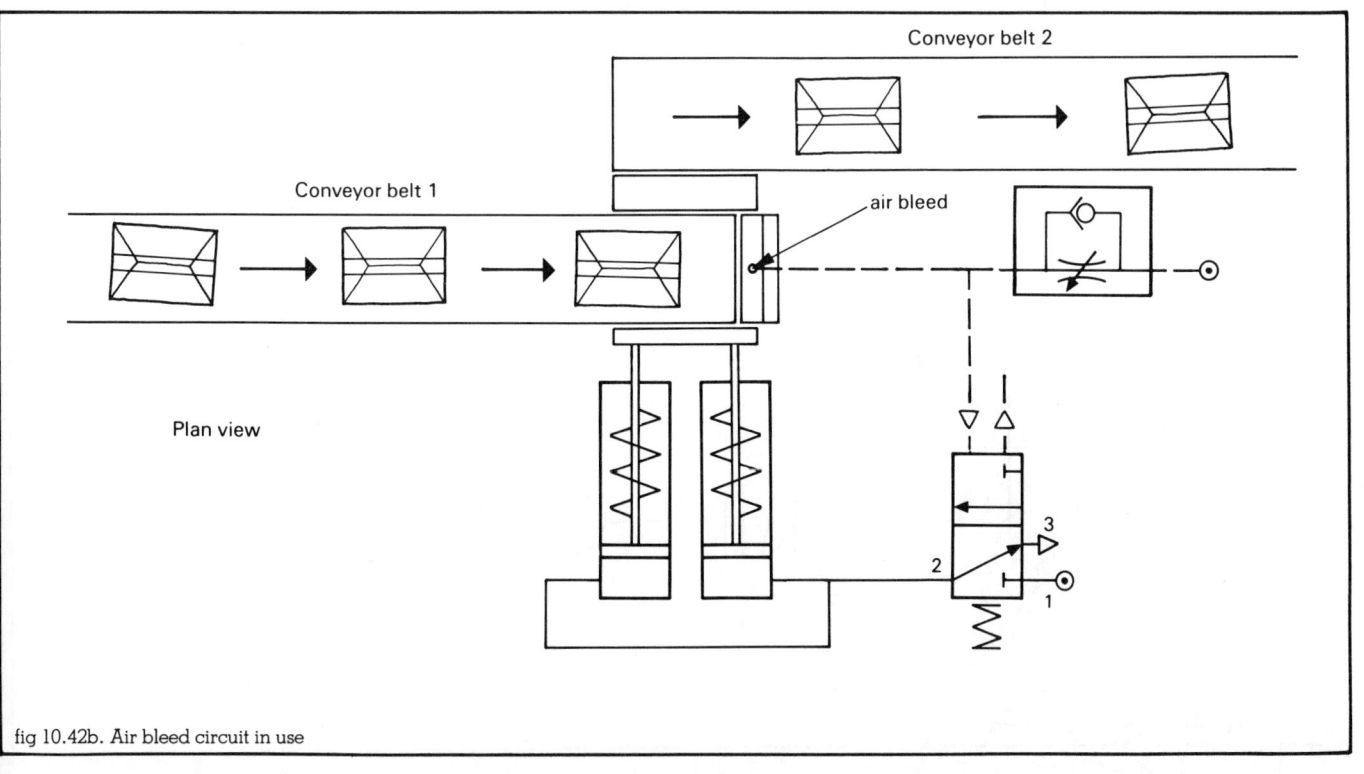

fig 10.42b. Air bleed circuit in use

TIME DELAYS

Time delays can be achieved in pneumatic circuits by using a **flow control valve and reservoir connected in series**. A reservoir is simply an empty container, rather like an empty bottle. The output from the reservoir is connected to an air operated valve (see fig 10.43).

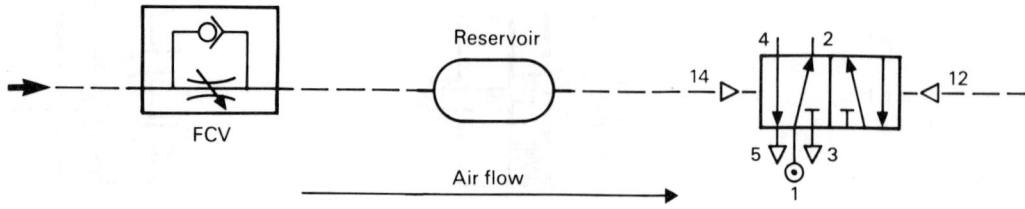

fig 10.43. Time delay

Air entering this arrangement at point A is restricted by the flow control valve. This restricted flow goes on to the reservoir, where pressure gradually builds up as it fills. When the pressure is high enough (around $0.2N/mm^2$), the air operated valve is switched. The length of the time delay depends on the adjustment of the flow control valve and the size of the reservoir. In the circuit shown in fig 10.44, there will be a delay of several seconds between valve X being switched on and the piston going positive. However, it will go negative immediately valve X is switched off.

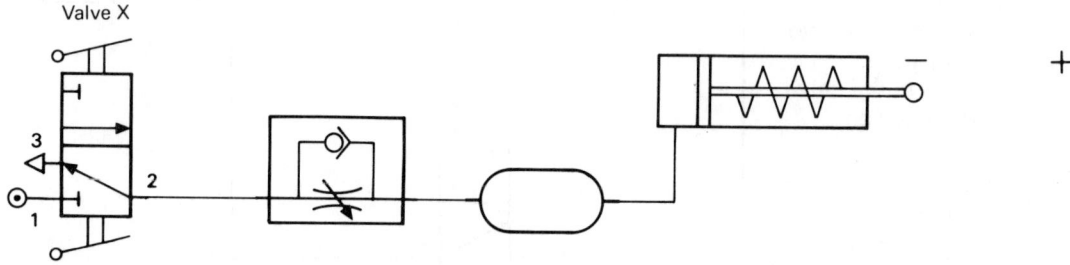

fig 10.44. Delay before SAC piston goes positive

In the semi-automatic circuit shown in fig 10.45, the piston will go positive when valve K is pressed. When it is fully positive, the piston rod will press valve L and, after a time delay, the piston will go negative.

Do not confuse time delays with speed control. Time delays give several seconds delay before a piston moves, whereas speed control slows down piston movement.

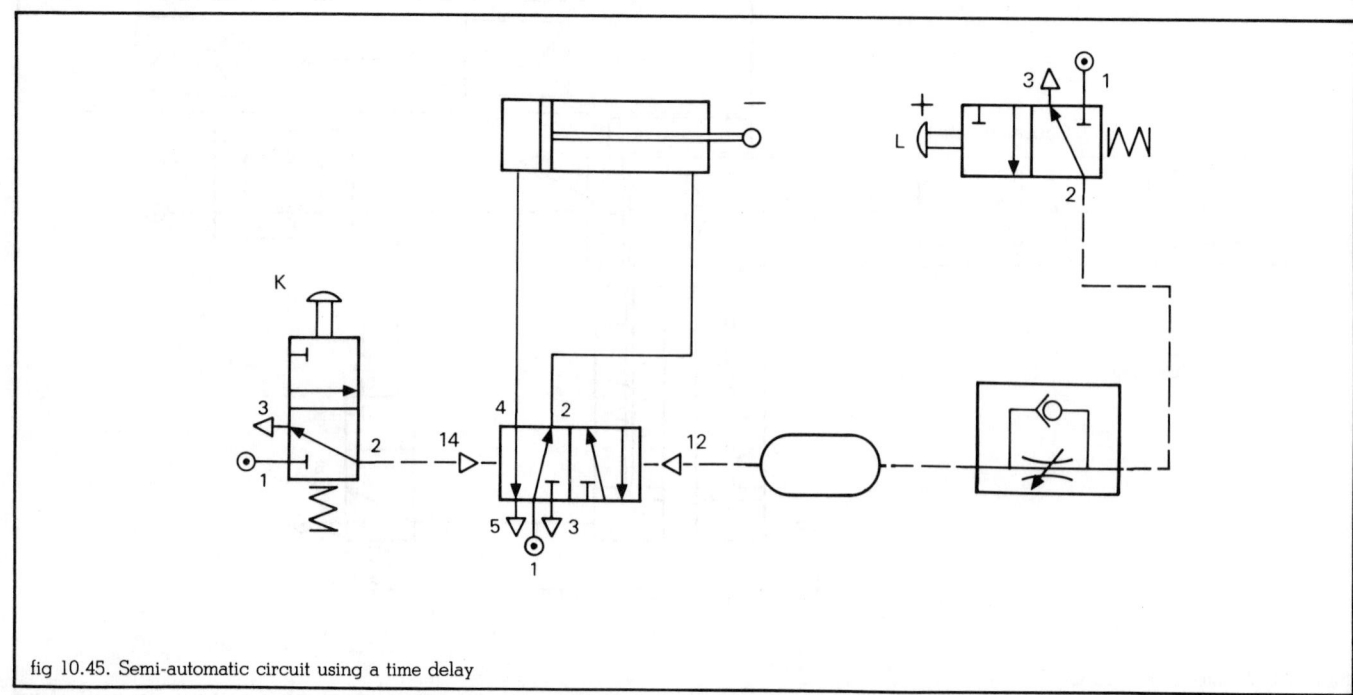

fig 10.45. Semi-automatic circuit using a time delay

AUTOMATIC CONTROL

There are several ways in which circuits can be made to operate automatically. The most obvious way is to place valves so that they are switched by the piston rod as it reaches the ends of its movement. Fig 10.46 shows a fully automatic circuit. As long as air is supplied to this circuit it will continue to work, and the piston will reciprocate automatically. The circuit can be switched on and off in a number of ways. One way would be to fit a lever operated 3PV at point Z in the circuit. With the valve fitted in that position the piston would always finish by going negative. Positioning it in the other signal line would make the piston finish in the positive position. Another way to switch the circuit on and off would be to fit a 3PV to the main air supply line to the 5PV. This would have the disadvantage that the piston could stop in any position.

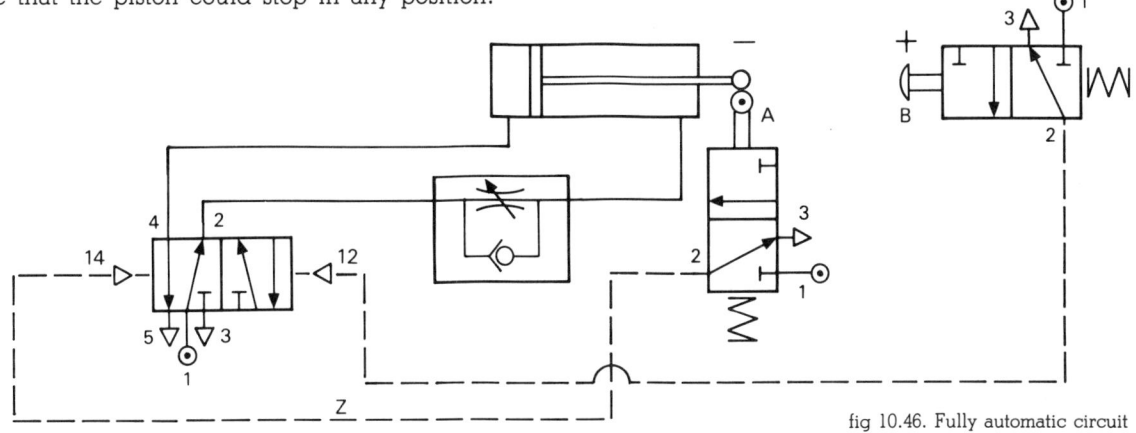

fig 10.46. Fully automatic circuit

IMPROVED AUTOMATIC CIRCUITS

The very real problem with the semi and fully automatic circuits we have looked at so far, is that they require valves to be placed where the piston rod can hit them. While this may be possible in a test situation, when it comes to doing a real job it is often very difficult or even impossible to place valves in the cylinder's **work area** (see fig 10.47). Anything placed in this area tends to get in the way of the work being done.

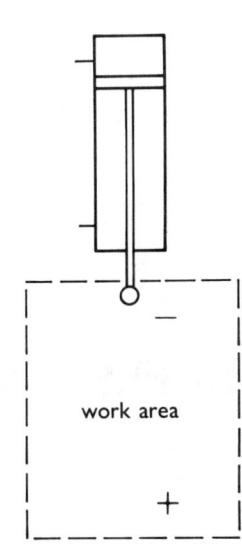

fig 10.47. Work area

USING AIR BLEEDS

One way to clear the work area is to use low pressure air operated valves and air bleeds. Fig 10.48 shows how this can be done. By positioning the air bleed so that it is blocked by the workpiece, the valve can be switched. All this requires in the work area, is a small hole or holes from which low pressure air is escaping. There are two disadvantages of this kind of circuit. First, producing compressed air costs money. To have it leaking away, even at low pressures, is not economic. Second, the small holes from which the air escapes can get blocked, causing one of the valves to be switched on all the time. This will stop the circuit working.

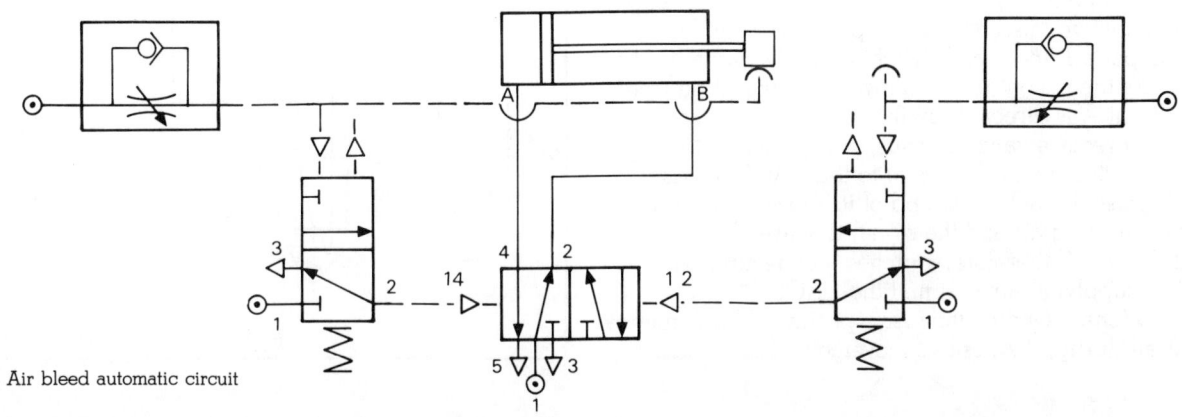

fig 10.48. Air bleed automatic circuit

USING TIME DELAYS

Another way to clear the work area is to use a time delay circuit as shown in fig 10.49. When valve A is pressed, it switches the DAO5PV, sending the piston positive. This also sends air, via the 'T' connector, to the flow control valve and reservoir. These two give a time delay of a few seconds while the pressure builds up to about $0.2N/mm^2$, enough to switch the DAO5PV and so send the piston negative. Once in the negative position, the piston will stay there until valve A is pressed again. This method leaves the work area totally clear.

The circuit shown in fig 10.49 can be made fully automatic by using another time delay in place of valve A. Fig 10.50 shows this circuit. The only air supply to the circuit is to port 1 of the DAO5PV. As long as valve X, a lever operated 3PV, is switched on, the piston will reciprocate. The two flow control valves have to be adjusted very carefully for this circuit to work reliably. This type of circuit is used where a continuous process is needed.

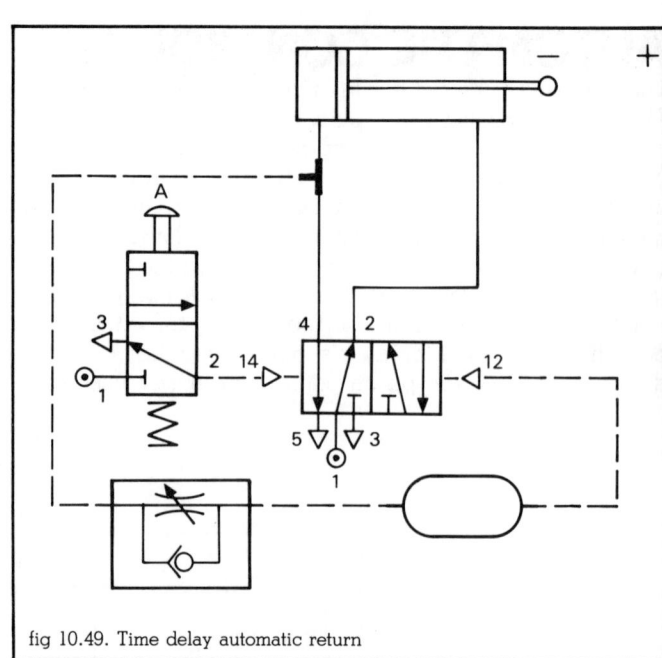

fig 10.49. Time delay automatic return

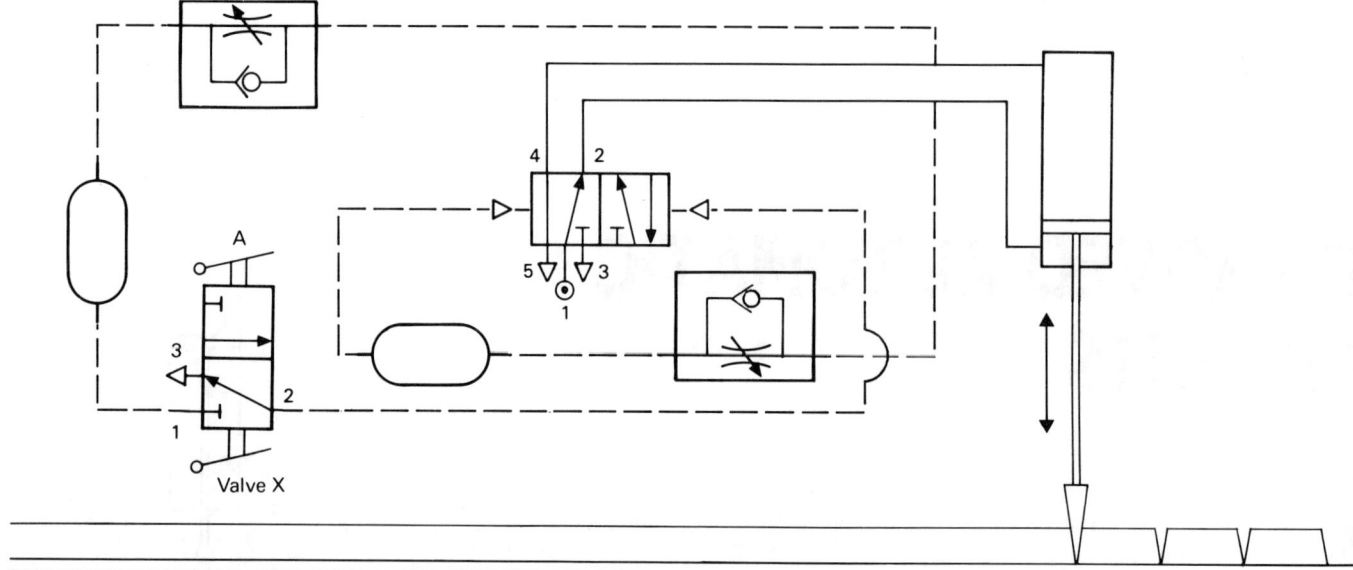

fig 10.50. Time delay fully automatic circuit

USING PRESSURE SENSING

Another way of achieving a fully automatic circuit is to use low pressure air operated valves to detect when a piston has come to the end of its movement. In order to get the valves to detect the end of the piston movement they have to be connected up in a rather unusual way. Fig 10.51 shows that instead of air being supplied to port 1 of the low pressure air operated valves, it is supplied to port 3.

It will help you to understand this circuit if you remember that for a low pressure air operated valve, with an air signal to the valve, ports 1 and 2 are connected and without an air signal to the valve, ports 2 and 3 are connected.

Air coming into the cylinder from the DAO5PV, to send the piston negative, also signals low pressure valve D, connecting ports 1 and 2 and switching off the signal to the DAO5PV. As the piston goes negative, air exhausting from the cylinder at A is directed, by the flow control valve, towards low pressure valve C. Thus, ports 1 and 2 are connected, switching off the signal to the DAO5PV. As soon as the piston reaches the end of its movement there is no more exhausting air and the signal to valve C is removed. Valve C, therefore, switches, connecting ports 2 and 3, thus supplying air to signal the DAO5PV. The DAO5PV switches, sending the piston positive. The sequence repeats itself during the positive movement of the piston.

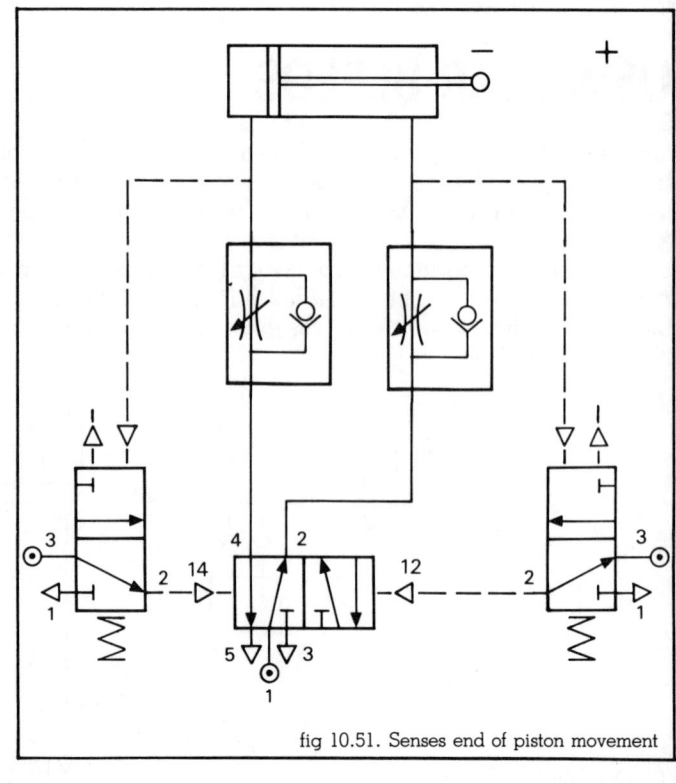

fig 10.51. Senses end of piston movement

SEQUENTIAL CONTROL

The circuits you have looked at so far have used either one cylinder only or two working in parallel. Many operations in industry require the use of two or more cylinders in sequence. Fig 10.52 shows a sequential circuit which uses mechanically operated valves. The sequence is:

1. Cylinder A goes positive
2. Cylinder B goes positive
3. Cylinder A goes negative
4. Cylinder B goes negative

This can be stated as **Sequence = A+, B+, A−, B−**

The circuit operates as follows:

1. Valve K is pressed to start the sequence. It signals control valve A to send piston A positive.

2. At the end of its positive stroke, piston A presses valve L, so signalling control valve B to send piston B positive.

3. At the end of its positive stroke, piston B presses valve N, so signalling control valve A to send piston A negative.

4. At the end of its negative stroke, piston A presses valve M, so signalling control valve B to send piston B negative.

Thus for one press of valve K the complete sequence is carried out. The circuit could be made fully automatic by moving the start valve, valve K, to a position next to the piston of control valve B. The sequence would then be triggered each time piston B went negative.

This circuit suffers from the same problem as some you looked at earlier, in that it has valves in the cylinder's work areas. This problem can be overcome in a number of ways. The more common ones are: using a **camshaft** to operate the valves; using **time delay circuits**; using **logic circuits**, either pneumatic or electronic.

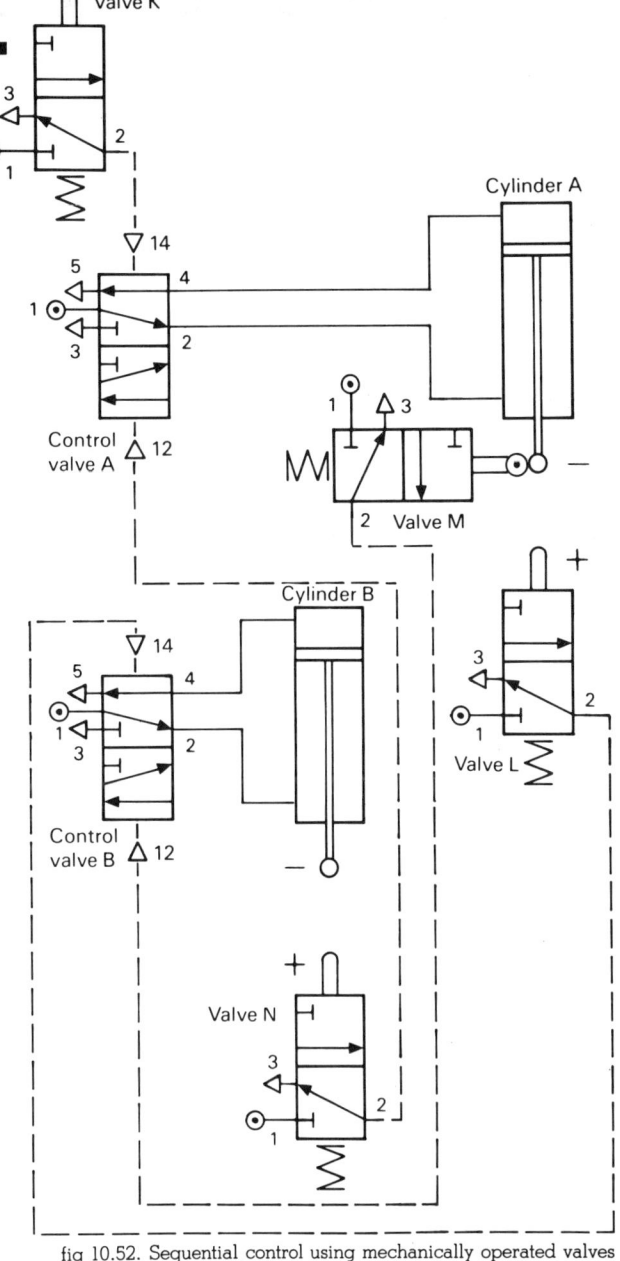

fig 10.52. Sequential control using mechanically operated valves

CAMSHAFT SEQUENTIAL CONTROL

The circuit shown in fig 10.52 could be operated by a camshaft, as shown in fig 10.53. The camshaft rotates, probably driven by an electric motor. As it rotates the cams press each of the valves in turn to complete the sequence.

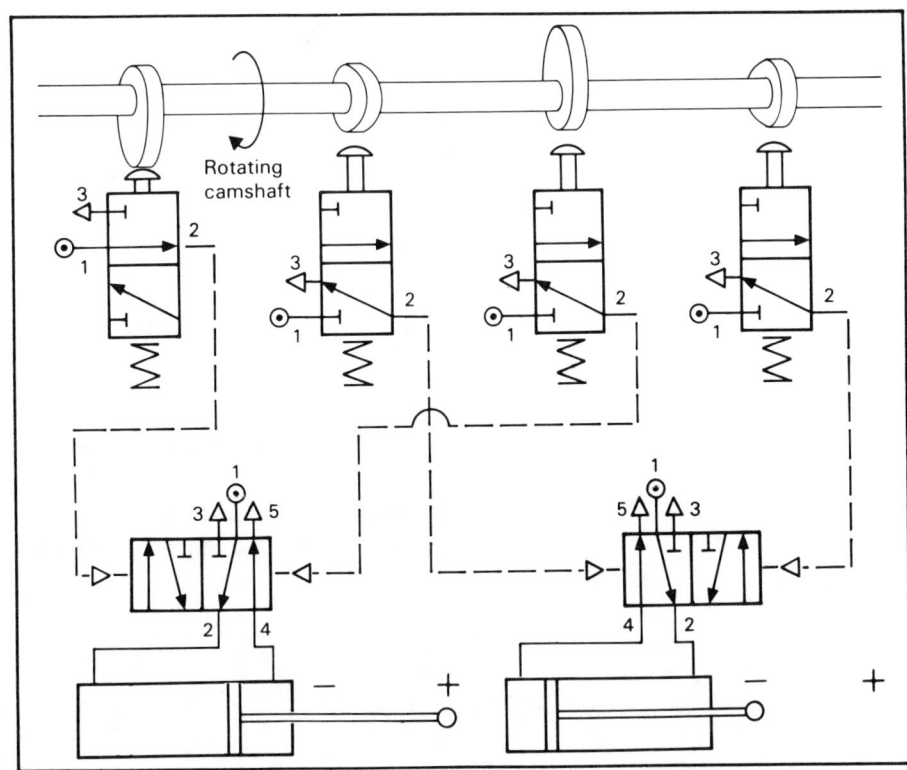

fig 10.53. Camshaft sequential control

TIME DELAY SEQUENTIAL CONTROL

Fig 10.54 shows two cylinders giving the sequence A+, B+, A−, B−. It makes use of three time delays to sequence the switching of the control valves.

It operates as follows:

1. The start valve, valve S, is pressed. This signals control valve A to send piston A positive. It also supplies air to 'time delay 1'.

2. Time delay 1, after a short delay, signals control valve B to send piston B positive. This also supplies air to 'time delay 2'.

3. Time delay 2, after a short delay, signals control valve A to send piston A negative. It also supplies air to 'time delay 3'.

4. Time delay 3, after a short delay, signals control valve B to send piston B negative. The sequence is now complete.

The circuit could be made fully automatic by removing valve S and connecting a fourth time delay between point X and control valve A.

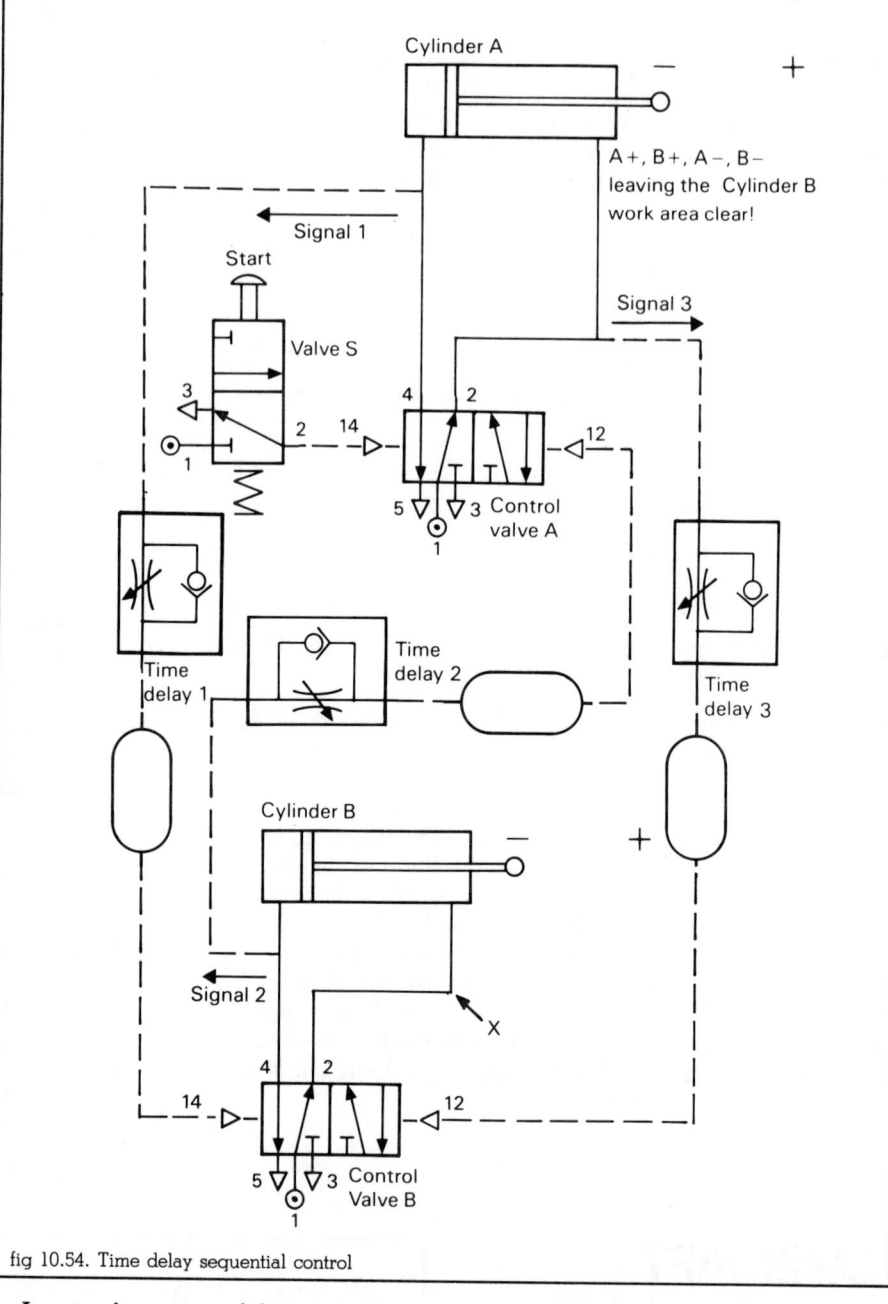

fig 10.54. Time delay sequential control

A major limitation of the sequential control circuits looked at so far, is that the sequence used must switch the cylinders alternately, i.e. A, B, A, B or B, A, B, A. Many operations require sequences other than this. Fig 10.55 shows a punching machine. It holds down a piece of metal with one piston, while a second piston punches a hole in the metal. The sequence is:

1. Metal is clamped
   = Clamping piston goes positive
   = A+

2. Punch goes down
   = Punching piston goes positive
   = B+

3. Punch pulls out
   = Punching piston goes negative
   = B−

4. Metal is released
   = Clamping piston goes negative
   = A−

The full sequence is A+, B+, B−, A−. Trying to produce this sequence using the circuits looked at so far would fail, because in trying to go from B+ to B−, control valve B would be getting a signal from both ends at the same time (see fig 10.56). When this happens the spool does not move, therefore the valve does not switch. The sequence would go A+, B+, then stop. The common way to overcome this problem using pneumatic components, is to use a **cascade system.**

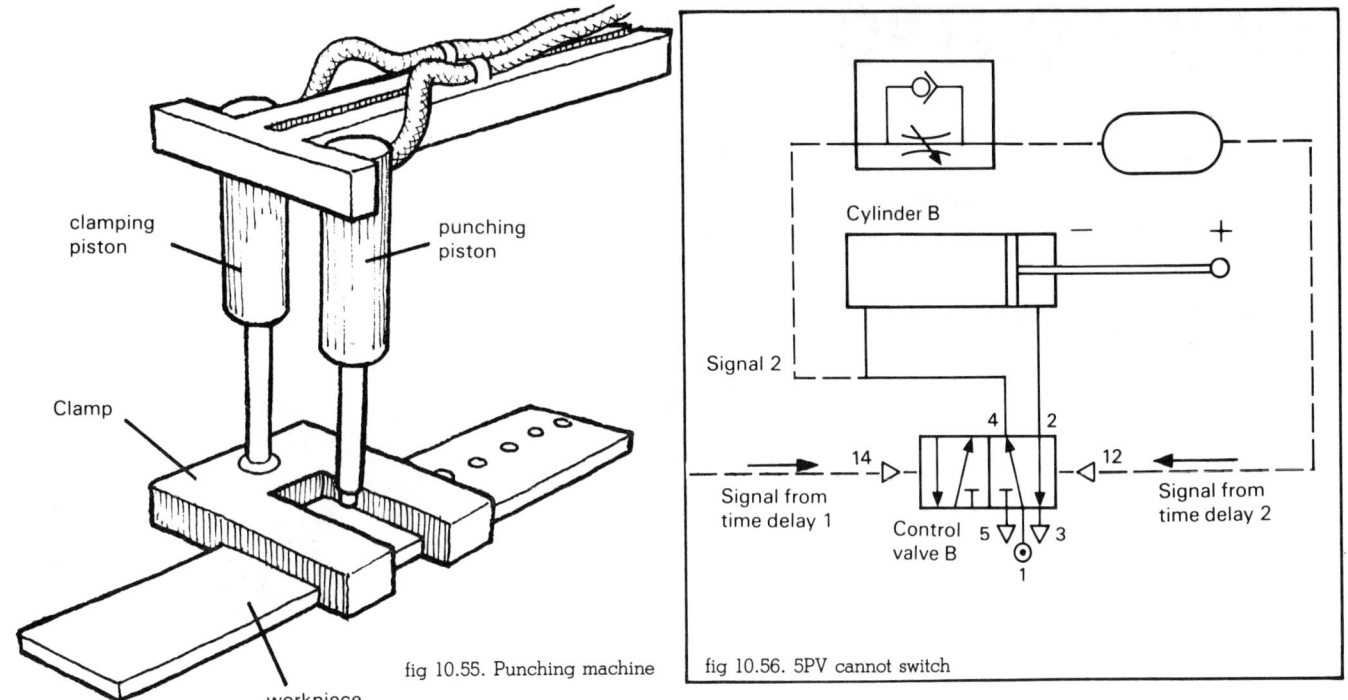

clamping piston

punching piston

Clamp

workpiece

fig 10.55. Punching machine

Cylinder B

Signal 2

Signal from time delay 1

14

4 2

12

Signal from time delay 2

5 3

Control valve B

1

fig 10.56. 5PV cannot switch

CASCADE SYSTEM FOR SEQUENTIAL CONTROL

This system allows sequences which would be 'inoperable' using the techniques looked at so far (e.g. A+, B+, B−, A−). This is done by splitting the air supply to the pilot valves into two groups. Only one group is supplied with air at any one time. The two groups for this sequence are **group I**, A+, B+ and **group II**, B−, A−.

The circuit is shown in fig 10.57. Note that the air supply to the pilot valves in both groups is controlled by the **group changeover valve**. The group I air supply lines are shown as thicker lines.

This system can be extended to control several cylinders in whatever sequence is required. However, it gets very complex and still has the disadvantage that the work area is not clear. In view of this and developments in electronics, most complex systems are controlled by microprocessors.

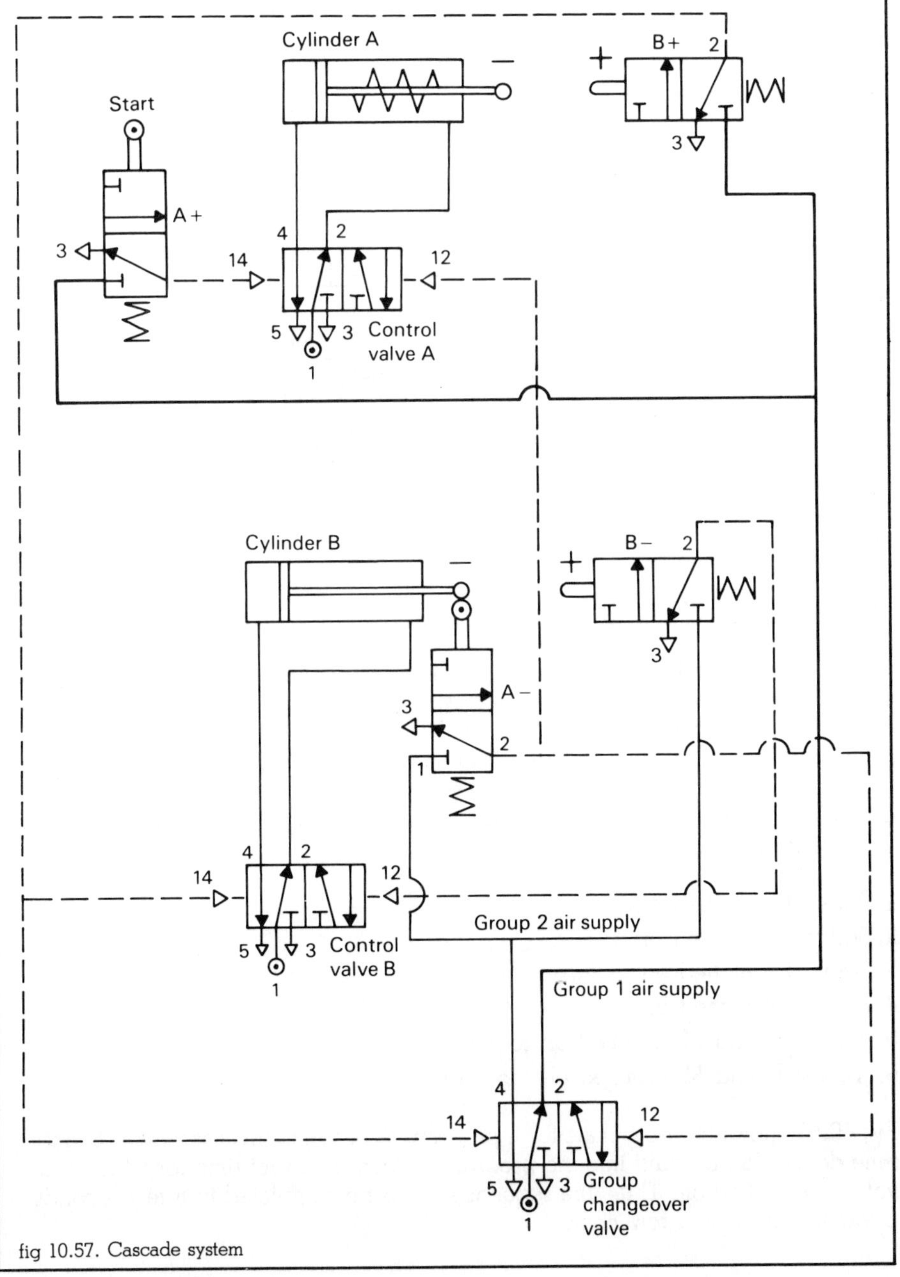

Cylinder A

Start

A+

3

14

4 2

12

5 3

1

Control valve A

B+ 2

3

Cylinder B

A−

3

2

1

4 2

14

12

5 3

1

Control valve B

B− 2

3

Group 2 air supply

Group 1 air supply

4 2

14

12

5 3

1

Group changeover valve

fig 10.57. Cascade system

173

LOGIC CONTROL

Pneumatic logic systems do exist, but they are rapidly being replaced by electronic systems. Electronic control systems are connected to pneumatic circuits using **solenoid operated valves** (fig 10.58). When energised the solenoid moves the valve spool, switching the valve. Solenoid operated valves have several advantages over other valves:

1. Electrical signals are **faster** than pneumatic signals. This means a faster response to remote signals.

2. Electrical signals can be transmitted over **longer distances** than pneumatic signals.

3. Electrical signals use **less energy** than pneumatic signals.

4. The components used in electrical signal circuits are **cheaper** and **smaller** than the equivalent pneumatic ones.

fig 10.58. Solenoid operated valve

Fig 10.59 shows a three port solenoid operated valve controlling a SAC. With the solenoid energised, ports 1 and 2 are connected, supplying air to send the piston positive. With it de-energised, ports 2 and 3 are connected, allowing the piston to go negative. Switch S can be any one of a number of different types of switch, for example, a simple hand operated on/off switch, a microswitch triggered by a moving piston, a reed switch positioned on a cylinder or a relay controlled by a computer.

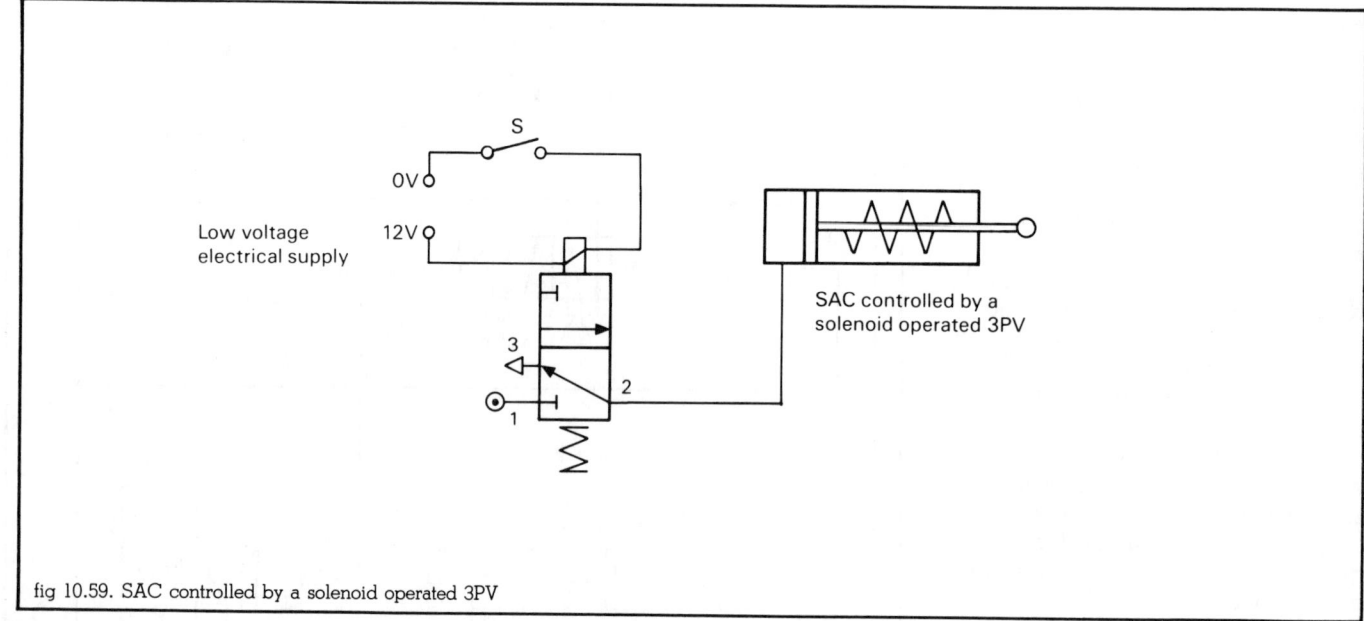

fig 10.59. SAC controlled by a solenoid operated 3PV

The last of these is the most versatile. The computer can be programmed to control a whole series of switches, **in any sequence**, thus controlling the cylinders. Fig 10.60 shows two cylinders being controlled via four three port solenoid operated valves. **Any required sequence can be programmed** because each of the four valves can be switched on and off independently. For example, sequence B+, A+, A- and B- together. The switching sequence of the solenoid valves for this is:

1. Valve L switched on.
2. Valve L switched off.
3. Valve J switched on.
4. Valve J switched off.
5. Valves K and M switched on together.
6. Valves K and M switched off together.

Fig 10.61 shows the flow chart for a program to produce this sequence. Exact time delays can be built into the program, as can the exact time for which the valve is switched on. Thus, the sequence could be completed in a few seconds, a few minutes, or a few hours.

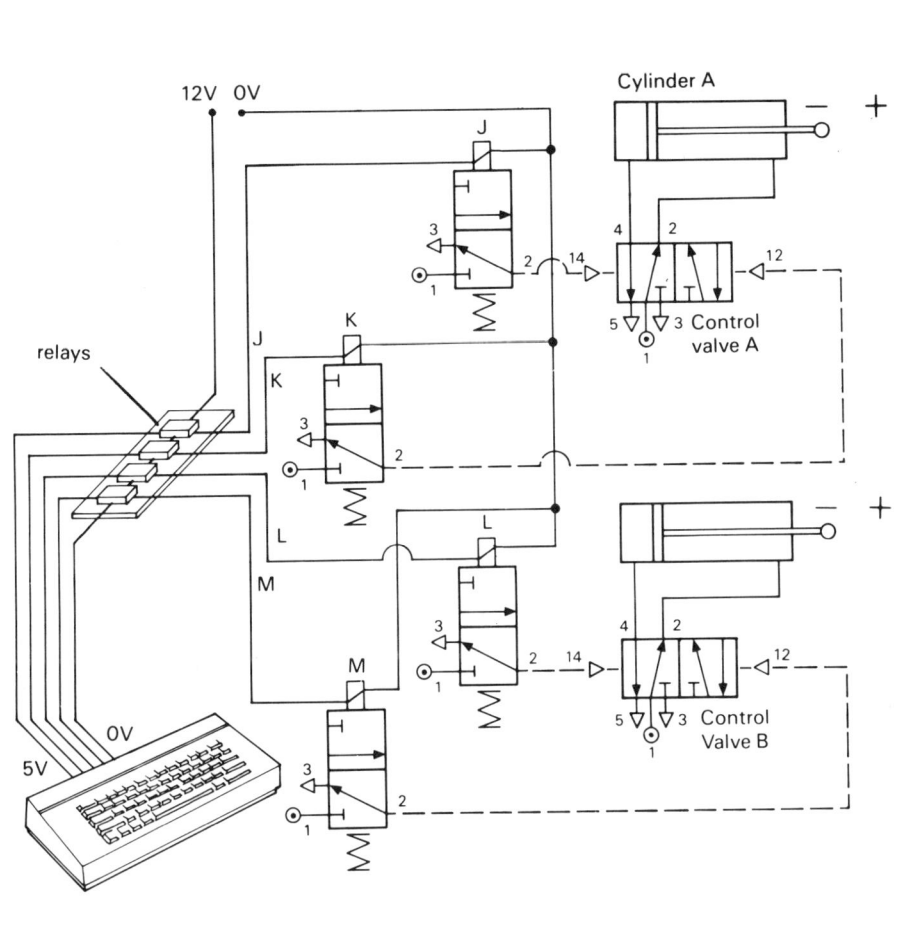

fig 10.60. Computer control

Industrial practice is not to use a full computer system as you would in school, but to use a programmable logic controller (PLC) (fig 10.62). These small, rugged controllers can be programmed, via a programming panel, to do a particular job. Once programmed, the PLC will do the same job as a full computer system, but at a much lower cost. Like a computer system it can be reprogrammed at any time to do another job.

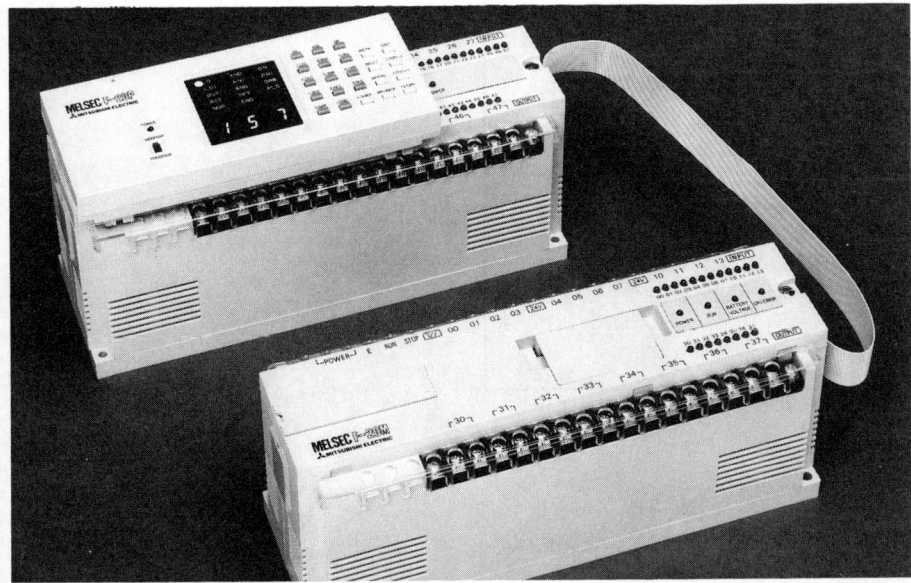

fig 10.62. PLC and programming unit

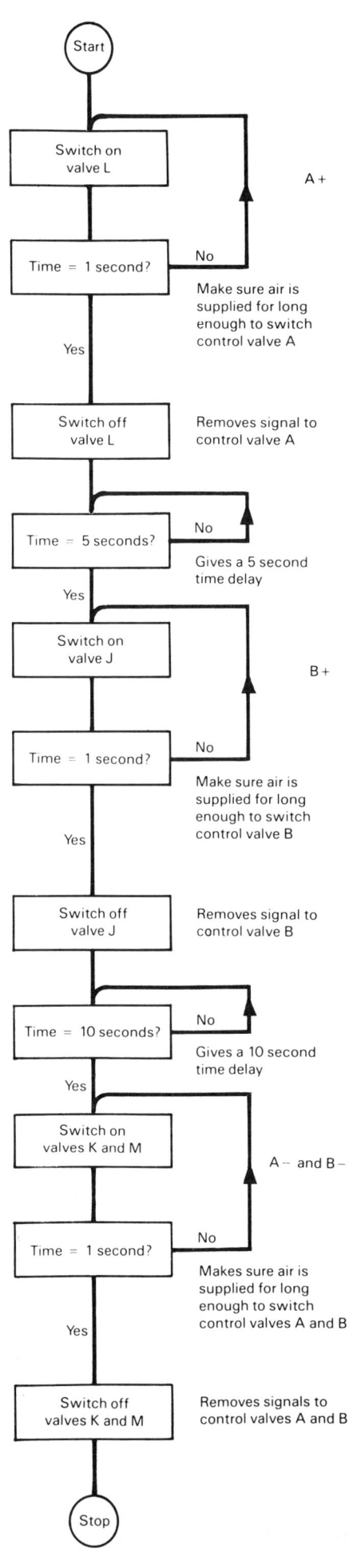

fig 10.61. Flow chart for sequence B+, A+, A− and B−

RELATED SYSTEMS — HYDRAULICS

Hydraulic systems are very closely related to pneumatic systems. They both use cylinders to produce linear motion at varying speeds and with varying forces. The basic difference between them is that hydraulics uses a fluid, hydraulic oil, as a means of transmitting motion and force instead of the compressed air used by pneumatics. The great advantage of using **the liquid**, is that it **cannot be compressed**, unlike compressed air which, as its name implies, can be compressed. Hydraulic pistons can be stopped at any point in their movement, and will stay there whatever happens to the load, whereas pneumatic pistons can only be used safely at the two ends of their movement. The basic principle of hydraulics is shown in fig 10.63.

As piston A is pressed down, hydraulic oil is pushed along the pipe to cylinder B, where it pushes up piston B. Because of the difference in the diameter of the two cylinders, a relatively small force pressing down on piston A will produce a large force pushing up on piston B. This is explained in fig 10.64.

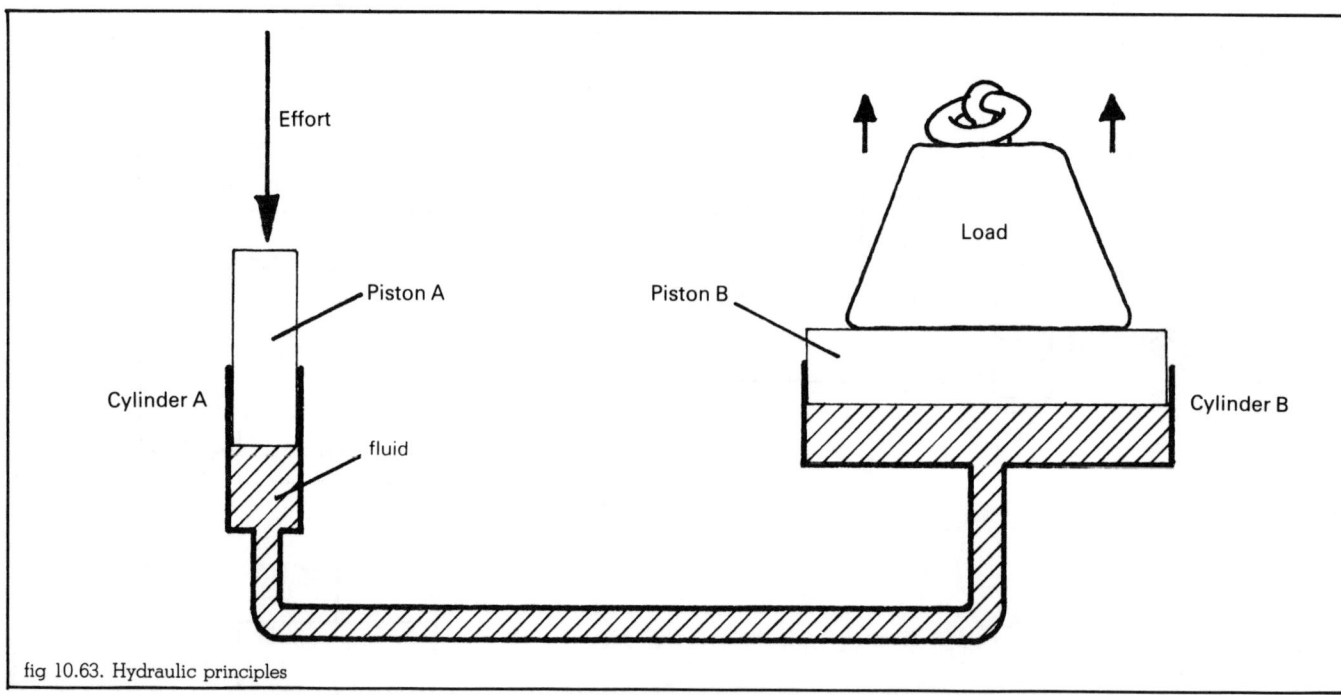

Effort

Piston A

Cylinder A

fluid

Load

Piston B

Cylinder B

fig 10.63. Hydraulic principles

The pressure produced in the fluid in cylinder A, by pressing down on piston A with an effort of 50N, is given by:

$$\text{Pressure} = \frac{\text{Force}}{\text{Area}} = \frac{50}{\pi R^2} = \frac{50}{3.14 \times 5 \times 5} = \frac{50}{78.5} = 0.6 \text{N/mm}^2$$

This pressure (here 0.6N/mm^2) is transmitted equally in all directions through the fluid. Therefore, the pressure pushing piston B upwards is also 0.6N/mm^2. The upward force this produces on piston B is given by:

Force = Pressure × Area

= 0.6 × Area

= 0.6 × 1962.25

= 1177.35N

Area = πR^2
= 3.14 × 25 × 25
= 1962.25mm^2

fig 10.64. Explanation of forces

Thus an effort of 50N is capable of moving a load of 1177N. The load is about 23 times greater than the effort. It is tempting to think that you are getting something for nothing, but of course you are not. What you must realise is that the effort has to move 23 centimetres to move the load 1 centimetre.

HYDRAULICS IN ACTION

Vehicle braking systems

Most vehicles use hydraulic systems to operate their brakes. Large forces are needed to press the brake shoes or pads against the revolving wheel hubs or discs. Fig 10.65 shows a basic hydraulic braking system. When the brake pedal is pressed, it pushes down the piston in the master cylinder, so creating pressure in the fluid. That pressure is transmitted to the wheel cylinder, which forces the brake shoes out against the revolving drum. An alternative to the drum brakes shown in fig 10.65 are disc brakes, which press on both sides of a disc fixed to the revolving wheel. Fig 10.66 shows this system on the front wheel of a motorbike. Disc brakes are better than drum brakes for several reasons, one being that because they use large diameter wheel cylinders, they apply larger braking forces to the wheels.

fig 10.65. Basic hydraulic brake system

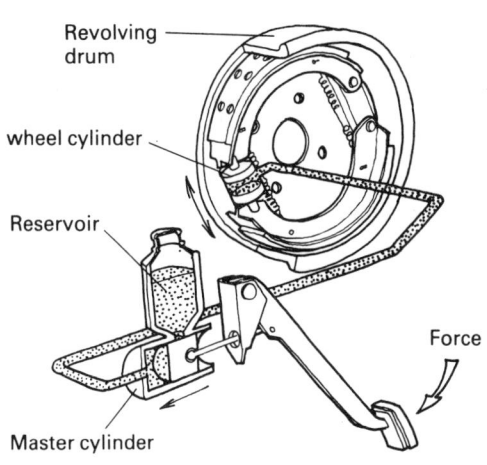

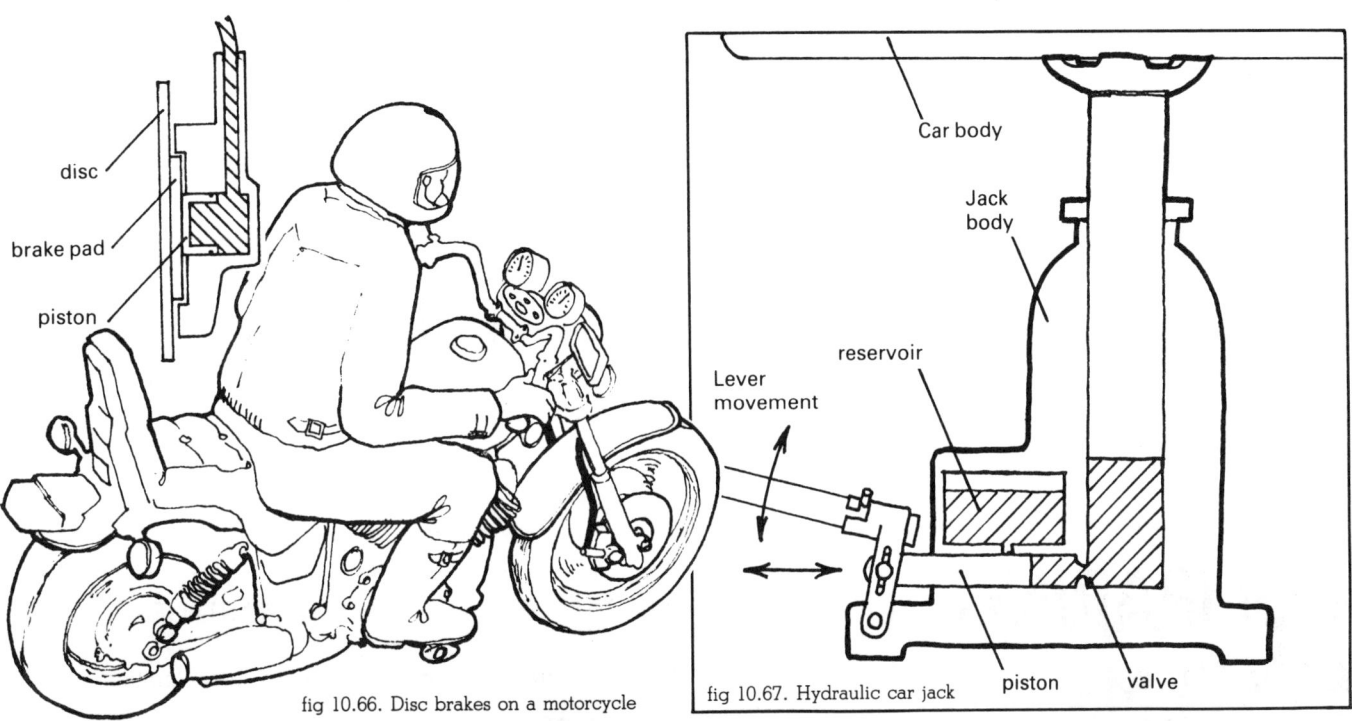

fig 10.66. Disc brakes on a motorcycle

fig 10.67. Hydraulic car jack

fig 10.68. Hydraulic systems on an excavator

Hydraulic jack

Garages and some motorists use hydraulic jacks to lift one wheel or one end of a car off the ground. They normally have a handle which has to be pumped to lift the car. Fig 10.67 shows the basic principles of a hydraulic jack. As the handle is moved backwards and forwards, fluid is pumped from a reservoir, through a one way valve, into the main cylinder. Each movement of the handle pumps only a small amount of fluid and so raises the car only a short distance. However, the lifting force is quite considerable; most cars weigh about 1000kg. To lower the car, the one way valve must be released, allowing the weight of the car to force fluid from the main cylinder back into the reservoir.

More advanced hyraulic systems use electrically operated pumps to pressurise the fluid and pump it around the system. Mechanical diggers, such as the one shown in fig 10.68, use this kind of system.

HYDRO-PNEUMATICS

Hydro-pneumatics is the name given to systems which use a combination of hydraulics and pneumatics. Compressed air is used to pressurise the hydraulic fluid which produces the output movement. Fig 10.69 shows the basic hydro-pneumatic system which is used to power the car servicing lifts used in garages.

The reservoir has a moveable piston. On one side there is air and on the other hydraulic fluid. When valve A is switched on, compressed air goes into the reservoir. This presses the piston down, so pressurising the hydraulic fluid. The pressure in the fluid will be exactly the same as that of the compressed air. When valve B is opened, the pressurised fluid flows into the hydraulic cylinder and raises the load. It will continue to rise until valve B is closed. With valve B closed, the flow of hydraulic fluid stops and the load stops. By careful use of valve B, the load can be positioned very accurately.

To lower the load, valve A must be switched off, allowing the air in the reservoir to escape, so depressurising the hydraulic fluid. The load will not move until valve B is opened, allowing fluid to flow back into the reservoir.

Circuits based on this principle have many uses. One of the more common ones is for 'power assisted' brakes on vehicles. These are fitted to all modern vehicles because they are so much easier to use, requiring much less effort from the driver than the purely hydraulic system shown in fig 10.67.

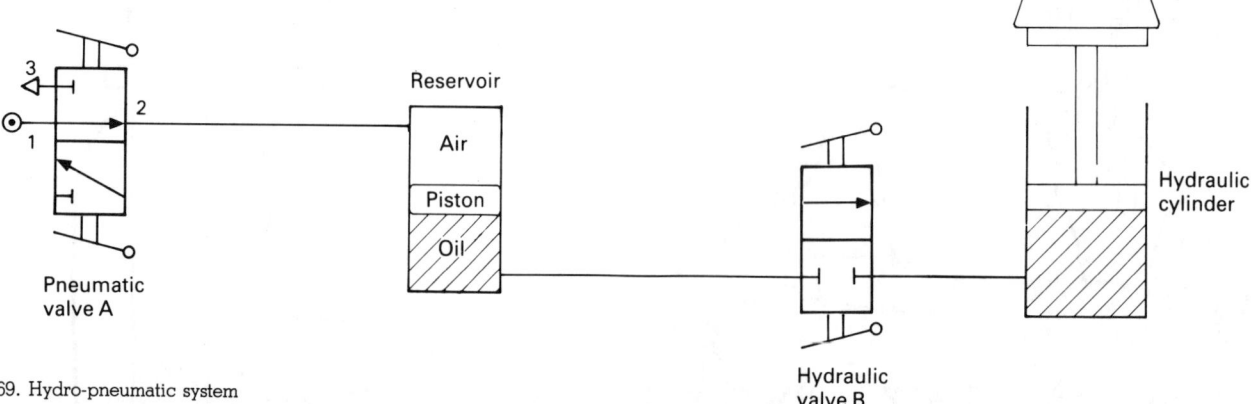

fig 10.69. Hydro-pneumatic system

COMPARING SYSTEMS

Industry uses three main systems to produce movement. They are electrical systems, pneumatic systems and hydraulic systems. They can be used individually or together, as in hydro-pneumatic or computer controlled pneumatic systems. Each has its advantages and disadvantages. The main ones are given in fig 10.70.

	ADVANTAGES	DISADVANTAGES
ELECTRICAL SYSTEMS	Good for **rotary** motion. Efficient, the cheapest to operate. Readily available. Excellent control systems.	Poor for linear motion. May cause sparks which are dangerous in some environments.
PNEUMATIC SYSTEMS	Good for **linear** motion. Can be used in dangerous environments. Cheapest way to get linear movements.	Only suitable for small rotary forces. Only suitable for relatively small linear forces. Needs a compressor to run the system. Approximately 3 times more expensive to run than electrical systems.
HYDRAULIC SYSTEMS	Excellent for precisely controlled linear motion. Can be used in some dangerous environments. Can produce very large forces.	The most expensive system. Can be dangerous because of the high pressure often used.

fig 10.70. Comparing systems

FAULT FINDING IN PNEUMATIC CIRCUITS

If your circuit fails to work as expected, there are a few simple checks you can make. Assuming that the compressor is switched on and working, and that the components are in good working order:

1. Check that the regulator is set correctly and providing air at the correct pressure. Many valves need a pressure of at least $0.2N/mm^2$ to operate reliably. Some, for example solenoid operated valves, may need more than this.

2. Check that air is being supplied to **all** necessary points in the circuit.

3. Check that the components are connected up as shown in your circuit diagram.

4. Make sure there are no major air leaks in your circuit, particularly in signal lines.

5. Check the adjustment of any flow regulators in the circuit.
   a) If they are completely closed they can stop the flow of air.
   b) Where they are being used in semi and fully automatic circuits their adjustment can be critical. Fig 10.50, on page 170 shows a circuit where this is very important.

SEQUENTIAL CIRCUITS

6. Check the sequence you are trying to operate! Does it alternate? (e.g. A, B, A, B). If it does not follow this kind of pattern, you need to be using a cascade system or computer control via solenoid operated valves.
7. If a circuit has carried out part of its sequence, note the point at which it has stopped.
   a) What signal is needed to continue the sequence?
   b) Has it stopped because the required signal is not getting to the control valve? This could be caused by a badly positioned pilot valve or a closed flow control valve.
   c) Has it stopped because the signal is having no effect on the control valve? This will happen if the control valve already has a signal on the other end. If this happens there is probably a fault in your circuit design!
   d) Has it stopped because the signal is going to the wrong end of the control valve? If so, is it an error in connecting up the circuit or is it a circuit design fault?

COMPONENT FAULTS

Component faults are quite rare, as pneumatic components are strong and hardwearing being designed to do real jobs in industry, hour after hour, day after day. In schools, however, the lubricating oil, which is normally carried as a fine mist in the air supply, is often not included. This avoids getting your hands and clothes oily, but it also means that components need to be lubricated from time to time in order to stop them sticking. The parts likely to stick are pistons and valve spools.

EXERCISES

1. Give three examples of the everyday use of compressed air.

2. Explain why a safety valve and drain tap are needed on a compressor.

3. Give two reasons why pneumatic systems are used in industry.

4. What three parts go to make up a regulator unit? Name the three parts and explain what each of them does.

5. Give one advantage and one disadvantage of including a lubricator unit in a pneumatic system.

6. Draw a cross-sectional view of a single acting cylinder and label four important parts.

7. Draw the symbol for a three port lever operated valve. Label each of the three ports to show what they are normally connected to.

8. What force will be produced by a 40mm diameter cylinder as it goes positive using a pressure of $0.4N/mm^2$.

9. Explain clearly why the forces produced by a double acting cylinder are different in each direction.

10. What air pressure needs to be supplied to a single acting cylinder of 56mm diameter to produce a force of 650N.

11. Calculate the positive and negative forces produced by a DAC of 30mm diameter using a pressure of $0.7N/mm^2$. The piston rod is 8mm diameter.

12. A cylinder is required to give a positive force of 800N using a supply pressure of $0.2N/mm^2$.
    a) What is the minimum diameter of cylinder needed to do this?
    b) Why, in reality, are you likely to need a slightly larger diameter?

13. Draw the symbol for a double air operated five port valve and indicate the connections to each of the ports. What sort of signal pressure is needed for these valves?

14. Draw the symbol for a flow control valve, showing the direction of free flow and the direction of restricted flow. Draw a circuit which uses one to control the speed of the positive movement of a DAC.

15. Draw the symbol for a low pressure air operated three port valve and explain how it is operated by low pressure air.

16. Draw a pneumatic AND circuit, using three valves, controlling a single acting cylinder.

17. Explain why a shuttle valve is used in a pneumatic OR circuit.

18. Use notes and sketches to explain how a three port valve can be connected up to give a NOT function.

19. Draw a circuit diagram showing a double acting cylinder, controlled by a lever operated five port valve, which will go positive slowly and negative quickly.

20. Use a diagram to show what is meant by the terms pilot valve and control valve. Give one advantage of this type of system.

21. Design a circuit which will control two single acting cylinders so that they both go slowly positive at the same time. It is important that they move exactly together.

22. Design a circuit which will control a double acting cylinder so that it reciprocates automatically once switched on. When switched off it should stop in the negative position.

STRUCTURES

When you think of structures you probably think of things like bridges, electricity pylons and tall buildings. These are very obvious structures, but there are other examples much closer to you, things such as stools, benches, cupboards and even doors. All these are examples of **man-made** structures.

Many made-made structures use the same principles as those found in **natural structures**, such as trees, plants and caves. Have you ever thought of yourself as a structure? You have a skeleton which forms a support for the softer parts of your body. It is just as much a structure as the poles that support a tent or the wooden frame inside a sofa.

Just from thinking about these few examples you can see that there are many different types of structure. Although evolved or designed to do different jobs, they do have some things in common. They all carry a load of some sort, for example tunnels support the rock above them, tree trunks carry branches and leaves. They all work to hold various parts in their correct position, for example, a bicycle frame holds the wheels, pedals and seat in place, tree roots hold a tree in place.

Once you start to think about it you will see that just about everything you can think of has a structure of some sort. It may be a very obvious man-made structure or a microscopic cell structure. This chapter is concerned largely with man-made structures.

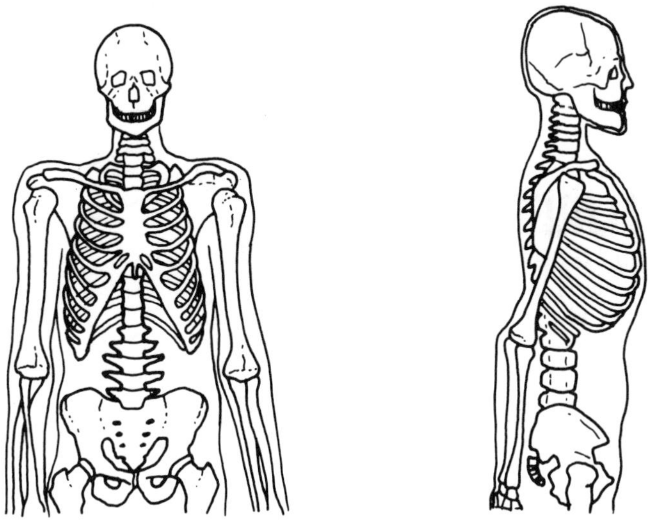

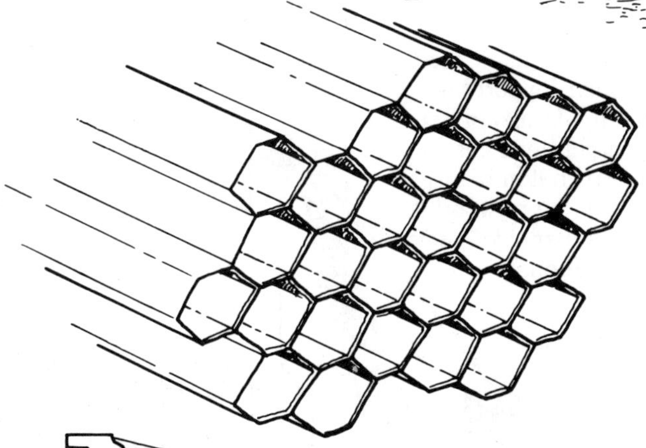

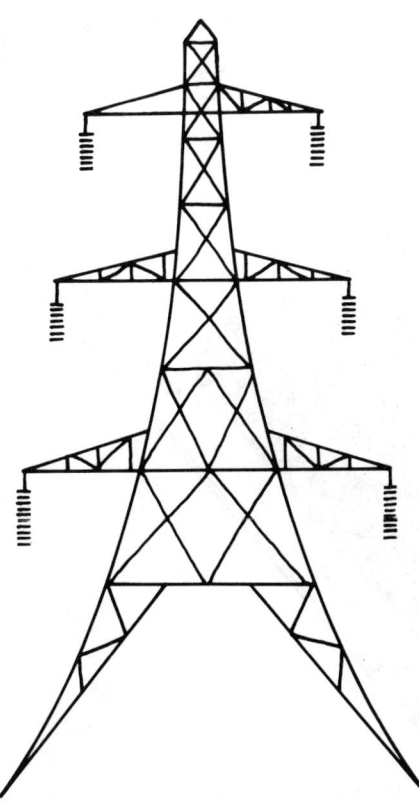

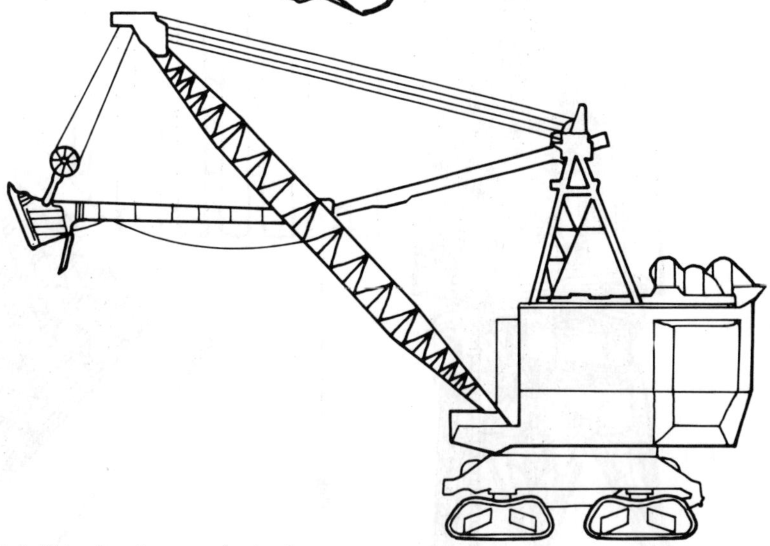

fig 11.1. Natural and man-made structures

FRAME AND SHELL STRUCTURES

All man-made structures are either frame structures or shell structures, or a mixture of the two.

FRAME STRUCTURES

The sledge in fig 11.2 is a typical frame structure, made by joining together a number of parts or **members**. Many frame structures are like this, with a very obvious structure. Frame structures make great use of triangles, simply because they are the strongest shape. Any other polygon can distort under load, but a triangle will not (fig 11.3). By adding other members to form triangles, other polygons can be made rigid. A rectangle can be made rigid by adding two diagonal members (11.4). However, adding two is not really necessary, one would be enough to **triangulate** it, making it rigid (fig 11.5). The frame structures shown in fig 11.6 each use triangulation to prevent distortion.

Other frame structures are not so obvious. They may have some kind of 'skin' over the framework. The internal door shown in fig 11.7 is a typical example of this. It has a wooden framework under its decorative plywood surface. The gaps in the framework are filled with interlocked cardboard strips which support the plywood. Note that the framework is rectangular. You might think this means it could distort, but in reality it doesn't, because once the plywood is glued into place on either side it acts in the same way as diagonal members would.

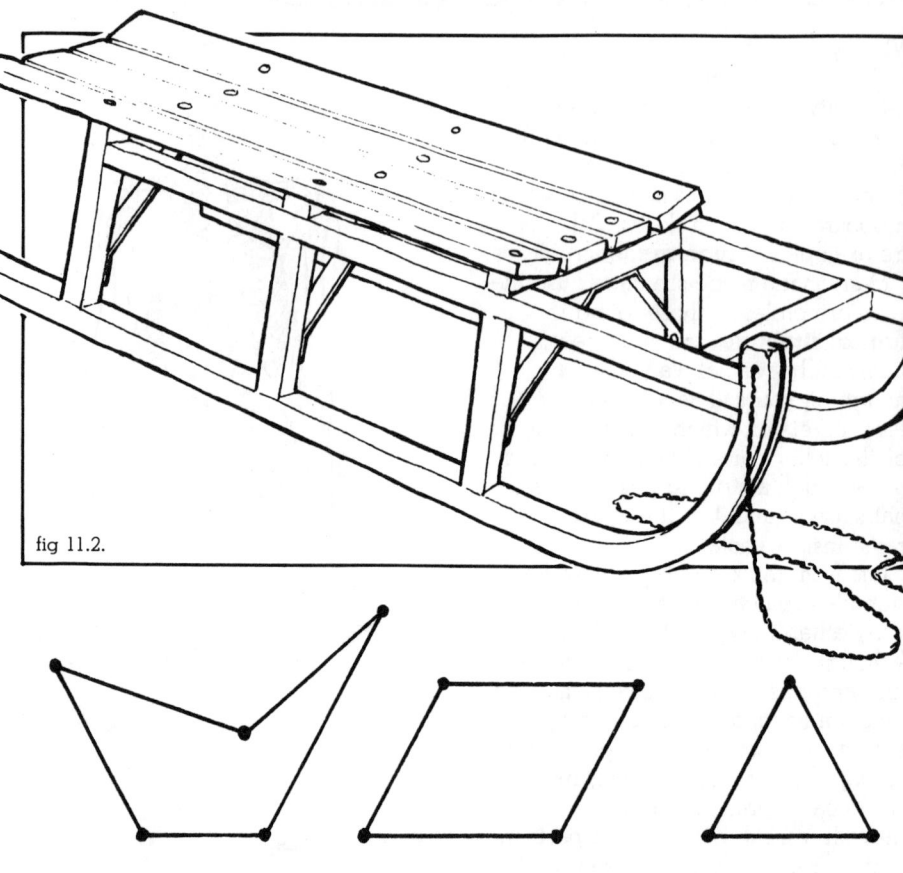

fig 11.2.

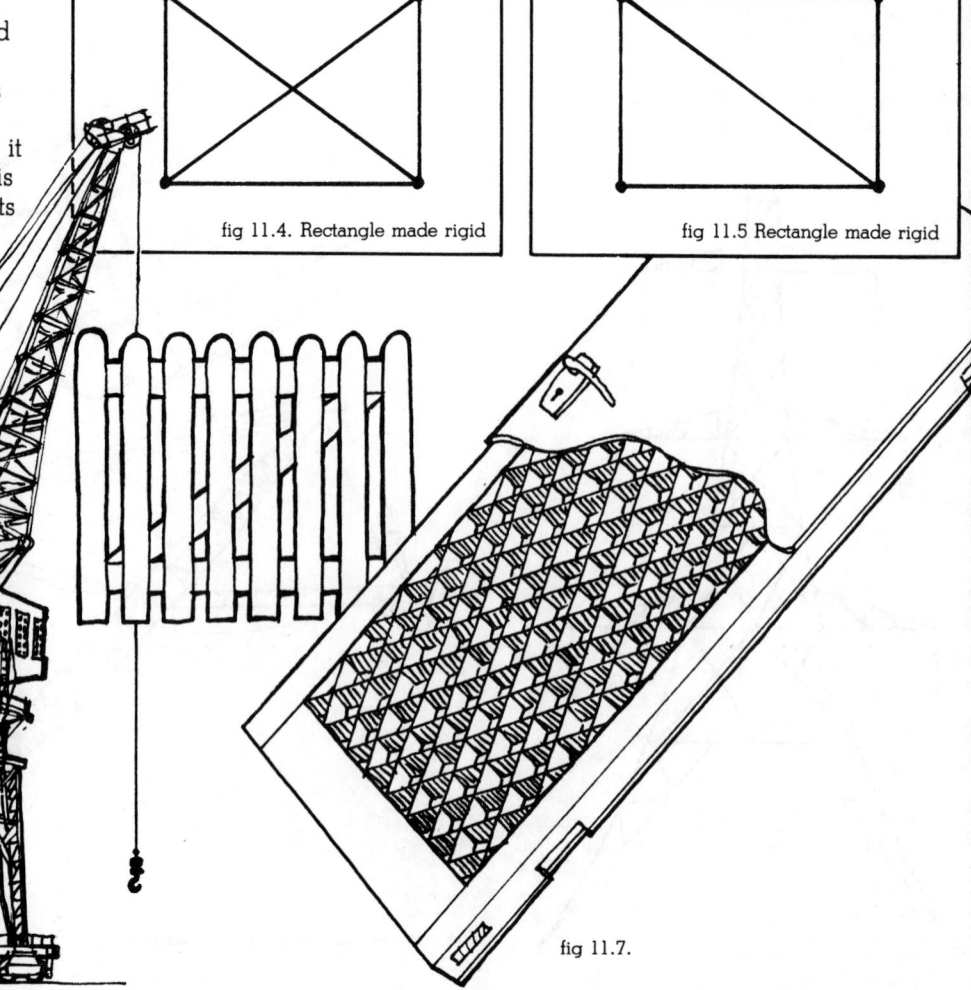

fig 11.3. Polygons

fig 11.4. Rectangle made rigid

fig 11.5 Rectangle made rigid

fig 11.6. Frame structures showing triangulation

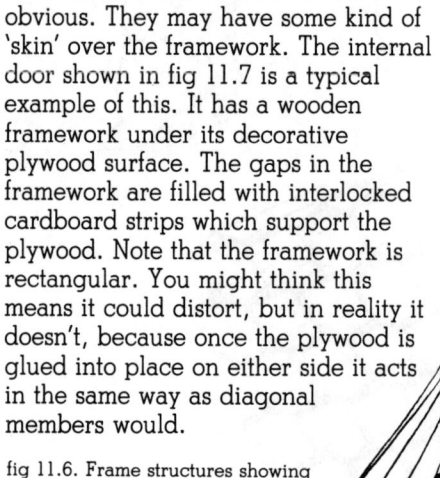

fig 11.7.

SHELL STRUCTURES

The sledge shown in fig 11.8 does not have a frame, it relies on the shape it has been moulded into for its strength. It is an example of a **shell** or **monocoque structure**. Other simple examples, shown in fig 11.9, are corrugated roofing sheets and egg boxes. Both are made from a thin, flexible, plastic sheet, yet are surprisingly strong. It is the folds and curves which give them their strength.

Monocoque structures are generally much lighter in weight than frame structures, which is why most car bodies are made this way. Sheet steel is pressed into the shapes of the various panels and welded together. The more curved or ridged a panel is, the stronger it will be. Large flat panels, such as the bonnet, are not very strong and often have to be supported by a framework (fig 11.10). It is interesting to see that the framework itself is manufactured from parts made by folding sheet metal, so each is, in fact, a monocoque structure. This kind of construction is used throughout a car body, giving a combination of light weight and strength, giving a good **strength to weight ratio**. Those areas which carry the greatest loads, such as the floor area, have the most folds and corrugations.

The frame and shell structures listed so far are man-made, but they are all based on structures found in nature. Trees, leaves and spiders webs are all examples of frame structures. An umbrella supports its load in a very similar way to a leaf. Egg shells, honeycombs and the hollow stems of many plants are all examples of shell structures. The hollow stems support their load in exactly the same way that a metal tube supports a TV aerial.

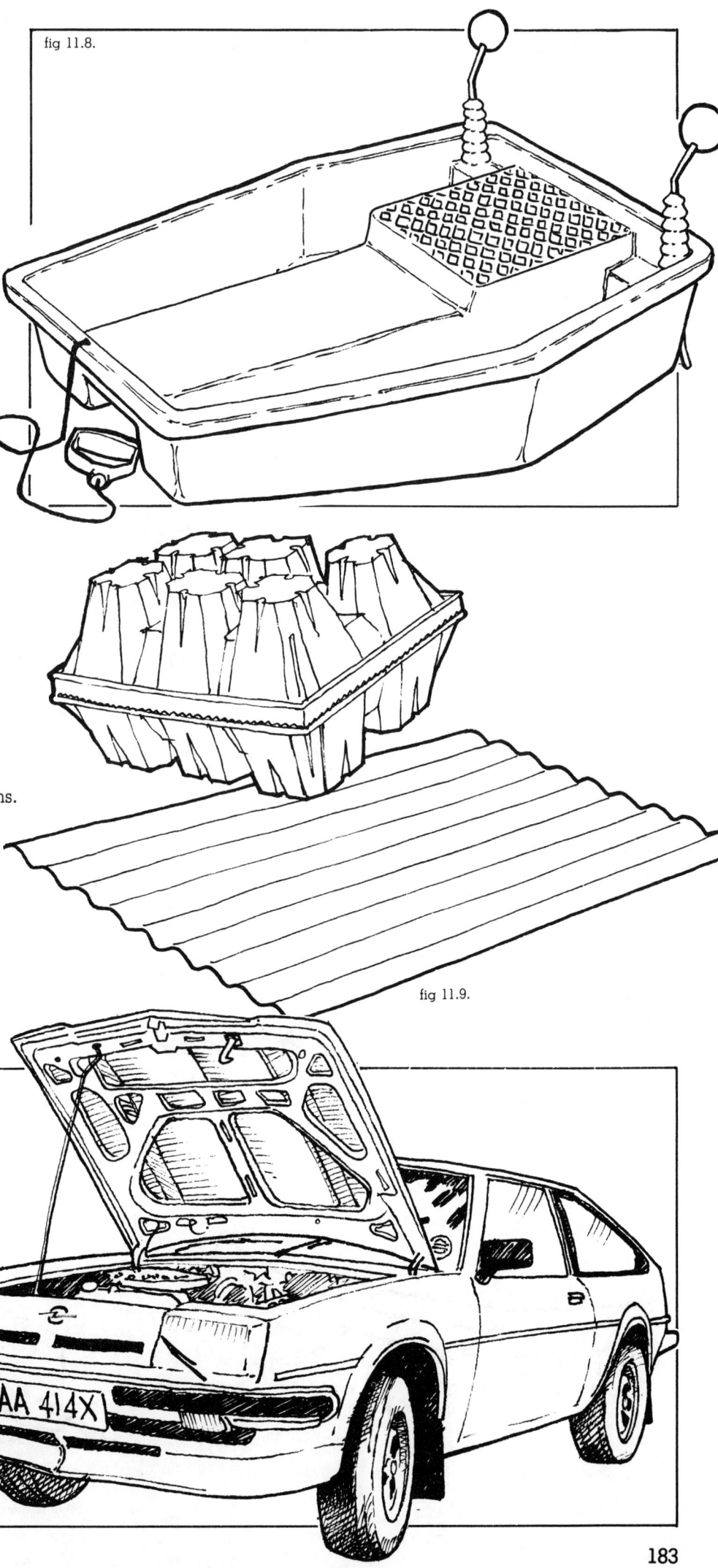

fig 11.8.

fig 11.9.

fig 11.10.

JAA 414X

FORCES

Structures are built to support a load. That load may be **static** or **dynamic**. Static loads are those which do not move, like a book on a shelf or a roof on a building. The forces acting on a structure can be calculated quite easily in this situation. Dynamic loads are those which move and change, like a diver on a springboard or a strong wind blowing on a tent (fig 11.11). Dynamic loads tend to produce much greater forces than static loads. For example, if a diver just stands on a springboard she will be producing a force equal to her own weight, but once she starts bouncing, the force acting on the board will be five or six times greater.

Many other structures have to be built to withstand dynamic loads even though they spend most of their time supporting static loads. The bunk beds in fig 11.12 have to be strong enough to take the dynamic forces created by children playing on them, even though their main purpose is to support a sleeping child, a fairly static load.

In order to be successful, all structures have to withstand the forces acting on them, whether they are static or dynamic forces. There are five basic types of force that act on a structure.

Compression forces act to **squash** a structure. Axle stands are compressed by the weight of a car (fig 11.13). A member in compression is called a **strut**.

Tension forces act to **stretch** a structure. The wires supporting a suspension bridge are in tension, as are the ropes of a swing (fig 11.14). A member in tension is called a **tie**.

Bending forces act to **bend** a structure. The beam in fig 11.15 is bending due to the weight of the engine. Closer study of the beam would show that it is in compression on its upper surface and in tension on its lower surface. Between the two is the **neutral axis** of the beam, which is neither in compression nor tension (fig 11.16). Concrete beams are strengthened by steel reinforcing near their bottom surface because concrete is weak in tension (fig 11.17).

fig 11.11.

fig 11.12.

fig 11.13. Axle stands

fig 11.14.

fig 11.15.

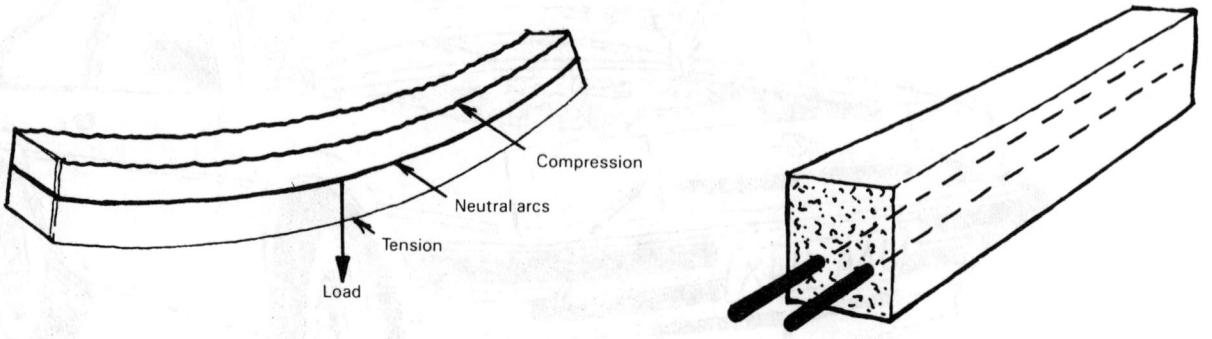

fig 11.16. Neutral axis

Compression

Neutral arcs

Tension

Load

fig 11.17. Steel reinforcing in a concrete beam

Torsion forces act to **twist** a structure, rather like you might twist a wet flannel to get the water out. They can be resisted by building a 'torsion box' (fig 11.18). Many structures use this idea, including aircraft, bridges and car bodies.

Shear forces act to **cut** a structure in two. Garden shears cut grass by forcing the two parts of the grass in opposite directions (fig 11.19). The forces acting in opposite directions on this bolt are trying to shear it (fig 11.21). By being in double shear it is much less likely to fail than a bolt in single shear (11.20).

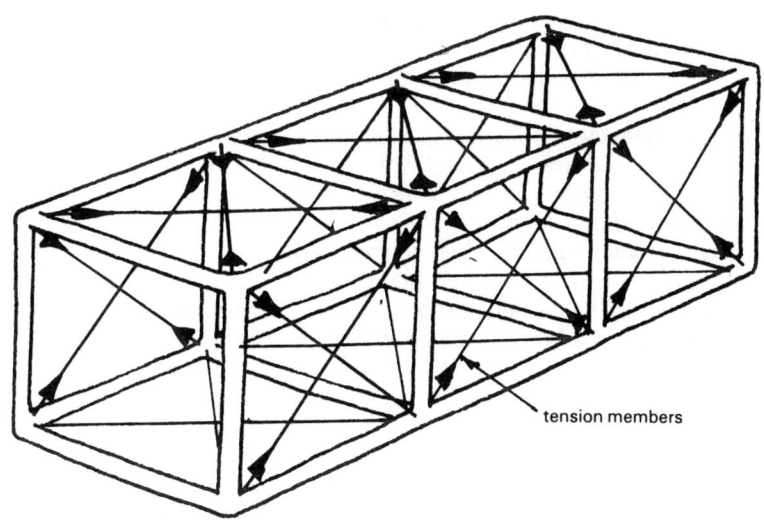
fig 11.18. Torsion box

tension members

fig 11.19. Shears cutting grass

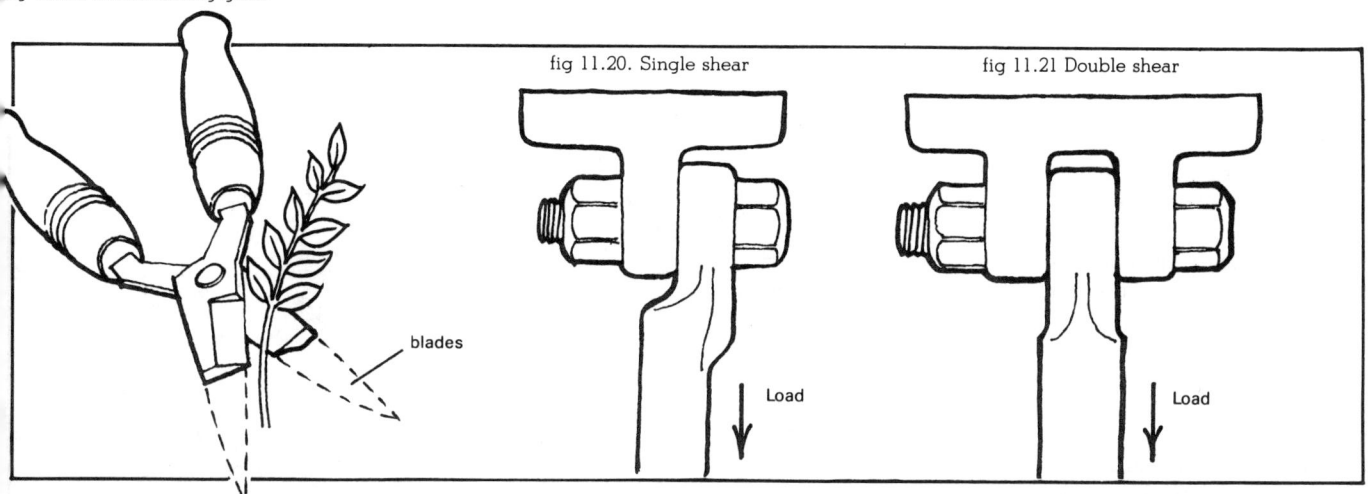
fig 11.20. Single shear

fig 11.21 Double shear

blades

Load

Load

In order to make a structure that will be strong enough, but will not be too expensive, designers have to try and calculate the forces that will act on it. Getting the balance right is very tricky. Even very skilled designers can never be absolutely sure they have planned for all the forces that might act on a structure. Who is to say that an 85kg man will not sit on a swing designed for 5-year-olds, or that a jack designed to lift loads of up to 500kg won't be used to try to lift 1000kg. Because of problems like this structures are designed with a **factor of safety**. The designer has to decide what that factor of safety will be. If he decides on a safety factor of 4 then, after calculating the load a member has to take, he would make it able to carry four times that load. So, if a swing frame is being designed to take loads of 400 newtons (N), using a safety factor of 4, it should in fact be capable of supporting loads of up to 1600N. In industry, all lifting devices have to be clearly labelled to show how much weight they can lift safely (fig 11.22). This **safe working load (SWL)** is the weight they were designed to lift. The factor of safety is over and above the SWL.

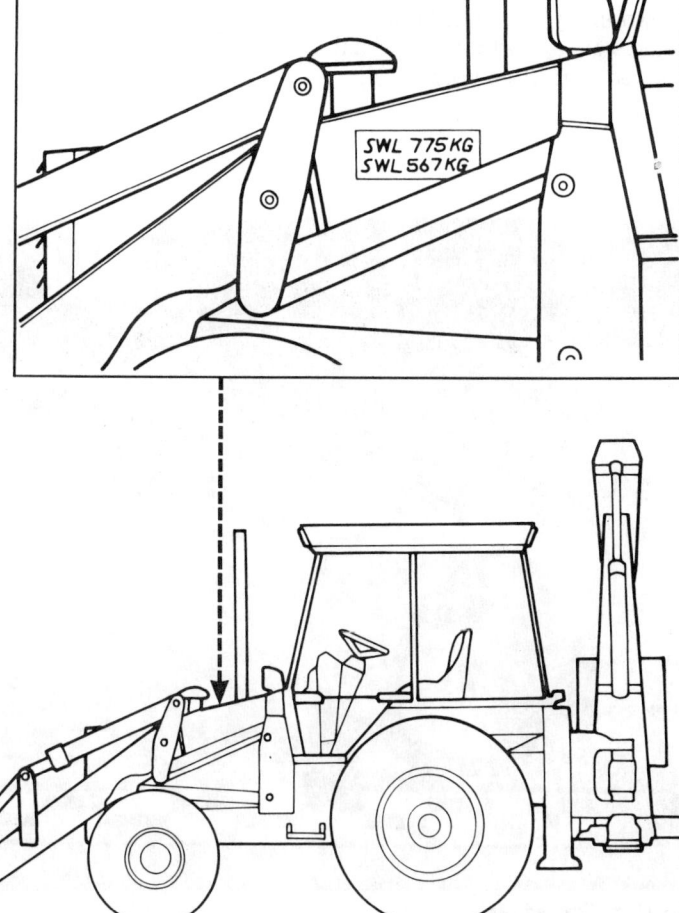
SWL 775 KG
SWL 567 KG

fig 11.22.

ACTION/REACTION

fig 11.23.

Newton's third law of motion states that **for every action there is an equal and opposite reaction**. This is as important for structures as for anything else. Structures have to resist the external forces acting on them by creating equal and opposite internal forces.

There is a tendency to think that structures, and many of the materials from which they are made, are absolutely rigid, and that no movement whatsoever takes place when they are loaded. This is just not true! In fact **all** materials and structures deflect under load. In some cases the deflections are large and obvious, like a plank bending when you walk on it (fig 11.23), or a cushion compressing when you sit on it. In other cases, when using materials like steel or concrete, the deflections are so small that you cannot see them, but they are still there. When you sit on a chair its legs get a little bit shorter. When you go upstairs your house walls really are a tiny bit shorter because of your weight. Although these movements are very tiny, they can be measured with the right equipment.

Not only is the whole structure of your house or chair deflected under load, but the material from which it is made also deflects. The molecules that go to make up the material, whether it be brick, concrete or wood, are either squashed closer together or pulled further apart (fig 11.24). It is the ability of a material to resist these molecular movements which allows it to produce the internal forces needed to counteract the external forces acting on it.

For example, fig 11.25 shows a car jack under compression. It is being compressed between two solid bodies, the car and the ground. In order to cope with the external forces, there must be equal and opposite forces acting within the jack.

This principle holds true not only for all structures, but also for all members within a structure, whatever types of external loads they have to take. If this did not happen the structure would fail in some way. Part of it would be permanently deformed, like the car shown in fig 11.26. Its body structure could not cope with the external forces applied to it when it hit a wall.

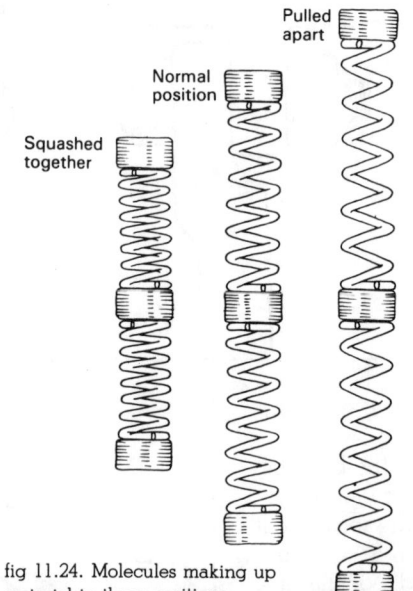

Pulled apart

Normal position

Squashed together

fig 11.24. Molecules making up material in three positions

fig 11.25. Hydraulic car jack taking a load

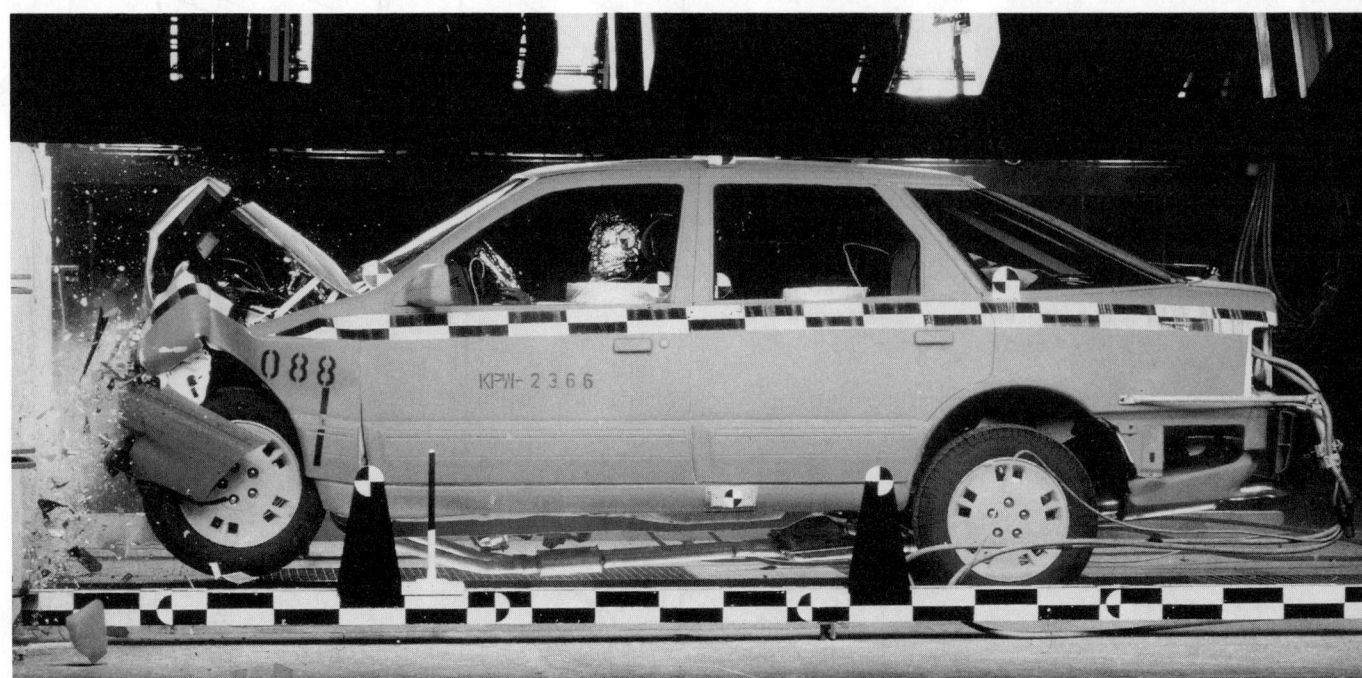

fig 11.26. Crashed car

EQUILIBRIUM

When the forces acting on a structure are balanced, it is in **equilibrium**. Fig 11.27 shows a radio mast. Its supporting wires are positioned so that they apply exactly equal forces on opposite sides of the mast. With other structures the balancing forces are not so obvious, but they are always there.

A force, and the direction in which it acts, can be represented by a line, called a **vector** (fig 11.28). The length of the line represents the size of the force and the arrowhead shows the direction in which it acts.

Vector diagrams can be used to find the size and direction of unknown forces acting in a structure. Fig 11.29 shows two forces acting on a bolt fastened to a wall. Their combined effect (resultant) and the balancing force (equilibrant) in the wall can be found using vectors. Fig 11.30 shows the stages involved in finding them.

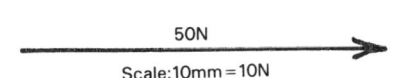

50N

Scale: 10mm = 10N

fig 11.28. Vector

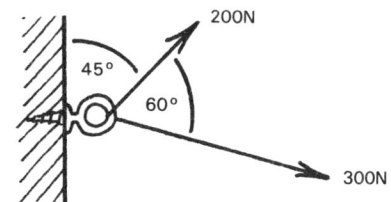

200N

45°

60°

300N

fig 11.29. Vector diagram

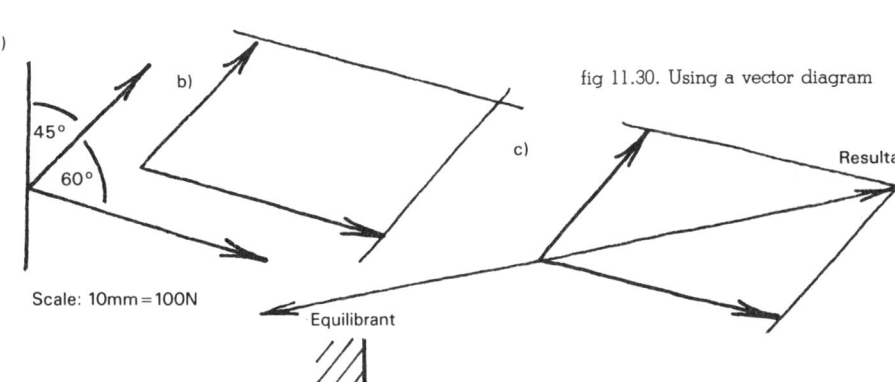

fig 11.27. TV mast with supporting wires

a)

45°

60°

b)

c)

Resultant

fig 11.30. Using a vector diagram

Scale: 10mm = 100N

Equilibrant

FORCE DIAGRAM

By drawing

Fig 11.31 shows a **space diagram** of a framework used to support a hanging basket. If the hanging basket, when full, weighs 60N, what forces will be created in the members supporting it? By drawing a **force diagram**, to a suitable suitable scale, you can find those forces. Fig 11.32 shows the stages involved in drawing the diagram, starting with the known force. The other two lines are drawn from the ends of the first line, parallel to the members on the space diagram. By measuring their lengths you can find the forces in them.

By calculation

The problem shown in fig 11.31 can also be solved using basic geometry based on right angled triangles. Fig 11.33 shows a sketch of the structure with the calculations alongside.

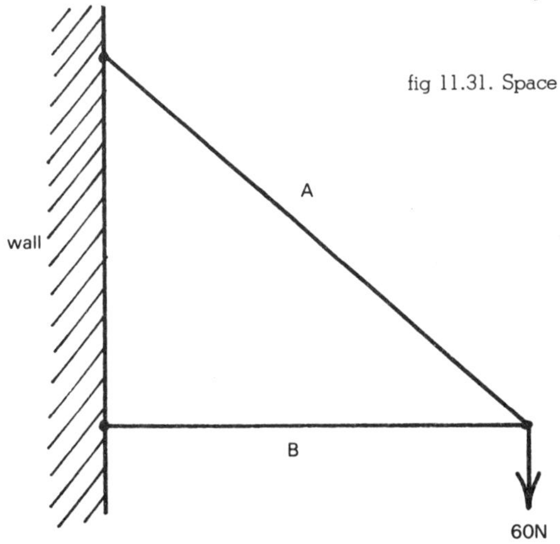

fig 11.31. Space diagram

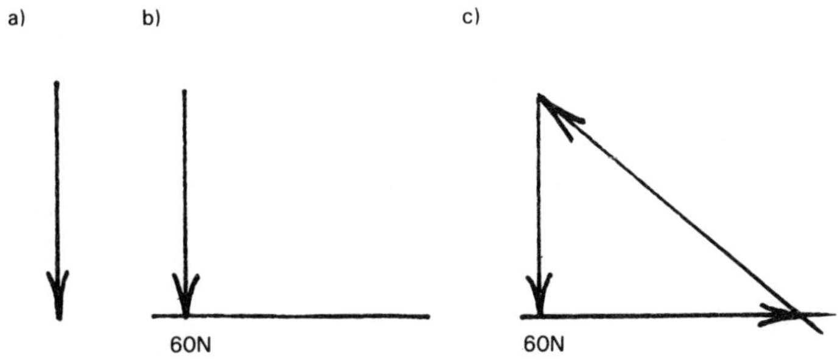

wall

A

B

60N

a)

b)

c)

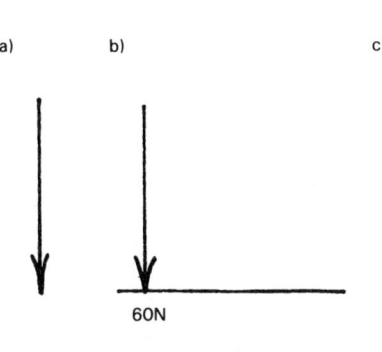

60N

60N

fig 11.32. Drawing a triangle of forces

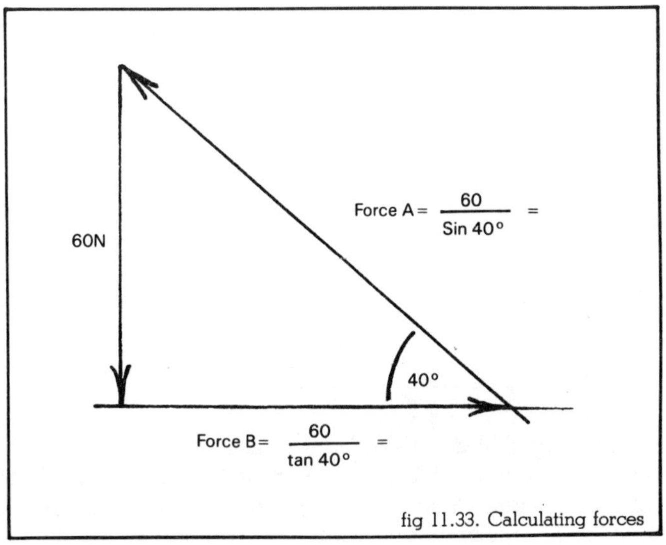

60N

$$\text{Force A} = \frac{60}{\sin 40°} =$$

40°

$$\text{Force B} = \frac{60}{\tan 40°} =$$

fig 11.33. Calculating forces

MATERIALS

STRESS AND STRAIN

Stress is a measure of **how hard** the molecules of a material are being pushed together or pulled apart by external forces. It tells you how much force is being applied to each unit of area. The greater the external forces, the greater the stress. Eventually, as the forces continue to increase, the material will break. The stress at which it breaks is called the **breaking stress** of that material. Fig 11.34 shows a steel rope with a cross sectional area of 20mm² being used to lift an engine weighing 2000N. The effects of that 2000N force are spread equally across the cross sectional area of the rope, so the stress in the rope can be found using the formula:

$$\text{Stress} = \frac{\text{Force}}{\text{Area}} = \frac{2000\text{N}}{20\text{mm}^2} = 100 \text{ N/mm}^2$$

This is a very low value for stress. Stress values for some materials can be very high. The units for stress are MN/m² (1 Meganewton = 1 million Newtons).

The concrete pillars in fig 11.35 are being compressed by the weight of the building. Each pillar is taking a load of 100,000N and has a cross sectional area of 0.1 metre². Therefore, the stress in each pillar is:

$$\frac{\text{Force}}{\text{Area}} = \frac{100,000\text{N}}{0.1} = 1\text{MN/m}^2$$

Fig 11.37 gives typical stress values for some common materials.

Strain is a measure of **how far** the molecules of a material are being pushed together or pulled apart. It compares the original distances with the new distances caused by the external forces. Fig 11.36 shows a rope supporting the weight of three children. Their weight stretches the rope. By comparing the change in length with the original length you can find the strain in the rope:

$$\text{Strain} = \frac{\text{Change in length}}{\text{Original length}} = \frac{30\text{mm}}{4\text{m}} = \frac{0.03\text{m}}{4\text{m}} = 0.0075 \text{ or } 0.75\%$$

This is a high value for strain. Many of the less flexible materials will have much smaller values. They are most often given as a percentage.

Material	Stress value in MN/m²
Wood	100
Aluminium	70
Cast Iron	150
Mild Steel	400
High Tensile Steel	1,000

fig 11.37. Typical stress values

STRENGTH

The **strength of a material** is a measure of how much stress is needed just to break the material, and the **strength of a structure** is a measure of how much external force is needed to break that structure.

fig 11.34. Steel rope

Cross Sectional area = 20mm²

fig 11.35. Stone pillars supporting heavy stone building

fig 11.36. Weight strain stretching rope

ELASTICITY

Elasticity is a measure of how **stiff** a material is, how much strain it will take before it is permanently distorted.

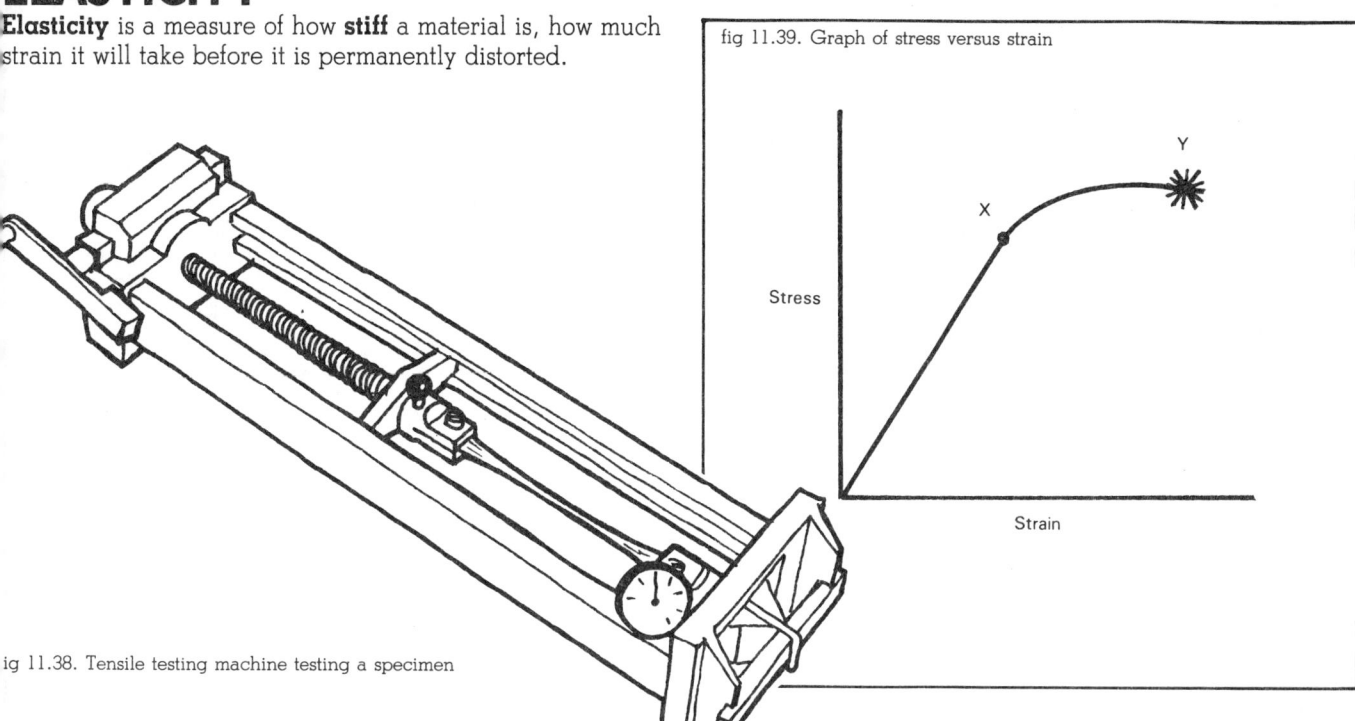

fig 11.39. Graph of stress versus strain

fig 11.38. Tensile testing machine testing a specimen

A specimen of any material can be tested in a tensile testing machine (fig 11.38). This pulls the two ends apart until the material breaks. Plotting stress against strain for the material will give you a graph something like fig 11.39. The end of the graph, point Y, is the point at which the material breaks. In structures, you are more concerned with the straight part of the graph. If the material is not stretched beyond point X then, when the external forces are removed, the material will return to its original shape and size, like a rubber band. If it is stretched beyond this point it will be permanently deformed.

By comparing the straight line parts of the graph for different materials you can compare their stiffness (fig 11.40). The steeper the slope, the stiffer a material is. These slopes can be represented by a single figure, known as the **modulus of elasticity** or **Young's modulus**. The following example is the calculation for mild steel.

$$\text{Modulus of elasticity (E)} = \frac{\text{Stress}}{\text{Strain}} = \frac{400\text{MN/m}^2}{0.002} = 200{,}000 \text{ MN/m}^2$$

Designers use this figure to help choose the right material for a particular structure. Fig 11.41 gives the E values for some common structural materials.

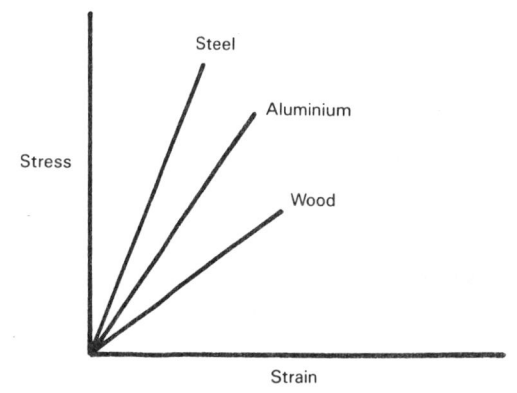

fig 11.40. Comparing different materials for stress/strain

Material	E value in MN/m^2
Steel	210,000
Aluminium	73,000
Concrete	17,000
Softwood	13,000

fig 11.41.

BEAMS

Beams of one sort or another are the most common kind of structural member. If designed and used correctly, they can cope with all types of forces. Originally all beams were made of solid material which made them very heavy and expensive. Over the years many different beam shapes have been developed which are much lighter, yet just as strong, giving them a much better **strength to weight ratio**. Fig 11.42 shows some of them. For maximum strength it is important that a beam is used the right way on, that is with the widest section taking the load (fig 11.43). You see many of these beams in use around you every day. Look at bridges, street lights, lorry chassis, bike frames and fence posts. Whether working horizontally like a bridge or vertically like a post, **the job of all beams is to resist bending!**

BEAM CALCULATIONS – MOMENTS

A **moment** is a turning force, like the force you use to turn a door handle or a spanner. The spanner in fig 11.44 is 200mm long and is being turned with a force of 50N. The **turning moment** acting on the nut is given by:

$$\text{Moment} = \text{Force(N)} \times \text{Distance(m)}$$
$$= 50 \times 0.2$$
$$= 10\text{Nm}$$

The nut will continue to turn until, as it tightens up, the friction force acting against it also equals 10Nm. At that point the forces acting on the nut will be balanced, in **equilibrium**.

The see-saw in fig 11.45 is also in equilibrium, the **clockwise** and **anti-clockwise** turning moments acting on it are equal. Taking moments about the fulcrum:

$$\text{Anti-clockwise moments} = \text{Clockwise moments}$$
$$270\text{N} \times 2\text{m} = 180\text{N} \times 3\text{m}$$
$$540\text{Nm} = 540\text{Nm}$$

The forces acting on a beam can be thought of in the same way. Fig 11.46 shows a beam in equilibrium. The downward forces acting on it are balanced by the upward forces, called **reactions**, at it ends. Because the load is in the centre of the beam, the reactions at X and Y are each half the value of the downward force. This can be proved by calculation, taking moments about one end of the beam.

You can calculate the reaction, R_2 by imagining that the support at end Y has been removed, so that the beam is free to rotate clockwise about end X. The force needed to stop this happening is found by:

Taking moments about end X,

$$\text{Reaction Y} \times 6 = 500 \times 3$$
$$\text{Reaction Y} = \frac{1500}{6}$$

$$\text{Reaction Y} = 250 \text{ N}$$

If Reaction X + Reaction Y = 500N then

$$\text{Reaction X} = 500 - \text{Reaction Y}$$
$$= 500 - 250$$
$$\text{Reaction X} = 250 \text{ N}$$

If the load is not in the centre of a beam then the reactions will not be equal. Fig 11.47 shows such a

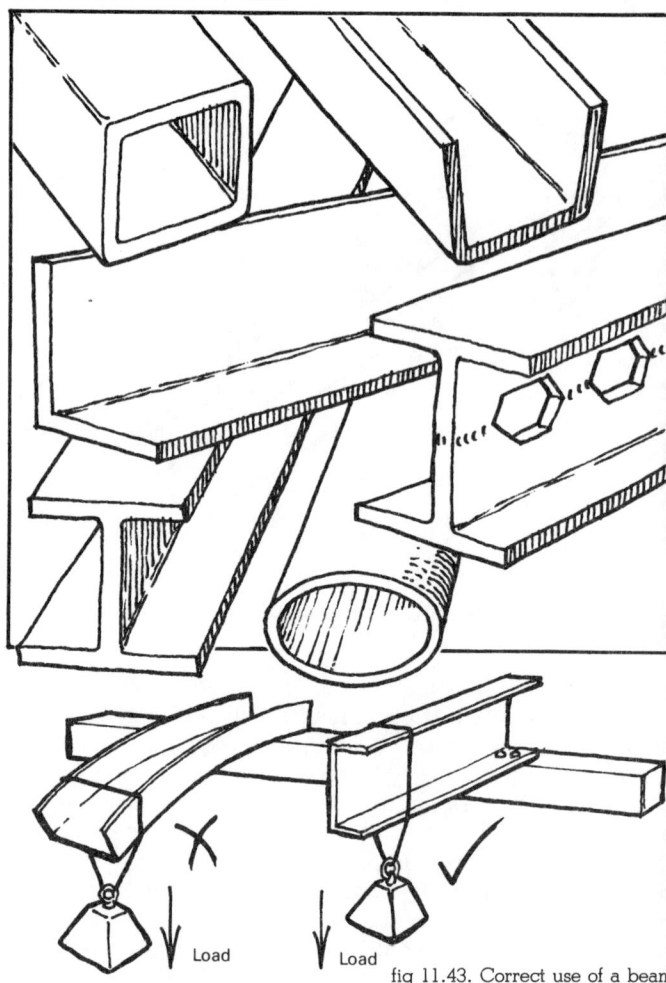

fig 11.42. Different beam shapes

fig 11.43. Correct use of a beam

Load   Load

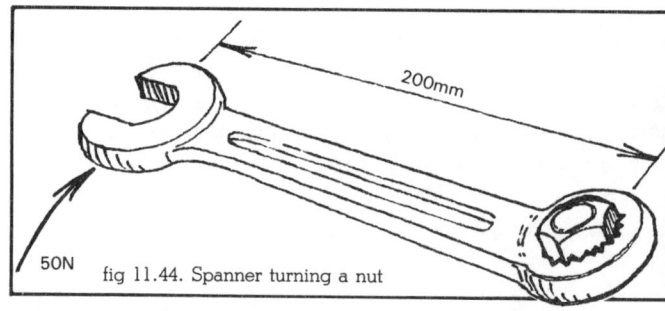

200mm

50N   fig 11.44. Spanner turning a nut

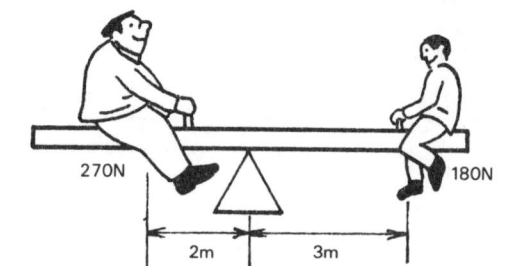

270N   180N

2m   3m

fig 11.45. See-saw in equilibrium

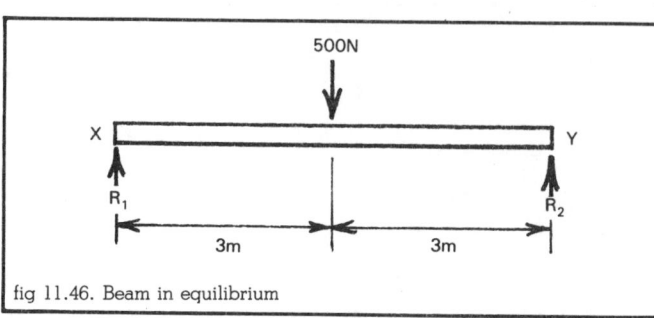

500N

X   Y

R_1 R_2

3m   3m

fig 11.46. Beam in equilibrium

situation. The reactions at X and Y can be calculated in just the same way as before.

Take moments about end X,

$$\text{Reaction } Y \times 6 = 500 \times 2$$
$$Y = \frac{1000}{6}$$

$$\text{Reaction } Y = 166.6\text{kN}$$

If Reaction X + Reaction Y = 500kN then

$$\text{Reaction } X = 500 - \text{Reaction } Y$$
$$= 500 - 166.6$$
$$\text{Reaction } X = 333.3\text{kN}$$

You can see that most of the load is taken by the support nearest to the load. The nearer the load gets to a support the more of the load that support is taking. So when a moving object, such as a car, goes over a bridge the load taken by each of the supports changes gradually as it crosses.

Sometimes a beam is loaded at two or more points, as in fig 11.48. The same method can be used to work out the reactions.

Taking moments about X,

$$\text{Reaction } Y \times 10 = (250 \times 3) + (450 \times 7) + (75 \times 8)$$
$$= 750 + 3150 + 600$$
$$= 45000$$

$$Y = \frac{4500}{10} = 450\text{kN}$$

Reaction X + Reaction Y = 775kN
therefore
Reaction X = 775 − 450 = 325kN

The loads looked at so far have been **point loads** acting on a beam, the weight of the beam itself has been ignored. Obviously the beam does have weight, which is spread evenly over the full length of the beam. A load like this is called a **uniformly distributed load (UDL)**. Another example of a UDL is shown in fig 11.49, a corrugated roof supported by a beam. Calculating the reactions on beams carrying UDLs is made simpler by taking the total load and treating it as a point load acting at the centre of the distributed load. For example, the load created by the beam itself is 20N/m and the load created by the roof is 5N/m, so the total load spread over a 12 metre span is (20N × 12) + (5N × 12) = 300N. The reactions can now be calculated as if it was a 300N point load acting at the centre of the beam.

CANTILEVERS

Some beams are supported at one end only. Fig 11.50 shows such a beam used to help unload lorries. When used like this it is called a **cantilever** beam. When loaded, it is in tension on the top surface and in compression on the lower surface. Cantilever beams are used either where it is possible to support one end only or where it is necessary to keep a large gap clear. Supermarket trolleys and diving boards are also examples of cantilevers. If supported at both ends the supermarket trolleys would not stack together. Supporting diving boards at both ends would be dangerous and, in the case of springboards, stop them working correctly.

Cantilever calculations are very simple. As the beam is only supported at one end, the reaction at that end must be equal to the load. For example in fig 11.51, Reaction A = 600kN.

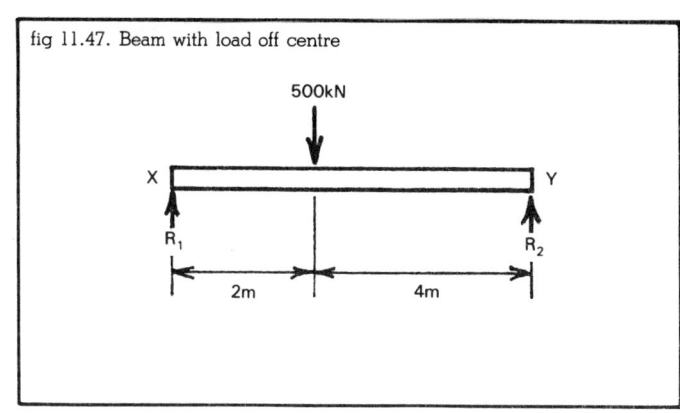

fig 11.47. Beam with load off centre

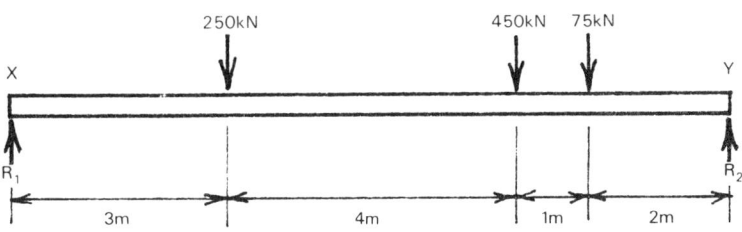

fig 11.48. Beam loaded at more than one point

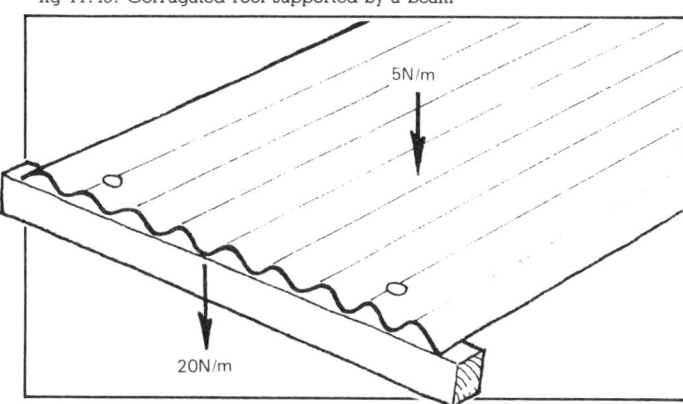

fig 11.49. Corrugated roof supported by a beam

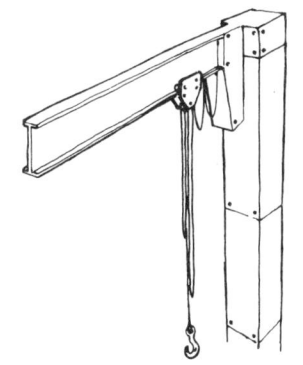

fig 11.50. Cantilever with load on the end of the beam

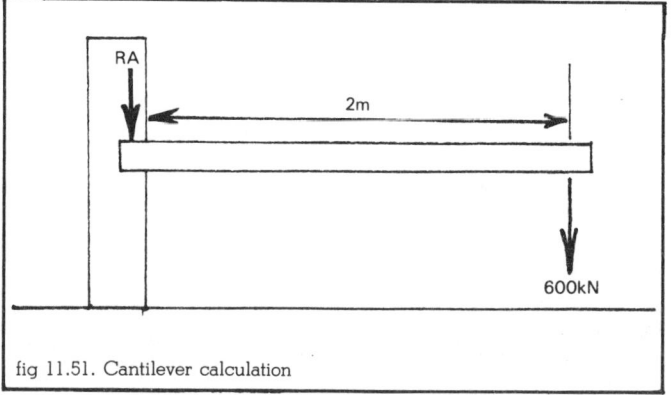

fig 11.51. Cantilever calculation

FORCES IN FRAMEWORKS

Calculating the actual forces acting in any framework is a very difficult and time consuming process, requiring powerful computers. Fortunately, there are other much simpler methods which give quite acceptable answers. You can find the forces acting in the members that go to make up a framework either by drawing or calculation. Both methods require you to make a number of assumptions which are obviously not true in real life:

1. That the members are light compared with the external loads they carry. In effect you ignore their own weight!

2. That the members are connected by pin-points, rather like fixing two bits of meccano together with a single bolt.

3. That the loads are only applied at joints in the framework, not between joints.

These assumptions mean that the forces in any member will always be acting directly along it. This means that a member will always be in either **compression** or **tension**, you can ignore other forces.

Bow's notation relies mainly on accurate drawing and measuring. Fig 11.52 shows a simple roof framework and the external loads it has to carry. The reactions at the walls are indicated, but their value is not given.

1. Calculate the reactions by taking moments about one end of the framework as if it were a simple beam. Downward forces equal 18kN, therefore R1 + R2 must equal 18kN.

Taking moments about the left hand end:
$$R2 \times 20 = (12 \times 5) + (6 \times 15)$$
$$R2 \times 20 = 150$$
$$R2 = \frac{150}{20} = 7.5kN$$

2. Draw a **space diagram** of the framework. It must be accurate and to scale (e.g. 10mm = 2m). Draw in all the external forces and number each of the joints (fig 11.53).

3. Letter each of the **spaces** between the external forces and also the spaces between the internal members. Notice that the letters go in a clockwise direction, first round the outside then the inside (fig 11.53). Each member is then referred to by the letters on either side of it (e.g. AE or FG).

4. Starting with space A construct a **vector diagram** to a suitable scale (e.g. 10mm = 2kN). This diagram will tell you what forces are acting in each member.

Between space A and space B, at joint 2, there is a force of 12kN acting **downwards**. So start the diagram by drawing a line 60mm long going straight down from 'a' (see fig 10.54a). Label the bottom of the line as point 'b'.

Between space B and space C, at joint 3, there is a force of 6kN also acting **downwards**. So from 'b', draw a line downwards, 30mm long, and label the bottom of it as point 'c' (fig 10.54b).

Between space C and space D, at joint 4, there is an **upward** acting force of 7.5kN. So from 'c', measure 37.5mm upwards to find point 'd' (fig 10.54c).

You now have a vector diagram representing the known external forces. To calculate the forces acting in each member of the framework, the rest of the diagram is completed by drawing lines from these points, **parallel to the lines on the space diagram.**

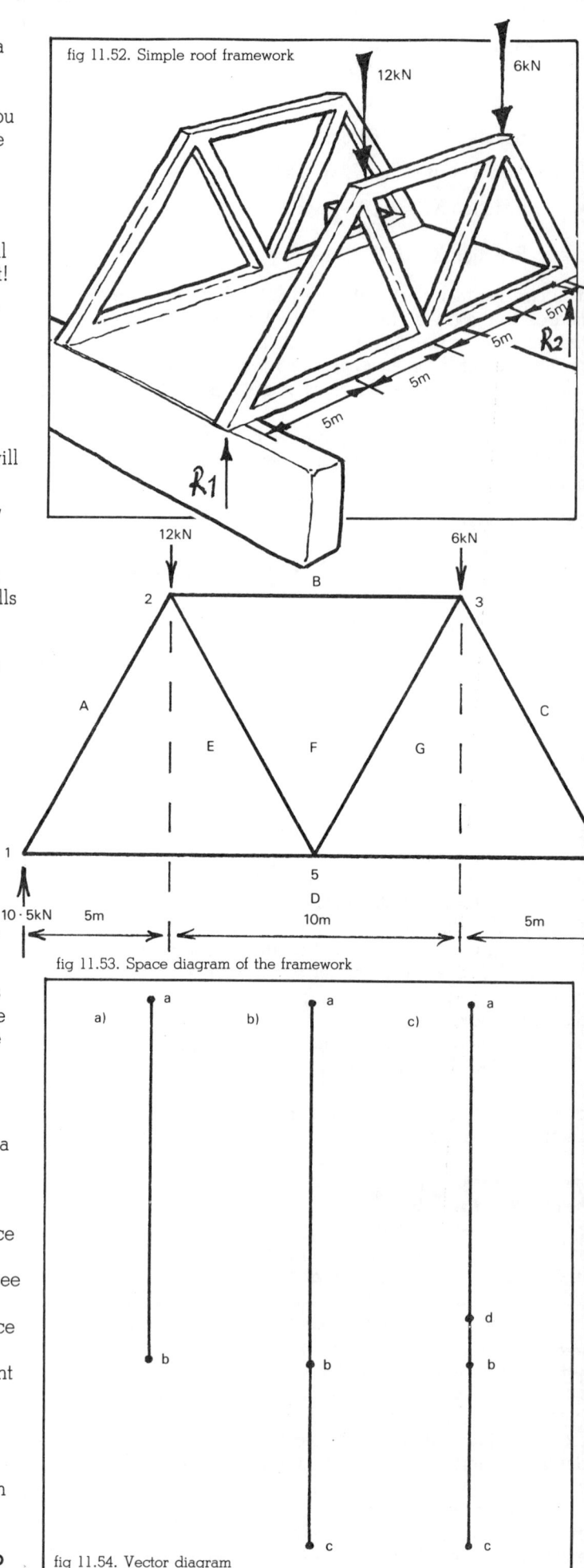

fig 11.52. Simple roof framework

fig 11.53. Space diagram of the framework

fig 11.54. Vector diagram

Members DE and DG are **horizontal**, so draw a horizontal line out from point 'd'. At this stage you don't know how long it needs to be.

Member BF is also horizontal, so draw a horizontal line out from point 'b'. Again you don't know its length yet.

Member AE is at 60° to the horizontal, so from point 'a' draw a line down at 60° to meet the horizontal line coming from point 'd'. Where the two lines meet is point 'e'. Members EF, FG and CG can be treated in the same way to complete the vector diagram (fig 11.55).

5. Each line on the vector diagram now represents, to scale, the forces acting in a particular member of the structure. For example, line 'ae' represents the force acting in member AE. To find its value, **measure its length**. It is 61mm long, representing a force of 12.2kN.

6. A little experience will often tell you whether a member is in compression or tension. You can check by examining each of the joints in turn, starting with joint 1. You must work round the joint in a **clockwise** direction.

For example, joint 1 (fig 11.56) has an external force of 10.5kN acting upwards. From the vector diagram you can see that the force in AE is acting downwards, from 'a' to 'e', towards the joint. So put an arrowhead on AE pointing towards the joint. Moving clockwise, the force in ED is acting, in the vector diagram, from 'e' to 'd', to the right, therefore away from the joint. So put an arrowhead pointing away from the joint. Now add the corresponding arrows on the other ends of members AE and DE (fig 11.56a).

fig 11.55. Complete vector diagram

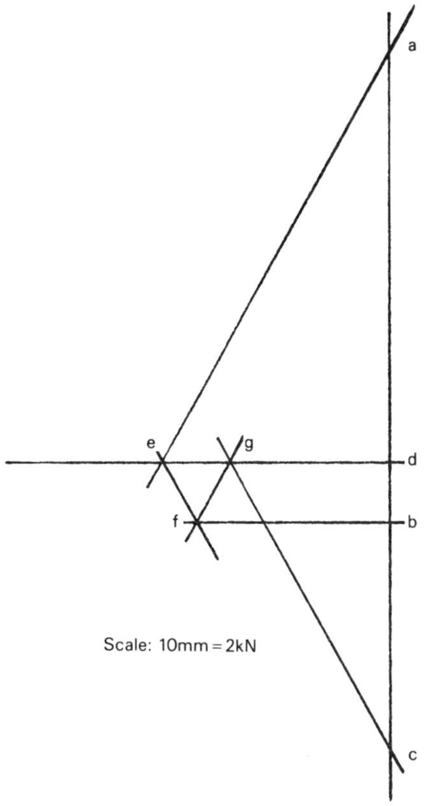

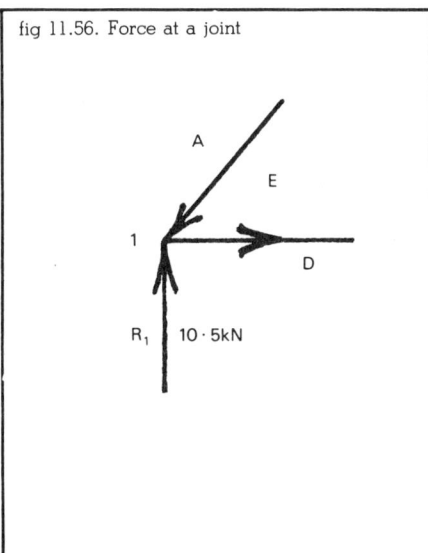

Scale: 10mm = 2kN

Having done this first joint you can soon complete the rest of the picture, each joint giving you more information. Fig 11.57 gives the complete picture. You can now give a complete list of the **type** and **value** of the forces acting in each member of the structure (fig 11.58).

Member	Force	Type
AE	12.2kN	Compression
BF	5.2kN	Compression
CG	8.7kN	Compression
DG	4.2kN	Tension
DE	6.0kN	Tension
EF	1.8kN	Compression
FG	1.8kN	Tension

fig 11.58. Type and value of forces acting

fig 11.56. Force at a joint

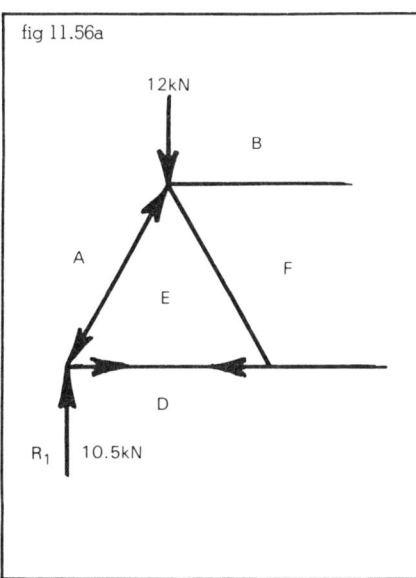

fig 11.56a

Sometimes, when working out the forces in frame structures, you will find that one or two of the members are not carrying any load, either compressive or tensile. These apparently useless members are known as **redundant members**. They are not always as useless as they appear. The assumptions you make in working out these forces do not hold true in real life. They do not allow for the weight of the members, fixed joints, loads between joints, materials corroding or the extra forces caused by strong winds. All this means that in reality these members may be taking loads, so it pays to think carefully before removing them.

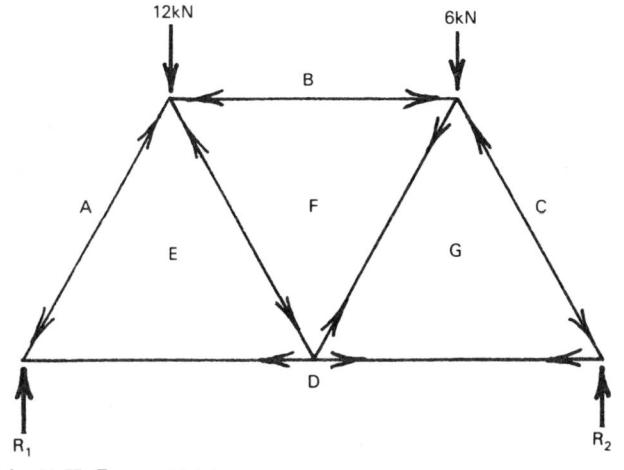

fig 11.57. Forces at joints

JOINTS

Ideally, joints should allow the forces acting on a structure to be spread evenly throughout it. A structure like this would have no strong or weak points and, if overloaded, could break at any point. For example, the welded joint shown in fig 11.59 should have exactly the same strength as the two pieces of mild steel it joins together, and the glued scarf joint in fig 11.60 should make the member just as strong as if it were made from one piece of wood. In reality, it is very difficult to achieve this. Most joints are far from perfect, and the stresses within them are not spread evenly.

GLUED JOINTS

As new, stronger glues have been developed, they have been used more and more, often being used instead of mechanical methods such as bolts or rivets. To understand how glue works, imagine wetting two pieces of wood, pressing them together and freezing them. This would fix them together, though it would not be a very strong joint because ice is brittle and would melt at temperatures above 0°C. However, if a better glue were used, a strong joint would result. Most glues work in this way, by penetrating the surfaces being joined, then setting (fig 11.61). When joining wood to wood, nails are often used as well as glue. The main purpose of the nails is to hold the two pieces together firmly while the glue sets. They do not add a great deal of strength to the joint. Many modern wood glues are so strong that close examination of failed joints will show that it is the wood on either side of the joint which has failed, not the glue.

The strength of a glued joint depends on several things:

1. The type of glue being used.
2. The type of joint being glued.
3. How well the parts fit together.
4. How well the parts are held while the glue sets.
5. The width of the joint.

It is often thought that it is the surface area being glued that is important, but tests show that most of the load is taken by the ends of the joint (fig 11.62), so it is the width of the joint which is important. The wider the joint, the greater the end area which takes most of the load.

Many specialist glues have been developed in recent years, not only for simple things like mending a broken vase, but also for things like sticking aircraft wings together. Some of the more common ones you are likely to come across are:

PVA: thick white liquid glue, used mainly for wood (e.g. Resin W).
Epoxy resin: separate resin and hardener which are mixed together in equal amounts (e.g. Araldite).
Cyanoacrylate: clear watery liquid, sets hard in seconds (e.g. Superglue).

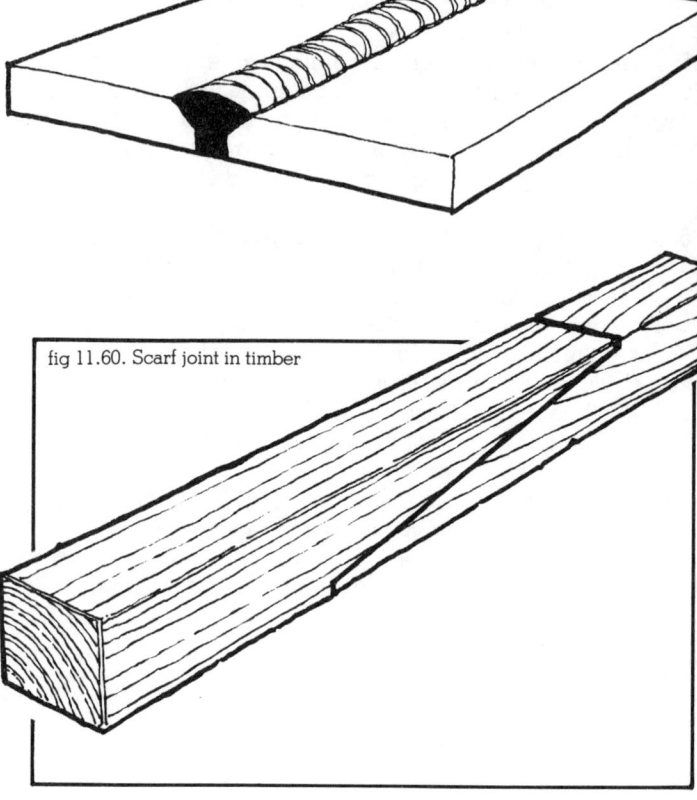

fig 11.59. Welded joint

fig 11.60. Scarf joint in timber

fig 11.61. How glue works

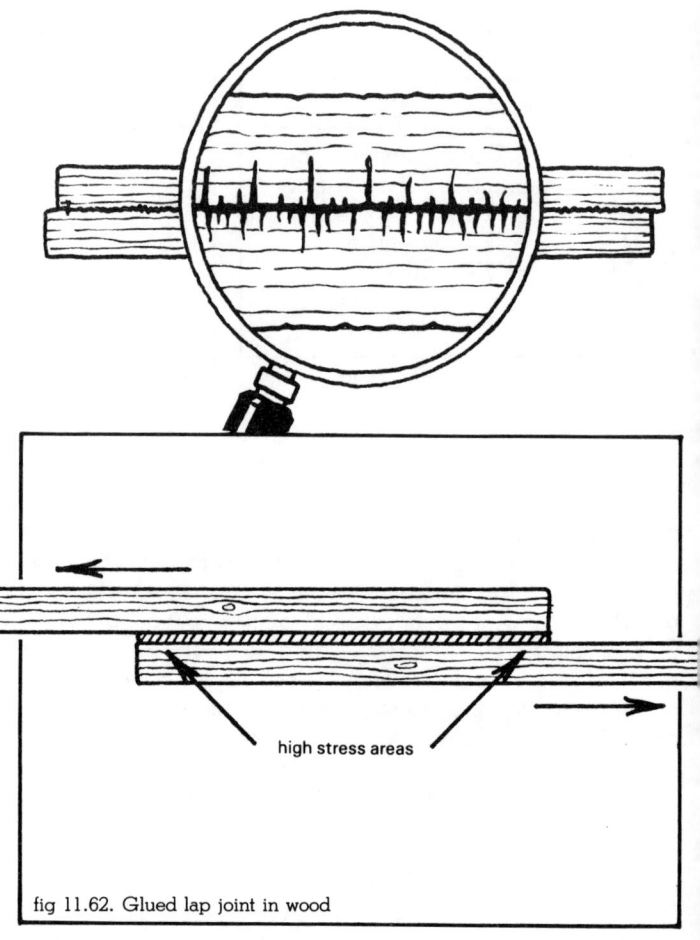

high stress areas

fig 11.62. Glued lap joint in wood

MECHANICAL JOINTS

Any joint which relies on nails, screws, bolts or rivets is called a mechanical joint.

Nails rely mainly on friction. They force their way into a piece of wood, bending the fibres on either side (11.63). The fibres try to straighten out again, so pressing onto the sides of the nail. Some wood fibres are very strong and refuse to bend, so the wood splits instead. Hardwoods are particularly likely to do this.

Screws of all types rely on the surfaces in contact being able to cope with the forces applied to them. Fig 11.64 shows a screw fitting taking the load of a hanging basket. The force created by the weight of the basket is taken by the thread in contact with the wood. Tests show that it is the first two or three threads that take the majority of the load, so using very long screws does little or nothing to improve the strength of a joint. This applies to all screws, whether going into wood or metal.

Rivets and **bolts** are simply metal pins used to hold metal plates together. They can be used to fix parts together firmly or to allow them to pivot (fig 11.65). Bolts have the disadvantage that unless held in some way (e.g. lock washers), they are likely to work loose. Experience has shown that for maximum strength, the positioning of the rivets or bolts and the distances between them is important. If they are too close together, too near to an edge or too far apart, the joint may fail. Fig 11.66 shows the common ways in which these joints fail.

Welded joints are, in theory, the ideal way of joining two pieces of metal together. A joint like the one shown in fig 11.59 should be just as strong as the material on the other side. In practice this is very difficult and expensive to achieve. It needs either very highly skilled welders and inspectors with X-ray machines, or automatic welding machines. Welding of this standard is often required on pipelines and oilrigs. Other welded joints (e.g. fig 11.67) may be weakened due to one or more of the following faults:

1. Getting impurities into the weld, usually 'slag' from the welding process.
2. Cracks and cavities inside the weld.
3. Distortion of the material on either side of the weld due to overheating. This not only causes changes in shape, but also leads to stress concentrations which may later cause corrosion and cracks to form.

fig 11.63. Cross section of nail in wood

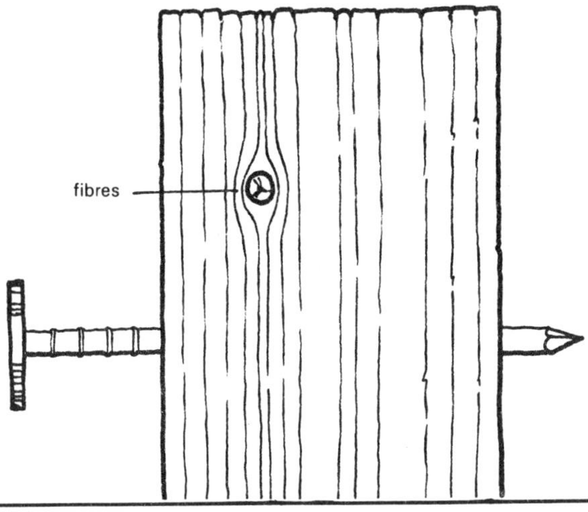

fibres

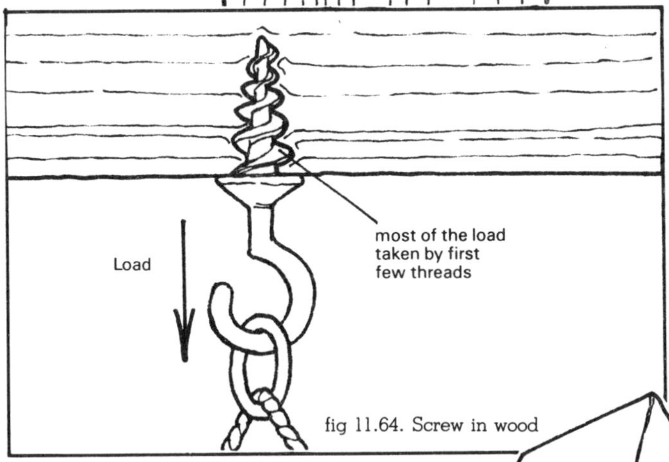

Load

most of the load taken by first few threads

fig 11.64. Screw in wood

fig 11.65. Rivets and bolts

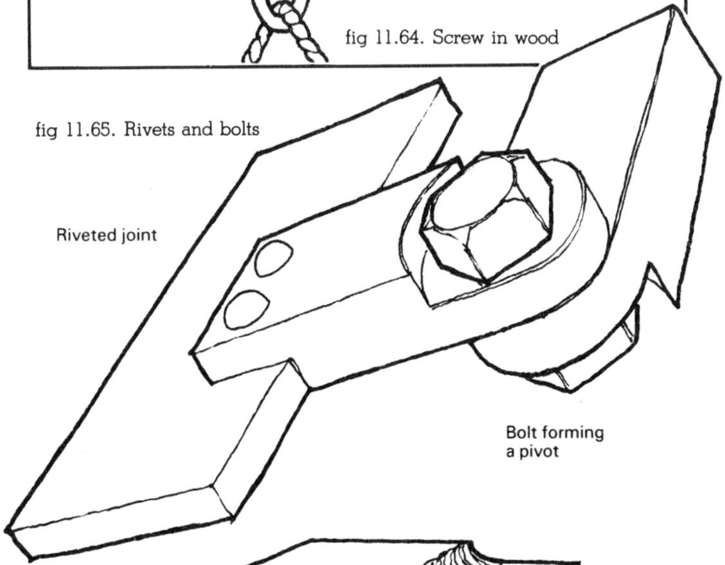

Riveted joint

Bolt forming a pivot

fig 11.66. Common causes of joint failure

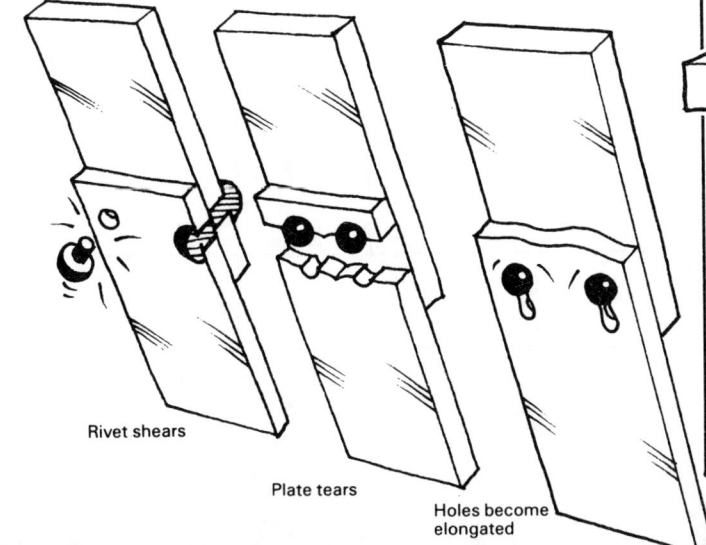

Rivet shears

Plate tears

Holes become elongated

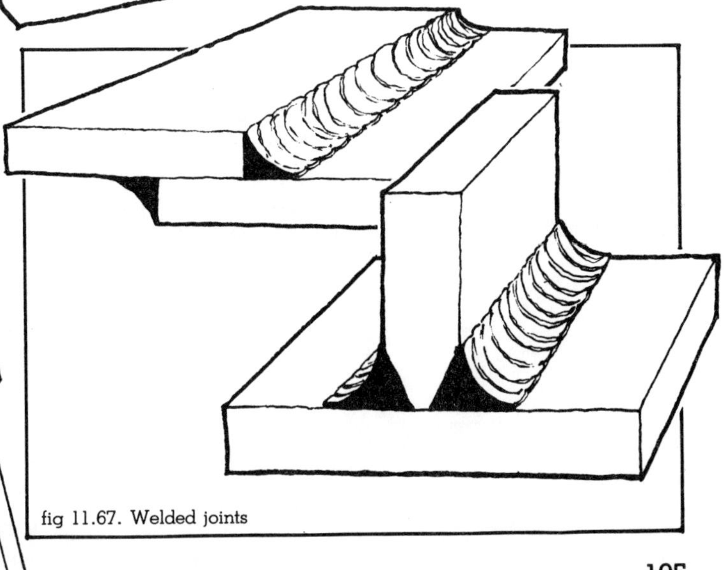

fig 11.67. Welded joints

TESTING

Once you get beyond the simple level of making model structures and watching what happens when they are loaded, it is important to find some way of recording accurately what is taking place when a structure is loaded. This can be achieved by very complex mathematical modelling, or more simply by two common methods which involve the use of **strain gauges** or **polarised light**.

STRAIN GAUGES

Strain gauges measure the amount of strain in a member. They work on the principle that the electrical resistance of a wire changes as it is stretched, becoming longer and thinner. The more it is stretched, the greater its resistance. Mathematically this is written as:

$$\text{Resistance} \; \alpha \; \frac{\text{Length}}{\text{Area}} \quad \text{or} \quad R \; \alpha \; \frac{L}{A}$$

By arranging the wire in tightly packed rows, quite long lengths can be fitted on to a small pad (fig 11.68). Modern strain gauges are made not of wire, but by etching a pattern into metal foil which is stuck to a polyester backing (fig 11.69).

In use, a gauge is stuck on to the surface of the member being tested. Its **active axis** is fixed along the direction in which you want to measure the strain. Movements on the **passive axis** will have no real effect on it. The gauge must then be connected to an electronic circuit. Fig 11.70 shows a block diagram of the complete circuit. The resistance of the gauge is compared with the resistance of fixed value resistors in the circuit. Any differences in resistance are converted into voltage differences. These very small changes in voltage are amplified before being displayed.

The final circuit, shown in fig 11.71 includes a **dummy gauge**. This compensates for any changes in the resistance of the active gauge caused by temperature changes. The active and dummy gauges form part of the **Wheatstone bridge**. With no forces applied to the active gauge the output from this part of the circuit should be zero. When forces are applied, the resistance of the active gauge changes so the output voltage to the amplifier changes. The amplifier magnifies that change so that it can be clearly seen on the meter. The three variable resistors in the circuit each allow different adjustments to be made. VR1 allows you to 'balance' the bridge, getting the resistances exactly equal. VR2 allows you to adjust the 'gain' of the amplifier, in other words how much the voltage is amplified. By adjusting VR3 the output can be adjusted to exactly zero before a load is applied to the member being tested.

In practice, strain gauges tend to be used in pairs or groups, often measuring the strain in various parts of a structure at the same time. When used like this they are often linked to a computer rather than a series of display meters. The computer keeps a constant check on the outputs from each of the strain gauges, making sure that no part of the structure is being loaded beyond normal limits.

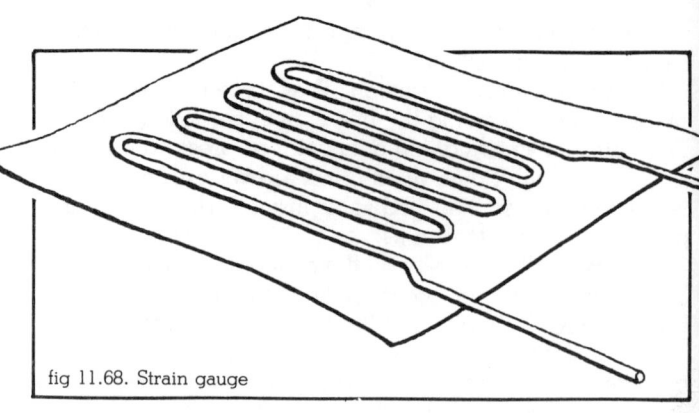

fig 11.68. Strain gauge

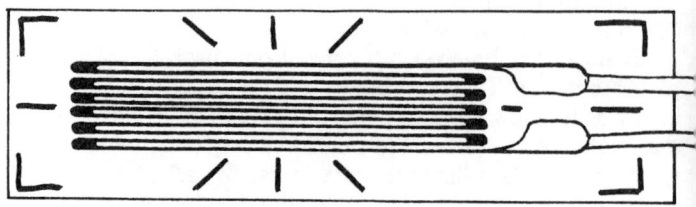

fig 11.69. Modern strain gauge

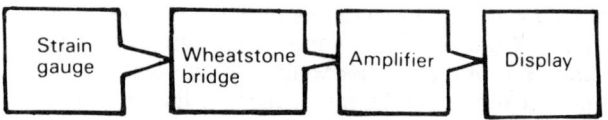

fig 11.70. Block diagram of the complete circuit

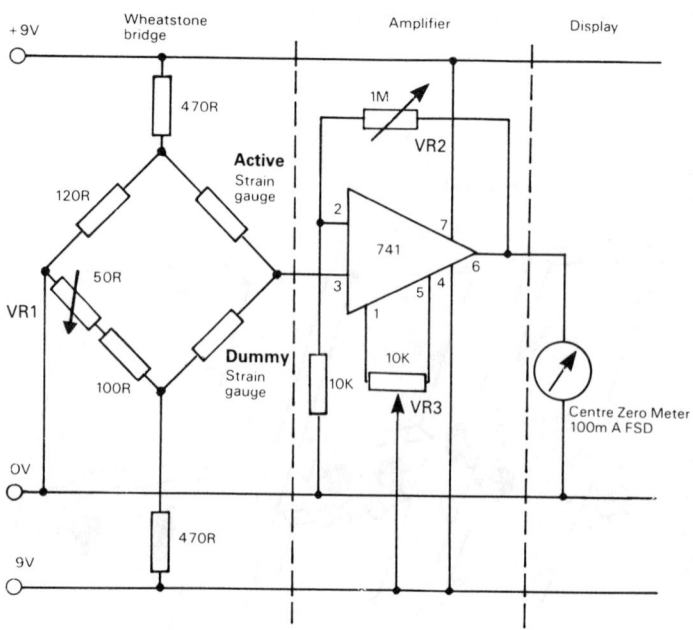

fig 11.71. Strain gauge in circuit

POLARISED LIGHT

This technique allows you to look at a scale model of a structure and actually see the stresses in it. For the best effect the model should be made of Araldite, an epoxy resin. Acrylic can also be used, but does not show the stresses so well. The model being tested has to be viewed through 'polarised' filters. A simple version of the system is shown in fig 11.72.

The darker an area, the greater the stress in that area. Stress concentrations are worst around cracks, holes and sharp corners (fig 11.74). They can be avoided by designing shapes which are more rounded or radiused (fig 11.75).

fig 11.73. Polariscope

fig 11.72. Polariscope and results

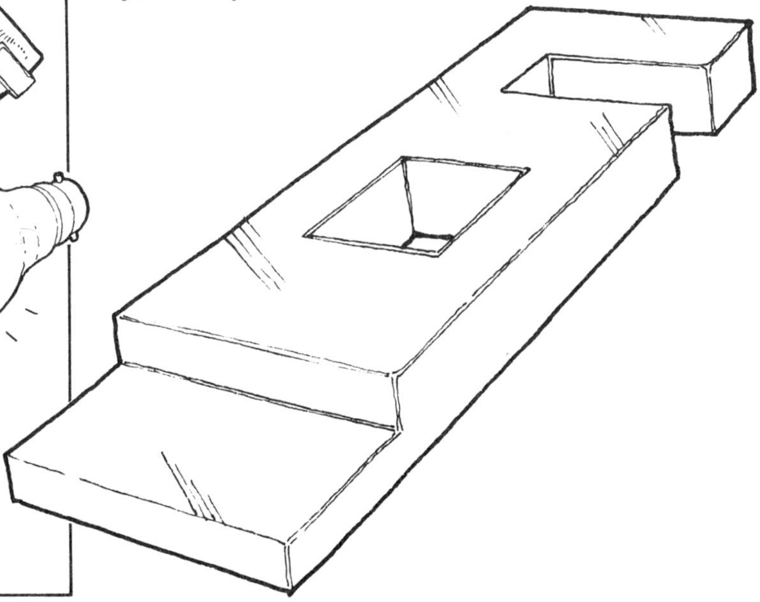

fig 11.74. Sharp corners cause stress concentrations

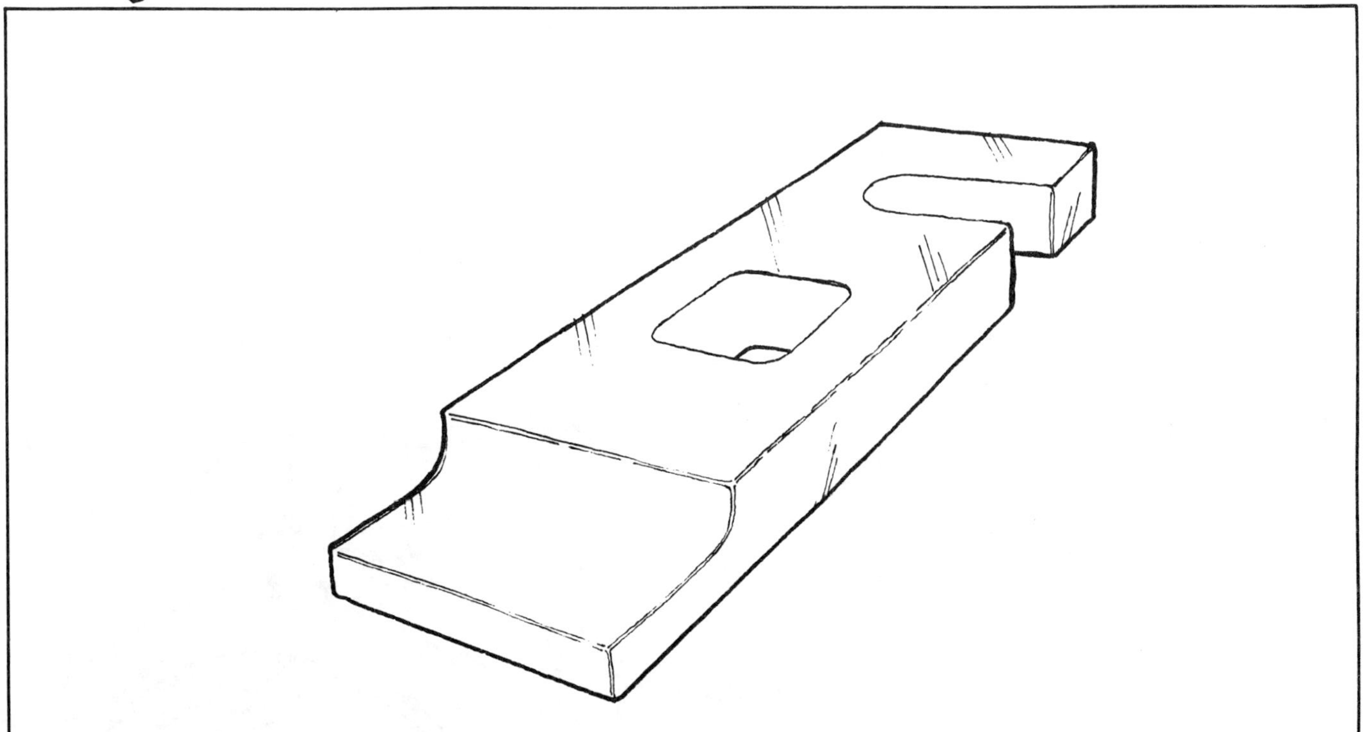

fig 11.75. Rounded or radiused shapes avoid stress

BRIDGES

All bridges are designed to allow loads to cross obstacles easily and safely. The obstacles may be ditches, rivers, valleys or swamps, while the loads may be people, animals, vehicles or trains.

There are four basic types of bridge, beam bridges, arch bridges, cantilever bridges and suspension bridges.

A common problem for all bridges is stiffness. If they are not stiff enough they will bend or twist when loaded, which can lead to disaster. Each of the members from which they are made must be strong enough to take the loads placed on them.

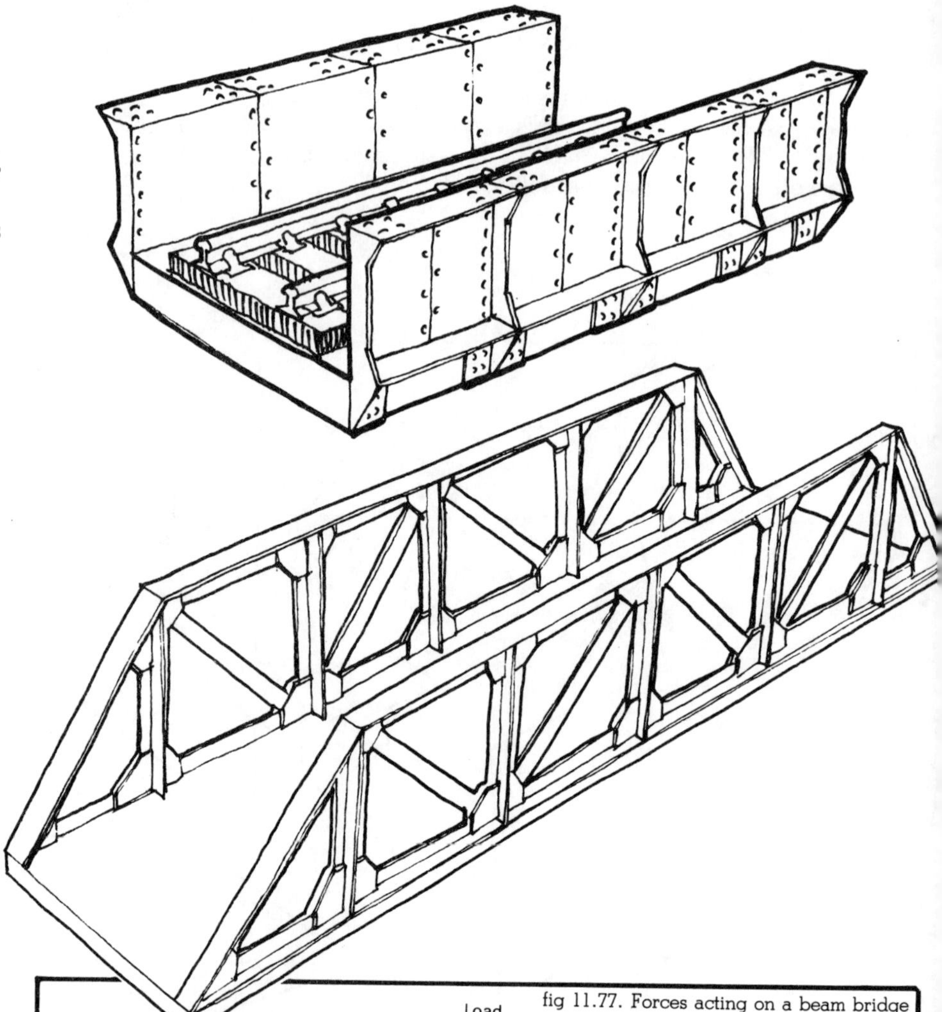

Beam bridges

Beam bridges are the simplest type. They have developed from a simple log across a stream to the large box girder bridges in use today. In between there have been many different types (fig 11.76). All are concerned with providing the necessary stiffness to prevent bending and twisting under load, which is what beams are good at doing. The forces acting on the supports of these bridges are directly downward forces caused by the load (fig 11.77).

Arch bridges

Arch bridges (fig 11.78) have been in use for thousands of years. Originally built in stone, and later brick, they now tend to be built in reinforced concrete or steel. The newer materials allow much longer, lower spans to be built. Arches carry their load by transferring it to the supports at either side (fig 11.79).

fig 11.76. Types of beam bridge

fig 11.77. Forces acting on a beam bridge

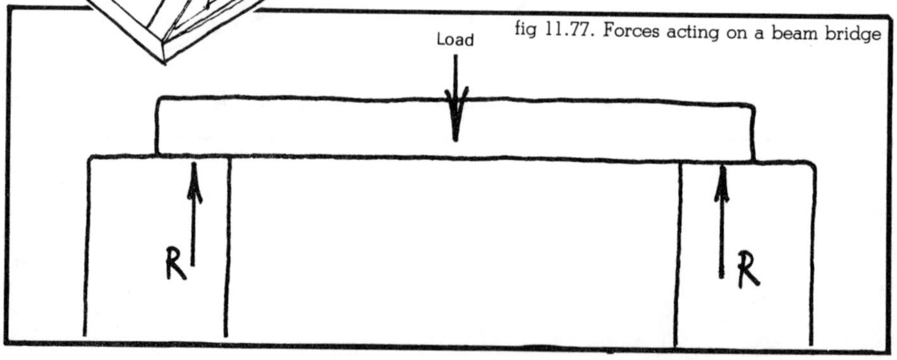

fig 11.78. Types of arch bridge

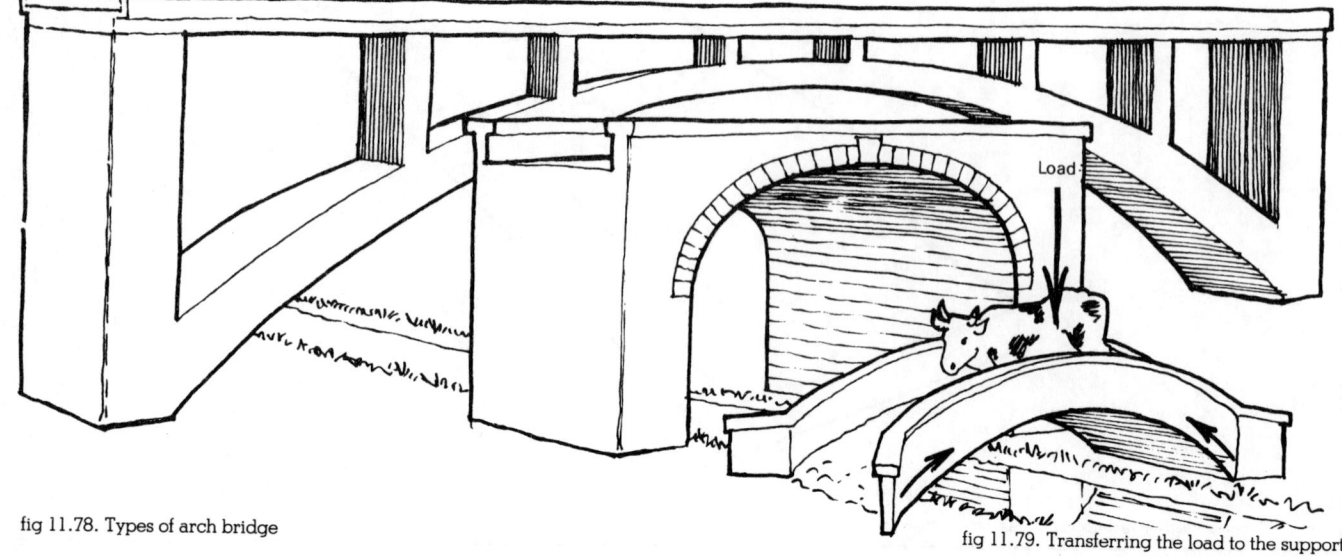

fig 11.79. Transferring the load to the supports

Cantilever bridges

Cantilever bridges normally use pairs of cantilevers. The Forth Railway bridge (fig 11.80) uses two cantilevers, back to back, with a short beam bridge between the sets of cantilevers. Many modern bridges over motorways also use cantilevers in pairs. They have a cantilever coming out from each side and a short beam between them. Fig 11.81 shows how they support a load.

Suspension bridges

Suspension bridges originated as rope and wood structures in various parts of the world. Modern suspension bridges use a box section roadway supported by high tensile steel cables (fig 11.82). The main cables, supported by the twin towers, have to be very securely anchored into the bank on either end of the bridge. Suspension bridges have the advantage that they are cheap and can give long clear spans, making them ideal for crossing busy waterways.

fig 11.80. Forth Railway bridge

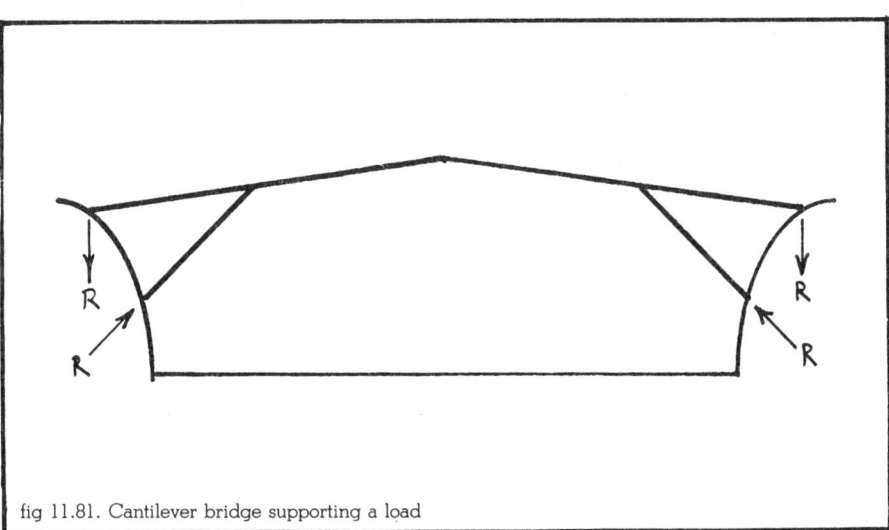

fig 11.81. Cantilever bridge supporting a load

fig 11.82. Suspension bridge

EXERCISES

1. Give two examples of natural structures and two examples of man-made structures.

2. Explain, using examples to make things clear, what is meant by 'shell structures' and 'frame structures'.

3. a) Fig 11.83 shows several pin-jointed frameworks which, when loaded as indicated, are likely to collapse. Draw each of them, adding as few extra members as possible to stop them distorting under load.
   b) Indicate on each of the members whether they are in tension or compression.

4. Give two examples of man-made shell structures and explain how they are made strong enough to cope with the forces acting on them.

5. Explain the difference between static and dynamic loads. Use examples to illustrate your answer.

6. Explain, with the aid of sketches, what forces are acting in a member which is bending.

7. Fig 11.84 shows a concrete beam which is to be used as a lintel above a window.
   a) Why is it likely to fail under load?
   b) How could it have been strengthened when being made?

8. Fig 11.85 shows a modified 'I' section beam.
   a) Explain how this type of beam is made.
   b) What are the advantages of this type of beam?

9. State Newton's third law and show how it applies to structures.

10. A steel wire of 10mm diameter is supporting a load of 1200N.
    a) What is the stress in the wire?
    b) If the wire has stretched from 1m to 1.02m what is the strain in the wire?

11. Fig 11.86 shows a graph of stress against strain for three materials.
    a) Which has the highest Young's modulus (E) value?
    b) Which one would you think is the most elastic?

12. Calculate the reactions R1 and R2 on each of the beams in fig 11.87.

13. a) By drawing or calculation find the forces acting on each of the members of the pin-joined framework shown in fig 11.88.
    b) State whether the load in each is compressive or tensile.

14. Explain how strain gauges are used to monitor the forces acting on a structure.

15. Explain how polarised light can be used to see stress concentrations.

16. Name two basic types of bridge and explain how each one carries a load.

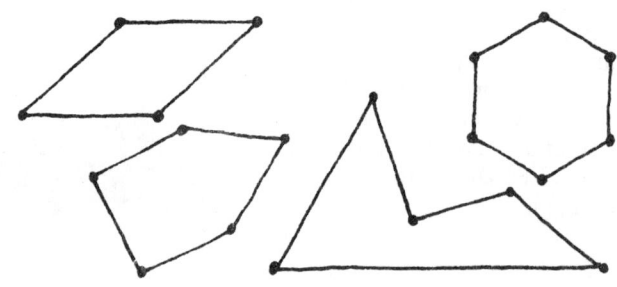
fig 11.83. Pin jointed frameworks

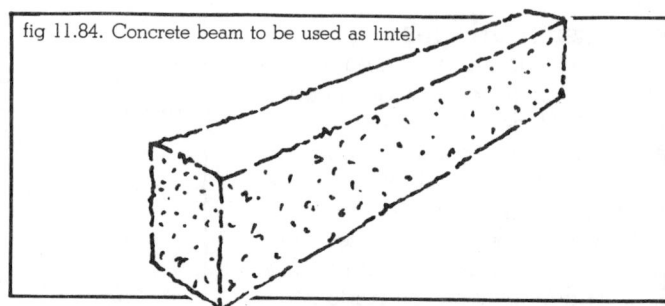
fig 11.84. Concrete beam to be used as lintel

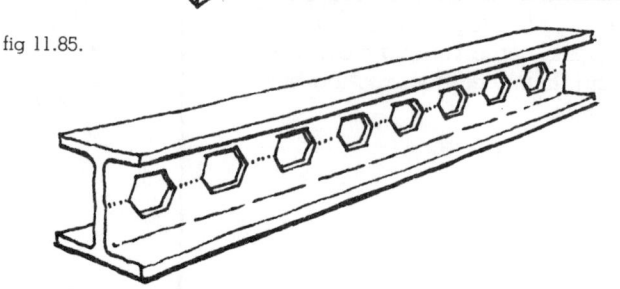
fig 11.85.

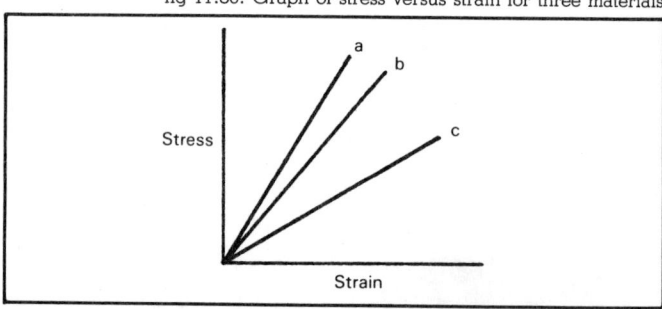
fig 11.86. Graph of stress versus strain for three materials

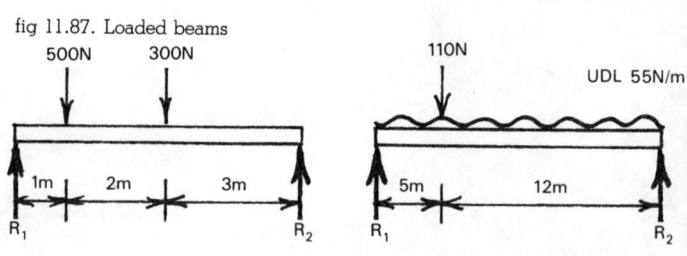
fig 11.87. Loaded beams

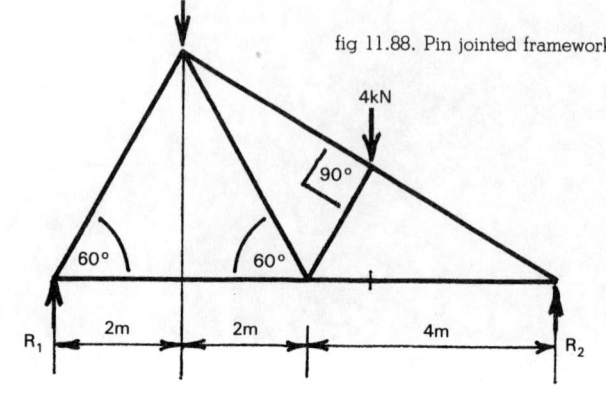
fig 11.88. Pin jointed framework

PROJECT WORK

Your design projects will carry a large percentage of the marks in any examination. In order to gain the best possible mark you must learn how to **manage** your project work. The points highlighted on this page, and examples of projects given later in this chapter, will help you develop project management skills.

Different examination syllabuses have different structures for their project work and it is wise to follow the structure set by the examining group if possible. Your teacher may provide you with a great deal of support by telling you exactly what he wishes to see in your design folder and the way it should be presented. However, if no structure is provided, the format set out on page 12 of this book may be helpful.

The nature of project work means that it is very easy to fall behind. Therefore, it is wise to devise a plan for the whole project as soon as possible. A sensible policy is to divide the overall time allocation for your project into small sections so that individual targets can be identified and met. This avoids the embarrassing situation of 'bottle necks' where many things are required at the same time. and you have little chance of producing any of them to the required standard.

At an early stage in the project things which may take a long time should be identified and acted upon. These may include letters requesting information or materials and components.

A consistent effort throughout the course nearly always produces the best results because very rarely can a project be neglected for a long period and then revived and completed in the final weeks of the course.

It is useful to lay down a few basic rules before beginning a project so that the time can be used as productively as possible. It will also help the finished folder to take on the appearance of one project and not a whole series of individual pieces of work assembled at the last minute for the examiner. A sensible list of rules might be as follows:

1. Establish the size of paper on which your project will be presented. A number of smaller drawings and photographs can be mounted to make complete sheets. Never throw any work away no matter how insignificant it may appear.

2. Decide upon a common layout for each sheet of paper. This may simply be a border around the edge of the paper but it can be developed to include a personal logo made from your initials.

3. Decide upon the colour, size and style of lettering for the page headings. If these are the same it will help to tie the project together.

4. Be aware of the sections that your project must include. You may use a chart similar to that shown below to plot your progress.

DATE	ACTIVITY	SIGNED
20/1/88	ANALYSIS	CMMA
15/2/88	RESEARCH	J.I.C.
2/3/88	SPECIFICATION	MMMA

5. Make sure all work is kept in a folder and this is stored in a safe place when not being worked upon.

Once the project is underway you may use a checklist similar to the one shown below to establish the level of progress that is being achieved.

PROGRESS CHECKLIST

- Do you know the date by which the completed project must be handed in?

- Have you divided the time allowed so that you have a realistic time scale for completing all sections of the work?

- Have you a folder in which all your design work is kept?

- Do you know in what format your project is to be presented?

- Are you regularly working on your design folder out of lesson time?

- Have you carried out sufficient research to allow a full range of solutions to the problem to be developed?

- Will the proposed solution satisfy your specification?

- Have you produced working drawings in sufficient detail that will allow the final design to be realised?

- Are the materials or components that you require to realise your design readily available?

- Is the equipment available in school that will allow the design to be realised?

- What progress did you make during the last lesson?

- What progress do you hope to make during the next lesson?

ANALYSIS

Child's Problems:—

For this age group the concepts of, left from right, names of colours, tall and short, fat and thin rough and smooth, are just a fraction of what they need to learn. Often they can not dress themselves or tie their laces, therefore a game to acquire these skills for when they are changing may be a great help. When starting school children are beginning to learn to read therefore a game might help them to recognize symbols, letters or numbers would be helpful.

Educational toy for a child aged 4 to 7 years
Rebecca Rigby

Function:—

an enjoyable game of which a child could pick up easily and could learn from it while still having fun. The game should not bore the child, or teach irrelevant things.

Ergonomics:—

The toy should not be fragile or easily broken.

Materials:—

wood — strong but heavy
cardboard — not strong enough
acrylic
rubber
fabrics
ceramics — too fragile
glass — far too dangerous.

Rebecca has clearly identified some of the important learning concepts for the age group. These helped her to come up with lots of good ideas later on. She has also realised that enjoyment is a great help to learning.

ERGONOMICS is about designing things to suit people. You should think about shapes, sizes, weights, movements and forces that will make things easy for people to use.

Which is the fattest?

Anthropometrics:—

The toy should not be too heavy because a child may like to have the object on her knee, it should also be easy to move it around.
It should be simple to operate and easy to understand.

Cost:—

A good child's toy can cost up to £30, more than this and it would be too expensive to break.

Safety:—

Being a child's game it must pass British Safety Standards. Loose equipment such as counters may be a bad idea because they are easily swallowed by young children

3

When doing her research, Rebecca investigated these areas in greater detail.

British Standards are a very useful source of information.

RESEARCH

Research - sensors thermistors

Thermistors are typically resistors made from a special heat sensitive composition material. Normally the higher the temperature, the more electrons become available, so its resistance falls.

Thermistors are produced in a varity of shapes and sizes, some in solid glass pellets, as shown on sheet overleaf. Precautions must be taken to avoid self heating effects due to the resistance measuring current and it is not clear how linear the resistance - temperature relationship is.

Remote temperature sensing device
Peter Crees

Supplier's catalogues can be a very useful source of information. You need to make some comment about any parts you use.

Temperature Sensing Resistors

These temperature sensing resistors are manufactured utilizing techniques proven in the manufacture of precision resistors. Advanced 'hick-film materials are formulated to ensure complete product control. The moulded case construction with axial leads permit normal mounting methods to be employed. The sensors are available in two different sizes having a standard resistance of either 1KΩ or 10KΩ at 25°C.

Specifications:

Standard Resistance @ 25°C	1KΩ or 10KΩ ± 2%
Temperature Coefficient	0.3%/°C @ 25°C (3000ppm/°C)
Resistance Ratio	R25°C/R125°C = 1.37
Linearity	< 1% deviation/100°C
Operating Temperature Range	−55°C to +175°C
High Temperature Stability	3.000 hours @ 175°C. < 0.8%ΔR

Description	Heat Dissipation mW/°C	Time Constant Seconds	Length	Diameter	Lead Length	Order Code	1+	25+	100+
1KΩ Sensing Resistor	8.1	7.4	6.4	2.3	38.1	143-584	£2.00	£1.81	£1.66
10KΩ Sensing Resistor	8.1	7.4	6.4	2.3	38.1	143-585	£2.00	£1.81	£1.66
1KΩ Sensing Resistor	4.7	2.9	3.8	1.7	25.4	143-586	£3.20	£2.88	£2.66
10KΩ Sensing Resistor	4.7	2.9	3.8	1.7	25.4	143-587	£3.20	£2.88	£2.66

Dimensions (mm) span Length/Diameter columns. *Price Each Per Quantity* spans 1+/25+/100+.

Platinum Resistance Element

Platinum Resistance Temperature Detectors meeting the requirements of BS1904 and DIN43760. They are available in two configurations, a flat substrate being ideal for air, gas and surface temperature measurement and rod shaped being designed to be a direct size replacement for the conventional wire wound elements.

Specifications	FLAT (146-884)	ROD (146-883)
Temperature Range	−70°C to +600°C	−70°C to +600°C
Ice Point Resistance	100 ± 0.1Ω	100 ± 0.1Ω
Fundamental Interval (0°−100°C)	38.5Ω (nominal)	38.5Ω (nominal)
Self Heating	< .005°C/mW	< 0.01°C/mW
Thermal Response	0.15 sec	0.3 sec
Stability	± 0.05%	± 0.05%

Length = 29mm. Dia = 3.1mm. Lead length = 8.0mm

Length = 32mm. Width = 4.7mm. Thickness = 0.8mm. Lead length = 8mm

Description	Order Code	1+	10+
Rod element	146-883	£5.98	£5.58
Flat element	146-884	£5.98	£5.58

Price Each Per Quantity spans 1+/10+.

Thermistor Probe

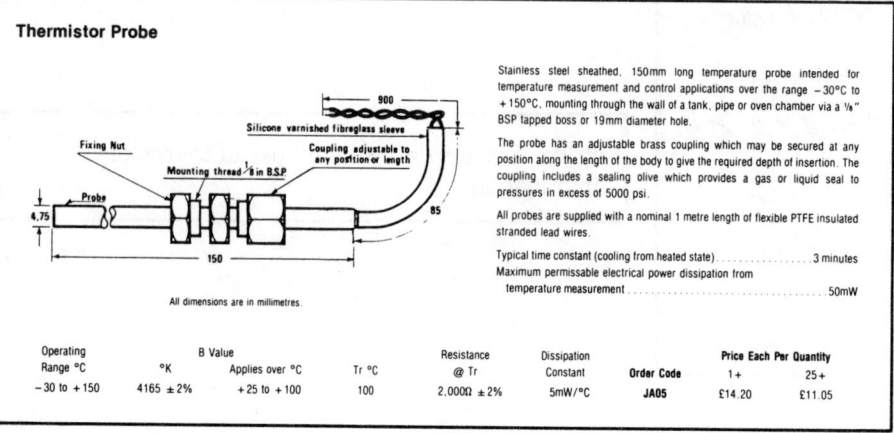

Stainless steel sheathed, 150mm long temperature probe intended for temperature measurement and control applications over the range −30°C to +150°C, mounting through the wall of a tank, pipe or oven chamber via a ⅛″ BSP tapped boss or 19mm diameter hole.

The probe has an adjustable brass coupling which may be secured at any position along the length of the body to give the required depth of insertion. The coupling includes a sealing olive which provides a gas or liquid seal to pressures in excess of 5000 psi.

All probes are supplied with a nominal 1 metre length of flexible PTFE insulated stranded lead wires.

Typical time constant (cooling from heated state) .... 3 minutes
Maximum permissable electrical power dissipation from
temperature measurement 50mW

900 · Silicone varnished fibreglass sleeve · Coupling adjustable to any position or length · Fixing Nut · Mounting thread ⅛ in B.S.P · Probe · 4.75 · 150 · 85

All dimensions are in millimetres.

Operating Range °C	B Value °K	B Value Applies over °C	Tr °C	Resistance @ Tr	Dissipation Constant	Order Code	1+	25+
−30 to +150	4165 ± 2%	+25 to +100	100	2.000Ω ± 2%	5mW/°C	JA05	£14.20	£11.05

Price Each Per Quantity spans 1+/25+.

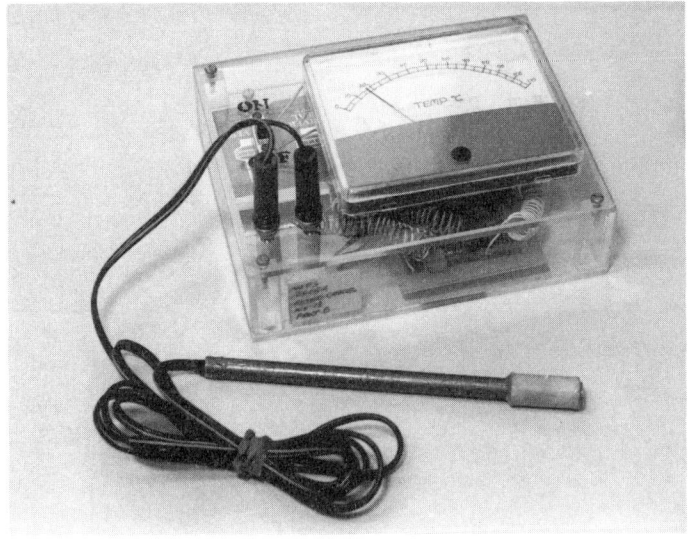

Peter has used his research to find details of the different types of temperature sensors and look at how they work.

Research - sensors p-n diode

The effect of a temperature change on a silicon p-n junction is to change the forward voltage at any fixed value of current. This is one problem that designers of power amplifiers, etc using transistors faced, which has now been virtually solved by the use of integrated circuits (IC's).

For a thermometer using a diode as the you want the biggest possible voltage difference for 2 different temperatures, this, as can be seen in Fig 16.1, occurs at low currents of about 0.25 amps or lower. Thus if a small current is passed through a forward biased silicon diode the temperature dependent forward voltage drop of the diode can be used to provide a sensitive temperature measurement method.

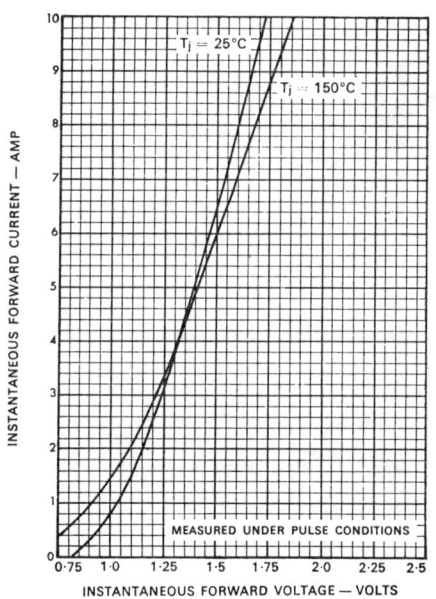

Fig 16.1: Forward silicon diode characteristics

Research - sensors thermocouples

Because of the Seebeck effect (discovered by Seebeck in 1821) a thermocouple consisting of two wires of different materials can be used to measure temperatures

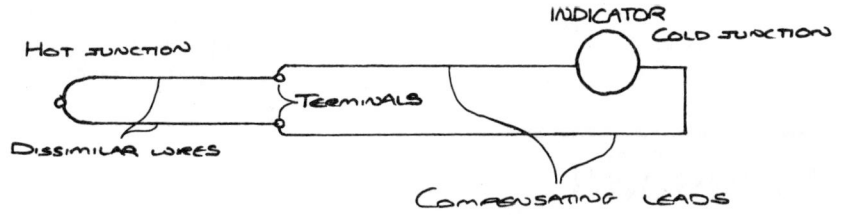

Fig 13.1: Thermoelectric Pyrometer

IDEAS

NOTE:— WHERE POSSIBLE EACH IDEA HAS BEEN EVALUATED. THIS IS TO BE FOUND IN THE BLACK BOXES WITH THE IDEAS

1

Egg timer
Jonathan Robinson

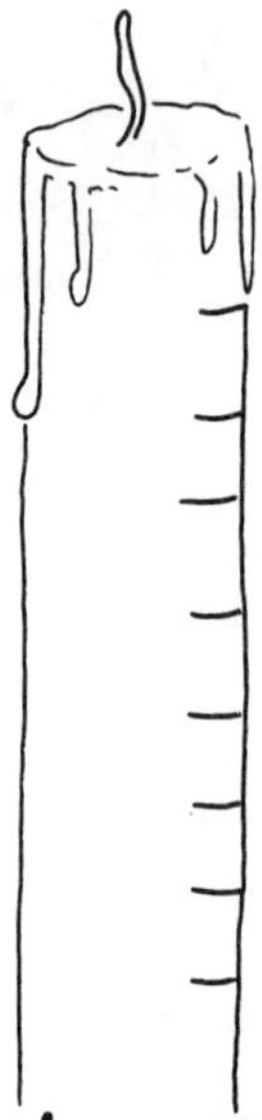

LEFT. A SIMPLE SOLUTION TO THE PROBLEM OF TIMING. A CANDLE IF IT BURNS AT A REASONABLE RATE, AND CONSTANT. THE ONLY PROBLEM WOULD BE TO MARK THE CANDLE IN EITHER 4 MINUTE OR 1 MINUTE DIVISIONS. THE LATTER ALLOWING FOR A WIDER SELECTION OF TIMES.

2

A GOOD LOOKING IDEA SIMPLE TO OPERATE AND LOOK AT. IT APPEARS FINE. HOWEVER THE TIMING CIRCUIT MAY NOT FIT IN WITH THE SIMPLE LAYOUT SUGGESTED. PLASTIC GIVES A GOOD EASY TO CLEAN BOX — PERSONALLY I WOULD PREFER A BOX IN WOOD - AS I DISLIKE USING ACRYLIC.

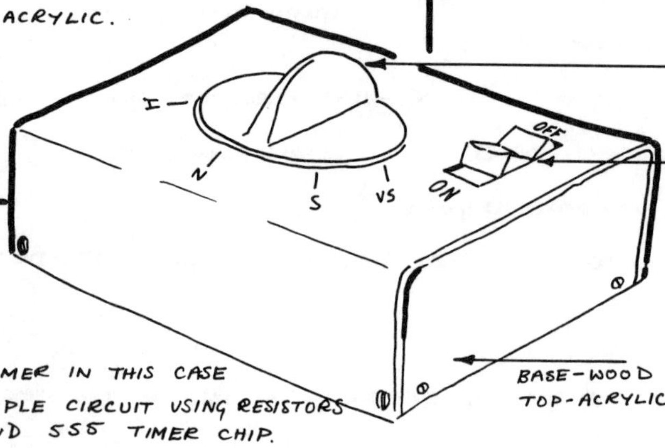

THE ABOVE NOTE SUGGESTS THIS TO BE AN IDEAL SOLUTION, REASONABLY ACCURATE, SIMPLE TO OPERATE, AND POSSIBLY CHEAP. UNFORTUNATELY THERE IS MORE TO IT THAN THAT. 1. CANDLES BURN OUT, OR DOWN AND WOULD NEED TO BE REPLACED REGULARLY. 2. THEY ARE DIRTY 3. THEY COULD BE A FIRE HAZARD.

VARIABLE RESISTOR

ON/OFF SWITCH

TIMER IN THIS CASE SIMPLE CIRCUIT USING RESISTORS AND 555 TIMER CHIP.

BASE — WOOD
TOP — ACRYLIC

Jonathan has lots of very different ideas for an egg timer. His sketches are simple yet clear. He has used lots of notes to explain his ideas, making it easy to follow his thinking.

J. J. ROBINSON

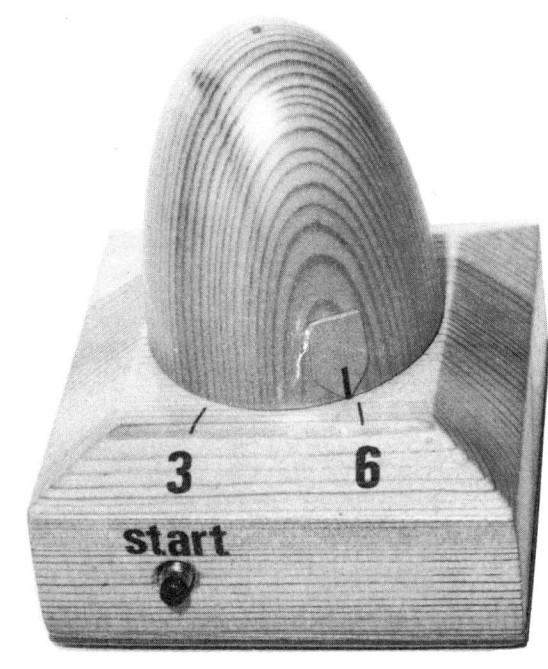

RIGHT. A SIMPLE BOX. A TIMER CIRCUIT WITH TWO TIME SETTINGS ACHIEVED BY HAVING TWO ALTERNATIVE RESISTORS FROM WHICH ONE CAN BE SELECTED. THE BOX SHOULD BE EASY TO MAKE IN HARDWOOD - MAHOGANY.

3

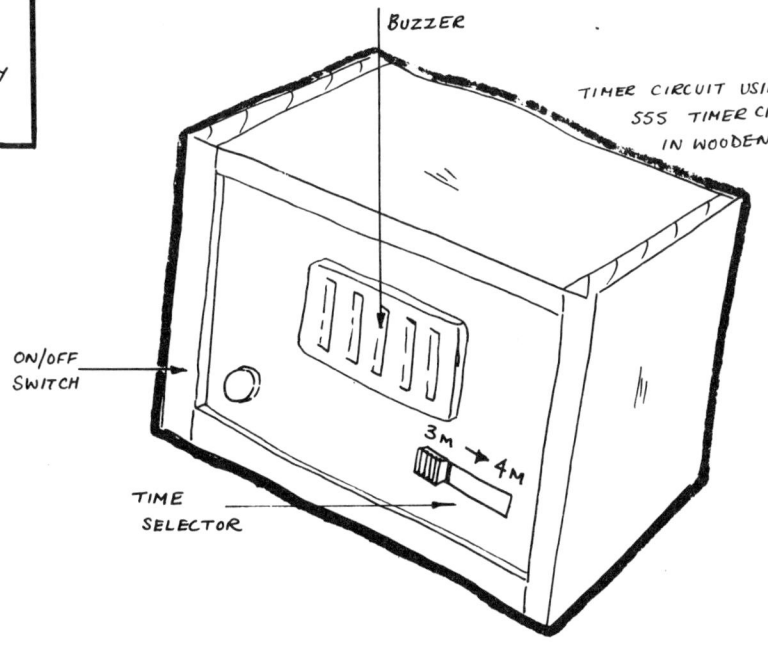

BUZZER

TIMER CIRCUIT USING 555 TIMER CHIP IN WOODEN BOX

ON/OFF SWITCH

3M → 4M

TIME SELECTOR

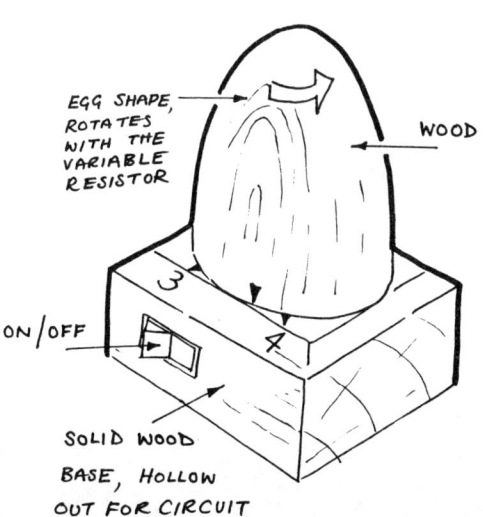

EGG SHAPE, ROTATES WITH THE VARIABLE RESISTOR

WOOD

ON/OFF

SOLID WOOD BASE, HOLLOW OUT FOR CIRCUIT

THIS IS AN INTERESTING PACKAGE. THE SHAPE SUGGESTS ITS USE AS AN EGG TIMER. THE TOP IS FIXED TO THE VARIABLE RESISTOR AND IS USED TO VARY THE TIME DELAY

4

IDEAS

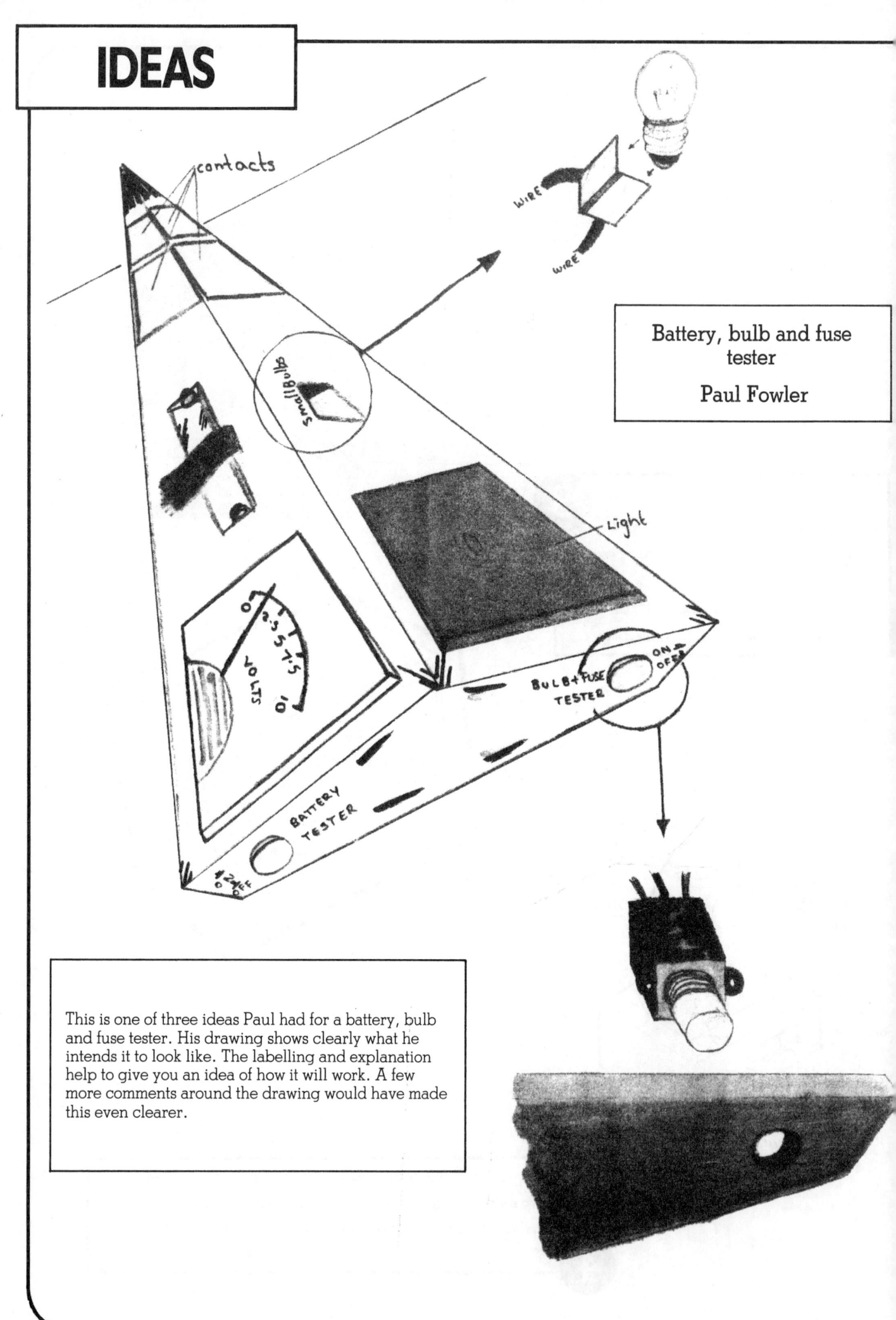

contacts

WIRE

WIRE

Small Bulbs

Light

0 2.5 5 7.5 10 VOLTS

BATTERY TESTER

4 2 0 OFF

BULB+FUSE TESTER

ON OFF

Battery, bulb and fuse tester

Paul Fowler

This is one of three ideas Paul had for a battery, bulb and fuse tester. His drawing shows clearly what he intends it to look like. The labelling and explanation help to give you an idea of how it will work. A few more comments around the drawing would have made this even clearer.

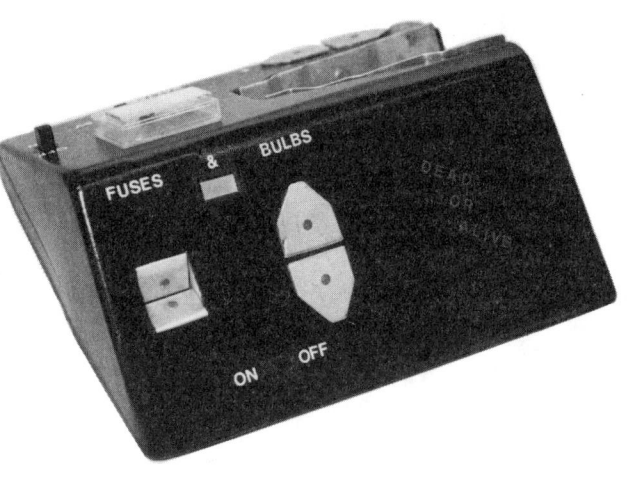

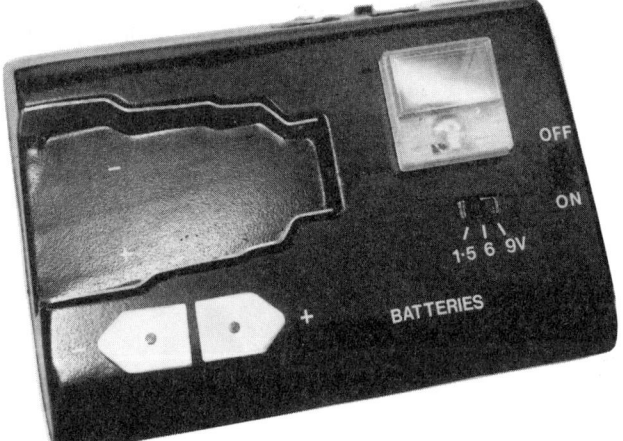

IDEA 3

<u>How It Works</u>

The bulb and fuse tester uses the end
contacts to check the Items. The small
bulb tester in two pieces of metal, one
touches the contact on the bottom, while
the other touches the side contact

The battery tester uses a voltmetre instead
of lights or a colour coded system This makes
it less complicated to make. The end contacts
are for 6 and 9 volt batteries while the slot
in for the 1.5 volt batteries.

Both the ON, OFF switches are simple
push switches that click in and out

8

DEVELOPMENT

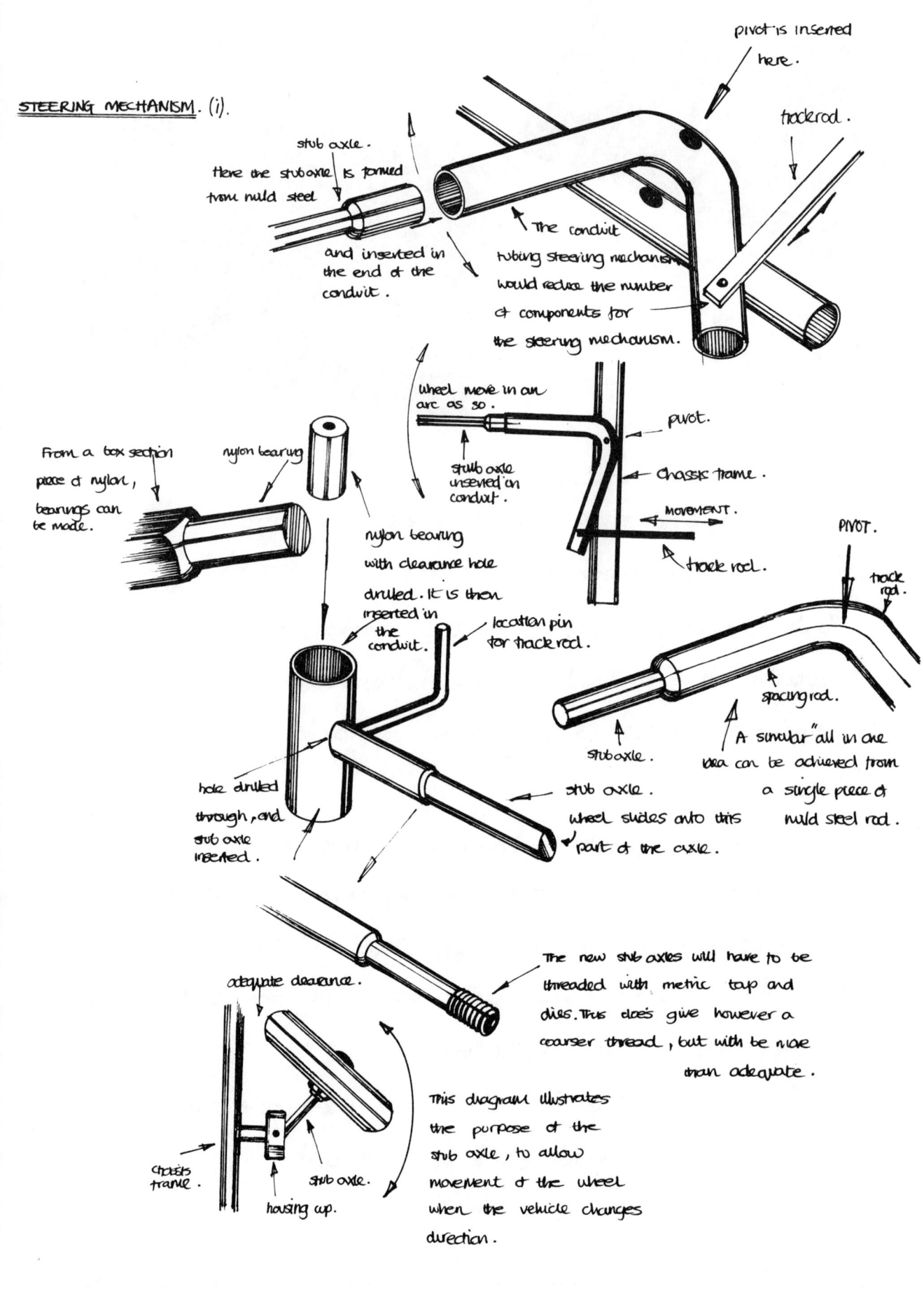

STEERING MECHANISM. (i).

stub axle.

Here the stuboxle is formed from mild steel

and inserted in the end of the conduit.

pivot is inserted here.

trackrod.

The conduit tubing steering mechanism would reduce the number of components for the steering mechanism.

wheel move in an arc as so.

stub axle inserted in conduit.

pivot.

Chassis frame.

MOVEMENT.

track rod.

From a box section piece of nylon, bearings can be made.

nylon bearing

nylon bearing with clearance hole drilled. It is then inserted in the conduit.

location pin for track rod.

PIVOT.

track rod.

spacing rod.

A similar "all in one" idea can be achieved from a single piece of mild steel rod.

stub axle.

hole drilled through, and stub axle inserted.

stub axle. wheel slides onto this part of the axle.

The new stub axles will have to be threaded with metric tap and dies. This does give however a coarser thread, but with be more than adequate.

adequate clearance.

chassis frame.

stub axle.

housing cup.

This diagram illustrates the purpose of the stub axle, to allow movement of the wheel when the vehicle changes direction.

Pedal-powered go-kart

John Powell

John examimed several models such as this to get ideas for his own development work.

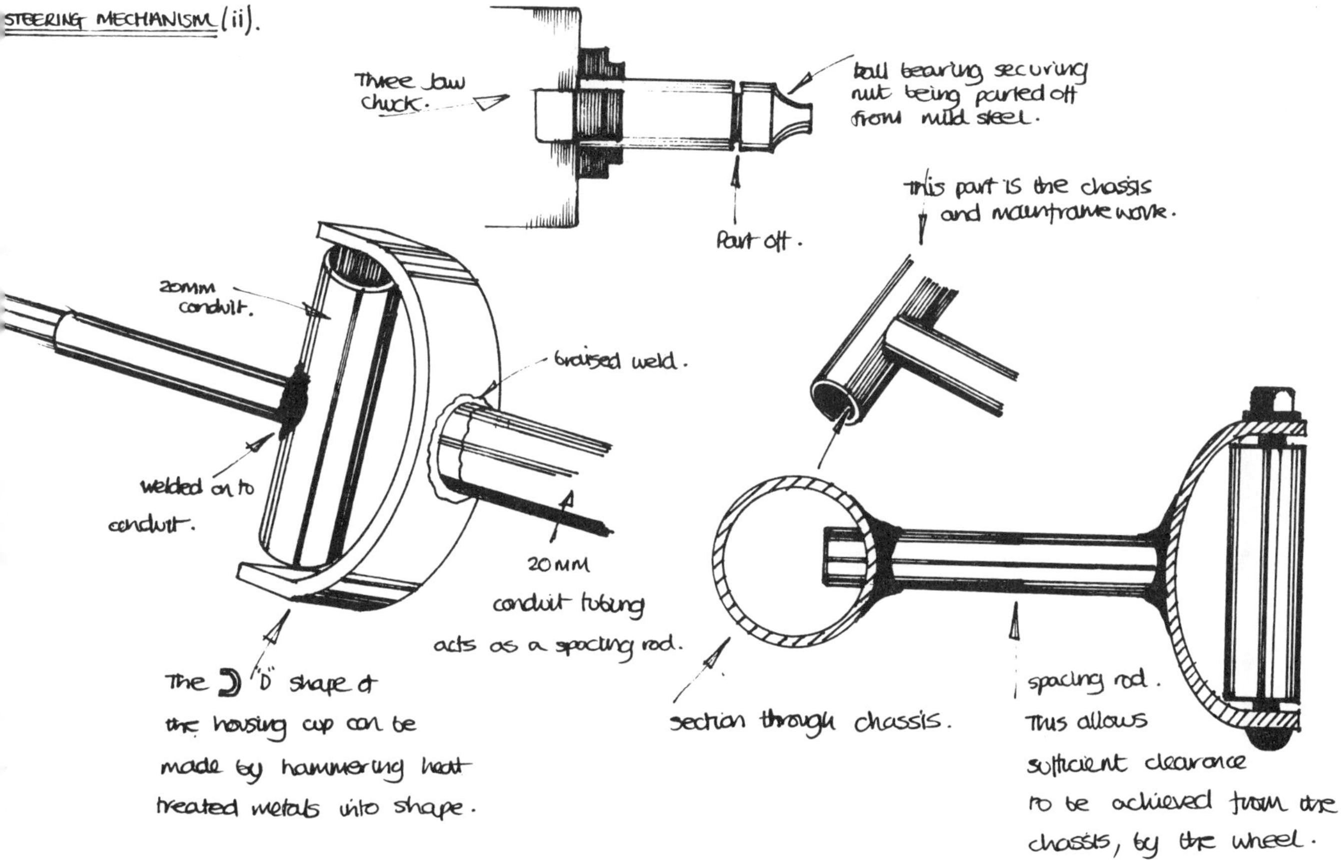

STEERING MECHANISM (ii).

Three jaw chuck.

ball bearing securing nut being parted off from mild steel.

Part off.

this part is the chassis and mainframe work.

20mm conduit.

braised weld.

welded on to conduit.

20 MM conduit tubing acts as a spacing rod.

section through chassis.

spacing rod. This allows sufficient clearance to be achieved from the chassis, by the wheel.

The 🌙 "D" shape of the housing cup can be made by hammering heat treated metals into shape.

This is just part of John's development work. As well as clear sketches, he has used lots of notes to explain what materials are used and how the parts fit together. He uses arrows well to show how the parts relate to each other.

WORKING DRAWING

SECTION VIEW AA'

Bird feeder

Tim Ward

R 10

R 10

R 5

R 10

R 10

190

5 | 40 | 5 | 35 | 5 | 20

5

Tim's working drawing of his bird feeder is clear and well presented. There are enough dimensions to enable him to make the final product. His parts list is on a separate sheet. A few labels indicating joining techniques would have been helpful.

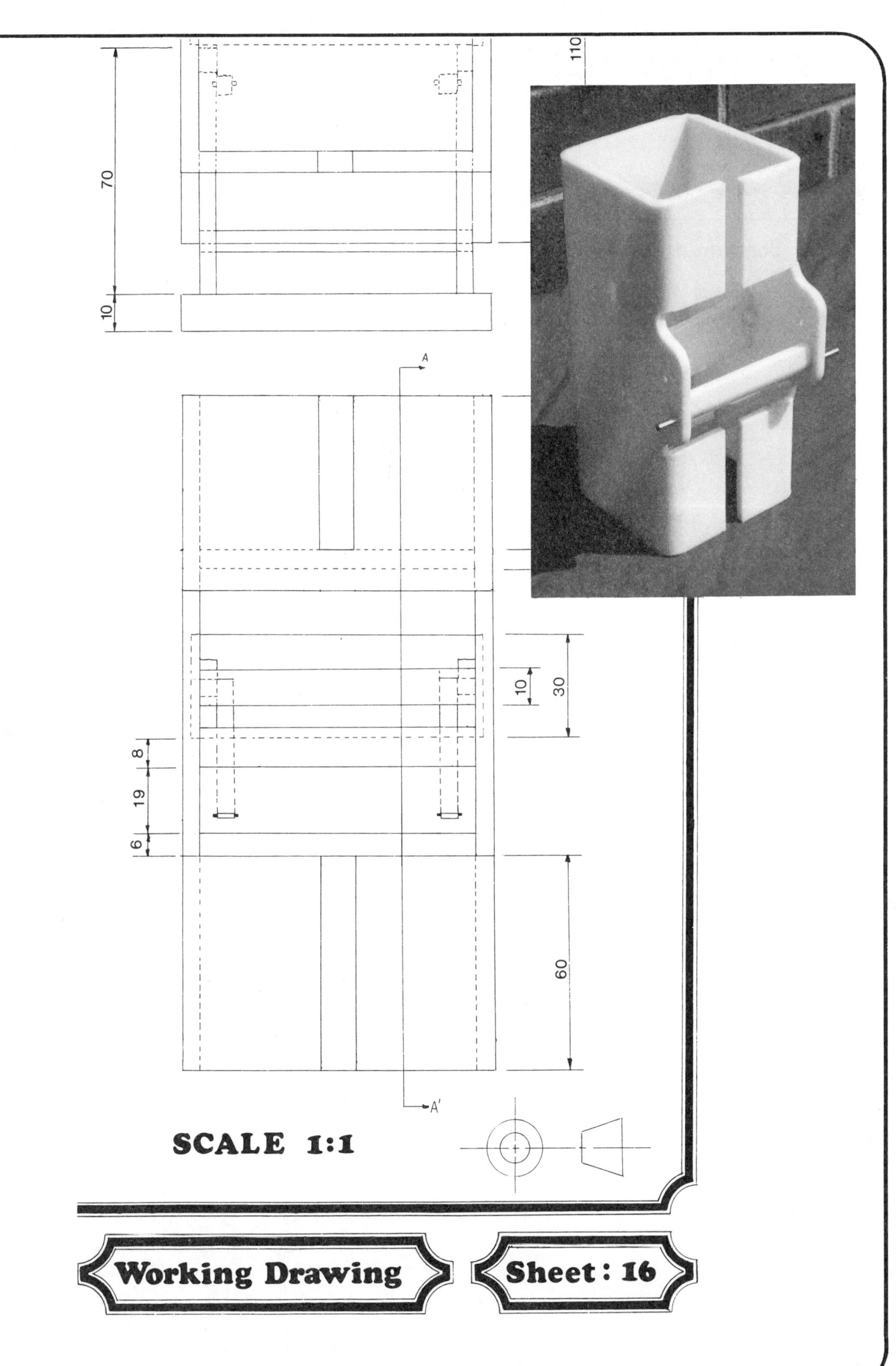

SCALE 1:1

70

10

110

A

10

30

8

19

6

60

A'

EVALUATION

Boat turner

Jon Fox

The front clamp fitted perfectly, and the securing pins were inserted with little difficulty. The clamp held the front of the boat very securely.

When it came to tightening the back clamp, a design flaw became apparent. Instead of the turning of the threaded bar tightening the clamps grip on the boat, the crossbars distorted under the strain, causing the rest of the back clamp to distort as well (DIAGRAM 2)

In order to rectify this problem I welded pieces of bar steel across the back of the distorting pieces to form much stronger angle bars (DIAGRAM 3). This solved the problem. The bars did not distort and the back clamp held the boat very securely.
Details of this modification have been added to the final drawing of the back clamp.

Jon's boat turner was designed to turn a boat in a confined space. This is just part of his evaluation. It shows that he tested his product well, even making changes to his design as a result. His drawings show details of some of the changes.

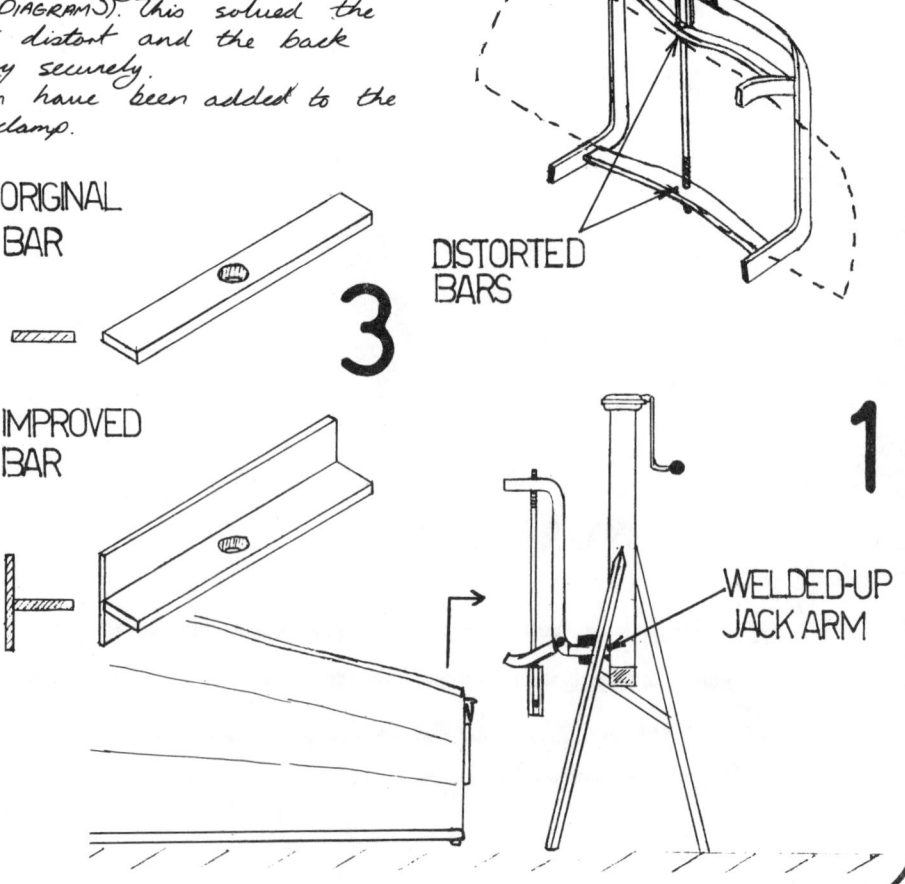

REFERENCE DATA

Prefix	Symbol	Meaning	Pronunciation
mega	M	$\times$ 1,000,000 (10^6)	megger
kilo	k	$\times$ 1000 (10^3)	keel-oh
		ONE WHOLE UNIT	
milli	m	$\div$ 1,000 (10^{-3})	milli
micro	μ	$\div$ 1,000,000 (10^{-6})	micro
nano	n	$\div$ 1,000,000,000 (10^{-9})	nar-no
pico	p	$\div$ 1,000,000,000,000 (10^{-12})	peeko

Energy

Work done = Force (Newtons) $\times$ Distance moved in direction of force (metres)

Units = Newton metres (Nm) or Joules (J)

e.g. Work done in lifting a load of 50N two metres off the ground
= 50N $\times$ 2m = 100 Nm or 100J

Efficiency $= \dfrac{\text{Useful energy out}}{\text{Energy in}} \times 100\%$

e.g. $\dfrac{430\,\text{J}}{1000\,\text{J}} \times 100\% = 43\%$ efficient

Rate of heat loss = Area (metres^2) $\times$ Temperature difference (°C) $\times$ 'U' value

e.g. $10\text{m}^2 \times 5°\text{C} \times 1.5 = 75$ Watts

Electronics

Resistor colour code

Number	Colour	Number	Colour
0	Black	5	Green
1	Brown	6	Blue
2	Red	7	Violet
3	Orange	8	Grey
4	Yellow	9	White

Resistors: preferred values

10, 12, 15, 18, 22, 27, 33, 39, 47, 56, 68, 82

from 10R to 8M2 (10 ohms to 8.2 million ohms)

Ohms Law : $V = I \times R$

where V = voltage across component (volts)
I = current through component (amps)
R = resistance of component (ohms)

e.g. When using a 6V battery what value resistor is needed to limit the current flow through an LED to 15 mA?

$$R = \frac{V}{I} = \frac{6}{0.015} = 400 \text{ ohms}$$

nearest available value is 390R

Resistors in series : R total = R1 + R2

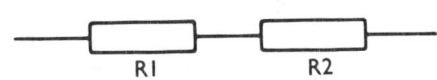

Resistors in parallel: with two resistors $R\ \text{total} = \dfrac{R1 \times R2}{R1 + R2}$

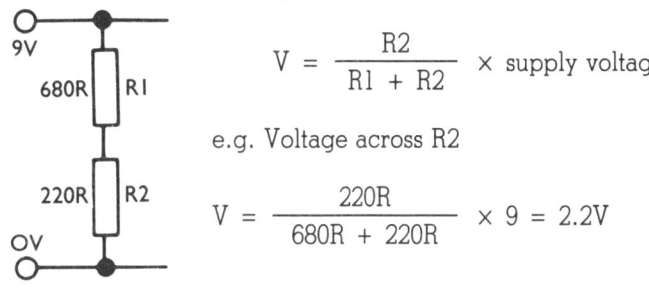

with two or more resistors $\dfrac{1}{R\ \text{total}} = \dfrac{1}{R1} + \dfrac{1}{R2} + \dfrac{1}{R3}$

Voltage dividers : finding voltages across components in a voltage divider.

$$V = \frac{R2}{R1 + R2} \times \text{supply voltage}$$

e.g. Voltage across R2

$$V = \frac{220R}{680R + 220R} \times 9 = 2.2V$$

Power : $P = I \times V$

where P = power used in component (watts)
I = current through component (amps)
V = voltage across component (volts)

Capacitors : measured in **farads**. Values tend to be in fractions of a farad (e.g. μF and nF)

In series : with two capacitors $C\ \text{total} = \dfrac{C1 \times C2}{C1 + C2}$

with two or more capacitors $\dfrac{1}{C\ \text{total}} = \dfrac{1}{C1} + \dfrac{1}{C2} + \dfrac{1}{C3}$

Capacitors in parallel : C total = C1 + C2

Time constant : time taken for a capacitor to charge up, via a resistor, to 2/3 supply voltage.

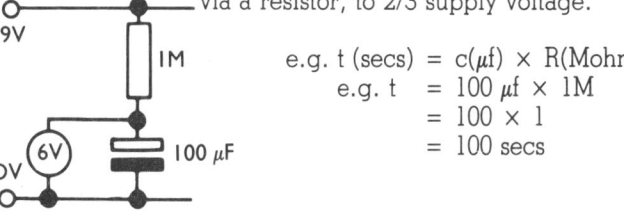

e.g. t (secs) = c(μf) $\times$ R(Mohms)
e.g. t = 100 μf $\times$ 1M
= 100 $\times$ 1
= 100 secs

Frequency of an astable : measured in Hertz (Hz)
1Hz = 1 cycle/second

$$f = \frac{1.44}{(R1 + 2R2)C} \quad \text{e.g.} \quad f = \frac{1.44}{(10K + 100K)0.001} = \frac{1.44}{1} = 1.44\text{Hz}$$

Transistor data

P_T = power rating at 25°C
I_C = current (max) through collector
V_{CEO} = voltage (max) across collector/emitter
V_{CBO} = voltage (max) across collector/base
hfe = gain of transistor
$F_T\text{typ}$ = operating frequency (max)

Counting systems

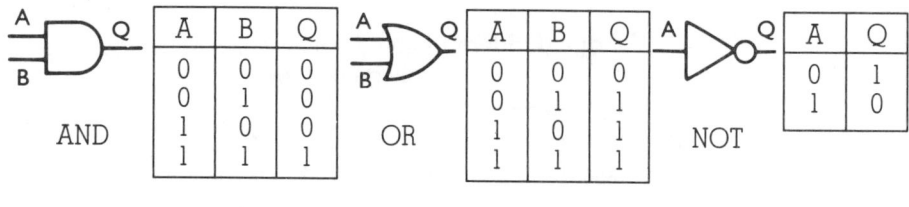

Decimal	0	1	2	3	4	5	6	7	8	9	10	11	12	13	14	15
Hexadecimal	0	1	2	3	4	5	6	7	8	9	A	B	C	D	E	F
Binary	0	1	10	11	100	101	110	111	1000	1001	1010	1011	1100	1101	1110	1111

Truth tables

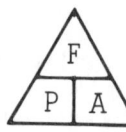

AND

A	B	Q
0	0	0
0	1	0
1	0	0
1	1	1

OR

A	B	Q
0	0	0
0	1	1
1	0	1
1	1	1

NOT

A	Q
0	1
1	0

Pneumatics

Pressure: Pressure (N/mm^2) = $\dfrac{\text{Force (N)}}{\text{Area (mm}^2)}$ e.g. P = $\dfrac{200}{1000}$ = 0.2 N/mm^2

gives F = P × A and A = $\dfrac{F}{P}$

Mechanisms

Mechanical advantage = $\dfrac{\text{Load force}}{\text{Effort force}}$

Velocity ratio = $\dfrac{\text{Distance moved by effort}}{\text{Distance moved by load}}$

Efficiency = $\dfrac{MA}{VR}$ × 100%

Weston differential pulley:

M.A. = $\dfrac{2R}{R-r}$

Rotary systems

VR = $\dfrac{\text{Driven}}{\text{Driver}}$

(Gear ratio)

For pulley systems use diameters. For chain and sprocket, and gear systems use number of teeth.

Output speed = $\dfrac{\text{Input speed}}{VR}$

Output torque = Input torque × VR

Structures

Stress = $\dfrac{\text{Force}}{\text{Area}}$ (MN/m^2)

Strain = $\dfrac{\text{Change in length}}{\text{Original length}}$

Moments:
ACM = CM (force × distance)

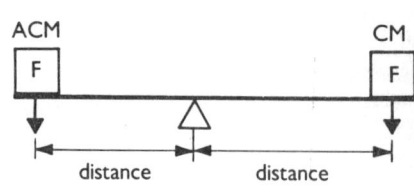

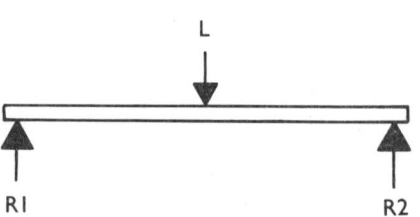

Reactions : R1 + R2 = Load

Bows Notation
1. Calculate reactions.
2. Draw space diagram to scale.
3. Put in reactions, label spaces A, B, etc. and number joints.
4. Construct vector diagram to scale.
5. Measure lengths and state forces acting in each member.
6. Go round each joint working out the type of force in each member.

Calculations: working out forces using geometry based on right angle triangles.

Sine X = $\dfrac{O}{H}$

Cos X = $\dfrac{A}{H}$

Tan X = $\dfrac{O}{A}$

Provided you know angle X and one of the forces you can calculate the other two.

INDEX